CARBON NANOMATERIALS FOR ADVANCED ENERGY SYSTEMS

CARBON NANOMATERIALS FOR ADVANCED ENERGY SYSTEMS

Advances in Materials Synthesis and Device Applications

Edited by

WEN LU
JONG-BEOM BAEK
LIMING DAI

Library of Congress Cataloging-in-Publication Data:

Carbon nanomaterials for advanced energy systems : advances in materials synthesis and device
applications / edited by Wen Lu, Jong-Beom Baek, Liming Dai.
    pages   cm
  Includes bibliographical references and index.
  ISBN 978-1-118-58078-3 (hardback)
1. Electric batteries–Materials.   2. Energy harvesting–Materials.   3. Fullerenes.
4. Nanostructured materials.   5. Carbon nanotubes.   I. Lu, Wen (Materials scientist)
II. Baek, Jong-Beom.   III. Dai, Liming, 1961–
  TK2945.C37C375 2015
  621.31′2420284–dc23

2015017196

Printed in the United States of America

10 9 8 7 6 5 4 3 2 1

CONTENTS

LIST OF CONTRIBUTORS

Nirupam Aich, Department of Civil, Architectural and Environmental Engineering, University of Texas, Austin, TX, USA

Paola Ayala, Faculty of Physics, University of Vienna, Vienna, Austria and Yachay Tech University, School of Physical Sciences and Nanotechnology, Urcuquí, Ecuador

Marcella Bini, Department of Chemistry, University of Pavia, Pavia, Italy

Doretta Capsoni, Department of Chemistry, University of Pavia, Pavia, Italy

Zhongfang Chen, Department of Chemistry, Institute for Functional Nanomaterials, University of Puerto Rico, San Juan, PR, USA

Wonbong Choi, Department of Materials Science and Engineering, University of North Texas, Denton, TX, USA

Vishnu T. Chundi, Department of Materials Science and Metallurgy, University of Cambridge, Cambridge, UK

Liming Dai, Center of Advanced Science and Engineering for Carbon (Case4Carbon), Department of Macromolecular Science and Engineering, Case Western Reserve University, Cleveland, OH, USA

Santanu Das, Department of Materials Science and Engineering, University of North Texas, Denton, TX, USA

Stefania Ferrari, Department of Chemistry, University of Pavia, Pavia, Italy

Michael W. Forney, NanoPower Research Laboratory, Rochester Institute of Technology, Rochester, NY, USA

Matthew J. Ganter, NanoPower Research Laboratory, Rochester Institute of Technology, Rochester, NY, USA

Lorenzo Grande, Helmholtz Institute Ulm, Ulm, Germany

Yong Soo Kang, Department of Energy Engineering, Hanyang University, Seoul, South Korea

Takahiro Kondo, Faculty of Pure and Applied Sciences, University of Tsukuba, Tsukuba, Japan

Brian J. Landi, NanoPower Research Laboratory, and Department of Chemical Engineering, Rochester Institute of Technology, Rochester, NY, USA

Fen Li, Laboratory of Materials Modification by Laser, Electron, and Ion Beams and College of Advanced Science and Technology, Dalian University of Technology, Dalian, China

Xinming Li, School of Materials Science and Engineering, Tsinghua University and National Center for Nanoscience and Technology, Beijing, China

Jun Liu, State Key Laboratory of Polymer Physics and Chemistry, Changchun Institute of Applied Chemistry, Chinese Academy of Sciences, Changchun, China

Piercarlo Mustarelli, Department of Chemistry, University of Pavia, Pavia, Italy

Junji Nakamura, Faculty of Pure and Applied Sciences, University of Tsukuba, Tsukuba, Japan

Vasile V.N. Obreja, National R&D Institute for Microtechnology (IMT Bucuresti), Bucharest, Romania

Toshiyuki Ohashi, Fundamental Technology Research Center, Honda R&D Co., Ltd., Wako, Japan

Jaime Plazas-Tuttle, Department of Civil, Architectural and Environmental Engineering, University of Texas, Austin, TX, USA

Reginald E. Rogers, NanoPower Research Laboratory, and Department of Chemical Engineering, Rochester Institute of Technology, Rochester, NY, USA

Navid B. Saleh, Department of Civil, Architectural and Environmental Engineering, University of Texas, Austin, TX, USA

P. Sudhagar, Department of Energy Engineering, Hanyang University, Seoul, South Korea

Toma Susi, Faculty of Physics, University of Vienna, Vienna, Austria

Di Wei, Nokia R&D UK Ltd, Cambridge, UK

Xiaowei Yang, School of Material Science and Engineering, Tongji University, Shanghai, China

Yuegang Zhang, Suzhou Institute of Nano-Tech and Nano-Bionics, Chinese Academy of Sciences, Suzhou, China

Jijun Zhao, Laboratory of Materials Modification by Laser, Electron, and Ion Beams and College of Advanced Science and Technology, Dalian University of Technology, Dalian, China

Tianshuo Zhao, Department of Material Science and Engineering, University of Pennsylvania, Philadelphia, PA, USA

Hongwei Zhu, School of Materials Science and Engineering, and Center for Nano and Micro Mechanics, Tsinghua University, Beijing, China

PREFACE

The global energy consumption has been accelerating at an alarming rate due to the rapid economic expansion worldwide, increase in world population, and ever-increasing human reliance on energy-based appliances. It was estimated that the world will need to double its energy supply by 2050. Consequently, the research and development of sustainable energy conversion and storage technologies have become more important than ever. Although the efficiency of energy conversion and storage devices depends on a variety of factors, their overall performance strongly relies on the structure and property of the materials used. The recent development in nanotechnology has opened up new frontiers by creating new nanomaterials and structures for efficient energy conversion and storage. Of particular interest, carbon nanomaterials have been cost-effectively structured into various nanostructures with a high surface area and energy conversion/storage capacities. This book will focus on advances in the research and development of carbon nanomaterials for advanced energy systems.

Carbon has long been known to exist in three forms: amorphous carbon, graphite, and diamond. However, the Nobel Prize-winning discovery of buckminsterfullerene C_{60} in 1985 has created an entirely new branch of carbon chemistry. The subsequent discoveries of carbon nanotubes in 1991 and graphene in 2004 opened up a new era in materials science and nanotechnology. Since then, carbon nanomaterials with unique size-/surface-dependent electrical, thermal, optical, and mechanical properties have been demonstrated to be useful as energy materials, and tremendous progress has been achieved in developing carbon nanomaterials for high-performance energy conversion and storage systems. This is a field in which a huge amount of literature has been rapidly generated with the number of publications continuing to increase each year. Therefore, it is very important to cover the most recent developments in this field in a timely manner.

This book deals with the synthesis, fundamentals, and device applications of a wide range of carbon nanomaterials. In order to cover the multidisciplinary field of such diversity, *Carbon Nanomaterials for Advanced Energy Systems* provides a collection of chapters written by top researchers who have been actively working in the field, and the text has been divided into three major parts. The first part consisting of Chapters 1–4, *Synthesis and Characterization of Carbon Nanomaterials*, deals with the synthesis and basic science of various carbon nanomaterials, including fullerenes, carbon nanotubes, graphene, and their multidimensional/multifunctional derivatives. In the second part, *Carbon Nanomaterials For Energy Conversion*, Chapters 5–8 present an overview of carbon nanomaterials for various energy conversion systems, such as solar cells, fuel cells, and thermoelectric devices. A large variety of carbon-based energy storage devices, ranging from supercapacitors through batteries to energy-related gas storage systems, are then described in the final part (Chapters 9–13), *Carbon nanomaterials for energy storage*, of the book. The above approach will allow the readers to first review the scientific basis of carbon nanomaterials and then extend the basic knowledge to the development, construction, and application of functional devices; many of them are of practical significance.

The readers who are new to the field will be exposed to many self-explanatory illustrations that could provide an overview understanding even before a serious reading. In the meantime, the large number of updated references cited in each of the chapters should enable advanced readers to quickly review the multidisciplinary and challenging field with information on the latest developments. Therefore, *Carbon Nanomaterials for Advanced Energy Systems* is an essential reference on carbon nanomaterials for energy systems to scientists, engineers, teachers, and students who are new to the field. Experienced academic and industrial professionals can use this book to quickly review the latest developments in this challenging multidisciplinary field and broaden their knowledge of carbon nanomaterials for developing novel devices/systems for advanced energy conversion and storage.

Finally, we wish to express our sincere thanks to Dr. Edmund H. Immergut, Ms. Anita Lekhwani, Ms. Cecilia Tsai, and their colleagues at Wiley and Wiley-VCH for their very kind and patient cooperation during the completion of this book project, without which this book would never have been appeared. We would also like to thank all of the chapter contributors, authors whose work was cited, and our colleagues who contributed in one way or the other to the book. Last, but not the least, we thank our families for their love, unceasing patience, and continuous support.

Wen Lu, Jong-Beom Baek, Liming Dai
November, 2014

PART I

SYNTHESIS AND CHARACTERIZATION OF CARBON NANOMATERIALS

1

FULLERENES, HIGHER FULLERENES, AND THEIR HYBRIDS: SYNTHESIS, CHARACTERIZATION, AND ENVIRONMENTAL CONSIDERATIONS

Nirupam Aich, Jaime Plazas-Tuttle and Navid B. Saleh

Department of Civil, Architectural and Environmental Engineering, University of Texas, Austin, TX, USA

1.1 INTRODUCTION

The search for alternative and renewable energy sources has become one of the major thrusts of the twenty-first-century researchers due to the increasing demand for energy. Innovations and development of photovoltaics, dye-sensitized or polymer solar cells, high-efficiency lithium ion batteries, supercapacitors, transparent conductors, hydrogen productions and storage systems, microbial fuel cells, catalyst-driven proton exchange membrane fuel cells, thermoelectric power generation, etc., have come to the forefront in alternative energy research [80, 149]. In quest of effective energy transfer, distribution, and storage, improved materials are being synthesized since the 1990s. Nanoscale manipulation of materials has fueled such development [11]. Improved surface area at the nanoscale and targeted molecular placement or alteration in nanomaterials resulted in desired band gap tuning and effective electron transfer, storage, and surface activity [111]. One of the key challenges that eluded energy researchers for decades was an efficient photoelectron acceptor with high structural stability and chemical reactivity; a spheroidal

Carbon Nanomaterials for Advanced Energy Systems: Advances in Materials Synthesis and Device Applications, First Edition. Edited by Wen Lu, Jong-Beom Baek and Liming Dai.
© 2015 John Wiley & Sons, Inc. Published 2015 by John Wiley & Sons, Inc.

carbon allotrope, known as fullerene, addressed this critical gap in alternative energy research and development [31, 111].

Sixty carbon atoms, organized following isolated pentagon rule (*i.e.*, 20 hexagons and 12 pentagons), forming a truncated icosahedron structure is known as buckminsterfullerene—the first member of the fullerene family, discovered by Sir Harry Kroto, Robert Curl, and Richard Smalley in 1985 [136]. Fullerenes' ability to effectively function in donor/acceptor heterojunctions has popularized its synthesis, derivatization, and supramolecular assembly for photovoltaic applications [31]. Later, detection and effective isolation of higher-order fullerenes [121], that is, C_{70}, C_{76}, C_{82}, C_{84}, etc., have encouraged their studies and uses in energy applications. Changes in hole/electron-pair generation ability and electronic band gap with the changing number of atoms in the fullerene structures have continued to evoke interest in these higher fullerenes [61, 169]. Electronic structure could be further tuned by conjugation of fullerenes with other carbon allotropes, for example, carbon nanotubes (CNTs), graphene, etc., which has encouraged synthesis of hierarchical assemblages of fullerenes with other nanoscale structures, resulting in nanoscale hybrid (NH) materials [3, 146, 201, 212, 272].

C_{60}s, especially its polymeric derivative [6,6]-phenyl-C_{61}-butyric acid methyl ester (PCBM), has been known to be the most effective electron acceptor for organic photovoltaics [31]. Recent advances in this field have proposed a novel donor/acceptor blend for hole/electron transfer. By photoexciting the donor, electron moves from the lowest unoccupied molecular orbital (LUMO) of the donor to the acceptor, where the hole gets transported to the donor. C_{60}'s excellent electron-accepting ability has presented it as an ideal candidate for photovoltaic solar cell construction. Their applications in organic field-effect transistors [9] and lithium or hydrogen storage [42] also depend on its high electron affinity and high charge transferability. C_{60}s also act as promising catalytic composites and electrode materials for Nafion-based proton exchange membrane fuel cells [243]. Similarly, higher-order fullerenes such as C_{70}, C_{76}, C_{84}, and C_{90} and their derivatives are also being utilized as higher-efficiency transistors and have shown promising solar cell efficiencies [128, 221, 244]. Moreover, hybridization of fullerenes to formulate concentric fullerene clusters or carbon nano-onions [90], fullerene nanopeapods [146] or nanobuds (fullerene–CNT hybrids) [245, 255], and endohedral metallofullerenes [263] enhances their promises in energy storage devices. However, such demand of fullerenes requires higher quantity to be synthesized and purified. Such high demand for this material requires unique synthesis and preparation processes, which in conjunction with fullerenes' inherent attributes can invoke toxic responses to the environment, hence necessitating careful consideration [200].

C_{60} and its derivatives such as C-3, fullerol $C_{60}(OH)_{24}$, bis-methanophosphonate fullerene, tris carboxyl fullerene adduct tris-C_{60}, dendritic C_{60} monoadduct, malonic acid C_{60} tris adduct, etc. are found to be responsible for inducing toxicological impacts in soil and aquatic microbes [41, 73, 114, 151, 259], invertebrates [276], and fish [276] as well as in human cell lines [78, 196, 205] and rats and mice [71]. Such environmental and biological toxic potentials are known to have resulted from fullerenes' ability to penetrate cell membranes and generate oxidative stresses. Similarly,

C_{70}s have also shown to adversely affect aquatic species, when C_{70}-gallic acid derivative at less than quantifiable concentration causes significant reduction in *Daphnia magna* fecundity after 21-day exposure. It has also demonstrated generation of oxidative stress through inhibition of enzymatic activities [211]. The demonstrated toxicity of fullerenes resulted in systematic evaluation of its fate, transport, and transformation in natural environment, which include fundamental aggregation [164], deposition and transport in porous media [270], photoinduced transformation [104], etc. C_{60}s, synthesized using different techniques [32, 150], have been studied to evaluate role of synthesis on their potential risk. However, very few studies have focused on systematic investigations of higher fullerenes and fullerene-based NH's fate, transport, transformation, and toxicity [2, 3, 200, 201].

This book chapter discusses synthesis, characterization, and application of fullerenes, higher fullerenes, and their NHs. The chapter will identify potential risk of these carbon allotropes when used in energy applications and discuss possible strategies for pursuing green synthesis of these materials. The discussion in this chapter will potentially highlight the relevant risk of using fullerenes in energy applications and help establish an understanding of environmental considerations.

1.2 FULLERENE, HIGHER FULLERENES, AND NANOHYBRIDS: STRUCTURES AND HISTORICAL PERSPECTIVE

1.2.1 C_{60} Fullerene

C_{60} fullerene is an all-carbon and perfectly symmetric molecule made from 60 carbon atoms (Fig. 1.1a). It was the first ever discovered regular truncated icosahedral molecule [197, 242]. The carbon atoms on the vertices of the polygons in C_{60}s possess sp^2 hybridization and become bonded through 6:6 double bond between hexagons and 6:5 bonds between hexagons and pentagons [87]. One carbon atom is bonded to 3 other carbon atoms with a bond length of 0.14 nm. The total spherical diameter of a C_{60} molecule becomes 0.71 nm, giving rise to the perfect symmetric cage [95]. Though such molecules possess high structural stability [162, 197, 228, 242], these were found to be highly reactive, where acceptance of electrons makes them strongly reductive [99, 100]. Such conjugate reactivity and structural stability help them to produce various derivatives as shown in Figure 1.1b.

The discovery of fullerene was rather extraordinary [130]. A research lab in Exxon group in 1984 had first seen carbon soot presenting similar time of flight (TOF) mass spectra for even numbers of carbon atoms starting from 40 to 200 [198]; however, they were unable to identify the abundance of C_{60}s in that mixture. In similar time frame, while searching for the mechanism of interstellar long-chain carbon molecule formation, an unusual TOF spectral signature of carbon soot was observed by Sir Kroto, Smalley, and Curl while synthesizing carbon soot through laser irradiation of graphite [136] at Rice University in 1985. The group hypothesized that the spectral signature was generated due to the formation of C_{60}s, the probable aromatic icosahedron structure with remarkable stability. Later on, nuclear magnetic resonance (NMR)

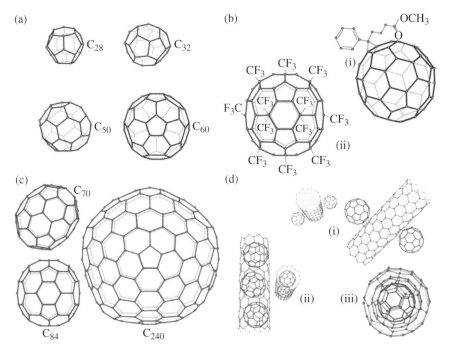

FIGURE 1.1 (a) Fullerenes. (b) Fullerene derivatives: (i) C_{60} derivative [6,6]-phenyl-C_{61}-butyric acid methyl ester (PCBM) and (ii) trifluoromethyl derivative of C_{84} ([C_{84}](CF$_3$)$_{12}$). (c) Higher-order fullerenes. (d) (i) Nanobud (fullerenes covalently bound to the outer sidewalls of single-walled carbon nanotube), (ii) peapod (fullerenes encapsulated inside a single-walled carbon nanotube), and (iii) nano-onion (multishelled fullerenes). (*See insert for color representation of the figure.*)

experiments were performed to conclude that the molecules obtained by Kroto and others truly resulted in C_{60} molecules [230]. Kratschmer, Huffman, and Fostiropoulos, on the other hand, came up with synthesis technique for macroscopic amount of C_{60} and C_{70} in 1989 [132]. In 1995, Harry Kroto, Richard Smalley, and Robert Curl were awarded Nobel Prize in Chemistry for the discovery of C_{60}. The aforementioned scientists later named the first discovered carbon allotrope as "buckminsterfullerene" or "fullerene" to pay homage toward the renowned American architect Buckminster Fuller, who designed geodesic dome-shaped structures resembling fullerenes.

1.2.2 Higher Fullerenes

Members of fullerene family possessing more than 60 carbon atoms are known as higher-order fullerenes (Fig. 1.1c). They are generally found in the same carbon soot obtained during C_{60} synthesis. C_{70}, being the first member of the higher-order fullerene family, is always found in abundance with the C_{60}s. However, the other members, that is, C_{76}, C_{78}, C_{84}, and C_{92} (up to fullerenes with more than a hundred carbon atoms), are found in much smaller quantities in the soot. Diederich *et al.* first found mass spectroscopic evidence for existence of C_{76}/C_{78} and C_{84} and isolated them through extraction technique employment during reproduction of the Kratschmer method for producing C_{60} and C_{70} fullerenes [59]. During the same time, theoretical prediction of their existence, isomerism, and chemical stability was presented by Fowler and Manolopoulos [20, 74, 155, 156]. With the help of the newer chromatographic techniques, fullerenes with a wide range of composition, that is, C_{20} to C_{400}, were extracted, isolated, and characterized alongside with identification of isomeric forms of several higher-order fullerenes [121, 189, 214].

1.2.3 Fullerene-Based Nanohybrids

When C_{60} and higher fullerenes are conjugated either exohedrally or endohedrally with carbon- and metal-based nanomaterials, the ensemble materials are known as fullerene-based nanohybrids (NHs) (Fig. 1.1d) [3, 201]. The overall scope of this book will limit the discussion to carbon NHs only. Endohedral NHs can be formed via fullerene and higher fullerene encapsulation within CNTs and larger fullerene structures. These structures are called peapods [218] and carbon nano-onions [219], respectively (Fig. 1.1d). Nanopeapods, prepared in 1998 by Luzzi et al., was one of the first NHs synthesized [218]. Growing interests in fullerene and NH chemistry encouraged development of other NH assemblages, either with CNTs [174] or graphene [194]. The conjugation is performed using both nonspecific short-ranged interaction [257] and via covalent bonding [148, 174] with the use of functional linking molecules or polymers.

1.3 SYNTHESIS AND CHARACTERIZATION

1.3.1 Fullerenes and Higher Fullerenes

Commercial production of C_{60}s and higher fullerenes involves a two-step process [63]. First, carbon soot containing fullerene mixtures is synthesized via carbon vapor generation methods. Second, fullerene separation and purification from the carbon soot are performed to obtain individual fractions of the carbon allotropes. Based on the raw materials and precursors, vaporization methods, and processing techniques, various soot generation processes have been developed. Most of the synthesis techniques were developed during the 1985–1995 time line, when fullerene discovery and techniques for primary isolations and separation were innovated [130]. Later, chemical synthesis processes to form fullerenes from aromatic hydrocarbons were developed [207]. A brief discussion on the major fullerene and higher fullerene production techniques is described in this section. Figure 1.2 shows a flow diagram demonstrating steps involved in carbon soot synthesis and fullerene extraction and purification.

1.3.1.1 Carbon Soot Synthesis

Arc Vaporization Methods Arc vaporization methods are the most effective ones for carbon soot synthesis. The process of resistive heating of graphite rods in helium environment, developed by Kratschmer et al., was the first step to produce carbon

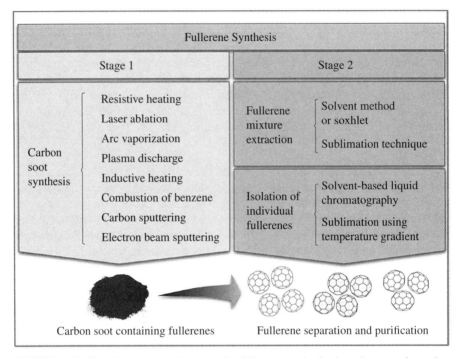

FIGURE 1.2   Flow diagram showing steps for fullerene synthesis via carbon soot formation and fullerene separation and purification.

soot containing fullerenes in macroscopic amounts [131]. The method was furthered into AC- or DC-arc-based carbon vaporization processes to produce gram quantities of fullerenes [4]; this technique reduced loss of carbon rods through complete heating of the electrodes. Figure 1.3 shows a typical arc process for fullerene soot generation. Two graphite rods are separated from each other by 1–10 mm in a helium-filled chamber under 100–200 torr pressure. An arc is discharged to generate 100–200 amp current at a voltage of 10–20 V. This process causes the graphite rods to evaporate and form soot containing fullerene products. Copper jacket covering the chamber wall and circulating cooling water control the temperature to allow carbon soot vapors to condense and deposit on the chamber walls, which can later be extracted for purification and processing. Modifications of these methods were performed to achieve several advantages. Such modifications include arc via contacting with the graphite [93, 214] or demineralized coal electrodes [186], employment of plasma discharge for high yield [189, 210], application of DC power rather than AC [93, 186, 214], and low current rather than high AC current [126], to achieve better fullerene yield, formation of tapered apparatus for gravity-based collection of carbon soot [126], etc.

Laser Ablation Method   This technique was first adopted by the Smalley group in 1985 [136], which involves a laser, such as neodymium-doped yttrium aluminum garnet (Nd:YAG), irradiated on a graphite rod causing the carbons to evaporate via

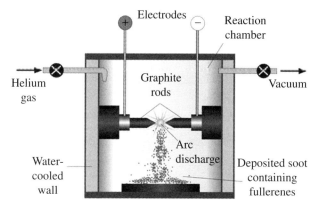

FIGURE 1.3 Arc discharge process for fullerene synthesis. Adapted and modified from Refs. 93 and 147.

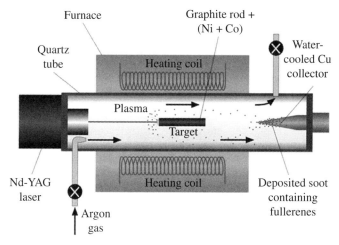

FIGURE 1.4 Laser ablation process for fullerene synthesis. Adapted and modified from Ref. 147.

heating and produce carbon plasma. Afterward, controlled cooling of the carbon plasma takes place to form fullerene clusters. Later, ablation at elevated temperature or inside of heated furnace resolved the issue by slowing down the cooling process [59, 147]. Around 1000–1200°C was found to be most efficient for fullerene cluster formation [147]. A high-temperature furnace containing such laser ablation arrangement is shown in Figure 1.4. Operating parameter modulation, such as changing laser intensity, wavelength, buffer gas pressure, and temperature in the furnace, can offer better control over fullerene formation and yield.

Other Methods   Versatility in fullerene synthesis processes has been achieved through adoption of different innovative approaches. A thermal vaporization method by inductively heating the graphite rods in a high-frequency furnace at 2700°C was developed for soot production at large quantities [190]. Moreover, combustion method was employed where laminar flames of premixed benzene and oxygen with argon diluents were used to produce C_{60} and C_{70} [102, 103]. Gram quantities of fullerenes were produced using this process with potential for easy scale-up, continuous process operations, easy dopant addition in the flame mixture, and changes in flame properties for controlling the fullerene size distribution [102, 147]. Efficient production of a large quantity of higher fullerene soot with minor C_{60} presence was developed by Bunshah *et al.* in 1992 [36]. Two different experimental setups were devised—one for carbon sputtering and the other for electron beam sputtering. In the carbon sputtering method, a magnetron sputtering cathode was attached to a graphite target, and carbon black was sputtered from the target by helium ions. In the other method, an electron beam was used to evaporate carbon from a graphite target. Efficiencies of fullerene synthesis methods along with their extraction methods, yields, and operating conditions, such as pressure, temperature, mode, etc., have been summarized in several review papers [188, 216].

1.3.1.2 *Extraction, Separation, and Purification*   Fullerene extraction and purification involve a two-step process [63]. First, fullerene mixtures are isolated from the carbon soot using a solvent extraction, followed by a separation of individual fullerene molecules using chromatography or sublimation processes. Details of these processes are described in the following.

Fullerene Mixture Extraction from Carbon Soot   In the solvent extraction method, fullerene mixtures along with some soluble hydrocarbons from the soot are dissolved in toluene or similar solvents and then filtered or decanted to remove the insoluble residue to recover extractable fullerenes at 10–44% mass [216]. Toluene-soluble extracts generally contain 65% C_{60}, 30% C_{70}, and 5% higher fullerenes [216]. Tetrahydrofuran (THF) is also used to ultrasonicate soot at room temperature, followed by filtration [58]. Evaporation of the filtrate in a rotary evaporator is employed to obtain fullerene powder mixture. A Soxhlet apparatus can also be used for efficient solubilization of fullerenes [63, 189]. In this process, the solvent is first boiled and evaporated, which is condensed down through a carbon soot matrix, extracting the fullerenes. The cycle is repeated for maximizing the fullerene extraction [189]. Sublimation process, in contrast, involves heating the raw soot in a quartz tube under helium gas or in vacuum followed by condensing the mixture [241]. Fullerene mixtures accumulate at the bottom, leaving the residue products from soot in the vapor phase.

Separation and Purification of Individual Fullerenes from Mixture   There are two major processes for purification or isolation of individual fullerenes. These are solvent-based liquid chromatography (LC) and sublimation using temperature gradient.

LC   This is the primary technique for separation of individual C_{60}, C_{70}, and other higher fullerenes from the extracted mixture [216]. In this process, a solution of fullerene mixture is passed through a packed porous column. The solution is known as the mobile phase, and the solvent as the eluent, while the solid surface is called the stationary phase. Based on the molecular weight, fullerenes undergo chromatographic separation [251]. Selective separation of individual fullerenes with highest purity can be achieved by changing the stationary phase and eluent constituent and compositions [216]. Historically, a wide variety of mobile phases and stationary phases have been used for successful separation of fullerenes. The eluents include toluene [86], hexane [113, 181], toluene–hexane mixture [4, 230], pi-basic groups, etc. On the other hand, alumina [7], silica gel [4], graphite [240], C_{18} reverse phase [89], pi-acidic Pirkle phase [192], etc., have been reported to be used as developed stationary phases. The Soxhlet method has also been combined with the purification process to have a one-step extraction–purification technique for fullerene separation from raw soot [119]. This process loads fullerene mixture from top and separates individual structures employing a chromatographic column. Higher fullerene separation is achieved using high-pressure liquid chromatography (HPLC), which involves repeated and reversed chromatographic methods. The process utilizes solvents like carbon disulfide for achieving enhanced solubility of higher fullerenes. Gel-permeation-based chromatographic techniques have also been used for fullerene separation [163]. Commercially available HPLC systems have been found to use pyrenylpropyl and pentabromobenzyl groups as stationary phases for fullerene isolations [172]. Toluene and toluene–acetonitrile mixture have been found to be the most commonly used mobile phases for C_{60} and C_{70} separation, while chlorobenzene and dichlorobenzene are used for higher fullerenes.

SUBLIMATION WITH TEMPERATURE GRADIENT   Differences in sublimation temperature and artificially created thermal gradient are used as the driving forces for the separation process [49, 260]. The raw soot containing fullerene mixture is directly added to a quartz tube under vacuum, and heat is applied at the center of the tube to raise the temperature to 900–1000°C. The tube containing the fullerene mixture has its one end at the center of the quartz tube, and the other is protruded outside of the tube in the ambient environment. Thereby, a temperature gradient is created from the center of the tube (hottest) to the outermost end. Individual fullerenes based on their sublimation temperatures deposit at different locations of the tube. Generally, higher fullerenes deposit closer to the center as they possess higher sublimation temperature compared to C_{60}s. Such spatial distance allows for purification of the individualized structures [22]. Several modifications have been performed on this method to improve separation efficiencies, which are elaborately described in other books and review articles [63].

1.3.1.3 *Chemical Synthesis Processes*

Fullerene synthesis using chemical methods has been sought for obtaining a large quantity of isomerically pure fullerenes. However, only a couple of attempts have been successful developing such method, with only one that has realized into large-scale production so far [168]. Pyrolysis of naphthalene and its derivatives to obtain C_{60}s and C_{70}s through patching of C_{10} fragments encouraged researchers to adopt chemical pathways for fullerene synthesis [232]. Inspired by Barth and Lawton [138], Scott and coworkers presented their pioneering work of chemically synthesizing bowl-shaped corannulene molecules ($C_{20}H_{10}$) using flash vacuum pyrolysis (FVP) process [209]. Later, the same group developed a rather ground-up chemical synthesis method where commercially available precursors, such as bromomethylbenzene and 2-naphthaldehyde, were used to form C_{60}s [28]. Successive chemical reactions and modifications of these precursors lead to formation of polyarenes and their derivatives, such as $C_{60}H_{30}$, $C_{80}H_{40}$, and $C_{60}H_{27}Cl_3$ [28, 208]. Finally, employing FVP at 1100°C can cause cyclodehydrogenation and cyclodehalogenation of these intermediates to produce notable quantities of isomerically pure C_{60}s [28, 208]. However, the yield of this process is typically low (i.e., <0.4%), which restricts the use of this method for commercial production of fullerenes [207]. Newer techniques for fullerene fragment production and their conjugation processes have been proposed, which require further investigations and research to fully realize [168].

1.3.1.4 Fullerene-Based Nanohybrids   Fullerenes can be hybridized with carbonaceous materials both endohedrally and exohedrally. Endohedral hybrid examples include nanopeapods [218] and nano-onions [219]. Fullerene molecules when entrapped within CNTs are called nanopeapods, while multilayered concentric fullerenes are known as carbon nano-onions. Such encapsulations are performed through thermal annealing [185, 218], carbon vapor deposition (CVD)-based growth processes [44], water-assisted electric arc [107, 185], or wet capillary filling processes [215]. Exohedral conjugations of fullerenes with CNTs [51] and graphene [194] require noncovalent functionalization; π–π interaction between functionalized molecules can hold the carbon allotropes together. However, covalent functionalization can also be performed. For example, porphyrin-derivatized fullerenes can provide bonding through amination reaction with the carboxy-functionalized CNTs [81]. Moreover, harsh chemical reactions forming seamless bonds can give rise to special form of hybrids named "nanobuds" [174, 255].

1.3.2   Characterization

With the advancement of nanotechnology, various techniques have been developed to characterize fullerenes and their hybrids [1–3, 201]. From production of soot to the formation of individual fullerenes, several characterization methods are employed to determine the composition, morphology, and concentration [63, 147]. Such techniques include mass spectroscopy, NMR, optical spectroscopy, HPLC, electron microscopy (EM), etc. We have limited our discussion to fullerene characterization only; NHs in the literature have mostly followed similar characterization tools. However, spectral, physical, and chemical signatures of NHs will differ significantly from fullerenes. This section will briefly describe key characterization techniques for fullerenes.

1.3.2.1 Mass Spectroscopy   Mass spec has been one of the primary identification tools of the first fullerenes. Several mass spec methods have evolved over the years [188]. The key process in this characterization involves ionization and charged separation of neutral molecules according to their mass-to-charge (m/z) ratio. It is a highly sensitive method that can detect as low as 10 ions, enabling detection of trace concentrations of fullerenes [35]. The ionization and desorption of molecules are generally done by laser-induced methods [38]. Other methods of ionization include thermal desorption [48], fast atomic bombardment [161, 189], electrospray ionization [98], etc. For detection of the ionized fullerene samples and their spectral recording, TOF [132] or Fourier transform mass spectrometry [63, 188] (FTMS) methods are used.

1.3.2.2 NMR NMR is also very useful in determination of fullerene and higher fullerene purity [121, 229–231]. 13C-NMR has been proven to bear the first evidences of fullerene structures, which led to the conclusion of C_{60}'s ability to follow the isolated pentagonal rule [132]. It is interesting to notice that the highest obtainable spherical symmetry of C_{60}, when produced or characterized in its purest form, presents with a singular peak around 142–143 ppm in the resonance spectra [147, 230]. On the other hand, five resonance peaks are obtained for the ellipsoidal shape of C_{70} fullerenes [115]. Stability of fullerenes in different reactive environments has also been understood employing NMR method [231].

1.3.2.3 Optical Spectroscopy The ability of light absorption by individual fullerenes differs based on the molecular weight and band structure of the fullerenes. While solubilized in toluene, C_{60} suspensions appear to be magenta or deep purple, whereas C_{70}s exhibit a color close to the red wine [63]. Other higher fullerenes show colors ranging from yellow to green with the increase in molecular weight of the fullerene molecules [68, 121]. Similarly, their light adsorption in infrared and UV region also differs that is utilized for spectral characterizations of fullerenes [147]. For example, Figure 1.5 shows UV–Vis spectral signatures of C_{60} and C_{70} aqueous suspensions. The suspensions were prepared by sonicating powdered fullerenes in a

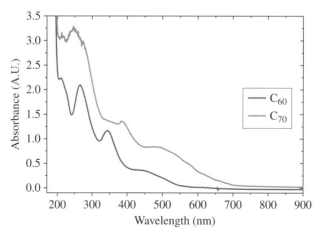

FIGURE 1.5 UV–visible spectra of Pluronic modified C_{60} and C_{70} aqueous suspensions. (*See insert for color representation of the figure.*)

biocompatible polymer solution. UV peaks appeared at 275 and 350 nm for C_{60}, while C_{70} showed widening and broad shoulders at those wavelengths, lacking in distinct peaking behavior. C_{70}s showed peaks close to 410 nm, consistent with the published literature [1, 2].

1.3.2.4 HPLC   HPLC is one of the key techniques for fullerene separation, as discussed earlier. Besides, HPLC can also serve as an effective analytical tool for purity assessment of C_{60}s, C_{70}s, and higher fullerenes [121, 147]. Moreover, HPLC can also be utilized for detection of fullerenes using the well-established elution times, proved to be reliable in the literature [147]. Such detection has been performed in commercial applications also. For example, in a 4.6 mm ID × 250 mm standard commercial column, using toluene as mobile phase at 1.0 ml/min flow rate, UV peaks at 312 nm wavelength for C_{60}, C_{70}, C_{76}, and C_{84} fullerenes can be observed at 8, 12.5, 17, and 23 min, respectively, allowing for their individual characterizations [172].

1.3.2.5 Electron Microscopy   Since the evolution of scanning tunneling microscopy, fullerene structures were confirmed through visual observation [91]. Development of electron microscopic (EM) techniques over the years has allowed for detailed characterization of fullerenes during synthesis and during their postproduction application. EM techniques enable evaluation of size and morphological characteristics of molecular and clustered fullerenes. For example, Figure 1.6 shows high-resolution transmission electron micrographs (HRTEM) obtained for aqueous fullerene clusters, solubilized via sonication in polymeric aqueous suspension as mentioned earlier [1]. Figure 1.6a, b shows C_{60} and C_{70} clusters, respectively. Their morphology appears to be spherical. Figure 1.6c, d presents higher magnified images, confirming fullerene lattice fringes, proving the crystalline nature of the clusters.

1.3.2.6 Static and Dynamic Light Scattering   Fullerene clusters in suspension are characterized using light scattering techniques. Dynamic light scattering (DLS) and static light scattering (SLS) are the most popular tools that are employed to evaluate time-dependent cluster size, fractal dimension, and aggregation propensity of fullerenes and other nanomaterials [1, 2, 117, 118]. Such methods are particularly useful for environmental implication studies, where interaction of fullerene clusters in water under varying chemical conditions can be systematically studied [1, 43]. Here, we will discuss measurement of aggregation kinetics of C_{60}s using DLS technique; detailed description of the SLS technique for determination of aggregate structure of carbonaceous nanomaterial is presented in a previous work by our group [118].

Time-dependent dynamic light scattering (TRDLS) intensity measurement can be performed on C_{60} aqueous suspension against different environmentally relevant concentrations of NaCl salt. The C_{60} aqueous suspension here was prepared by a well-established solvent exchange method [2]. An ALV/CGS-3 compact goniometer system (ALV-Laser Vertriebsgesellschaft m-b.H., Langen/Hessen, Germany) equipped with 22 mW HeNe laser at 632 nm (equivalent to 800 mW laser at 532 nm) and high QE APD detector with photomultipliers of 1:25 sensitivity was used for this purpose. The obtained scattering data for each condition were used to profile

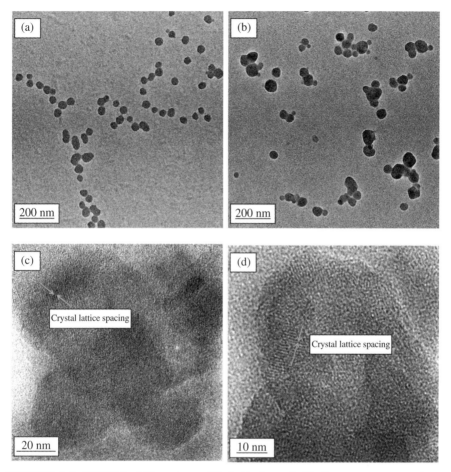

FIGURE 1.6 HRTEM of Pluronic modified (a) C_{60} and (b) C_{70} aqueous suspensions. Zoomed-in micrographs showing (c) C_{60} and (d) C_{70} crystalline features.

time-dependent aggregation of fullerene nanoparticles at each electrolyte condition as shown in Figure 1.7a. It is observed that with no salt addition and at low ionic concentration of NaCl (up to 10 mM), the hydrodynamic radius of C_{60} clusters remained unchanged over time. However, increased aggregation is observed at higher salt concentrations. The initial slope of this profile is the initial rate of aggregation that is proportional to the initial rate constant (k_{in}) and also to the initial concentration of the fullerene suspension (Eq. 1.1) [202]:

$$k \propto \frac{1}{N_0} \left[\frac{dR_h(t)}{dt} \right]_{t \to 0} \tag{1.1}$$

Attachment efficiency (α) of fullerene clusters at each solution condition can then be obtained through dividing the initial aggregation rate at each solution condition by the initial aggregation rate at favorable condition for aggregation (which is obtained at high salt concentration). The theoretical formulation is expressed in Equation 1.2 [202]:

$$\alpha = \frac{\left[dR_h(t)/dt\right]_{t\to 0}}{\left[dR_h(t)/dt\right]_{t\to 0,\text{fav}}} \tag{1.2}$$

The attachment efficiencies can then be plotted against corresponding salt concentrations (Fig. 1.7b), known as stability plot. Figure 1.7b shows that C_{60} aqueous suspension follows classical Derjaguin–Landau–Verwey–Overbeek (DLVO) behavior [43, 202]. Further, quantitation of the aggregation propensity of the fullerenes can be obtained by analyzing the stability plot.

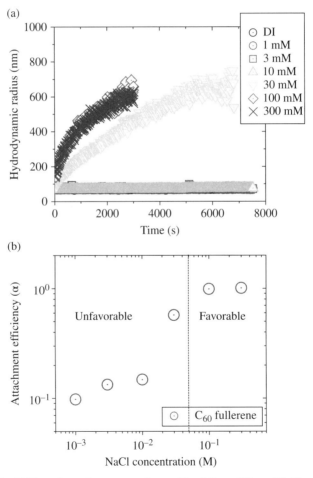

FIGURE 1.7 (a) Time-dependent aggregation profile of C_{60} at different NaCl concentrations. (b) Stability plot for C_{60} aqueous suspension at different NaCl concentrations.

1.4 ENERGY APPLICATIONS

A large number of fullerene-related publications offering insights into energy applications can be found in the literature. A recent literature search in Web of Science® has resulted in a total of 1626 publications from 1991 to 2012 that concern energy applications. The search was performed using a glossary of energy terms and a search algorithm designed with wild cards and Boolean operators. Title field tag and article-only document type were also combined in the search criteria to limit the obtained results. The literature search reveals that the energy application sector of fullerenes and related materials is at an early stage; however, there is a rather rapid increase in the fullerene energy application literature over the past decade (Fig. 1.8a). Most publications focus on fullerenes (89.2%), while less than 10% of the yearly publications are devoted to HOFs (0.3%), fullerene derivatives (6.1%), and/or hybrids (2.6%). The advantages of fullerenes and related materials on the energy application sector are derived from their fascinating characteristics: good acceptors of electrons and exceptionally low reorganization energies in electron transfer [7], superconductivity [94], absorption of light throughout the visible region [157], and their stability due to rigid spherical carbon framework [106].

The retrieved publications have also provided information regarding the various practical applications, which can be generally categorized as follows: (i) solar cells and photovoltaic materials, (ii) hydrogen storage materials, and (iii) electronic components. The technical literature also contains information on properties of fullerenes and related materials relevant to energy applications: (iv) superconductivity, electrical, and electronic properties and (v) photochemical, photophysical, and photocatalytic studies (Fig. 1.8b). The following section will briefly describe the different aspects and relevant properties of energy applications with fullerenes.

1.4.1 Solar Cells and Photovoltaic Materials

Some of the most promising applications for fullerene-related materials are *solar cells and photovoltaic materials* (~47% of the publications; Fig. 1.8b). The increased demand for low-cost renewable energy sources and the photoexcitation properties of C_{60}s and related materials has generated interest for their application as novel photovoltaic materials and has motivated new approaches to production of efficient and inexpensive solar cells and photovoltaic devices.

Solar cells convert the energy of light into electricity by photovoltaic effect and consist of an electron donor and an acceptor material arranged in a bilayer structure of interpenetrating network. Organic materials, for example, conjugated polymers, have been explored as economic alternatives to inorganic semiconductors (silicon, amorphous silicon, gallium arsenide, selenide, etc.) currently used [266]. The discovery of photoinduced electron transfer from conjugated conducting polymers (as donors) and C_{60}s (as acceptors) provided the first highly efficient plastic photovoltaic cell [77, 266]. C_{60}-doped polymers, for example, polyvinylcarbazole (PVK), poly(paraphenylene vinylene) (PPV), and phenylmethylpolysilane (PMPS), have been reported to exhibit exceptionally good photoconductive properties [235, 246]. Organic photovoltaic materials of poly[2-methoxy-5-(2′-ethylhexyloxy)-1,4-phenylenevinylene]

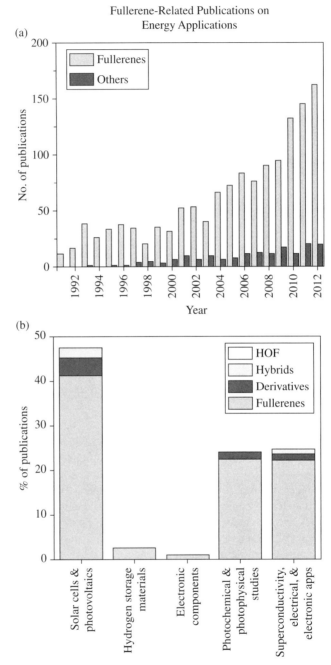

FIGURE 1.8   (a) Total number of publications on fullerene and related materials on energy topics. Note: Others correspond to HOFs, derivatives, and hybrids. (b) Energy applications of fullerenes, HOFs, hybrids, and its derivatives. Source: ISI Web of Science, September 2013.

(MEH-PPV)/C_{60} exhibit an enhancement in the photovoltaic effect with increasing C_{60} concentration [122]. Several studies report the use of other C_{60}-doped polymer combinations for photovoltaic cells: methyl-ethyl-hydroxyl-polypropylvinyl (MEH-PPV)/ C_{60} thin film [204], poly(4-vinyl pyridinated) fullerenes (PVPyF) [137], ITO/polyalkylthiophene (PAT)/C_{60}/Al [182], and poly(3-alkylthiophenes)/C_{60} [37]. However, material stability was found to be a persistent problem for applications of conjugated polymers that are simultaneously exposed to light and oxygen, causing rapid degradation of the materials [176]. Fullerene and high fullerene derivatives (*e.g.*, oligophenylenevinylene (OPV) group attached to C_{60} through a pyrrolidine ring [65], mono- and multiadducts of C_{60} derivative PCBM and MDMO-PPV [75], PPV and PCBM [183], C_{70}/poly(2-methoxy-5-(3,7'-dimethyloctyloxy)-*p*-phenylenevinylene) (MDMO-PPV) [250], poly(2,7-(9-(2'-ethylhecyl)-9-hexyl-fluorene)-alt-5,5-(4'7'-di-2-thienyl2',1',3'-benzothiadiazole)) (PFDTBT) and PCBM [222], C_{70}-PCBM [141]) have also been studied for incorporation in photovoltaic devices; however, such research is only in preliminary stages.

The use of hybrids for efficient solar energy conversion is also emerging. Studies have looked at chemically linked CdSe quantum dots (QDs) with thiol-functionalized C_{60} hybrids. The photoinduced charge separation between CdSe QDs and C_{60}s opens up new design strategies for developing light harvesting assemblies [23]. Other studies have looked into the effects of incorporating CNTs in a polymer–fullerene blend host. Nanobuds (C_{60}-functionalized CNTs) were found to be disadvantageous and somewhat detrimental to overall photovoltaic device performance [8].

The use of fullerenes and other related materials has also been focused on enhancing the thermal stability of solar cells [217], improving the performance and efficiency of polymer–fullerene conjugates [154] and photovoltaic properties of new blends [112], and optimizing polymer–fullerene solar cells [120]. Driven by technology advances, a better understanding of fullerenes and their synthesis and processing techniques will likely allow to lower the cost of the material to meet the exponential demand of the energy industry [223].

1.4.2 Hydrogen Storage Materials

Hydrogen is a clean and renewable source of energy that could be generated by electrolysis of water. Only a small percentage of the publications surveyed here report on fullerene and related materials as storage devices for molecular hydrogen (~3%; Fig. 1.8b). However, it appears that fullerene and related materials may be promising toward storage capacities for hydrogen. In its gaseous form, hydrogen has a low specific volumetric energy density, compared to other liquid fuel sources. To increase its energy density, compressed hydrogen should be stored in a hydrogen storage material such as hydrogen storage alloys and CNTs [62]. Because of low efficiency from high frictional power loss, an electrochemical compressor using a membrane electrode assembly (MEA) film of proton (H^+) conductor is more effective than the conventional mechanical compressing methods [158]. However, humidity affects the proton conductivity and has to be removed from the compressed hydrogen. New fullerene composite membranes have been synthesized and have demonstrated enhanced proton conductivity under low relative humidity conditions [187, 227, 243]. $C_{60}H_{36}$ has been under scrutiny as the source of hydrogen for the *in situ* hydrogenation of $(C_{59}N)_2$. It has led to $C_{59}NH_5$ as the main reaction product, identified by negative-ion mass spectrometry and providing evidence of the usage of C_{60}s as a storage device for hydrogen [239]. The electrochemical compression and hydrogen storage capacity using the MEA of fullerene-related materials (hydrogensulfated fullerenol) have been confirmed [158]. Studies also show that hydrogenation of carbon materials (fullerenes) requires activation centers [203, 239]. While considering these aspects, heteroatoms such as N, P, and S seem to be promising to behave as activators in heteroatom containing carbon materials for hydrogen storage applications. Boron atoms have also been identified for low-energy hydrogenation [203].

It has been demonstrated that coated fullerenes are ideal for many practical hydrogen storage applications. A single Ni-coated fullerene can store up to three H_2 molecules (storage capacity up to 6.8 wt%) [213]. The capacity of charged fullerenes C_n ($20 \leq n \leq 82$) as hydrogen storage media has been found to be up to 8.0 wt% [262]. Hydrogenated silicon fullerene has also been proposed for hydrogen storage with up to 9.48 wt% storage capacity [268]. Calcium has been proposed as a desirable metal coating to functionalize fullerenes and obtain high-capacity hydrogen storage materials with a hydrogen uptake up to 8.4 wt% [261]. Ti-decorated-doped silicon fullerene, Ca-coated boron fullerenes, and Mg-decorated boron fullerenes with storage capacities up to 5.23, 8.2, and 14.2 wt%, respectively, have also been reported [24, 143, 144].

1.4.3 Electronic Components (Batteries, Capacitors, and Open-Circuit Voltage Applications)

The incorporation of fullerene and related materials to improve the electrochemical performance of electronic components is scarcely reported by specific research groups, with publications in the last 10 years, dealing mostly with batteries [12–19, 193], capacitors [66, 116, 125, 253], and open-circuit voltage studies [50, 57, 79, 84, 129, 142, 166, 170, 177, 220, 236–238, 256, 267]. However, the demand of these components with higher capacity will likely increase to meet future demands.

1.4.4 Superconductivity, Electrical, and Electronic Properties Relevant to Energy Applications

Superconductivity is the event of exactly zero electrical resistance and expulsion of magnetic fields, occurring in certain materials when cooled below a critical temperature T_c [92]. Zero resistance and magnetic field exclusion have a major impact on electric power transmission and also enable the development of much smaller electronic components that are more reliable, efficient, and environmentally benign for energy applications [85, 92]. Fullerene-based superconductors have drawn enormous scientific interest toward energy applications (~25% of the publications; Fig. 1.8b). In 1991, research on semiconducting technology found that alkali metal-doped films of C_{60} lead to metallic behavior [88]. Shortly thereafter, these alkali-doped C_{60}s are found to be superconducting at T_c that is only exceeded by the cuprates [70, 94, 101]. It was also found that potassium-doped C_{60} becomes superconducting at 18 K, making it the highest transition temperature for a molecular superconductor [94]. It has been discovered that the superconducting transition temperature in alkaline metal-doped fullerene increases with the unit-cell volume [275]. Cesium-doped fullerene (Cs_3C_{60}) has been reported to lead to superconductivity at 38 K under applied pressure in 1998 [76], but the highest superconducting transition temperature of 33 K at ambient pressure was reported for cesium–rubidium-doped fullerene (Cs_2RbC_{60}) in 1991 [225].

One of the biggest limitations of superconducting fullerenes is their instability in air; exposing the materials to air for a fraction of a second can completely compromise the superconductivity. Investigations in superconducting fullerenes continue as new combinations of surface and other fullerene-related materials are synthesized [27]. HOF analogues of the alkali-doped fullerenes have also been investigated; however, results indicate absence of superconductivity above 5 K [55]. A vast body of literature can be found on superconducting fullerenes and on the electrical and electronic properties of these materials. However, more research on superconductors using other fullerene-related materials is necessary.

1.4.5   Photochemical and Photophysical Properties Pertinent for Energy Applications

The photoactivity and the ability to "tune" fullerenes [40] and related material properties (*i.e.*, band gap, chemical environment, conductance, thermal storage, etc.) are fundamentally important to fabricate devices for the collection, conversion, and storage of renewable energy (solar energy). Photochemical and photophysical properties of fullerenes and related materials result in the distinctive switching of chemical reactions, electrical energy, luminescence, degradation, absorption, and thermal and electrical properties of functional composites, which is crucial for novel devices with excellent performance.

In general terms, research in photochemical and photophysical properties of fullerenes revolves around optical absorption [82, 173, 191], photoluminescence and fluorescence [247, 265], excited state dynamics and properties of the singlet and triplet states [25, 72, 110, 152, 184], photochemical reactions [5, 6, 224], synthesis [21, 83, 105, 108], photocatalyst degradation [6, 134, 135, 165], singlet oxygen production, and charge transfer reactions [26, 30, 109, 178, 252, 264]. A fair number (24.2%) of the retrieved publications deal with photochemical and photophysical information on fullerenes (Fig. 1.8b). However, additional progress is required given the diversity of fullerene and related materials and the necessity to functionalize and "tune" their properties for specific energy applications.

1.5   ENVIRONMENTAL CONSIDERATIONS FOR FULLERENE SYNTHESIS AND PROCESSING

Sustainable use of materials for energy applications not only demands for a renewable alternative with a small energy footprint but also necessitates low-risk involvement in the usage and disposal of such materials. Fullerenes, one of the most attractive nanomaterials for energy applications, should present minimum environmental risk to be considered truly sustainable. However, fullerenes' unique electronic properties are also known to be responsible for reactive oxygen species (ROS) generation, resulting in environmental toxicity. Moreover, synthesis and solubilization process of fullerenes and the soft polymeric and surfactant surface coatings (used for processing) will likely contribute to altered environmental risk. Thus, synthesis and processing of fullerenes, higher fullerenes, and their hybrids necessitate careful consideration for choosing potentially greener options [1, 3, 200, 201].

For example, organic photovoltaics, a major fullerene-based device, is an assemblage of multiple layers containing electron donors and acceptors, electrodes, and hole/electron transporters packaged within plastic materials such as poly(ethylene terephthalate) or PET [133]. These organic photovoltaics at the end of their usage will likely be disposed off to landfills [160, 175]. Though PET packagings are non-biodegradable, their fragmentation through abrasion and photodegradation under long-term exposure to sunlight are likely [69, 249]. Such degradation can lead to potential release of fullerene derivatives PCBM from the interior of these solar cells [277]. During their residence in landfills and the material exposure to soil surfaces and water (after being carried via surface runoffs), these fullerene, their derivatives, and associated solvents will inevitably interact with the aquatic environment and terrestrial ecosystems. Thus, true sustainable energy generation requires a thorough understanding of material energy cost as well as environmental risks associated with environmental fate, transport, and exposure.

1.5.1   Existing Environmental Literature for C_{60}

C_{60} and its several derivatives, including PCBMs, have been systematically studied to better understand their fate and transport, environmental transformations, and toxicity toward aquatic species. It is important to note here that fullerene preparation methods and the chemical identity of the surface moieties of their derivatives can play a significant role in their environmental behavior. Fullerene cluster size and surface chemistry are known to impact their environmental behavior [43, 150, 248]. Importantly, the initial cluster size and surface chemistry are found to differ based on fullerene processing techniques; while extended mixing of water results in larger fullerene clusters with rough irregular edges, solvent exchange using toluene or THF as intermediate can produce smooth round edge crystalline structures [33]. Similarly, surface charge of the aqueous suspension using THF as intermediate was found to be significantly more negative compared to that produced via extended mixing [32]. Systematic evaluation of aqueous C_{60} and their derivatives showed such technique-dependent behavior; the OPV-containing PCBM and corresponding butyl (PCBB) and octyl (PCBO) esters showed exceptionally high stability compared to pristine aqueous C_{60} [29]. Mobility of fullerenes in porous media and their interfacial interaction are also highly dependent on the preparation methods [41], stabilizing agents [248], and particle size [41]. For example, toluene-dissolved aqueous C_{60} was found to show higher toxicity to Japanese medaka fish compared to C_{60}s solubilized with dimethyl sulfoxide (DMSO) any extended stirring [123]. Toluene-dissolved C_{60}

suspension exhibited spherical aggregates, while the other methods formed mesoscale aggregates. Thus, synthesis and preparation techniques of fullerenes can allow for environmentally friendly alternatives.

C_{60} aqueous suspensions are known to show toxicity to microbial entities in aquatic [73, 150] and soil [114] media, other invertebrates [180], and fish [180]. There have been evidences of genotoxicity [56] and developmental toxicity [226] induced by C_{60} aqueous suspensions. Several mechanisms for fullerene toxicity are postulated that include ROS-mediated oxidative damage [205], lipid peroxidation of cell organelles [206], direct contact with the cell membrane, and consequent membrane protein oxidation. Such toxic potential depends on the physicochemical characteristics such as surface charge and aggregate sizes [41, 150] as well as on preparation methods [73, 150] or solvents used [150]. For example, continuous sonication and separation through filtration of fullerene suspensions yield smaller aggregates that tend to produce more ROS leading to increased antibacterial effects [41]. Solvent effects on C_{60} toxicity evaluations, on the other hand, have created controversies in the literature [96, 97]. Toluene- and THF-based fullerene suspensions have shown significantly higher toxicity compared to solvent-free ones [96, 97, 276]. In some cases residual toluene or THF and their degradation by-products were also found to be more responsible for the enhanced toxicity than the pristine C_{60} aqueous suspension [96, 97]. It is also important to note that fullerene and its derivatives are often coated with polymers or surfactants to attain desired properties. Many of such synthetic molecules possess either nonbiodegradable polyaromatics or cyclic organic compounds. Such moieties have already reported to exhibit toxic behavior. For example, gamma-cyclodextrin–C_{60} aqueous suspension showed higher photodynamic activity under UV illumination than polyvinyl pyrrollidone (PVP)-dissolved C_{60} [67]. Similarly, Tween-80 induced higher toxicity to E. coli when compared to N,N-dimethylformamide in solubilizing C_{60} [47]. Thus, not only solvents used for fullerene synthesis but also chemical moieties used to functionalize these carbon allotrope surfaces need careful evaluation to reduce environmental risk. However, it has been found that the solvents' inherent toxicity is increased when it is associated with fullerenes [52].

Environmental transformation of C_{60}s through reaction with atmospheric oxygen [234] or ozone or via photochemical reactions under sunlight or UV irradiation or by adsorption of bio- and geomacromolecules (e.g., humic and fulvic acids) is also inevitable. Such transformations, which are likely going to be influenced by their synthesis and processing techniques, will also influence their subsequent environmental interactions. Sunlight or UV exposure to fullerenes is inevitable in the natural environment, which is known to cause chemical transformation of fullerenes by surface alteration via oxidation [139]. Such functionalization can enhance their stability in water, thus making them more persistent [145] in aqueous environment. Similarly, dissolved organic matter (NOM), generated from degradation of flora and fauna, can coat fullerene surfaces and stabilize their colloidal presence in water [271]. Dissolved organic matter in wastewater effluent was also found to inhibit fullerene aggregation by providing similar steric stabilization [258]. Thus, alteration of fullerene interaction in water can occur as a result of a combined presence of sunlight and NOMs [195]; where sunlight-induced functionalization can inhibit humic adsorption of fullerenes [195], humics can reduce UV-inflicted oxidation via scavenging of ROS [104]. Transformations of C_{60}s by ozonation or UV irradiation have shown to increase ROS production and subsequent *E. coli* inactivation [45, 46]. NOMs when interacting with fullerene suspensions can affect the triplet excited state, an intermediate state responsible for ROS generation [127]. While NOMs can quench such photoactivity of pristine C_{60} suspensions, they can enhance it for fullerenols [127]. Therefore, a complex interplay of both NOM and sunlight exposure will determine the environmental fate of the fullerenes and necessitate systematic evaluation.

1.5.2   Environmental Literature Status for Higher Fullerenes and NHs

Unlike C_{60}s, environmental considerations of higher fullerenes (*e.g.*, C_{70}, C_{76}, C_{78}, C_{84}, C_{90}, etc.) and NHs have mostly been ignored [2, 3, 200, 201]. A limited number of studies evaluated colloidal properties [2, 53, 153] and toxicity [211] of C_{70}s. One of the previous studies from the authors' group incorporated additional higher-order fullerene (*i.e.*, C_{76} and C_{84}) colloidal property evaluations upon aqueous solubilization [2]. These studies show that higher fullerene hydrodynamic radii and surface potential differ substantially from those of C_{60}s. The enhanced surface potential measured in higher fullerene suspensions appeared to have originated from differences in their molecular structures and their enhanced electronegativity [199]. It is probable that such differences in electronic properties will alter their interfacial interaction in the environment; predicting such behavior of these higher fullerenes from C_{60}s is thus unrealistic. It is likely that stabilization achieved via increased electron density will make the higher fullerenes more mobile in the aqueous media. Our unpublished work with four different fullerenes (C_{60}, C_{70}, C_{76}, and C_{84}) shows evidence of such behavior.

Toxicity studies on *D. magna*, an aquatic organism, showed acute toxicity in presence of gallic acid stabilized C_{70} suspensions; mechanism identified was oxidative stress generated from fullerenes [211]. It is also to be noted that higher fullerene isomers possess smaller band gap, which can make them more reactive [61] compared to C_{60}s and C_{70}s, influencing ROS generation and subsequent toxic potential of the materials. Such band gap-modulated toxicity mechanism is already demonstrated in the case of metal nanoparticles [269]. NHs, on the other hand, have not yet been studied for environmental fate or toxicological implications. However, it can be safely argued that fullerene, their derivatives, and higher fullerenes when conjugated to form NH ensembles will likely present altered electron charge transfer, band gap, photoactivity, sorption properties, morphology, etc., which will result in a unique environmental behavior. The state-of-the-art literature shows a major gap in knowledge for risk and safety evaluation of higher fullerenes and fullerene-based NHs [2, 3, 200, 201].

1.5.3   Environmental Considerations

Based on the existing literature, it is clearly evident that advantages of fullerenes do not come without associated environmental risk. Figure 1.9 presents a schematic showing probable release of fullerene and its homologues from energy applications

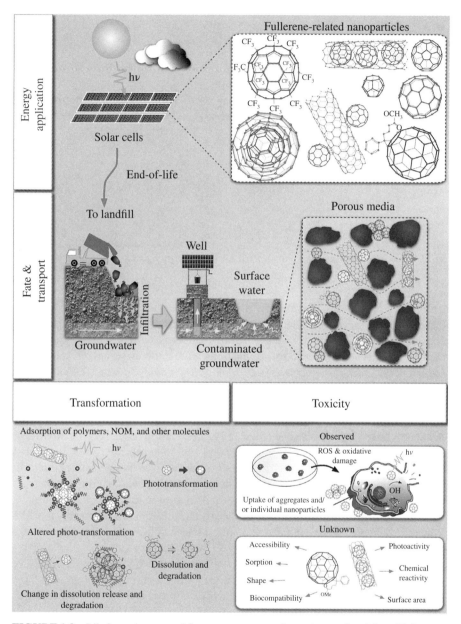

FIGURE 1.9   Likely environmental fate, transport, transformation, and toxicity of fullerenes and related nanomaterials. (*See insert for color representation of the figure.*)

to the environment as well as their fate, transport, and transformation, leading to potential environmental risk. Therefore, choice of synthesis and processing techniques as well as chemical functionalization of fullerenes should be considered keeping environmental risk factors in mind. C_{60}s, higher fullerenes, and NHs will

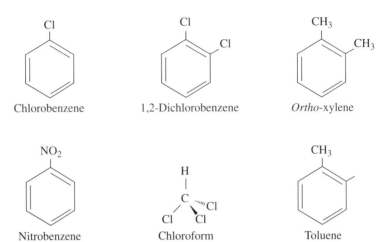

Chlorobenzene   1,2-Dichlorobenzene   *Ortho*-xylene

Nitrobenzene   Chloroform   Toluene

FIGURE 1.10   Structures of commonly used solvents for fullerene synthesis and processing.

likely to be, if not already, used in commercial energy applications. Exposure of these materials during manufacture, use, or end of life is thus likely and should be considered for their design and production. This section discusses key considerations of fullerene use in energy applications and identifies some relevant aspects critical to ensure environmental safety.

1.5.3.1   Consideration for Solvents   Solvents used to formulate the electron donor/acceptor (P3HT:PCBM) blends or thin films generally consist of chlorobenzene [124, 171], dichlorobenzene [124, 171], orthoxylene [171], mixture of chloro- and nitrobenzene [171], chloroform [167], and toluene [274] (Fig. 1.10). Most of these solvents possess aromatic groups and are halogenated, which are known as nonbiodegradable as well as toxic [64]. Thus, choosing these solvents may compromise environmental safety due to the residual in the solar cells. Similarly, solvents used for fullerene processing, for example, aqueous solubilization, may also pose such risks; for example, THF or toluene. For greener synthesis of energy devices, comparatively benign, short-chain substituted alkanes (*e.g.*, chloroform) can be opted [64]; or novel solvent-free methods for blend formation can also be developed [39]. Thermal annealing process for morphological control of the P3HT:PCBM film can be chosen over solvent evaporation annealing to reduce solvent use. However, such alternative decisions require systematic studies comparing solvent effects with energy efficiencies of synthesized and/or processed fullerenes.

1.5.3.2   Considerations for Derivatization   Fullerene derivatives, for example, PCBM, can have C_{60} and C_{70} as origins, which are mostly employed as the electron acceptors [31]. Recently, higher fullerenes, that is, C_{84}-based PCBM derivatives, have also been studied for solar cell applications [128]. Methanofullerenes like

PCBM have shown toxic effects to *D. magna* [34], which can be reduced through substitution with low toxicity derivatives. Substitutions of phenyl group with thienyl groups and other alkyl analogues can also present lower-risk alternatives. Other fullerene derivatives used should also be carefully considered, since many of such possess aromatic rings and/or cyclic chemical structures with less biodegradability. For example, indene fullerene [273], 1,4-di(organo)fullerenes [159], dihydronaphthyl fullerenes [54], penta(organo)fullerenes [179], etc. contain such nondegradable chemical structures, whereas fulleropyrrolidines [10] can be more bio-friendly. Moreover, derivatization is also performed using certain chromophores to enhance photoinduced charge transfers, such as porphyrin, phthalocyanines, etc. [254]. Such chromophores with potential safe usage for fullerene derivatization should be carefully evaluated for their biodegradability.

1.5.3.3 *Consideration for Coatings*   Moreover, fullerene processing involves use of polymers or surfactants to enhance their dispersion as well as photophysical properties. However, it has been determined in studies that biocompatibility and aggregation behavior depend significantly on the coating characteristics on fullerenes or other nanomaterials [233]. For example, fullerene suspensions stabilized with sodium dodecyl sulfate (SDS) and Triton X 100 produced higher ROS compared to pure fullerene water suspensions [140]. Furthermore, transformation processes in the environment can become complex as overcoating of these coated fullerenes with geo- and biomacromolecules will alter their environmental persistence as well as potential risk [60].

Fullerene synthesis, processing, and separation thus require an underlying risk consideration. Properties of the solvents used, relative degradability of the derivatives and coatings, contribution of coatings on environmental safety, and such similar issues should be considered for safer usage of fullerenes in energy applications. A few critical questions that should be asked to pursue lower risk in fullerene's energy applications are listed below:

1. Are the solvents chosen for fullerene synthesis and separation relatively less toxic?
2. Are there environmentally safer alternatives while derivatizing or hybridizing fullerenes?
3. Are the chemical moieties used to coat fullerene surfaces environmentally benign?
4. Can the fullerenes be immobilized to reduce their release from the devices or processes?

However, environmental considerations should pose these questions at a minimum, encouraging a more systematic and complete environmental study. Effective and sustainable use of fullerenes in energy applications requires attaining environmentally safe usage of these materials, which has been a missing link in most material science research and development.

REFERENCES

[1] Aich N, Boateng L K, Flora J R V and Saleh N B. 2013. Preparation of non-aggregating aqueous fullerenes in highly saline solutions with a biocompatible non-ionic polymer. *Nanotechnology* **24**, 395602.

[2] Aich N, Flora J R V and Saleh N B. 2012. Preparation and characterization of stable aqueous higher-order fullerenes. *Nanotechnology* **23**, 055705.

[3] Aich N, Plazas-Tuttle J, Lead J R and Saleh N B. 2014. A critical review of nanohybrids: synthesis, applications and environmental implications. *Environmental Chemistry* **11**, 609–23.

[4] Ajie H, Alvarez M M, Anz S J, Beck R D, Diederich F, Fostiropoulos K, Huffman D R, Kratschmer W, Rubin Y, Schriver K E, Sensharma D and Whetten R L. 1990. Characterization of the soluble all-carbon molecules C_{60} and C_{70}. *The Journal of Physical Chemistry* **94**, 8630–3.

[5] Akselrod L, Byrne H J, Thomsen C, Mittelbach A and Roth S. 1993. Raman studies of photochemical-reactions in fullerene films. *Chemical Physics Letters* **212**, 384–90.

[6] Akselrod L, Byrne H J, Thomsen C and Roth S. 1993. Reversible photochemical processes in fullerenes—a Raman-study. *Chemical Physics Letters* **215**, 131–6.

[7] Allemand P M, Koch A, Wudl F, Rubin Y, Diederich F, Alvarez M M, Anz S J and Whetten R L. 1991. Two different fullerenes have the same cyclic voltammetry. *Journal of the American Chemical Society* **113**, 1050–1.

[8] Alley N J, Liao K S, Andreoli E, Dias S, Dillon E P, Orbaek A W, Barron A R, Byrne H J and Curran S A. 2012. Effect of carbon nanotube-fullerene hybrid additive on P3HT:PCBM bulk-heterojunction organic photovoltaics. *Synthetic Metals* **162**, 95–101.

[9] Anthopoulos T D, Singh B, Marjanovic N, Sariciftci N S, Ramil A M, Sitter H, Colle M and de Leeuw D M. 2006. High performance n-channel organic field-effect transistors and ring oscillators based on C-60 fullerene films. *Applied Physics Letters* **89**, 213504.

[10] Antonietta Loi M, Denk P, Hoppe H, Neugebauer H, Winder C, Meissner D, Brabec C, Serdar Sariciftci N, Gouloumis A, Vazquez P and Torres T. 2003. Long-lived photoinduced charge separation for solar cell applications in phthalocyanine-fulleropyrrolidine dyad thin films. *Journal of Materials Chemistry* **13**, 700–4.

[11] Arico A S, Bruce P, Scrosati B, Tarascon J M and Van Schalkwijk W. 2005. Nanostructured materials for advanced energy conversion and storage devices. *Nature Materials* **4**, 366–77.

[12] Arie A A, Chang W and Lee J K. 2010. Effect of fullerene coating on silicon thin film anodes for lithium rechargeable batteries. *Journal of Solid State Electrochemistry* **14**, 51–6.

[13] Arie A A and Lee J K. 2009. Fullerene coated silicon electrodes prepared by a plasma-assisted evaporation technique for the anodes of lithium secondary batteries. *Journal of Ceramic Processing Research* **10**, 614–17.

[14] Arie A A and Lee J K. 2010. A study of Li-ion diffusion kinetics in the fullerene-coated Si anodes of lithium ion batteries. *Physica Scripta* **T139**, 014013. 10.1088/0031-8949/2010/T139/014013.

[15] Arie A A and Lee J K. 2011. Effect of boron doped fullerene C-60 film coating on the electrochemical characteristics of silicon thin film anodes for lithium secondary batteries. *Synthetic Metals* **161**, 158–65.

[16] Arie A A and Lee J K. 2011. Nano-carbon coating layer prepared by the thermal evaporation of fullerene C-60 for lithium metal anodes in rechargeable lithium batteries. *Journal of Nanoscience and Nanotechnology* **11**, 6569–74.

[17] Arie A A and Lee J K. 2012. Fullerene C-60 coated silicon nanowires as anode materials for lithium secondary batteries. *Journal of Nanoscience and Nanotechnology* **12**, 3547–51.

[18] Arie A A, Song J O and Lee J K. 2009. Structural and electrochemical properties of fullerene-coated silicon thin film as anode materials for lithium secondary batteries. *Materials Chemistry and Physics* **113**, 249–54.

[19] Arie A A, Vovk O M, Song J O, Cho B W and Lee J K. 2009. Carbon film covering originated from fullerene C-60 on the surface of lithium metal anode for lithium secondary batteries. *Journal of Electroceramics* **23**, 248–53.

[20] Austin S J, Fowler P W, Hansen P, Monolopoulos D E and Zheng M. 1994. Fullerene isomers of C_{60}. Kekulé counts versus stability. *Chemical Physics Letters* **228**, 478–84.

[21] Averdung J and Mattay J. 1996. Syntheses of aziridino-60 fullerenes via photochemically induced conversions of 1,2,3-triazolino-60 fullerenes and azafulleroids. *Journal of Information Recording* **22**, 577–80.

[22] Averitt R D, Alford J M and Halas N J. 1994. High-purity vapor phase purification of C[sub 60]. *Applied Physics Letters* **65**, 374–6.

[23] Bang J H and Kamat P V. 2011. CdSe quantum dot-fullerene hybrid nanocomposite for solar energy conversion: electron transfer and photoelectrochemistry. *ACS Nano* **5**, 9421–7.

[24] Barman S, Sen P and Das G P. 2008. Ti-decorated doped silicon fullerene: a possible hydrogen-storage material. *Journal of Physical Chemistry C* **112**, 19963–8.

[25] Benasson R V, Schwell M, Fanti M, Wachter N K, Lopez J O, Janot J M, Birkett P R, Land E J, Leach S, Seta P, Taylor R and Zerbetto F. 2001. Photophysical properties of the ground and triplet state of four multiphenylated 70 fullerene compounds. *Chemphyschem* **2**, 109–14.

[26] Bernstein R and Foote C S. 1999. Singlet oxygen involvement in the photochemical reaction of C-60 and amines. Synthesis of an alkyne-containing fullerene. *Journal of Physical Chemistry A* **103**, 7244–7.

[27] Bhuiyan K H and Mieno T. 2003. Effect of oxygen on electric conductivities of C-60 and higher fullerene thin films. *Thin Solid Films* **441**, 187–91.

[28] Boorum M M, Vasil'ev Y V, Drewello T and Scott L T. 2001. Groundwork for a rational synthesis of C_{60}: cyclodehydrogenation of a $C_{60}H_{30}$ polyarene. *Science* **294**, 828–31.

[29] Bouchard D, Ma X and Isaacson C. 2009. Colloidal properties of aqueous fullerenes: isoelectric points and aggregation kinetics of C_{60} and C_{60} derivatives. *Environmental Science & Technology* **43**, 6597–603.

[30] Bourdelande J L, Font J and Gonzalez-Moreno R. 2001. Fullerene C-60 bound to insoluble hydrophilic polymer: synthesis, photophysical behavior, and generation of singlet oxygen in water suspensions. *Helvetica Chimica Acta* **84**, 3488–94.

[31] Brabec C J, Gowrisanker S, Halls J J M, Laird D, Jia S J and Williams S P. 2010. Polymer-fullerene bulk-heterojunction solar cells. *Advanced Materials* **22**, 3839–56.

[32] Brant J, Lecoanet H, Hotze M and Wiesner M. 2005. Comparison of electrokinetic properties of colloidal fullerenes (n-C-60) formed using two procedures. *Environmental Science & Technology* **39**, 6343–51.

[33] Brant J A, Labille J, Bottero J Y and Wiesner M R. 2006. Characterizing the impact of preparation method on fullerene cluster structure and chemistry. *Langmuir* **22**, 3878–85.

[34] Brausch K A, Anderson T A, Smith P N and Maul J D. 2010. Effects of functionalized fullerenes on bifenthrin and tribufos toxicity to Daphnia magna: survival, reproduction, and growth rate. *Environmental Toxicology and Chemistry* **29**, 2600–6.

[35] Buchanan M V and Hettich R L. 1993. Fourier transform mass spectrometry of high-mass biomolecules. *Analytical Chemistry* **65**, 245A–59A.

[36] Bunshah R F, Jou S, Prakash S, Doerr H J, Isaacs L, Wehrsig A, Yeretzian C, Cynn H and Diederich F. 1992. Fullerene formation in sputtering and electron beam evaporation processes. *The Journal of Physical Chemistry* **96**, 6866–9.

[37] Camaioni N, Garlaschelli L, Geri A, Maggini M, Possamai G and Ridolfi G. 2002. Solar cells based on poly(3-alkyl)thiophenes and 60 fullerene: a comparative study. *Journal of Materials Chemistry* **12**, 2065–70.

[38] Campbell E E B and Hertel I V. 1992. Molecular beam studies of fullerenes. *Carbon* **30**, 1157–65.

[39] Casalegno M, Zanardi S, Frigerio F, Po R, Carbonera C, Marra G, Nicolini T, Raos G and Meille S V. 2013. Solvent-free phenyl-C_{61}-butyric acid methyl ester (PCBM) from clathrates: insights for organic photovoltaics from crystal structures and molecular dynamics. *Chemical Communications* **49**, 4525–7.

[40] Cates N C, Gysel R, Beiley Z, Miller C E, Toney M F, Heeney M, McCulloch I and McGehee M D. 2009. Tuning the properties of polymer bulk heterojunction solar cells by adjusting fullerene size to control intercalation. *Nano Letters* **9**, 4153–7.

[41] Chae S R, Badireddy A R, Budarz J F, Lin S H, Xiao Y, Therezien M and Wiesner M R. 2010. Heterogeneities in fullerene nanoparticle aggregates affecting reactivity, bioactivity, and transport. *ACS Nano* **4**, 5011–18.

[42] Chandrakumar K R S and Ghosh S K. 2008. Alkali-metal-induced enhancement of hydrogen adsorption in C-60 fullerene: an ab initio study. *Nano Letters* **8**, 13–19.

[43] Chen K L and Elimelech M. 2009. Relating colloidal stability of fullerene (C-60) nanoparticles to nanoparticle charge and electrokinetic properties. *Environmental Science & Technology* **43**, 7270–6.

[44] Chikkannanavar S B, Luzzi D E, Paulson S and Johnson A T. 2005. Synthesis of peapods using substrate-grown SWNTs and DWNTs: an enabling step toward peapod devices. *Nano Letters* **5**, 151–5.

[45] Cho M, Fortner J D, Hughes J B and Kim J H. 2009. *Escherichia coli* inactivation by water-soluble, ozonated C-60 derivative: kinetics and mechanisms. *Environmental Science & Technology* **43**, 7410–15.

[46] Cho M, Snow S D, Hughes J B and Kim J-H. 2011. *Escherichia coli* inactivation by UVC-irradiated C_{60}: kinetics and mechanisms. *Environmental Science & Technology* **45**, 9627–33.

[47] Cook S M, Aker W G, Rasulev B F, Hwang H-M, Leszczynski J, Jenkins J J and Shockley V. 2010. Choosing safe dispersing media for C_{60} fullerenes by using cytotoxicity tests on the bacterium *Escherichia coli*. *Journal of Hazardous Materials* **176**, 367–73.

[48] Cox D M, Behal S, Disko M, Gorun S M, Greaney M, Hsu C S, Kollin E B, Millar J and Robbins J. 1991. Characterization of C_{60} and C_{70} clusters. *Journal of the American Chemical Society* **113**, 2940–4.

[49] Cox D M, Sherwood Rexford D, Tindall P, Creegan Kathleen M, Anderson W and Martella David J. 1992. Mass Spectrometric, Thermal, and Separation Studies of Fullerenes. In: *Fullerenes: Synthesis, Properties, and Chemistry of Large Carbon Clusters*, ed G S Hammond and V J Kuck (ACS Symposium Series, Washington, DC: American Chemical Society) pp 117–25.

[50] Cremer J, Bauerle P, Wienk M M and Janssen R A J. 2006. High open-circuit voltage poly(ethynylene bithienylene): fullerene solar cells. *Chemistry of Materials* **18**, 5832–4.

[51] D'Souza F, Chitta R, Sandanayaka A S D, Subbaiyan N K, D'Souza L, Araki Y and Ito O. 2007. Supramolecular carbon nanotube-fullerene donor–acceptor hybrids for photoinduced electron transfer. *Journal of the American Chemical Society* **129**, 15865–71.

[52] Dai J, Wang C, Shang C I, Graham N and Chen G H. 2012. Comparison of the cytotoxic responses of *Escherichia coli* (*E. coli*) AMC 198 to different fullerene suspensions (nC(60)). *Chemosphere* **87**, 362–8.

[53] Deguchi S, Alargova R G and Tsujii K. 2001. Stable dispersions of fullerenes, C_{60} and C_{70}, in water. Preparation and characterization. *Langmuir* **17**, 6013–17.

[54] Deng L-L, Feng J, Sun L-C, Wang S, Xie S-L, Xie S-Y, Huang R-B and Zheng L-S. 2012. Functionalized dihydronaphthyl-C_{60} derivatives as acceptors for efficient polymer solar cells with tunable photovoltaic properties. *Solar Energy Materials and Solar Cells* **104**, 113–20.

[55] Denning M S, Dennis T J S, Rosseinsky M J and Shinohara H. 2001. K3+delta C_{84}. Higher fullerene analogues of the A(3)C(60) superconductors. *Chemistry of Materials* **13**, 4753–9.

[56] Dhawan A, Taurozzi J S, Pandey A K, Shan W Q, Miller S M, Hashsham S A and Tarabara V V. 2006. Stable colloidal dispersions of C_{60} fullerenes in water: evidence for genotoxicity. *Environmental Science & Technology* **40**, 7394–401.

[57] Di Nuzzo D, Wetzelaer G, Bouwer R K M, Gevaerts V S, Meskers S C J, Hummelen J C, Blom P W M and Janssen R A J. 2013. Simultaneous open-circuit voltage enhancement and short-circuit current loss in polymer: fullerene solar cells correlated by reduced quantum efficiency for photoinduced electron transfer. *Advanced Energy Materials* **3**, 85–94.

[58] Diack M, Hettich R L, Compton R N and Guiochon G. 1992. Contribution to the isolation and characterization of buckminsterfullerenes. *Analytical Chemistry* **64**, 2143–8.

[59] Diederich F, Ettl R, Rubin Y, Whetten R L, Beck R, Alvarez M, Anz S, Sensharma D, Wudl F, Khemani K C and Koch A. 1991. The higher fullerenes: isolation and characterization of C_{76}, C_{84}, C_{90}, C_{94}, and $C_{70}O$, an oxide of D_{5h}-C_{70}. *Science* **252**, 548–51.

[60] Diegoli S, Manciulea A L, Begum S, Jones I P, Lead J R and Preece J A. 2008. Interaction between manufactured gold nanoparticles and naturally occurring organic macromolecules. *Science of the Total Environment* **402**, 51–61.

[61] Diener M D and Alford J M. 1998. Isolation and properties of small-bandgap fullerenes. *Nature* **393**, 668–71.

[62] Dillon A C, Jones K M, Bekkedahl T A, Kiang C H, Bethune D S and Heben M J. 1997. Storage of hydrogen in single-walled carbon nanotubes. *Nature* **386**, 377–9.

[63]   Dresselhaus M S. 1996. *Science of Fullerenes and Carbon Nanotubes* (San Diego, CA: Academic Press).

[64]   Duchowicz P, Vitale M and Castro E. 2008. Partial Order Ranking for the aqueous toxicity of aromatic mixtures. *Journal of Mathematical Chemistry* **44**, 541–9.

[65]   Eckert J F, Nicoud J F, Nierengarten J F, Liu S G, Echegoyen L, Barigelletti F, Armaroli N, Ouali L, Krasnikov V and Hadziioannou G. 2000. Fullerene-oligophenylenevinylene hybrids: synthesis, electronic properties, and incorporation in photovoltaic devices. *Journal of the American Chemical Society* **122**, 7467–79.

[66]   Egashira M, Okada S, Korai Y, Yamaki J and Mochida I. 2005. Toluene-insoluble fraction of fullerene-soot as the electrode of a double-layer capacitor. *Journal of Power Sources* **148**, 116–20.

[67]   Eropkin M I, Piotrovskii L B, Eropkina E M, Dumpis M A, Litasova E V and Kiselev O I. 2011. Effect of polymer carrier origin and physical state on fullerene C_{60} phototoxicity in vitro. *Eksperimental'naia i klinicheskaia farmakologiia* **74**, 28–31.

[68]   Ettl R, Chao I, Diederich F and Whetten R L. 1991. Isolation of C_{76}: a chiral D_2 allotrope of carbon. *Nature* **353**, 149–53.

[69]   Fechine G J M, Souto-Maior R M and Rabello M S. 2002. Structural changes during photodegradation of poly(ethylene terephthalate). *Journal of Materials Science* **37**, 4979–84.

[70]   Fleming R M, Ramirez A P, Rosseinsky M J, Murphy D W, Haddon R C, Zahurak S M and Makhija A V. 1991. Relation of structure and superconducting transition-temperatures in A_3C_{60}. *Nature* **352**, 787–8.

[71]   Folkman J K, Risom L, Jacobsen N R, Wallin H, Loft S and Moller P. 2009. Oxidatively damaged DNA in rats exposed by oral gavage to C-60 fullerenes and single-walled carbon nanotubes. *Environmental Health Perspectives* **117**, 703–8.

[72]   Foote C S. 1994. Photophysical and photochemical properties of fullerenes. *Electron Transfer I* **169**, 347–63.

[73]   Fortner J D, Lyon D Y, Sayes C M, Boyd A M, Falkner J C, Hotze E M, Alemany L B, Tao Y J, Guo W, Ausman K D, Colvin V L and Hughes J B. 2005. C-60 in water: nano-crystal formation and microbial response. *Environmental Science & Technology* **39**, 4307–16.

[74]   Fowler P W, Manolopoulos D E and Ryan R P. 1992. Isomerisations of the fullerenes. *Carbon* **30**, 1235–50.

[75]   Fromherz T, Padinger F, Gebeyehu D, Brabec C, Hummelen J C and Sariciftci N S. 2000. Comparison of photovoltaic devices containing various blends of polymer and fullerene derivatives. *Solar Energy Materials and Solar Cells* **63**, 61–8.

[76]   Ganin A Y, Takabayashi Y, Khimyak Y Z, Margadonna S, Tamai A, Rosseinsky M J and Prassides K. 2008. Bulk superconductivity at 38[thinsp]K in a molecular system. *Nature Materials* **7**, 367–71.

[77]   Gao J, Hide F and Wang H L. 1997. Efficient photodetectors and photovoltaic cells from composites of fullerenes and conjugated polymers: photoinduced electron transfer. *Synthetic Metals* **84**, 979–80.

[78]   Gao J, Wang H L, Shreve A and Iyer R. 2010. Fullerene derivatives induce premature senescence: a new toxicity paradigm or novel biomedical applications. *Toxicology and Applied Pharmacology* **244**, 130–43.

[79] Garcia-Belmonte G and Bisquert J. 2010. Open-circuit voltage limit caused by recombination through tail states in bulk heterojunction polymer-fullerene solar cells. *Applied Physics Letters* **96**, 113301.

[80] Gillett S I.. 2002. Nanotechnology: Clean Energy and Resources for the Future. In: *White Paper for Forsight Institute*: Forsight Institute; University of Nevada: Reno, NV.

[81] Giordani S, Colomer J-F, Cattaruzza F, Alfonsi J, Meneghetti M, Prato M and Bonifazi D. 2009. Multifunctional hybrid materials composed of 60 fullerene-based functionalized-single-walled carbon nanotubes. *Carbon* **47**, 578–88.

[82] Glenis S, Cooke S, Chen X and Labes M M. 1994. Photophysical properties of fullerenes prepared in an atmosphere of pyrrole. *Chemistry of Materials* **6**, 1850–3.

[83] Glenis S, Cooke S, Chen X and Labes M M. 1995. Photophysical properties of nitrogen substituted fullerenes. *Synthetic Metals* **70**, 1313–16.

[84] Gong X, Yu T Z, Cao Y and Heeger A J. 2012. Large open-circuit voltage polymer solar cells by poly(3-hexylthiophene) with multi-adducts fullerenes. *Science China Chemistry* **55**, 743–8.

[85] Grant P M. 1997. Superconductivity and electric power: promises, promises … past, present and future. *IEEE Transactions on Applied Superconductivity* **7**, 112–32.

[86] Gügel A and Müllen K. 1993. Separation of C_{60} and C_{70} on polystyrene gel with toluene as mobile phase. *Journal of Chromatography. A* **628**, 23–9.

[87] Haddon R C. 1997. C_{60}: sphere or polyhedron? *Journal of the American Chemical Society* **119**, 1797–8.

[88] Haddon R C, Hebard A F, Rosseinsky M J, Murphy D W, Duclos S J, Lyons K B, Miller B, Rosamilia J M, Fleming R M, Kortan A R, Glarum S H, Makhija A V, Muller A J, Eick R H, Zahurak S M, Tycko R, Dabbagh G and Thiel F A. 1991. Conducting films of C_{60} and C_{70} by alkali-metal doping. *Nature* **350**, 320–2.

[89] Haddon R C, Schneemeyer L F, Waszczak J V, Glarum S H, Tycko R, Dabbagh G, Kortan A R, Muller A J, Mujsce A M, Rosseinsky M J, Zahurak S M, Makhija A V, Thiel F A, Raghavachari K, Cockayne E and Elser V. 1991. Experimental and theoretical determination of the magnetic susceptibility of C_{60} and C_{70}. *Nature* **350**, 46–7.

[90] Han F D, Yao B and Bai Y J. 2011. Preparation of carbon nano-onions and their application as anode materials for rechargeable lithium-ion batteries. *Journal of Physical Chemistry C* **115**, 8923–7.

[91] Hashizume T, Wang X D, Nishina Y, Shinohara H, Saito Y, Kuk Y and Sakurai T. 1992. Field ion-scanning tunneling microscopy study of C_{60} on the Si (100) surface. *Japanese Journal of Applied Physics. Part 2-Letters* **31**, L880–L3.

[92] Hassenzahl W V, Hazelton D W, Johnson B K, Komarek P, Noe M and Reis C T. 2004. Electric power applications of superconductivity. *Proceedings of the IEEE* **92**, 1655–74.

[93] Haufler R E, Conceicao J, Chibante L P F, Chai Y, Byrne N E, Flanagan S, Haley M M, O'Brien S C and Pan C. 1990. Efficient production of C_{60} (buckminsterfullerene), $C_{60}H_{36}$, and the solvated buckide ion. *The Journal of Physical Chemistry* **94**, 8634–6.

[94] Hebard A F, Rosseinsky M J, Haddon R C, Murphy D W, Glarum S H, Palstra T T M, Ramirez A P and Kortan A R. 1991. Superconductivity at 18-K in potassium-doped C-60. *Nature* **350**, 600–1.

[95]   Hedberg K, Hedberg L, Bethune D S, Brown C A, Dorn H C, Johnson R D and Vries
       M D. 1991. Bond lengths in free molecules of buckminsterfullerene, C_{60} from gas-
       phase electron diffraction. *Science* **254**, 410–12.

[96]   Henry T B, Menn F-M, Fleming J T, Wilgus J, Compton R N and Sayler G S. 2007.
       Attributing effects of aqueous C-60 nano-aggregates to tetrahydrofuran decomposi-
       tion products in larval zebrafish by assessment of gene expression. *Environmental
       Health Perspectives* **115**, 1059–65.

[97]   Henry T B, Petersen E J and Compton R N. 2011. Aqueous fullerene aggregates
       (nC(60)) generate minimal reactive oxygen species and are of low toxicity in fish: a
       revision of previous reports. *Current Opinion in Biotechnology* **22**, 533–7.

[98]   Hiraoka K, Kudaka I, Fujimaki S and Shinohara H. 1992. Observation of the fullerene
       anions C 60– and C 70– by electrospray ionization. *Rapid Communications in Mass
       Spectrometry* **6**, 254–6.

[99]   Hirsch A. 1999. Principles of Fullerene Reactivity. In: *Fullerenes and Related
       Structures*, ed A Hirsch (Berlin/New York: Springer) pp 1–65.

[100] Hirsch A, Li Q and Wudl F. 1991. Globe-trotting hydrogens on the surface of the ful-
       lerene compound $C_{60}H_6(N(CH_2CH_2)_2O)_6$. *Angewandte Chemie International Edition
       in English* **30**, 1309–10.

[101] Holczer K, Klein O, Huang S M, Kaner R B, Fu K J, Whetten R L and Diederich F.
       1991. Alkali-fulleride superconductors—synthesis, composition, and diamagnetic
       shielding. *Science* **252**, 1154–7.

[102] Howard J B, McKinnon J T, Johnson M E, Makarovsky Y and Lafleur A L. 1992.
       Production of C_{60} and C_{70} fullerenes in benzene-oxygen flames. *The Journal of
       Physical Chemistry* **96**, 6657–62.

[103] Howard J B, McKinnon J T, Makarovsky Y, Lafleur A L and Johnson M E. 1991.
       Fullerenes C_{60} and C_{70} in flames. *Nature* **352**, 139–41.

[104] Hwang Y S and Li Q. 2010. Characterizing photochemical transformation of aqueous
       nC_{60} under environmentally relevant conditions. *Environmental Science & Technology*
       44, 3008–13.

[105] Imahori H, Hagiwara K, Akiyama T, Taniguchi S, Okada T and Sakata Y. 1995.
       Synthesis and photophysical property of porphyrin-linked fullerene. *Chemistry Letters*
       265–6.

[106] Imahori H and Sakata Y. 1997. Donor-linked fullerenes: photoinduced electron
       transfer and its potential application. *Advanced Materials* **9**, 537–46.

[107] Imasaka K, Kanatake Y, Ohshiro Y, Suehiro J and Hara M. 2006. Production of carbon
       nanoonions and nanotubes using an intermittent arc discharge in water. *Thin Solid
       Films* **506**, 250–4.

[108] Inoue M, Machi L, Brown F, Inoue M B and Fernando Q. 1995. Photochemical
       syntheses of fullerene-amine adducts and their characterization with H-1-NMR
       spectroscopy. *Journal of Molecular Structure* **345**, 113–17.

[109] Itou M, Fujitsuka M, Araki Y, Ito O and Kido H. 2003. Photophysical properties of
       self-assembling porphinatozinc and photoinduced electron transfer with fullerenes.
       Journal of Porphyrins and Phthalocyanines **7**, 405–14.

[110] Janot J M, Eddaoudi H, Seta P, Ederle Y and Mathis C. 1999. Photophysical properties
       of the fullerene C-60 core of a 6-arm polystyrene star. *Chemical Physics Letters* **302**,
       103–7.

[111] Jariwala D, Sangwan V K, Lauhon L J, Marks T J and Hersam M C. 2013. Carbon nanomaterials for electronics, optoelectronics, photovoltaics, and sensing. *Chemical Society Reviews* **42**, 2824–60.

[112] Jin H, Hou Y B, Meng X G and Teng F. 2007. Concentration dependence of photovoltaic properties of photodiodes based on polymer-fullerene blends. *Materials Science and Engineering: B-Solid State Materials for Advanced Technology* **137**, 5–9.

[113] Jinno K, Saito Y, Chen Y L, Luehr G, Archer J, Fetzer J C and Biggs W R. 1993. Separation of C_{60} and C_{70} fullerenes on methoxyphenylpropyl bonded stationary phases in microcolumn liquid-chromatography. *Journal of Microcolumn Separations* **5**, 135–40.

[114] Johansen A, Pedersen A L, Jensen K A, Karlson U, Hansen B M, Scott-Fordsmand J J and Winding A. 2008. Effects of C(60) fullerene nanoparticles on soil bacteria and protozoans. *Environmental Toxicology and Chemistry* **27**, 1895–903.

[115] Johnson R D, Meijer G, Salem J R and Bethune D S. 1991. 2D Nuclear magnetic resonance study of the structure of the fullerene C_{70}. *Journal of the American Chemical Society* **113**, 3619–21.

[116] Kavak P, Menda U D, Parlak E A, Ozdemir O and Kutlu K. 2012. Excess current/capacitance observation on polymer-fullerene bulk heterojunction, studied through I-V and C/G-V measurements. *Solar Energy Materials and Solar Cells* **103**, 199–204.

[117] Khan I A, Afrooz A R M N, Flora J R V, Schierz P A, Ferguson P L, Sabo-Attwood T and Saleh N B. 2013. Chirality affects aggregation kinetics of single-walled carbon nanotubes. *Environmental Science & Technology* **47**, 1844–52.

[118] Khan I A, Aich N, Afrooz A R M N, Flora J R V, Schierz P A, Ferguson P L, Sabo-Attwood T and Saleh N B. 2013. Fractal structures of single-walled carbon nanotubes in biologically relevant conditions: role of chirality vs. media conditions. *Chemosphere* **93**, 1997–2003.

[119] Khemani K C, Prato M and Wudl F. 1992. A simple Soxhlet chromatographic method for the isolation of pure fullerenes C_{60} and C_{70}. *The Journal of Organic Chemistry* **57**, 3254–6.

[120] Khlyabich P P, Burkhart B, Rudenko A E and Thompson B C. 2013. Optimization and simplification of polymer-fullerene solar cells through polymer and active layer design. *Polymer* **54**, 5267–98.

[121] Kikuchi K, Nakahara N, Wakabayashi T, Honda M, Matsumiya H, Moriwaki T, Suzuki S, Shiromaru H, Saito K, Yamauchi K, Ikemoto I and Achiba Y. 1992. Isolation and identification of fullerene family—C_{76}, C_{78}, C_{82}, C_{84}, C_{90} and C_{96}. *Chemical Physics Letters* **188**, 177–80.

[122] Kim H, Kim J Y, Lee K, Shin J, Cha M S, Lee S E, Suh H and Ha C S. 2000. Conjugated polymer/fullerene composites as a new class of optoelectronic material: application to organic photovoltaic cells. *Journal of the Korean Physical Society* **36**, 342–5.

[123] Kim K T, Jang M H, Kim J Y and Kim S D. 2010. Effect of preparation methods on toxicity of fullerene water suspensions to Japanese medaka embryos. *The Science of the Total Environment* **408**, 5606–12.

[124] Kim Y, Choulis S A, Nelson J, Bradley D D C, Cook S and Durrant J R. 2005. Device annealing effect in organic solar cells with blends of regioregular poly(3-hexylthiophene) and soluble fullerene. *Applied Physics Letters* **86**, 063502.

[125] Klocek J, Henkel K, Kolanek K, Zschech E and Schmeisser D. 2012. Spectroscopic and capacitance-voltage characterization of thin aminopropylmethoxysilane films

doped with copper phthalocyanine, tris(dimethylvinylsilyloxy)-POSS and fullerene cages. *Applied Surface Science* **258**, 4213–21.

[126] Koch A S, Khemani K C and Wudl F. 1991. Preparation of fullerenes with a simple benchtop reactor. *The Journal of Organic Chemistry* **56**, 4543–5.

[127] Kong L, Mukherjee B, Chan Y F and Zepp R G. 2013. Quenching and sensitizing fullerene photoreactions by natural organic matter. *Environmental Science & Technology* **47**, 6189–96.

[128] Kooistra F B, Mihailetchi V D, Popescu L M, Kronholm D, Blom P W M and Hummelen J C. 2006. New C-84 derivative and its application in a bulk heterojunction solar cell. *Chemistry of Materials* **18**, 3068–73.

[129] Koster L J A, Mihailetchi V D, Ramaker R and Blom P W M. 2005. Light intensity dependence of open-circuit voltage of polymer: fullerene solar cells. *Applied Physics Letters* **86**, 123509.

[130] Krätschmer W. 2011. The story of making fullerenes. *Nanoscale* **3**, 2485–9.

[131] Krätschmer W, Fostiropoulos K and Huffman D R. 1990. The infrared and ultraviolet absorption spectra of laboratory-produced carbon dust: evidence for the presence of the C_{60} molecule. *Chemical Physics Letters* **170**, 167–70.

[132] Krätschmer W, Lamb L D, Fostiropoulos K and Huffman D R. 1990. Solid C_{60}: a new form of carbon. *Nature* **347**, 354–8.

[133] Krebs F C, Gevorgyan S A and Alstrup J. 2009. A roll-to-roll process to flexible polymer solar cells: model studies, manufacture and operational stability studies. *Journal of Materials Chemistry* **19**, 5442–51.

[134] Krishna V, Noguchi N, Koopman B and Moudgil B. 2006. Enhancement of titanium dioxide photocatalysis by water-soluble fullerenes. *Journal of Colloid and Interface Science* **304**, 166–71.

[135] Krishna V, Yanes D, Imaram W, Angerhofer A, Koopman B and Moudgil B. 2008. Mechanism of enhanced photocatalysis with polyhydroxy fullerenes. *Applied Catalysis B-Environmental* **79**, 376–81.

[136] Kroto H W, Heath J R, Obrien S C, Curl R F and Smalley R E. 1985. C_{60}—Buckminsterfullerene. *Nature* **318**, 162–3.

[137] Kuo C S, Kumar J, Tripathy S K and Chiang L Y. 2001. Synthesis and properties of 60 fullerene-polyvinylpyridine conjugates for photovoltaic devices. *Journal of Macromolecular Science-Pure and Applied Chemistry* **38**, 1481–98.

[138] Lawton R G and Barth W E. 1971. Synthesis of corannulene. *Journal of the American Chemical Society* **93**, 1730–45.

[139] Lee J, Cho M, Fortner J D, Hughes J B and Kim J-H. 2009. Transformation of aggregated C_{60} in the aqueous phase by UV irradiation. *Environmental Science & Technology* **43**, 4878–83.

[140] Lee J, Yamakoshi Y, Hughes J B and Kim J H. 2008. Mechanism of C-60 photoreactivity in water: fate of triplet state and radical anion and production of reactive oxygen species. *Environmental Science & Technology* **42**, 3459–64.

[141] Lenes M, Shelton S W, Sieval A B, Kronholm D F, Hummelen J C and Blom P W M. 2009. Electron trapping in higher adduct fullerene-based solar cells. *Advanced Functional Materials* **19**, 3002–7.

[142] Lenes M, Wetzelaer G, Kooistra F B, Veenstra S C, Hummelen J C and Blom P W M. 2008. Fullerene bisadducts for enhanced open-circuit voltages and efficiencies in polymer solar cells. *Advanced Materials* **20**, 2116–19.

[143] Li J L, Hu Z S and Yang G W. 2012. High-capacity hydrogen storage of magnesium-decorated boron fullerene. *Chemical Physics* **392**, 16–20.

[144] Li M, Li Y F, Zhou Z, Shen P W and Chen Z F. 2009. Ca-coated boron fullerenes and nanotubes as superior hydrogen storage materials. *Nano Letters* **9**, 1944–8.

[145] Li Q, Xie B, Hwang Y S and Xu Y. 2009. Kinetics of C_{60} fullerene dispersion in water enhanced by natural organic matter and sunlight. *Environmental Science & Technology* **43**, 3574–9.

[146] Li Y F, Kaneko T and Hatakeyama R. 2008. Electrical transport properties of fullerene peapods interacting with light. *Nanotechnology* **19**, 415201.

[147] Lieber C M and Chen C C. 1994. Preparation of fullerenes and fullerene-based materials. *Solid State Physics: Advances in Research and Applications* 48, 109–48.

[148] Liu Z B, Xu Y F, Zhang X Y, Zhang X L, Chen Y S and Tian J G. 2009. Porphyrin and fullerene covalently functionalized graphene hybrid materials with large nonlinear optical properties. *The Journal of Physical Chemistry. B* **113**, 9681–6.

[149] Luther W. 2008. Application of Nanotechnologies in the Energy Sector. In: *Actionslinie Hessen-Nanotech* (Wiesbaden, Germany: Hessen Ministry of Economy, Transport, Urban and Regional Development).

[150] Lyon D Y, Adams L K, Falkner J C and Alvarez P J J. 2006. Antibacterial activity of fullerene water suspensions: effects of preparation method and particle size. *Environmental Science & Technology* **40**, 4360–6.

[151] Lyon D Y, Fortner J D, Sayes C M, Colvin V L and Hughes J B. 2005. Bacterial cell association and antimicrobial activity of a C-60 water suspension. *Environmental Toxicology and Chemistry* **24**, 2757–62.

[152] Ma B, Riggs J E and Sun Y P. 1998. Photophysical and nonlinear absorptive optical limiting properties of 60 fullerene dimer and poly 60 fullerene polymer. *Journal of Physical Chemistry B* **102**, 5999–6009.

[153] Ma X and Bouchard D. 2009. Formation of aqueous suspensions of fullerenes. *Environmental Science & Technology* **43**, 330–6.

[154] Maeda R, Fujita K and Tsutsui T. 2007. Improved performance in polymer-fullerene blend photovoltaic cells by insertion of C_{60} interlayer. *Molecular Crystals and Liquid Crystals* **471**, 123–8.

[155] Manolopoulos D E and Fowler P W. 1992. Molecular graphs, point groups, and fullerenes. *The Journal of Chemical Physics* **96**, 7603–14.

[156] Manolopoulos D E, Fowler P W, Taylor R, Kroto H W and Walton D R M. 1992. An end to the search for the ground-state of C_{84}. *Journal of the Chemical Society, Faraday Transactions* **88**, 3117–18.

[157] Martin N, Sanchez L, Illescas B, Gonzalez S, Herranz M A and Guldi D M. 2000. Photoinduced electron transfer between C_{60} and electroactive units. *Carbon* **38**, 1577–85.

[158] Maruyama R. 2002. Electrochemical hydrogen storage into $LaNi_5$ using a fullerene-based proton conductor. *Electrochemical and Solid State Letters* **5**, A89–A91.

[159] Matsuo Y, Iwashita A, Abe Y, Li C-Z, Matsuo K, Hashiguchi M and Nakamura E. 2008. Regioselective synthesis of 1,4-di(organo)[60]fullerenes through DMF-assisted monoaddition of silylmethyl grignard reagents and subsequent alkylation reaction. *Journal of the American Chemical Society* **130**, 15429–36.

[160] McDonald N C and Pearce J M. 2010. Producer responsibility and recycling solar photovoltaic modules. *Energy Policy* **38**, 7041–7.

[161] McElvany S and Ross M. 1992. Mass spectrometry and fullerenes. *Journal of the American Society for Mass Spectrometry* **3**, 268–80.

[162] McKee D W. 1991. The thermal-stability of fullerene in air. *Carbon* **29**, 1057–8.

[163] Meier M S and Selegue J P. 1992. Efficient preparative separation of C_{60} and C_{70}. Gel permeation chromatography of fullerenes using 100% toluene as mobile phase. *The Journal of Organic Chemistry* **57**, 1924–6.

[164] Meng Z, Hashmi S M and Elimelech M. 2013. Aggregation rate and fractal dimension of fullerene nanoparticles via simultaneous multiangle static and dynamic light scattering measurement. *Journal of Colloid and Interface Science* **392**, 27–33.

[165] Meng Z D, Zhang F J, Zhu L, Park C Y, Ghosh T, Choi J G and Oh W C. 2012. Synthesis and characterization of M-fullerene/TiO_2 photocatalysts designed for degradation azo dye. *Materials Science & Engineering C-Materials for Biological Applications* **32**, 2175–82.

[166] Mihailetchi V D, Blom P W M, Hummelen J C and Rispens M T. 2003. Cathode dependence of the open-circuit voltage of polymer: fullerene bulk heterojunction solar cells. *Journal of Applied Physics* **94**, 6849–54.

[167] Miller S, Fanchini G, Lin Y-Y, Li C, Chen C-W, Su W-F and Chhowalla M. 2008. Investigation of nanoscale morphological changes in organic photovoltaics during solvent vapor annealing. *Journal of Materials Chemistry* **18**, 306–12.

[168] Mojica M, Alonso J A and Méndez F. 2013. Synthesis of fullerenes. *Journal of Physical Organic Chemistry* **26**, 526–39.

[169] Morvillo P. 2009. Higher fullerenes as electron acceptors for polymer solar cells: a quantum chemical study. *Solar Energy Materials & Solar Cells* **93**, 1827–32.

[170] Morvillo P and Bobeico E. 2008. Tuning the LUMO level of the acceptor to increase the open-circuit voltage of polymer-fullerene solar cells: a quantum chemical study. *Solar Energy Materials and Solar Cells* **92**, 1192–8.

[171] Moulé A J and Meerholz K. 2008. Controlling morphology in polymer–fullerene mixtures. *Advanced Materials* **20**, 240–5.

[172] Nacalaitesque I. HPLC column for fullerene separation: cosmosil buckyprep cosmosil PBB (Kyoto, Japan: Nacalaitesque, Inc). http://www.nacalai.co.jp/global/download/pdf/COSMOSIL_Buckyprep_PBB.pdf. Accessed March 17, 2015.

[173] Nakamura Y, Taki M, Tobita S, Shizuka H, Yokoi H, Ishiguro K, Sawaki Y and Nishimura J. 1999. Photophysical properties of various regioisomers of 60 fullerene-o-quinodimethane bisadducts. *Journal of the Chemical Society. Perkin Transactions* **2**, 127–30.

[174] Nasibulin A G, Pikhitsa P V, Jiang H, Brown D P, Krasheninnikov A V, Anisimov A S, Queipo P, Moisala A, Gonzalez D, Lientschnig G, Hassanien A, Shandakov S D, Lolli G, Resasco D E, Choi M, Tomanek D and Kauppinen E I. 2007. A novel hybrid carbon material. *Nature Nanotechnology* **2**, 156–61.

[175] Nath I. 2010. Cleaning up after clean energy: hazardous waste in the solar industry. *Stanford Journal of International Relations* **11**, 6–15.

[176] Neugebauer H, Brabec C, Hummelen J C and Sariciftci N S. 2000. Stability and photodegradation mechanisms of conjugated polymer/fullerene plastic solar cells. *Solar Energy Materials and Solar Cells* **61**, 35–42.

[177] Ng T W, Lo M F, Fung M K, Lai S L, Liu Z T, Lee C S and Lee S T. 2009. Electronic properties and open-circuit voltage enhancement in mixed copper phthalocyanine: fullerene bulk heterojunction photovoltaic devices. *Applied Physics Letters* **95**, 203303.

[178] Nierengarten J F, Eckert J F, Felder D, Nicoud J F, Armaroli N, Marconi G, Vicinelli V, Boudon C, Gisselbrecht J P, Gross M, Hadziioannou G, Krasnikov V, Ouali L, Echegoyen L and Liu S G. 2000. Synthesis and electronic properties of donor-linked fullerenes towards photochemical molecular devices. *Carbon* **38**, 1587–98.

[179] Niinomi T, Matsuo Y, Hashiguchi M, Sato Y and Nakamura E. 2009. Penta(organo) [60]fullerenes as acceptors for organic photovoltaic cells. *Journal of Materials Chemistry* **19**, 5804–11.

[180] Oberdorster E, Zhu S Q, Blickley T M, McClellan-Green P and Haasch M L. 2006. Ecotoxicology of carbon-based engineered nanoparticles: effects of fullerene (C-60) on aquatic organisms. *Carbon* **44**, 1112–20.

[181] Ohta H, Saito Y, Jinno K, Nagashima H and Itoh K. 1994. Temperature effect in separation of fullerene by high-performance liquid-chromatography. *Chromatographia* **39**, 453–9.

[182] Oumnov A G, Mordkovich V Z and Takeuchi Y. 2001. Polythiophene/fullerene photo-voltaic cells. *Synthetic Metals* **121**, 1581–2.

[183] Padinger F, Brabec C J, Fromherz T, Hummelen J C and Sariciftci N S. 2000. Fabrication of large area photovoltaic devices containing various blends of polymer and fullerene derivatives by using the doctor blade technique. *Opto-Electronics Review* **8**, 280–3.

[184] Palit D K, Sapre A V, Mittal J P and Rao C N R. 1992. Photophysical properties of the fullerenes, C-60 and C-70. *Chemical Physics Letters* **195**, 1–6.

[185] Palkar A, Melin F, Cardona C M, Elliott B, Naskar A K, Edie D D, Kumbhar A and Echegoyen L. 2007. Reactivity differences between carbon nano onions (CNOs) prepared by different methods. *Chemistry--An Asian Journal* **2**, 625–33.

[186] Pang L S K, Vassallo A M and Wilson M A. 1991. Fullerenes from coal. *Nature* **352**, 480.

[187] Park C, Anderson P E, Chambers A, Tan C D, Hidalgo R and Rodriguez N M. 1999. Further studies of the interaction of hydrogen with graphite nanofibers. *Journal of Physical Chemistry B* **103**, 10572–81.

[188] Parker D H, Chatterjee K, Wurz P, Lykke K R, Pellin M J, Stock L M and Hemminger J C. 1993. Fullerenes and Giant Fullerenes: Synthesis, Separation, and Mass Spectrometric Characterization. In: *The Fullerenes*, ed H W Kroto (Oxford: Pergamon) pp 29–44.

[189] Parker D H, Wurz P, Chatterjee K, Lykke K R, Hunt J E, Pellin M J, Hemminger J C, Gruen D M and Stock L M. 1991. High-yield synthesis, separation, and mass-spectrometric characterization of fullerenes C_{60} to C_{266}. *Journal of the American Chemical Society* **113**, 7499–503.

[190] Peters G and Jansen M. 1992. A new fullerene synthesis. *Angewandte Chemie International Edition in English* **31**, 223–4.

[191] Pichler K, Graham S, Gelsen O M, Friend R H, Romanow W J, McCauley J P, Coustel N, Fischer J E and Smith A B. 1991. Photophysical properties of solid films of fullerene C-60. *Journal of Physics-Condensed Matter* **3**, 9259–70.

[192] Pirkle W H and Welch C J. 1991. An unusual effect of temperature on the chromatographic behavior of buckminsterfullerene. *The Journal of Organic Chemistry* **56**, 6973–4.

[193] Qiao L, Sun X L, Yang Z B, Wang X H, Wang Q and He D Y. 2013. Network structures of fullerene-like carbon core/nano-crystalline silicon shell nanofibers as anode material for lithium-ion batteries. *Carbon* **54**, 29–35.

[194] Qu S, Li M, Xie L, Huang X, Yang J, Wang N and Yang S. 2013. Noncovalent functionalization of graphene attaching 6,6-phenyl-C_{61}-butyric acid methyl ester (PCBM) and application as electron extraction layer of polymer solar cells. *ACS Nano* **7**, 4070–81.

[195] Qu X, Hwang Y S, Alvarez P J J, Bouchard D and Li Q. 2010. UV irradiation and humic acid mediate aggregation of aqueous fullerene (nC_{60}) nanoparticles. *Environmental Science & Technology* **44**, 7821–6.

[196] Rancan F, Rosan S, Boehm F, Cantrell A, Brellreich M, Schoenberger H, Hirsch A and Moussa F. 2002. Cytotoxicity and photocytotoxicity of a dendritic C-60 mono-adduct and a malonic acid C-60 tris-adduct on Jurkat cells. *Journal of Photochemistry and Photobiology B-Biology* **67**, 157–62.

[197] Randić M, Kroto H W and Vukičević D. 2007. Numerical Kekulé structures of fullerenes and partitioning of π-electrons to pentagonal and hexagonal rings†. *Journal of Chemical Information and Modeling* **47**, 897–904.

[198] Rohlfing E A, Cox D M and Kaldor A. 1984. Production and characterization of supersonic carbon cluster beams. *The Journal of Chemical Physics* **81**, 3322–30.

[199] Ruoff R S, Kadish K M, Boulas P and Chen E C M. 1995. Relationship between the electron affinities and half-wave reduction potentials of fullerenes, aromatic hydrocarbons, and metal complexes. *The Journal of Physical Chemistry* **99**, 8843–50.

[200] Saleh N, Afrooz A, Bisesi, Joseph, Aich N, Plazas-Tuttle J and Sabo-Attwood T. 2014. Emergent properties and toxicological considerations for nanohybrid materials in aquatic systems. *Nanomaterials* **4**, 372–407.

[201] Saleh N B, Aich N, Plazas-Tuttle J, Lead J R and Lowry G V. 2015. Research strategy to determine when novel nanohybrids pose unique environmental risks. *Environmental Science: Nano* **2**, 11–18.

[202] Saleh N B, Pfefferle L D and Elimelech M. 2008. Aggregation kinetics of multiwalled carbon nanotubes in aquatic systems: measurements and environmental implications. *Environmental Science & Technology* **42**, 7963–9.

[203] Sankaran M, Muthukumar K and Viswanathan B. 2005. Boron-substituted fullerenes—can they be one of the options for hydrogen storage? *Fullerenes, Nanotubes, and Carbon Nanostructures* **13**, 43–52.

[204] Sariciftci N S, Braun D, Zhang C, Srdanov V I, Heeger A J, Stucky G and Wudl F. 1993. Semiconducting polymer-buckminsterfullerene heterojunctions—diodes, photodiodes, and photovoltaic cells. *Applied Physics Letters* **62**, 585–7.

[205] Sayes C M, Fortner J D, Guo W, Lyon D, Boyd A M, Ausman K D, Tao Y J, Sitharaman B, Wilson L J, Hughes J B, West J L and Colvin V L. 2004. The differential cytotoxicity of water-soluble fullerenes. *Nano Letters* **4**, 1881–7.

[206] Sayes C M, Gobin A M, Ausman K D, Mendez J, West J L and Colvin V L. 2005. Nano-C-60 cytotoxicity is due to lipid peroxidation. *Biomaterials* **26**, 7587–95.

[207] Scott L T. 2004. Methods for the chemical synthesis of fullerenes. *Angewandte Chemie, International Edition* **43**, 4994–5007.

[208] Scott L T, Boorum M M, Brandon J M, Hagen S, Mack J, Blank J, Wegner H and Meijere A D. 2002. A rational chemical synthesis of C_{60}. *Science* **295**, 1500–3.

[209] Scott L T, Hashemi M M, Meyer D T and Warren H B. 1991. Corannulene—a convenient new synthesis. *Journal of the American Chemical Society* **113**, 7082–4.

[210]   Scrivens W A and Tour J M. 1992. Synthesis of gram quantities of C_{60} by plasma discharge in a modified round-bottomed flask. Key parameters for yield optimization and purification. *The Journal of Organic Chemistry* **57**, 6932–6.

[211]   Seda B C, Ke P C, Mount A S and Klaine S J. 2012. Toxicity of aqueous C_{70}-gallic acid suspension in Daphnia magna. *Environmental Toxicology and Chemistry* **31**, 215–20.

[212]   Shimada T, Ohno Y, Okazaki T, Sugai T, Suenaga K, Kishimoto S, Mizutani T, Inoue T, Taniguchi R, Fukui N, Okubo H and Shinohara H. 2004. Transport properties of C-78, C-90 and Dy@C-82 fullerenes-nanopeapods by field effect transistors. *Physica E* **21**, 1089–92.

[213]   Shin W H, Yang S H, Goddard W A and Kang J K. 2006. Ni-dispersed fullerenes: hydrogen storage and desorption properties. *Applied Physics Letters* **88**, 53111.

[214]   Shinohara H, Sato H, Saito Y, Takayama M, Izuoka A and Sugawara T. 1991. Formation and extraction of very large all-carbon fullerenes. *The Journal of Physical Chemistry* **95**, 8449–51.

[215]   Simon F, Kuzmany H, Rauf H, Pichler T, Bernardi J, Peterlik H, Korecz L, Fülöp F and Jánossy A. 2004. Low temperature fullerene encapsulation in single wall carbon nanotubes: synthesis of $N@C_{60}@SWCNT$. *Chemical Physics Letters* **383**, 362–7.

[216]   Singh H and Srivastava M. 1995. Fullerenes: synthesis, separation, characterization, reaction chemistry, and applications—a review. *Energy Sources* **17**, 615–40.

[217]   Sivula K, Luscombe C K, Thompson B C and Frechet J M J. 2006. Enhancing the thermal stability of polythiophene: fullerene solar cells by decreasing effective polymer regioregularity. *Journal of the American Chemical Society* **128**, 13988–9.

[218]   Smith B W, Monthioux M and Luzzi D E. 1998. Encapsulated C-60 in carbon nanotubes. *Nature* **396**, 323–4.

[219]   Sonkar S K, Ghosh M, Roy M, Begum A and Sarkar S. 2012. Carbon nano-onions as nontoxic and high-fluorescence bioimaging agent in food chain—an in vivo study from unicellular *E. coli* to multicellular *C. elegans*. *Materials Express* **2**, 105–14.

[220]   Stevens D M, Qin Y, Hillmyer M A and Frisbie C D. 2009. Enhancement of the morphology and open circuit voltage in bilayer polymer/fullerene solar cells. *Journal of Physical Chemistry C* **113**, 11408–15.

[221]   Suglyama H, Nagano T, Nouchi R, Kawasaki N, Ohta Y, Imai K, Tsutsui M, Kubozono Y and Fujiwara A. 2007. Transport properties of field-effect transistors with thin films of C-76 and its electronic structure. *Chemical Physics Letters* **449**, 160–4.

[222]   Svensson M, Zhang F L, Veenstra S C, Verhees W J H, Hummelen J C, Kroon J M, Inganas O and Andersson M R. 2003. High-performance polymer solar cells of an alternating polyfluorene copolymer and a fullerene derivative. *Advanced Materials* **15**, 988–91.

[223]   Swanson R M. 2009. Photovoltaics power up. *Science* **324**, 891–2.

[224]   Tajima Y, Arai H, Tezuka Y, Ishii T and Takeuchi K. 1997. Photochemical reaction of furans in the presence of 60 fullerene. *Fullerene Science and Technology* **5**, 1531–44.

[225]   Tanigaki K, Ebbesen T W, Saito S, Mizuki J, Tsai J S, Kubo Y and Kuroshima S. 1991. Superconductivity at 33-K in $Cs_xRb_yC_{60}$. *Nature* **352**, 222–3.

[226]   Tao X, Fortner J D, Zhang B, He Y, Chen Y and Hughes J B. 2009. Effects of aqueous stable fullerene nanocrystals (nC_{60}) on Daphnia magna: evaluation of sub-lethal reproductive responses and accumulation. *Chemosphere* **77**, 1482–7.

[227] Tasaki K, DeSousa R, Wang H B, Gasa J, Venkatesan A, Pugazhendhi P and Loutfy R O. 2006. Fullerene composite proton conducting membranes for polymer electrolyte fuel cells operating under low humidity conditions. *Journal of Membrane Science* **281**, 570–80.

[228] Taylor R. 1991. A valence bond approach to explaining fullerene stabilities. *Tetrahedron Letters* **32**, 3731–4.

[229] Taylor R, Avent A G, Birkett P R, Dennis T J S, Hare J P, Hitchcock P B, Holloway J H, Hope E G, Kroto H W, Langley G J, Meidine M F, Parsons J P and Walton D R M. 1993. Isolation, characterization, and chemical-reactions of fullerenes. *Pure and Applied Chemistry* **65**, 135–42.

[230] Taylor R, Hare J P, Abdul-Sada A K and Kroto H W. 1990. Isolation, separation and characterisation of the fullerenes C_{60} and C_{70}: the third form of carbon. *Journal of the Chemical Society, Chemical Communications* 1423–5.

[231] Taylor R, Langley G J, Avent A G, Dennis T J S, Kroto H W and Walton D R M. 1993. C-13 NMR-spectroscopy of C_{76}, C_{78}, C_{84} and mixtures of C_{86}–C_{102}—anomalous chromatographic behavior of C_{82} and evidence for $C_{70}H_{12}$. *Journal of the Chemical Society. Perkin Transactions* 2, 1029–36.

[232] Taylor R, Langley G J, Kroto H W and Walton D R M. 1993. Formation of C_{60} by pyrolysis of naphthalene. *Nature* **366**, 728–31.

[233] Tejamaya M, Römer I, Merrifield R C and Lead J R. 2012. Stability of citrate, PVP, and PEG coated silver nanoparticles in ecotoxicology media. *Environmental Science & Technology* **46**, 7011–17.

[234] Tiwari A J and Marr L C. 2010. The role of atmospheric transformations in determining environmental impacts of carbonaceous nanoparticles. *Journal of Environmental Quality* **39**, 1883–95.

[235] Tong Q Y, Eom C B, Gosele U and Hebard A F. 1994. Materials with a buried C(60) layer produced by direct wafer bonding. *Journal of the Electrochemical Society* **141**, L137–L8.

[236] Vandewal K, Gadisa A, Oosterbaan W D, Bertho S, Banishoeib F, Van Severen I, Lutsen L, Cleij T J, Vanderzande D and Manca J V. 2008. The relation between open-circuit voltage and the onset of photocurrent generation by charge-transfer absorption in polymer: fullerene bulk heterojunction solar cells. *Advanced Functional Materials* **18**, 2064–70.

[237] Vandewal K, Ma Z F, Bergqvist J, Tang Z, Wang E G, Henriksson P, Tvingstedt K, Andersson M R, Zhang F L and Inganas O. 2012. Quantification of quantum efficiency and energy losses in low bandgap polymer: fullerene solar cells with high open-circuit voltage. *Advanced Functional Materials* **22**, 3480–90.

[238] Vandewal K, Tvingstedt K, Gadisa A, Inganas O and Manca J V. 2009. On the origin of the open-circuit voltage of polymer-fullerene solar cells. *Nature Materials* **8**, 904–9.

[239] Vasil'ev Y V, Hirsch A, Taylor R and Drewello T. 2004. Hydrogen storage on fullerenes: hydrogenation of $C_{59}N$ center dot using $C_{60}H_{36}$ as the source of hydrogen. *Chemical Communications* 1752–3.

[240] Vassallo A M, Palmisano A J, Pang L S K and Wilson M A. 1992. Improved separation of fullerene-60 and -70. *Journal of the Chemical Society, Chemical Communications* 60–1.

[241] Vaughan G B M, Heiney P A, Fischer J E, Luzzi D E, Rickettsfoot D A, McGhie A R, Hui Y W, Smith A L, Cox D E, Romanow W J, Allen B H, Coustel N, McCauley J P and Smith A B. 1991. Oriental disorder in solvent-free solid C_{70}. *Science* **254**, 1350–3.

[242] Vukičević D and Randić M. 2005. On Kekulé structures of buckminsterfullerene. *Chemical Physics Letters* **401**, 446–50.

[243] Wang H, DeSousa R, Gasa J, Tasaki K, Stucky G, Jousselme B and Wudl F. 2007. Fabrication of new fullerene composite membranes and their application in proton exchange membrane fuel cells. *Journal of Membrane Science* **289**, 277–83.

[244] Wang L, Xu M, Ying L, Liu F and Cao Y. 2008. [70] Fullerene-based efficient bulk heterojuntion solar cells. *Acta Polymerica Sinica* 993–7.

[245] Wang X, Ma C, Chen K, Li H and Wang P. 2009. Interaction between nanobuds and hydrogen molecules: a first-principles study. *Physics Letters A* **374**, 87–90.

[246] Wang Y. 1992. Photoconductivity of fullerene-doped polymers. *Nature* **356**, 585–7.

[247] Wang Y. 1992. Photophysical properties of fullerenes and fullerene/N,N-diethylaniline charge-transfer complexes. *Journal of Physical Chemistry* **96**, 764–7.

[248] Wang Y G, Li Y S, Costanza J, Abriola L M and Pennell K D. 2012. Enhanced mobility of fullerene (C-60) nanoparticles in the presence of stabilizing agents. *Environmental Science & Technology* **46**, 11761–9.

[249] Wegelin M, Canonica S, Alder A C, Marazuela D, Suter M J F, Bucheli T D, Haefliger O P, Zenobi R, McGuigan K G, Kelly M T, Ibrahim P and Larroque M. 2001. Does sunlight change the material and content of polyethylene terephthalate (PET) bottles? *Journal of Water Supply Research and Technology-Aqua* **50**, 125–33.

[250] Wienk M M, Kroon J M, Verhees W J H, Knol J, Hummelen J C, van Hal P A and Janssen R A J. 2003. Efficient methano 70 fullerene/MDMO-PPV bulk heterojunction photovoltaic cells. *Angewandte Chemie-International Edition* **42**, 3371–5.

[251] Willard H H, Merritt Jr L L, Dean J A and Settle Jr F A. 1988. *Instrumental Methods of Analysis* (Belmont, CA: Wadsworth Publishing Company).

[252] Williams R M, Zwier J M and Verhoeven J W. 1995. Photoinduced intramolecular electron-transfer in a bridged C-60 (acceptor) aniline (donor) system—photophysical properties of the first active fullerene diad. *Journal of the American Chemical Society* **117**, 4093–9.

[253] Winkler K, Grodzka E, D'Souza F and Balch A L. 2007. Two-component films of fullerene and palladium as materials for electrochemical capacitors. *Journal of the Electrochemical Society* **154**, K1–K10.

[254] Wrobel D and Graja A. 2011. Photoinduced electron transfer processes in fullerene-organic chromophore systems. *Coordination Chemistry Reviews* **255**, 2555–77.

[255] Wu X and Zeng X C. 2009. Periodic graphene nanobuds. *Nano Letters* **9**, 250–6.

[256] Yang B, Guo F W, Yuan Y B, Xiao Z G, Lu Y Z, Dong Q F and Huang J S. 2013. Solution-processed fullerene-based organic Schottky junction devices for large-open-circuit-voltage organic solar cells. *Advanced Materials* **25**, 572–7.

[257] Yang J, Heo M, Lee H J, Park S M, Kim J Y and Shin H S. 2011. Reduced graphene oxide (rGO)-wrapped fullerene (C-60) wires. *ACS Nano* **5**, 8365–71.

[258] Yang Y, Nakada N, Nakajima R, Yasojima M, Wang C and Tanaka H. 2013. pH, ionic strength and dissolved organic matter alter aggregation of fullerene C_{60} nanoparticles suspensions in wastewater. *Journal of Hazardous Materials* **244–245**, 582–7.

[259] Yao L, Song G G, Huang C and Yang X L. 2011. Inhibitory effects of aqueous nanoparticle suspensions of 60 fullerene derivatives on bacterial growth. *Chemical Journal of Chinese Universities* **32**, 885–90.

[260] Yeretzian C, Wiley J B, Holczer K, Su T, Nguyen S, Kaner R B and Whetten R L. 1993. Partial separation of fullerenes by gradient sublimation. *The Journal of Physical Chemistry* **97**, 10097–101.

[261] Yoon M, Yang S Y, Hicke C, Wang E, Geohegan D and Zhang Z Y. 2008. Calcium as the superior coating metal in functionalization of carbon fullerenes for high-capacity hydrogen storage. *Physical Review Letters* **100**, 206806.

[262] Yoon M, Yang S Y, Wang E and Zhang Z Y. 2007. Charged fullerenes as high-capacity hydrogen storage media. *Nano Letters* **7**, 2578–83.

[263] Yoon M, Yang S Y and Zhang Z Y. 2009. Interaction between hydrogen molecules and metallofullerenes. *Journal of Chemical Physics* **131**, 064707.

[264] Yoshino K, Akashi T, Morita S, Yoshida M, Hamaguchi M, Tada K, Fujii A, Kawai T, Uto S, Ozaki M, Onoda M and Zakhidov A A. 1995. Photophysical properties of fullerene-conducting polymer system. *Synthetic Metals* **70**, 1317–20.

[265] Yoshino K, Yin X H, Akashi T, Yoshimoto K, Morita S and Zakhidov A A. 1994. Novel photophysical properties of fullerene doped conducting polymers. *Molecular Crystals and Liquid Crystals Science and Technology Section A-Molecular Crystals and Liquid Crystals* **255**, 197–211.

[266] Yu G, Gao J, Hummelen J C, Wudl F and Heeger A J. 1995. Polymer photovoltaic cells—enhanced efficiencies via a network of internal donor-acceptor heterojunctions. *Science* **270**, 1789–91.

[267] Zhang C F, Tong S W, Zhu C X, Jiang C Y, Kang E T and Chan D S H. 2009. Enhancement in open circuit voltage induced by deep interface hole traps in polymer-fullerene bulk heterojunction solar cells. *Applied Physics Letters* **94**, 103305.

[268] Zhang D J, Ma C and Liu C B. 2007. Potential high-capacity hydrogen storage medium: hydrogenated silicon fullerenes. *Journal of Physical Chemistry C* **111**, 17099–103.

[269] Zhang H, Ji Z, Xia T, Meng H, Low-Kam C, Liu R, Pokhrel S, Lin S, Wang X, Liao Y-P, Wang M, Li L, Rallo R, Damoiseaux R, Telesca D, Mädler L, Cohen Y, Zink J I and Nel A E. 2012. Use of metal oxide nanoparticle band gap to develop a predictive paradigm for oxidative stress and acute pulmonary inflammation. *ACS Nano* **6**, 4349–68.

[270] Zhang L L, Hou L, Wang L L, Kan A T, Chen W and Tomson M B. 2012. Transport of fullerene nanoparticles (nC_{60}) in saturated sand and sandy soil: controlling factors and modeling. *Environmental Science & Technology* **46**, 7230–8.

[271] Zhang W, Rattanaudompol U S, Li H and Bouchard D. 2013. Effects of humic and fulvic acids on aggregation of aqu/nC_{60} nanoparticles. *Water Research* **47**, 1793–802.

[272] Zhang X, Huang Y, Wang Y, Ma Y, Liu Z and Chen Y. 2009. Synthesis and characterization of a graphene–C_{60} hybrid material. *Carbon* **47**, 334–7.

[273] Zhao G, He Y and Li Y. 2010. 6.5% Efficiency of polymer solar cells based on poly (3-hexylthiophene) and indene-C_{60} bisadduct by device optimization. *Advanced Materials* **22**, 4355–8.

[274] Zhao J, Swinnen A, Van Assche G, Manca J, Vanderzande D and Van Mele B. 2009. Phase diagram of P3HT/PCBM blends and its implication for the stability of morphology. *The Journal of Physical Chemistry. B* **113**, 1587–91.

[275] Zhou O, Vaughan G B M, Zhu Q, Fischer J E, Heiney P A, Coustel N, McCauley J P and Smith A B. 1992. Compressibility of M_3C_{60} fullerene superconductors—relation between T_c and lattice-parameter. *Science* **255**, 833–5.

[276] Zhu S, Oberdorster E and Haasch M L. 2006. Toxicity of an engineered nanoparticle (fullerene, C-60) in two aquatic species, Daphnia and fathead minnow. *Marine Environmental Research* **62**, S5–S9.

[277] Zimmermann Y-S, Schäffer A, Hugi C, Fent K, Corvini P F X and Lenz M. 2012. Organic photovoltaics: potential fate and effects in the environment. *Environment International* **49**, 128–40.

2

CARBON NANOTUBES

Toshiyuki Ohashi

Fundamental Technology Research Center, Honda R&D Co., Ltd., Wako, Japan

2.1 SYNTHESIS OF CARBON NANOTUBES

2.1.1 Introduction and Structure of Carbon Nanotube

Single-walled carbon nanotube (SWCNT) is made by rolling a graphene sheet, which consists of hexagonal carbon rings, in the form of seamless cylinder. In general, both ends of SWCNT are capped by fullerene-like structures such as pentagons. The structure of SWCNT is geometrically defined by the chiral vector, C_h.

$$C_h = na_1 + ma_2$$

where a_1 and a_2 are unit vectors in the two-dimensional hexagonal lattice on the graphene sheet and the pair of integers (n, m) is known as the chiral index [1–3], as shown in Figure 2.1. The chiral angle, θ, is the angle between C_h and a_1. Each pair of integers (n, m) defines a unique way of rolling the graphene sheet to form a carbon nanotube (CNT), and there are three different ways. When $n = m$ and $\theta = 30°$, armchair nanotube is formed. When either n or $m = 0$ and $\theta = 0°$, zigzag nanotubes are formed. All the other nanotubes have intermediate chiral angle between $0°$ and $30°$ and are chiral nanotubes, as shown in Figure 2.2. Based on the two-dimensional dispersion function of graphene, the function of electronic density of states (DOS), which is defined as the electronic state of SWCNT, is estimated using both periodic boundary condition of circumferential direction and periodicity of axial direction to form geometrical structure of SWCNT. Sharp peaks (van Hove singularity) in DOS are

Carbon Nanomaterials for Advanced Energy Systems: Advances in Materials Synthesis and Device Applications, First Edition. Edited by Wen Lu, Jong-Beom Baek and Liming Dai.
© 2015 John Wiley & Sons, Inc. Published 2015 by John Wiley & Sons, Inc.

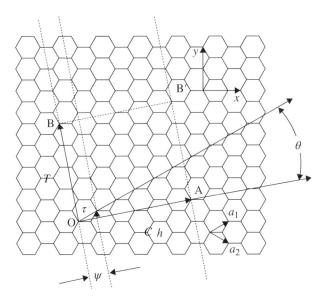

FIGURE 2.1   The 2D graphene sheet is shown along with the vector that specifies the chiral nanotubes. The chiral vector OA or $C_h = na_1 + ma_2$ is defined on the honeycomb lattice by unit vectors a_1 and a_2 and chiral angle θ is defined with respect to the zigzag axis. Along the zigzag axis $\theta = 0°$. Also shown are lattice vector OB $= T$ of the 1D tubule unit cell, and the rotation angle ψ and the transition τ which constitute the basic symmetry operation $R = (\psi/\tau)$. The diagram is constructed for $(n, m) = (4, 2)$. Reprinted from Ref. 2 with permission from American Carbon Society. © PERGAMON.

appeared only if the boundary condition of circumferential direction is satisfied. For example, either semiconducting or metallic behavior of CNTs depends on their chiral indexes. It is generally known that SWCNT behaves like a metal when $(n - m)/3$ is integer, whereas the others behave like a semiconductor. The geometrical ratio of semiconductor is 2/3 and that of metal is 1/3. Accurately, armchair behaves as a metal only when $(n = m)$, whereas when $(n - m)/3 =$ integer and $n \neq m$ it behaves as a zero-bandgap semiconductor.

Multi-walled carbon nanotube (MWCNT) consists of more than one layer of graphene roll in a concentric manner. Based on the number of wall, they are known as double-walled carbon nanotube or few-walled carbon nanotube. Wide application is expected from these CNTs regardless of their use as functional and structural materials.

In this chapter, synthesis and characterization of CNTs are discussed. Firstly, in the section that deals with synthesis, arc-discharge and laser ablation methods which were intensively used in the early period are noted, and subsequently chemical vapor deposition (CVD), a method widely used today, is described. In the point of view of assembly and selectivity of CNTs, the important criteria for device application, synthesis of vertically and horizontally aligned nanotube and selective synthesis of metallic or semiconducting of nanotubes are reviewed. Moreover, recent progress of chirality selective growth, an ultimate goal for CNT synthesis, is discussed.

(a)

(b)

(c)

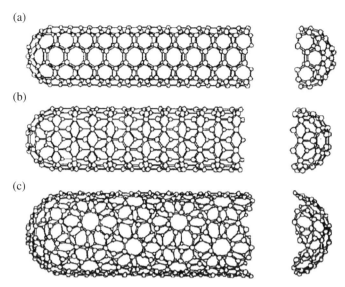

FIGURE 2.2   By rolling up a graphene sheet (a single layer of carbon atoms from a 3D graphite crystal) as a cylinder and capping each end of the cylinder with half of a fullerene molecule, a "fullerene-derived tubule," one layer in thickness, is formed. Shown here is a schematic theoretical model for a single-wall carbon tubule with the tubule axis OB (see Fig. 2.1) normal to (a) the $\theta = 30°$ direction (an "armchair" tubule), (b) the $\theta = 0°$ direction (a "zigzag" tubule), and (c) a general direction B with $0 < \theta < 30°$ (a "chiral" tubule). The actual tubules shown in the figure correspond to (n, m) values of: (a) (5, 5), (b) (9, 0), and (c) (10, 5). Reprinted from Ref. 2 with permission from American Carbon Society. © PERGAMON.

In the succeeding section, characterization methods of nanotubes are reviewed. Especially two representative methods are noted. One is spectroscopic techniques such as Raman spectroscopy, optical absorption spectroscopy (using ultraviolet-, visible-, and near infrared-lights (UV-vis-NIR)), and photoluminescence spectroscopy. Another is microscopic techniques such as transmission electron microscopy (TEM) and scanning tunneling microscopy (STM), the methods used for direct observation of the structure of nanotubes.

2.1.2   Arc Discharge and Laser Ablation

In both arc discharge and laser ablation methods, condensation of vapor-phase atomic carbon produced by vaporization of solid carbon takes place. These methods were mainly used in the early period of research of CNTs. The arc discharge is well known as the method Iijima used to discover MWCNT in 1991 [4]. In 1992, Ebbesen and Ajayan reported the growth of MWCNT at gram level [5]. In general, the synthesis of CNT by arc discharge is carried out as follows: Firstly, arc discharge is generated between two opposite graphite electrodes immersed in an inert gas. Then, carbon is vaporized at 3000–4000°C, and a carbon soot is formed during cooling down. Soot that is attached mainly to anode contains CNTs. MWCNT can be prepared by this process if both electrodes are pure graphite. On the other hands, if a graphite anode containing metallic particles (e.g., Fe, Co, Ni, Cu) is used with a graphite cathode, SWCNTs are generated [6–8]. The yield of nanotubes by this method is relatively high, but as-grown nanotubes contain an abundance of by-products like amorphous carbons and fullerenes. Therefore, purification is necessary to obtain only nanotubes [9–13].

Arc discharge method made evaluation of SWCNTs possible. However, low yield and inhomogeneous nature of SWCNT prepared by arc discharge method restricted them from further experimental studies using individual nanotubes to provide theoretical estimation. By making improvements to existing vaporization method of carbon, Thess et al. successfully synthesized high-purity SWCNT in 1996 [14]. This is so-called laser ablation method. This method is superior and yields high-purity SWCNTs. Many researchers today use this method to experimentally validate their theoretical studies. In general, the synthesis of CNT using laser ablation is carried out as follows: graphite along with Ni/Co is heated to 1200°C. Then, the graphite is vaporized using a pulsed laser under flowing argon at 500 Torr. When the vaporized carbon is crystallized again, SWCNT and fullerene are synthesized with carbon soot. This method was originally developed to synthesize fullerenes with high efficiency. Different from the synthesis for fullerene, in order to obtain SWCNT, a small amount (ca. 1 atomic %) of metallic catalyst is usually added to graphite material. If pure graphite is used, fullerene is synthesized. This method yields nearly 60% of SWCNT.

The SWCNTs formed by the aforementioned methods are generally bundled together and are highly tangled. Also they cannot be produced at higher scale, due to limitation of the methods.

2.1.3   Chemical Vapor Deposition

CVD is similar to pyrolysis of gas accompanying catalytic reaction by metallic nanoparticles or carbon volatile compounds. Here, the nanoparticles play the role of nucleation site initiating the growth of CNT. As the method is simple and produce high yield of CNT, it is used for higher scale mass production. Hence, at present, this method is considered as the most important method for the synthesis of CNT.

The synthesis of CNT using catalytic CVD (CCVD) as most general method is carried out as follows: metallic fine particles used as catalyst (Fe, Co, etc.) coexist in a heated reactor (typically, 800–1000°C). A gas mixture containing hydrocarbon gas (e.g., CH_4, C_2H_4, and C_2H_2), which acts as carbon source, and carrier gas (Ar, He, etc.) is flowed into the reactor. The catalyst and source gas react to form CNT.

Synthesis of CNT by CCVD was first reported by Endo et al. in 1993. They improved the CVD method used to produce vapor-grown carbon fiber and synthesized MWCNT [15]. During that time, it was thought that it is difficult to make SWCNT. However, Smalley et al. in 1996 reported that it is possible to grow SWCNT using carbon monoxide (CO) as carbon source in the presence of Mo as catalyst at 1200°C [16]. Moreover, Dai et al. succeeded in synthesizing SWCNT using CH_4 and various catalysts [17]. Later, CVD became the primary method to synthesize SWCNT and many follow-up studies were carried out [18–24]. The key for the synthesis of

SWCNT is to make fine particles of metallic catalysts. Hence, in most cases, metallic particles (e.g., Fe, Ni, Co, Mo, or their alloy) were supported to materials like alumina (Al_2O_3), silica (SiO_2), magnesia (MgO), zeolite, which are generally stable to high temperature. By using these supported particles, metallic fine particles in nanometer level were obtained. Consequently, through the combination of carbon source and these catalysts, the synthesis of high-yield SWCNT was embodied.

As previously described, CO was the first gas used for the synthesis of SWCNT in CVD. The advantage of using CO as carbon source than that of hydrocarbon is the level of amorphous carbon produced as a by-product could be comparatively minimized when CO is used. An important progress in CO CVD method that made it potentially commercial was by Resasco et al., which is known as CoMoCAT process [24, 25]. They combined Co/Mo binary catalyst and a flowing bed CVD to make large amount of SWCNTs. Another method is the high-pressure catalytic decomposition of CO gas (HiPco process). This method was originally developed by Smalley et al. [26, 27]. Figure 2.3 shows the experimental setup of HiPco process [26]. It is possible to synthesize SWCNT with almost no amorphous CNTs, based on the disproportionation reaction of CO:

$$2CO \rightarrow C + CO_2$$

The catalysts used in HiPco process are formed in gas phase from volatile organometallic precursors. These precursors decompose at high temperature, forming metal clusters on which SWCNT nucleate and grow. Currently, the HiPco process is one of the processes that can prepare SWCNT on a kilogram per day scale, and widely used as reference sample in many studies.

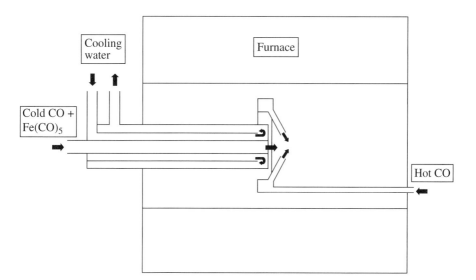

FIGURE 2.3 Layout of CO flow-tube reactor, showing water-cooled injector and "showerhead" mixer. Reprinted from Ref. 26 with permission from Elsevier.

Plasma-enhanced CVD (PECVD) [28–34] is also widely used for the growth of aligned CNT (discussed in the following) especially. Early in this study, with the field effect originated from plasma sheath, aligned CNT has been grown perpendicular to substrate [35, 36]. Using this method, the research of growing MWCNT for application as an emitter of field emission display was intensively conducted. On the other hand, the inhibitory effect by the reactive species derived from plasma to the growth of small-diameter tubes was of more concern. However, since the successes of the growth of SWCNT using radio-frequency plasma enhanced CVD by Dai et al. [28] and using antenna-type micro-wave plasma CVD (APCVD) by Kawarada et al. [29], many groups have reported about the advantage of plasma process such as the controllability of active species. In addition to this, many recent studies have reported about the growth of high-quality SWCNT by PECVD [37–43]. Figure 2.4 shows the experimental setup of a remote plasma CVD (APCVD) that was used by Kawarada et al. [41].

2.1.4   Aligned Growth

One of the advantages of the CVD method is that it allows good controllability of the morphology and structure of CNTs grown by them. Actually, it is possible to grow aligned CNTs or isolated individual tube on a substrate. These CNTs are directly adaptable to a functional material or device. Recent research has shown that structures made using nanodevice that uses horizontally or vertically aligned CNTs have advanced characterizations and superior properties.

The aligned growth of CNTs on a substrate has much developed with vertical alignment. In the beginning, the control of alignment using template or assistance by external force (e.g., magnetic force) was studied. Actually, a method using a porous

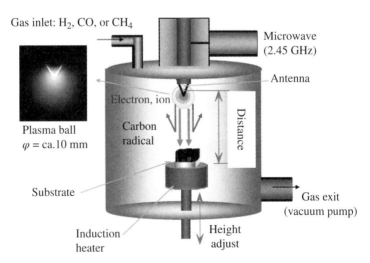

FIGURE 2.4   Antenna-type remote plasma CVD. Reprinted from Ref. 41 with permission from Elsevier.

membrane was reported [44–46]. However, in 1998, Rao et al. grew some aligned structures by decomposing ferrocene at high temperature [47] and Ren using PECVD [48]; since then the growth of aligned CNTs without a template has become an important method. Within few years, many other groups also reported about the success of the growth of aligned CNTs without a template [49–54]. In many cases, PECVD was used, and consequently vertically aligned CNTs (va-CNTs) to the substrate were successfully grown by the plasma field effect. These va-CNTs are also called CNT arrays [48]. For some time, these va-CNTs have been considered as only va-MWCNTs. However, Murakami et al. successfully grew va-SWCNTs in 2004 [55], as shown in Figure 2.5. They prepared high-quality SWCNTs using alcohol as

FIGURE 2.5 SEM micrographs of vertically aligned SWCNT bundles grown on quartz surface with different magnifications, taken at 20° from the horizon. Fractured substrate edges were observed to study cross-sectional morphology. The inset shows the substrate without CVD (left) and that with CVD used for synthesis of vertically aligned SWCNT film (right); its upper left corner is blank because it was covered during the dip-coat process. Reprinted from Ref. 55 with permission from Elsevier.

 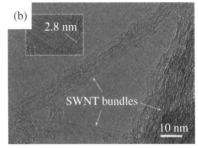

FIGURE 2.6 Typical characterizations of the as-grown densely packed and vertically aligned SWCNTs: (a) the cross-sectional FE-SEM morphology and (b) high-resolution TEM images; the inset shows an individual SWCNT with a diameter of about 2.8 nm. Reprinted from Ref. 31 with permission from American Carbon Society. © PERGAMON.

carbon source in CVD (alcohol CVD). Moreover, they pointed out that hydroxyl group present in alcohol is effective in removing amorphous carbon. They also suggested that the formation of amorphous carbon on a catalyst can avoid the degradation and thus enhance the activity of the catalyst. This formed the basis for "super growth (SG)" method [56]. Using thermal CVD with ppm level of water, SG realized significant enhancement for growth rate and height of va-SWCNTs (SWCNT forest).

Long va-SWCNT can be synthesized by this method; this method is simple because it could be achieved by just adding hundreds ppm of water to conventional thermal CVD. Therefore, this method has been extensively studied to grow not only va-SWCNTs but also va-DWCNTs and va-MWCNTs [57–62].

On the other hand, techniques such as APCVD [29] and hot filament CVD [63] were also reported to obtain long va-SWCNTs. In these methods, the growth of long va-SWCNTs was achieved without addition of water during the growth phase. Especially, in 2005, Kawarada et al. successfully grew va-SWCNTs using APCVD [29]. In 2007, they also reported 5-mm-long va-SWCNTs [34]. Figure 2.6 shows va-SWCNT that Kawarada et al. produced using APCVD [31]. To synthesize SWCNT forest without water using thermal CVD, it was reported that parameters such as thickness of the catalyst (linear to the size of catalyst particles) and concentration of synthesis gas should be optimized [64].

The length of these long SWCNT forests measured in millimeters; they are not only aligned but also visible enough to be handled, significantly benefiting the research for device and functional materials using CNTs.

It is pointed out that the purity of SWCNT decreases markedly as the nanotubes increase in length [34]. It is also reported that millimeter-long SWCNT forest synthesized using SG method is of low purity [65]. Moreover, SWCNTs of average diameter prepared using SG method is ca. 3 nm, whereas it is ca. 2.5 nm for the one synthesized using APCVD. These average diameters are somewhat large for SWCNTs. In general, the size of the catalyst corresponds with the diameter of CNTs; therefore the thickness of the catalyst should be prepared as thin as possible to keep control of the diameter of CNTs [66]. Optimized temperature and gas concentration

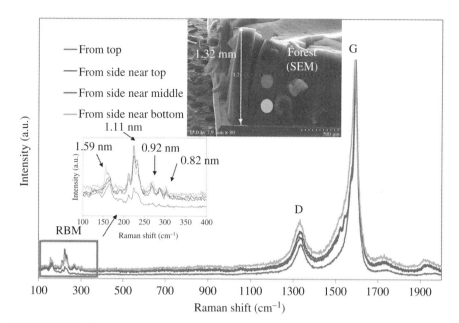

FIGURE 2.7   Raman spectra along the height of a forest grown on a top Al_2O_3/Fe/bottom Al_2O_3 sandwich catalytic substrate with a Ti layer on the top, with nanotube diameters estimated via $d\,(n,m) = 248/x$ (cm^{-1}) [69]. The upper inset is an SEM image of the forest. The colored circles on the forest show the measurement points of Raman spectroscopy. The lower inset is a magnified image of RBM (for Raman spectra shown in the red square). The laser excitation wavelength was 532 nm. Reprinted from Ref. 68 with permission from Elsevier. (*See insert for color representation of the figure.*)

can also control the diameter of CNTs [32, 67]. In general, these methods are applied to change the size of catalyst particle or the amount of carbon dissolving in catalyst by temperature and gas pressure. However, the growth window for SWCNTs is narrower than that of MWCNTs; hence, it is difficult to get a reproducible result, if the size of catalyst and gas concentration both are not correctly optimized. Consequently, conventional methodologies used to obtain SWCNT forests could not been applied to SWCNT forest with nanotubes of small diameter less than 2 nm. However, recently, 1.3-mm-long SWCNT forest with an average diameter of 1.32 nm has been synthesized successfully [68]. SEM image and Raman spectra of the SWCNT forest are shown in Figure 2.7. This result is obtained by adding titanium onto the top alumina layer of a top Al_2O_3/Fe/bottom Al_2O_3 sandwich catalyst on Si substrate, which has been proved effective for the growth of CNT forests [70]. Adding titanium layer further increases the ionization potential above that of Al (Ti > Al), which most likely reinforces the function of Al_2O_3; hence, nanoparticles of Fe is more stabilized.

Besides the development of the synthesis of va-CNTs, the synthesis of horizontally aligned CNTs (ha-CNTs) is also widely studied. The growth density of ha-CNTs is lower than that of va-CNTs, which is believed to be due to the sparse deposition of catalysts used for the growth of ha-CNTs. Nevertheless, this arrangement of catalysts presents some advantages for ha-CNTs including, for example, growth of ultra-long CNTs having no contact between nanotubes [71–74], alignment control of CNTs along the direction of gas flow [75, 76], alignment control of CNTs by utilizing an edge or a step on the surface of single crystal [77–81], as shown in Figure 2.8, and

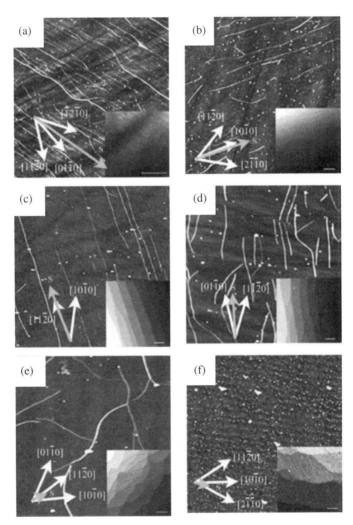

FIGURE 2.8 SWCNTs on miscut sapphire. Comparative analysis of representative samples with different miscut inclination and azimuth angles: (a) $\theta = 3.4 \pm 0.3°$, $\phi = 42 \pm 5°$; (b) $\theta = 2.3 \pm 0.2°$, $\phi = -33 \pm 5°$; (c) $\theta = 2.1 \pm 0.2°$, $\phi = 0 \pm 5°$; (d) $\theta = 1.7 \pm 0.1°$, $\phi = 18 \pm 5°$; (e) $\theta = 0.4 \pm 0.2°$, $\phi = -5 \pm 5°$; and (f) $\theta = 0.3 \pm 0.2°$, $\phi = -50 \pm 5°$ (image sizes are 2.5 µm, except (e), 5 µm). The vectors indicate the relevant lattice directions and the step vector **s** (black) obtained from XRD (except in (e) and (f), where s is from AFM). Insets show AFM topographic images of the respective annealed samples (inset scale bars 100 nm) with macrosteps. In (f), the atomic steps are spaced enough to be observed and are decorated with inactive catalyst nanoparticles. Reprinted from Ref. 78 with permission from Gesellschaft Deutscher Chemiker. © Wiley.

moreover, growth of cross-shaped CNTs using two isolated nanotubes from different growth directions [82]. Making full use of the characteristics of ha-CNTs that va-CNTs lacks, the synthesis of ha-CNTs is mainly applied for application to electronic device.

2.1.5 Selective Synthesis of Carbon Nanotubes

The growth of CNTs using transition metals is generally explained using vapor–liquid–solid (VLS) model [83–85]. Based on the observed shapes of catalyst particle, it was reported that particles in the liquid state were the active sites in the process of catalytic CNT growth [84]. This is the same for Si nanowires [83]. Carbon is dissolved in a melting metallic catalyst, and eventually oversaturated metallic particles are formed. Then, solid carbon (as an origin of nanotubes) is precipitated on the surface of the metallic particle. The melting point of Fe is about 1500°C, but it decreases as the size of Fe particles decreases and when carbon is dissolved into Fe particles (forming iron carbide). The interfacial energy contribution to the total free energy of nanoparticles was assumed to be responsible for the significant lowering of the melting temperature of the catalytic particles. VLS model is widely used for the general growth of CNTs using relatively large particles of Fe catalyst operated at 700–1000°C. This VLS growth mechanism is observed mainly for MWCNTs using environmental TEM (E-TEM) [86–93]. Figure 2.9 shows the images observed during CNT growth using E-TEM [93]. On the other hand, quantum mechanical molecular dynamics (QM/MD) methods were also used to simulate nucleation and growth process of SWCNT on transition metal catalysts [85]. With respect to the nucleation of a SWCNT cap precursor, it was shown that the presence of a transition metal carbide particle is not a necessary prerequisite for SWCNT nucleation, contrary to conventional experimental assumptions.

The controllability of the diameter of CNTs grown by VLS model depends on the size of catalyst. Not only the size of catalyst but also the addition of other elements (e.g., Cu) to transition metal catalyst [94] has been reported to obtain SWCNTs of

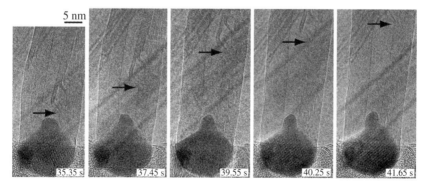

FIGURE 2.9 Environmental TEM images showing a growing MWCNT from a nanoparticle of $(Fe, Mo)_{23}C_6$. The arrow indicates the bend of the graphitic walls. Reprinted from Ref. 93 with permission from Elsevier.

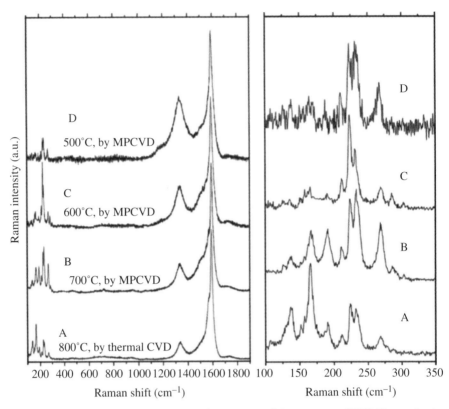

FIGURE 2.10 Resonant Raman scattering spectra of the as-grown SWCNT samples by thermal CVD at 800°C (A), and by Microwave-PCVD at 700°C (B), 600°C (C), and 500°C (D), respectively, by using the same batch of catalyst. The right panel is a magnified view of the low-frequency RBM region. The excitation laser line was 532 nm. The spectra in the left panel are normalized to the G-band at ca. 1593 cm^{-1}, while the magnified RBM signals in the right panel are normalized to the strongest peak. Reprinted from Ref. 32 with permission from Elsevier.

small diameter. Moreover, the optimization of the process condition such as temperature [32] and gas pressure [95] is also effective. Figure 2.10 shows the report of making SWCNTs of small diameter by using optimized temperature (especially, in lower temperature). In general, selection of existing (n, m) becomes narrow as the diameter of SWCNTs decreases. Hence, this approach is also the first step for the selective synthesis of SWCNTs.

About the selective synthesis of either metallic (m-) or semiconducting (sc-) SWCNTs, already some reports in the early stage of CNT synthesis are available. One is based on selective etching of either m-SWCNTs or sc-SWCNTs using oxidative atmospheric gas [22, 96, 97], ultra violet (UV) light [98, 99], and weak acid [100, 101]. Another is optimization of the process conditions such as using PECVD [28, 102] and changing the ratio of carbon sources [103–105].

Abovementioned selective method is mainly available to selectively keep sc-SWCNTs or synthesize them. This is based on the concept that m-SWCNTs is more reactive and can be selectively etched or difficult to grow. However, this method has limitations. As is obvious, owning to their reactivity, some sc-SWCNTs are also removed by this etching process. It is unavoidable that impure SWCNTs are present after the process. Therefore, it is ideal to develop a direct selective synthesis based on catalyst control. However, not so many reports are found in the literature about the direct selective synthesis. For example, there are methods such as the following: method in which the composition ratio of Fe/Ni on the Fe/Ni binary catalyst could be changed [106], approach based on the stabilization of Fe crystal facet using water in less than 10 ppm level [107], and synthesis of m-SWCNTs by the addition of sulfur to Fe using flowing bed CVD [108], as shown in Figure 2.11. The purity of m-SWCNTs obtained by selective synthesis is more than 90% [107].

However, in order to evaluate the quantification of selectivity, each group uses a combination of methods such as Raman, electric conductivity, and UV absorption, as there is no standard method. Moreover, as there is no follow-up study that could be reproduced, it is expected that the versatile method to evaluate the selectivity has been established.

Chirality control is the ultimate goal in the synthesis of CNTs. For the existing ratios of sc-SWCNTs (2/3) and m-SWCNTs (1/3) which are geometrically assigned, the difference in existence distribution by various synthesis methods is reported. Especially, about the chirality control, some studies report that (6, 5) tube is synthesized at a higher percentage than usual [23, 94, 106, 109–111]. To synthesize a higher percentage of (6, 5) tube, bimetallic catalysts such as Co/Mo [23], Fe/Ru [109], and Fe/Cu [94] are used. Figure 2.12 shows contour plots fluorescence spectra of SWCNTs synthesized using Fe/Co bimetallic catalyst and alcohol as carbon source, where (6, 5) and (7, 5) tubes are enriched in the case of alcohol compared to chirality distribution in HiPco sample [110].

Theoretical calculations predict that the chiral-selective growth of SWCNTs is more likely to be achieved on (n, m) tubes with low energy barriers of growth reactions such as (6, 5) and (7, 5) [112]. Actually, as mentioned earlier, in as-grown samples, abundant large chiral-angle tubes containing (6, 5) and (7, 5) tubes are observed. As far as large chiral-angle tubes are considered, theoretical calculation predicted that armchair tubes grow faster [113], which has been confirmed experimentally [114]. As armchair tubes seem to be more abundant than other tubes, it is suggested that armchair and other large chiral-angle nanotubes kinetically grow faster than zigzag and other small chiral species.

On the other hand, (6, 5) tube is near the chirality of armchair but is a chiral tube. Moreover, at present, species other than (6, 5) tube (or neighborhood tubes) cannot be synthesized selectively, but accidentally these (6, 5) tubes are left in high percentage. Discussions are available about the reason why the products after synthesis contain highly pure (6, 5) tube but is still an open question.

Recently, the growth of CNT based on vapor–solid–solid (VSS) model has attracted attention compared to VLS model [115]. Page et al. [85] indicated that the VSS mechanism rather than VLS mechanism is responsible for the growth of CNTs

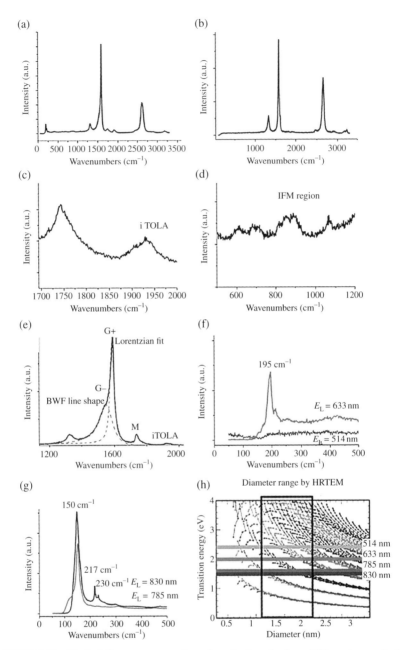

FIGURE 2.11 Typical Raman spectra (wavelength, λ excitation = 633 nm in panels (a–e)) from fibers obtained with (a) carbon disulfide and (b) thiophene as the sulfur precursors. (c) M, iTOLA, and (d) IFM regions in the Raman spectra of the carbon disulfide fibers. (e) The internal structure of the G band, with the Lorentzian G+ and the G− exhibiting the Fano line shape with fit parameters I_0, ω, T, and $q = 2256$, 1556, 49.5, and −0.20, respectively. (f) RBM regions with a peak at 195 cm^{-1} with $\lambda = 633$ nm and the absence of the RBM peak with $\lambda = 514$ nm. (g) RBM regions for $\lambda = 785$ nm and 830 nm. (h) Kataura plot with mapping of the four wavelengths and the measured diameter range from high-resolution TEM. Black points represent families of metallic nanotubes while red and blue represent semiconducting nanotubes. Reprinted from Ref. 108 with permission from Wiley. (*See insert for color representation of the figure.*)

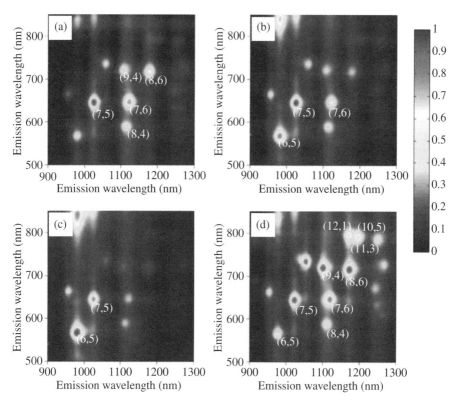

FIGURE 2.12 Contour plots of fluorescence spectra as a function of excitation wavelength and the resultant emission: (a) 850°C, (b) 750°C, (c) 650°C, and (d) HiPco. Reprinted from Ref. 110 with permission from Elsevier.

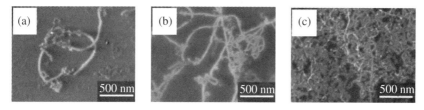

FIGURE 2.13 SEM images of carbon nanotubes grown after annealing treatment of SiO_2 in Ar/H$_2$ for different time: (a) 1, (b) 5, and (c) 30 min. Reprinted from Ref. 122 with permission from American Carbon Society. © PERGAMON.

on SiO_2 nanoparticles. There are many reports about the growth of CNT using catalyst particles such as SiC [116, 117], Ge [118, 119], noble metals [120, 121], SiO_2 [122, 123], and nanodiamond [124], which were thought as inert for the growth so far. Moreover, reports show that apart from catalyst, the growth of CNTs could be achieved even from a surface of a substrate [125] or a scratch-like surface bump on a substrate [126]. Figure 2.13 shows CNTs grown on SiO_2 particles [122].

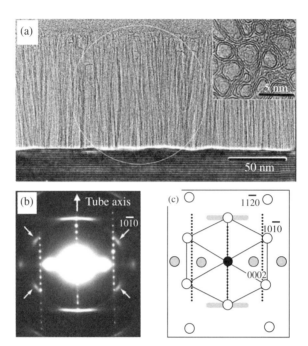

FIGURE 2.14 Aligned carbon nanotube film on the SiC (0 0 0 −1) C-face formed by surface decomposition of SiC. (a) TEM micrographs taken from cross-sectional and plan-view directions (upper right). (b) A selected-area electron diffraction pattern obtained from the circled area shown in (a). (c) A schematic of the diffraction pattern shown in (b). The intensity distribution is classified into three types of reflections: solid circles—an SiC crystal (electron beam/ [1 −1 0 2]$_{SiC}$); small dots—horizontal plane of CNTs; and gray circles—vertical planes of CNTs. Reprinted from Ref. 128 with permission from Elsevier.

In common, as growth could be achieved with elements having high melting point or via only substrates (no catalyst), where the dissolution of carbon in the catalyst or the substrate cannot be expected. In this synthesis, it is assumed that vapor carbon gets directly converted to CNTs on the solid surface of the catalyst or the substrate. In VSS, it is possible to control the structure of CNTs, whereas in VLS, only the diameter of CNTs can be controlled. To control the chirality of CNT synthesis, the important point to note is the precise control of catalyst structure based on VSS mechanism. No catalytic synthesis of CNTs was demonstrated already in 1997. A forest of well-oriented DWCNTs were produced by sublimation decomposition of SiC nanoparticles at high temperatures over 1700°C [127, 128], as shown in Figure 2.14. However, due to the huge success of CNT growth using transition metal catalysts, a large number of researchers working on CVD at that time ignored this work. However, with recent developments in novel catalysts, interest in the use of no transition metal or even no metal particles for CNT growth is gradually rising. Actually, most recently, selective growth of SWCNTs is reported with the surface of boron nitride [129] or from the cloning of nanotubes [130]. In the latter case, they used a seed of an isolated known chirality SWCNT prepared by wet separation using density gradient ultra-centrifugation (DGU) after growth of SWCNTs, and successfully synthesized SWCNT with a structure similar to that of the seed, based on VSS mechanism. This attracted a huge attention recently from the point of view that both low yield and short SWCNTs, a disadvantage of DGU, could be overcome and an intended length of SWCNTs with same chirality be obtained.

2.1.6 Summary

The growth of CNTs has been described form its early stage to the latest topics. There is a general consensus about the intensive use of CVD method covering from mass production to catalyst control on a substrate. However, research question such as chirality control is yet to be resolved. In the true sense, for application of CNTs, it is ideal to make intended structure of CNT in desired amount. With a focus on VSS mechanism, this question of structural control of SWCNTs will be resolved by precisely controlling the catalyst on a substrate with keeping up the standard of the growth mechanism of CNTs.

2.2 CHARACTERIZATION OF NANOTUBES

2.2.1 Introduction

It is already noted that the chirality control in CNT synthesis is an unresolved issue. However, to further discuss the issue, not only the synthesis but also the characterization for clarifying the structure of CNTs and growth mechanism is important. Techniques that are capable of characterizing the detail of CNTs are scarce. The commercially available techniques that characterize the structures of CNTs fall roughly into two categories. One is spectroscopic techniques such as containing Raman spectroscopy, optical absorption (UV-vis-NIR) spectroscopy, and photoluminescence spectroscopy. Another is microscopic techniques such as transmission electron microscopy (TEM) and scanning tunneling microscopy (STM) used for direct observation of the structure of nanotubes. In this section, these methods are described.

2.2.2 Spectroscopy

2.2.2.1 Raman Spectroscopy Scattering of light from a solid is due to the interaction between incident light on a substance and lattice vibration (phonon). By observing Raman shift, which is the energy difference between incident light and scattering light, it is possible to get the information about lattice structure of a substance. The first measurement of CNT using Raman spectroscopy was done by Rao et al. in 1997 [131]. Raman spectroscopy is noncontact, nondestructive, simple, and fast characterization method, which also possibly measures at atmospheric pressure and room temperature. Raman spectroscopy is widely used for the characterization

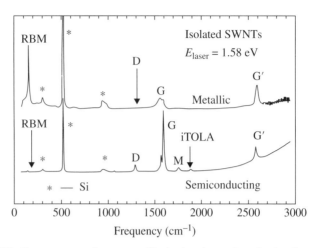

FIGURE 2.15 Raman spectra from a metallic (top) and a semiconducting (bottom) SWCNT at the single nanotube level. The spectra show the radial breathing modes (RBM), D-band, G-band, and G′ band features, in addition to weak double resonance features associated with the M-band and the iTOLA second-order modes. The isolated carbon nanotubes are present on an oxidized silicon substrate that contributes to the Raman spectra denoted by "*." Reprinted from Ref. 133 with permission from Elsevier.

of CNTs as the most powerful tool. All carbon allotropes are Raman active [132]. Following characteristics are especially distinctive for CNTs. Figure 2.15 shows representative Raman spectra [133] of CNTs, in which the upper spectrum and lower one are *m*-SWCNTs and *sc*-SWCNTs, respectively.

G band: It is a strong signal caused from sp^2 vibration of graphite structure observed between 1500 and 1600 cm^{-1}. While it is only one peak from in-plane vibration in the case of graphite, a peak is split into two bands, G+ and G−, by the displacement of longitudinal optical mode (LO) and transverse optical mode (TO) originated from the curved surface of CNTs. This low-frequency component (G−) and high-frequency component (G+) correspond to atomic vibration vertical to tube axis (TO) and parallel to that (LO), respectively. Jorio et al. followed-up their earlier work by studying G-band frequencies of 62 different individual tubes. With that array of data on different (*n, m*) identified species, they attempted to extract an empirical relationship describing the diameter dependence of the lower energy G− peak relative to the G+ peak, which was claimed to have no diameter dependence [134]. For *m*-SWCNTs, it is known that this dissociative width is about three times compared to that of *sc*-SWCNTs in the case when same diameter tubes are compared [135]. Moreover, the line shape of G-band is changed by the difference in *sc*-SWCNTs or *m*-SWCNTs [134, 135]. For *m*-SWCNTs, the shape of spectrum changes from Lorentz to asymmetric. Brown et al. assigned this asymmetric, broad G− peak to a Breit–Wigner–Fano (BWF) resonance between the discrete circumferential vibrational mode and the continuum of electronic states [135]. Recently, the change in shape to BWF is attributed to phonon softening due to Kohn anomaly of LO

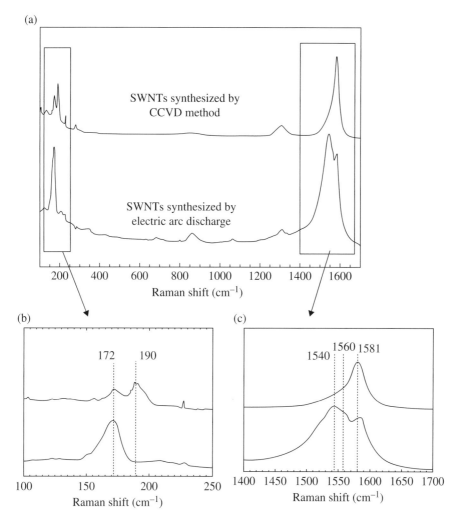

FIGURE 2.16 Superposition of SWCNT Raman spectra produced by CCVD method and ones synthesized by electric arc discharge, using an excitation wavelength of 676.4 nm. (a) Complete Raman spectrum, (b) detail of the RBM frequency range, and (c) detail of the TM frequency range. Reprinted from Ref. 138 with permission from Elsevier.

phonon mode [136]. More recently, the comparative study of Raman spectroscopy for various condensed SWCNT by post-growth separation techniques such as DGU-based and DNA-based separation has reported that the broad G− peak (BWF) is only found for zigzag m-SWCNT and not for armchair SWCNTs [137]. Figure 2.16 shows these distinctive G-band shapes for sc-SWCNTs (upper spectrum) and m-SWCNTs (lower spectrum) [138].

D band: It is a signal that appears around 1350 cm^{-1}. It is closely related to crystalline CNTs. CNTs containing many defects has strong D band signal.

G′ band: It is a signal that appears around $2700\,cm^{-1}$. This is overtone of D band. This signal is not related to crystalline CNTs and always have intensity equal to G band.

Radial breathing mode (RBM): This is a peak that appears mainly at the region of low wave number. Especially, this is distinctive for small-diameter tubes like SWCNTs and information of the diameter of CNTs is acquired. Here, the RBM frequency (ω) is inversely proportional to the diameter of CNTs (d). The expression is as follows:

$$\omega = A \,/\, d + B$$

Both A and B parameters are experimentally obtained constants. Based on this expression, it is possible to assign the chirality. RBM frequency is not related to the chiral angle θ of the nanotubes [139–141]. Many expressions for various conditions of CNTs such as "being bundle" are presented [139–147]. For example, for an isolated SWCNT on Si substrate, the value of A is reported to be 248 and $B=0$ [142]. Bandow et al. calculated the value of $A=223.75$ [131, 143] using a force constant model with interactions to the fourth neighbor. A calculated value of RBM parameters ($A=232$ and $B=6.5$) is presented by Alvarez et al. using a crystal of identical and infinite nanotubes and taking into account the Van der Waals interactions between tubes [145]. According to Bandow et al. [146], values of parameter A depend on the models used: zone folding method ($A=223.75$) [131], force constant model ($A=218$) [139], local density approximation ($A=234$) [140], pseudopotential density functional theory ($A=236$) [141], and elastic deformation model ($A=227$) [147]. Moreover, for CNT forests that have strong interaction between a tube and a tube, up shift of RBM is reported [148, 149].

As described earlier, Raman spectroscopy gives us much information of CNTs. However, following issues should be considered. Firstly, it is difficult to get absolute intensity. The factors that define the intensity are the probability of scattering, time of measurement, and size of sample. Therefore, it is difficult to discuss the intensity per unit weight of a specific spectrum. In most cases, observed intensity is converted using a certain amount of known sample and unit time of the measurement, or relative intensity is compared for the sample having two and more spectra. As example, non-scattering spectrum of Si is often used for calibration of the intensity.

Next, Raman spectrum of SWCNTs is resonant in nature; therefore, RBM shifts as excitation laser wavelength changes. For example, to assign specific chirality, measurement should be carried out with plural lasers having different wave number.

Because of the resonant nature and the unique band structure of CNTs, only phonons corresponding to species electronically resonant with the excitation energy should appear. In 1999, Kataura et al. plotted the optical transition using energy difference of plural laser wavelength on the vertical axis and wave number for each chiral on horizontal axis [150], which is so-called Kataura plot. Based on Kataura plot, RBM obtained using Raman spectroscopy is assigned for specific chirality [69].

2.2.2.2 *Optical Absorption (UV-Vis-NIR)*   Optical absorption of a substance is calculated using Lambert–Beer law. According to Lambert–Beer law, when a monochromic light passes through a uniform absorption layer with concentration C ($mol\,l^{-1}$)

and thickness b (cm), the following expression is derived from the relationship between the intensity of incident light I_0 and that of transmitted light I.

$$A = -\log\left(\frac{I}{I_0}\right) = \varepsilon C b$$

where I/I_0 is transmittance, A is absorbance, ε (1 mol^{-1}/cm^{-1}) is a molar absorption coefficient, which is inherence constant for each substance. Optical absorption spectrum is normally plotted with absorbance (A) on the vertical axis and the wavenumber of incident light (or energy) on horizontal axis.

To obtain optical absorption spectrum, firstly sample should be prepared. The optical absorption spectrum of nanotubes was first reported by Chen et al. using a solution of chemically modified SWCNTs in water [151]. Kataura et al. clarified the relationship between the absorption peak of SWCNTs and band gap by comparing the optical absorption peaks of SWCNTs and energy band using Tight-binding model. They measured optical absorption using air-sprayed SWCNTs on quartz. However, these initial works found that absorption features could not be attributed to specific chiral structures. Although liquid suspensions of nanotubes could be produced using ultrasonication or chemical functionalization, absorption spectra remained poorly defined due to the reaggregation of SWCNTs in suspension or modification of band structure due to functionalization. However, in 2002, O'Connell et al. [152] reported optical absorption spectra with strong peaks reflecting each electronic structure of SWCNTs. They firstly developed the method to obtain isolated SWCNTs using ultrasonication with sodium dodecyl sulfate (SDS) surfactant in deuterium water and a continuing separation by an ultracentrifuge.

In general, in optical absorption spectrum, distinctive peaks S_{11} (first transition of sc-SWCNTs), S_{22} (second transition of sc-SWCNTs), and M_{11} (first transition of m-SWCNTs), which are originated from inter-band transitions appeared in the region from infrared light to visible light. For example, in the case of CoMoCAT with tube diameters of 0.7–1.3 nm, the S_{11}, S_{22}, and the M_{11} transitions are usually observed in the ranges 800–1400, 550–800, and 450–650 nm, respectively [150]. At a time, π-plasmon originated from both carbon impurities and SWCNTs is generally observed. Especially, it becomes dominant in the region of high wave numbers. Hence, aforementioned plural peaks from band transition of SWCNTs are observed as overlapping on background of π-plasmon [24, 153]. Figure 2.17 shows a typical spectrum obtained by optical absorption [154]. The purity of SWCNTs can be estimated from the peaks on the background [153]. Nair et al. suggested a calculation algorism from a baseline of the spectrum and made peak height and area calculation for each (n, m) chirality possible [155].

SWCNT samples that should be measured for optical absorption should be dispersed and centrifuged. This method is useful for observing chirality controllability rather than total purity. With optical absorption, it is possible to obtain the information for M_{11} transition which is unavailable using later described PLE. However, it is difficult to distinguish between M_{11} and S_{22}, because S_{22} sometimes overlaps with M_{11}. Moreover, M_{11} of m-SWCNTs belong to the same family

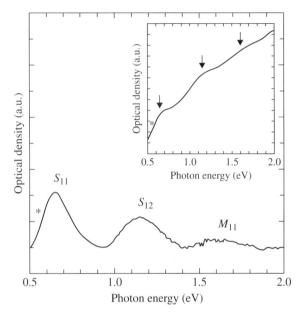

FIGURE 2.17 UV–Vis–NIR spectrum of the soot synthesized at Ce concentration of 1.7 at.%, He 500 Torr, and arc-discharge current of 70 A and subsequently burned in the air at 533 K for 1 h. Inset shows the raw UV–Vis–NIR spectrum. Asterisk (*) indicates absorption peak due to the quartz substrate. Reprinted from Ref. 154 with permission from Elsevier.

($2n + m$ = constant) that has close value, hence it is hard to distinguish chirality of m-SWCNTs from a spectrum.

2.2.2.3 Photoluminescence Spectroscopy
Photoluminescence spectroscopy (PL) is a method to observe photoluminescence spectrum of a substance excited by incident light. The principle of luminescence and optical absorption is the same and is attributed to the energy level of a substance. When a substance absorbs light, the electrons of the substance transit from the ground state to an excited state of higher level. Then, the electrons move back toward the ground state, during which time, they emit light. This light-emitting phenomenon is known as photoluminescence. As each substance has an inherent spectrum, important information about each substance is obtained.

In the case of SWCNTs, the relationship between excited energy and emitted energy for photoluminescence peaks is intrinsic for each chiral vector. Therefore, based on scanning excited wavelength at intervals of few nm, photoluminescence excitation spectroscopy (PLE) is developed. The representative is near infrared spectra mapped as a contour plot [156]. From the peak position on PLE mapping, the combination between excited energy and emission energy is identified, then (n, m) of sc-SWCNTs existing in a sample is confirmed. Figure 2.18 shows a representative contour plot obtained using isolated SWCNTs [157].

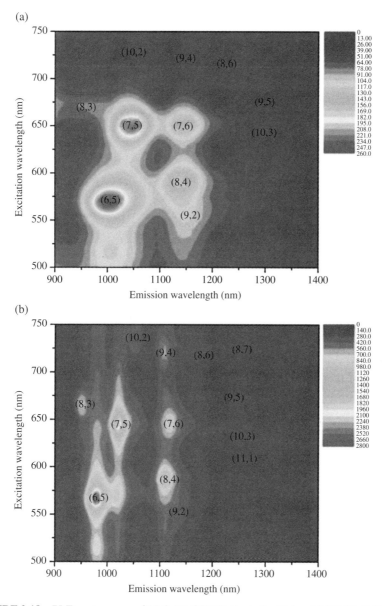

FIGURE 2.18 PLE contour map of (a) CoMoCAT-IL (ionic liquids) and (b) CoMoCAT-SDS versus excitation and emission wavelengths with corresponding chiral assignments. Reprinted from Ref. 157 with permission from Elsevier.

As is described in the previous section, while the research of selective synthesis of SWCNTs is pushed ahead, the development of separation techniques after synthesis has been rapidly progressed [158–173]. Methods used to separate fine particles and biomaterials are applied to separate SWCNTs, which are based on geometrical and electronic structures and polarizability. In 2006, Arnold et al. first applied density gradient ultracentrifugation (DGU) to separate SWCNTs using aqueous suspensions of surfactant-wrapped HiPco- and CoMoCAT-SWCNT materials [161]. They used anionic surfactants like SDS for wrapping. This method separates different molecular species by exploiting differences in their respective mass densities as they travel through a viscous medium under a large applied centrifugal field. Actually, this method is effective when separation is done based on the difference in the diameter and electronic character (*sc* or *m*), and high selectivity is confirmed [165–173]. Moreover, DNA-based ion-exchange chromatography (IEX) has been developed as an alternative to DGU with a focus on chirality-specific interactions to purify bulk, unsorted nanotube materials into single-chirality SWCNTs [170].

It is not overwhelming to say that isolation of SWCNTs by DGU and IEX made it possible to clearly observe photophysical behaviors of *sc*-SWCNTs by PL and PLE.

However, while both DGU and DNA-based separation methods can produce highly enriched specific chiral SWCNT aqueous suspensions, it should be noted that only sub-microgram to as much as single milligram quantities of unseparated SWCNTs can be produced and usually only after combining multiple applications of either separation technique. This is a consequence not of the chirality composition of the initial SWCNTs but rather of the overall efficiency of the separation technique. Although these techniques are not ideal for large-scale separation (even at gram level), they can produce sufficient amount for spectroscopic measurement.

It is thought that luminescence intensity corresponding to each chirality is affected by the factors such as abundance of each (*n*, *m*) of SWCNTs, internal relaxation, and transition probability of luminescence [169]. Therefore, by combining appropriate theoretical estimation for each transition probability of luminescence, it is expected that PL method will be a powerful tool for the measurement of chirality distribution of SWCNTs. However, at present, the relationship between chirality distribution of actual sample of SWCNTs and the observed luminescence spectra is not completely clarified. Moreover, in PLE mapping, various irretraceable subpeaks still exist. No detailed explanation of PL of SWCNTs is available. Moreover, in principle, precise information of *m*-SWCNT could not be acquired by PL method because no excitonic luminescence is produced in *m*-SWCNT.

2.2.3 Microscopy

2.2.3.1 Scanning Tunneling Microscopy and Transmission Electron Microscopy

The main factors that define the physical properties of CNTs are structural parameters such as diameter, chirality, and defects like atomic defects, and a hetero atom which exists inter lattice. It is important to observe and detect the morphologies and location precisely for each tube.

First, the images of CNTs in atomic level should be captured using scanning tunneling microscopy (STM) [174, 175],which are shown in Figure 2.19. Moreover, the curvature-induced gap and trigonal warping splitting for various zigzag nanotubes [176] were experimentally confirmed by measuring zigzag tubes using STM

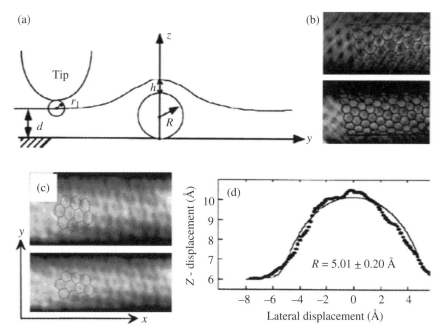

FIGURE 2.19 (a) A model situation of STM tip and SWCNT. (b) A high-resolution SWCNT STM image and projection of ideal tube. (c) A low-resolution distorted STM image and correction of the inflation effect. (d) Average cross-section of scan line of SWCNT STM image and fitting of the equation for a phenomenological form of the tunneling current. Reprinted from Ref. 175 with permission from American Carbon Society. © PERGAMON.

[177], where a splitting of the van Hove singularities and gap opening at the Fermi energy were observed in the electronic density of states.

Since the early stage of CNT research, TEM has been widely used for the structural measurement. TEM is effective in observing diameter and the number of wall of CNTs but it is difficult to obtain chirality structure directly. Therefore, to obtain chirality, TEM is performed combined with electron diffraction. It was the reason that CNTs, which are formed by only light carbon elements, are easy to destroy by electron beam irradiation for capture of high resolution image. However, recent development of aberration correction in TEM made high-resolution observation possible in low electron acceleration voltage [178–182]. Figure 2.20 shows a representative image of SWCNTs captured using aberration-corrected high-resolution TEM (ac-HRTEM) with corresponding electron diffraction patterns [183]. Direct observation of chirality using ac-HRTEM, along with its comparison with the result obtained from Raman spectroscopy, has been also carried out [180].

TEM is also used to evaluate the purity of SWCNTs. However, attention is needed to estimate the purity quantitatively. To get a significant result on total purity of CNTs on typical TEM, it is necessary to analyze many images captured randomly from total macro sample. However, the concern is that appropriate algorism to objectively measure relative quantity of different carbon species present in unpurified

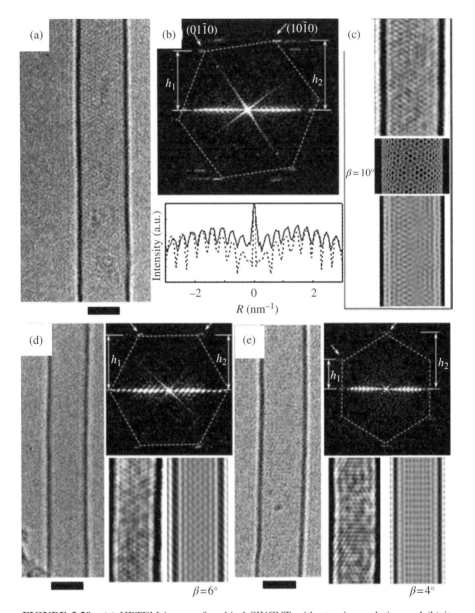

FIGURE 2.20 (a) HRTEM image of a chiral SWCNT with atomic resolution and (b) its corresponding optical diffraction (top) with profile of the equatorial line (solid line) and simulated result for solution (27, 17) (dot line) (bottom). (c) Reconstructed image and simulation image of SWCNT (27, 17) with tilting angle $\beta = 10°$. (d) An armchair tube (17, 17). $\beta = 6°$. (e) A zigzag tube (34, 0). $\beta = 4°$. Scale bar is 2 nm. Reprinted from Ref. 183 with permission from Elsevier.

typical SWCNTs is not established. TEM provides with valuable information about the morphologies and structures of products, but when used to evaluate purity, it should be used with caution.

2.3 SUMMARY

In this section, characterization methods, especially Raman and TEM, which are widely used for CNT research, are reviewed. In Raman spectroscopy, for assignment of chirality, it was necessary to use different plural excited lasers, and time-consumed measurements were so far unavoidable. In TEM, it was difficult to capture the structure of SWCNTs in atomic level so far. However, both rapid progression of a CCD detector in Raman spectroscopy and technique of aberration correction in TEM decreased the measurement time significantly in Raman and made possible the direct observation of SWCNT structure in TEM. Future development of these methods will make possible the characterization of CNTs such as location of atomic defect or quantification evaluation, which is not possible by conventional methods. For this, there should be a significant contribution toward CNT research.

REFERENCES

[1] Dresselhaus, M. S.; Dresselhaus, G.; Eklund, P. C. *Science of Fullerenes and Carbon Nanotubes*; Academic Press, New York: 1996.

[2] Dresselhaus, M. S.; Dresselhaus, G.; Saito, R. Physics of carbon nanotubes. *Carbon* 1995, **33**, 883–891.

[3] Saito, R.; Dresselhaus, G.; Dresselhaus, M. S. Trigonal warping effect of carbon nanotubes. *Phys. Rev. B* 2000, **6**, 2981–2990.

[4] Iijima, S. Helical microtubules of graphitic carbon. *Nature* 1991, **354**, 56–58.

[5] Ebbesen, T. W.; Ajayan, P. M. Large-scale synthesis of carbon nanotubes. *Nature* 1992, **358**, 220–222.

[6] Bethune, D. S.; Kiang, C. H.; DeVries, M.; Gorman, G.; Savoy, R.; Vazquez, J.; Beyers, R. Cobalt-catalyzed growth of carbon nanotubes with single-atomic-layer walls. *Nature* 1993, **363**, 605–607.

[7] Saito, Y.; Yoshikawa, T.; Okuda, M.; Fujimoto, N.; Sumiyama, K.; Suzuki, K.; Kasuya, A.; Nishina, Y. Carbon nanocapsules encaging metals and carbides. *J. Phys. Chem. Solids* 1993, **54**, 1849–1860.

[8] Iijima, S.; Ichihashi, T. Single-shell carbon nanotubes of 1-nm diameter. *Nature* 1993, **363**, 603–605.

[9] Ebbesen, T. W.; Ajayan, P. M.; Hiura, H.; Tanigaki, K. Purification of nanotubes. *Nature* 1994, **367**, 519.

[10] Zimmerman, J. L.; Bradley, R. K.; Huffman, C. B.; Hauge, R. H.; Margrave, J. L. Gas-phase purification of single-wall carbon nanotubes. *Chem. Mater.* 2000, **12**, 1361–1366.

[11] Moon, J. M.; An, K. H.; Lee, Y. H.; Park, Y. S.; Bae, D. J.; Park, G. S. High-yield purification process of single-walled carbon nanotubes. *J. Phys. Chem. B* 2001, **105**, 5677–5681.

[12]   Haddon, R. C.; Sippel, J.; Rinzler, A. G.; Papadimitrakopoulos, F. Purification and separation of carbon nanotubes. *MRS Bull.* 2004, **29**, 252–259.

[13]   Park, T. J.; Banerjee, S.; Hemraj-Benny, T.; Wong, S. S. Purification strategies and purity visualization techniques for single-walled carbon nanotubes. *J. Mater. Chem.* 2006, **16**, 141–154.

[14]   Thess, A.; Lee, R.; Nikolaev, P.; Dai, H. J.; Petit, P.; Robert, J.; Xu, C. H.; Lee, Y. H.; Kim, S. G.; Rinzler, A. G.; Colbert, D. T.; Scuseria, G. E.; Tomanek, D.; Fischer, J. E.; Smalley, R. E. Crystalline ropes of metallic carbon nanotubes. *Science* 1996, **273**, 483–487.

[15]   Endo, M.; Takeuchi, K.; Igarashi, S.; Kobori, K.; Shiraichi, M.; Kroto, H. W. The production and structure of pyrolytic carbon nanotubes (PCNTs). *J. Phys. Chem. Solids* 1993, **54**, 1841–1848.

[16]   Dai, H.; Rinzler, A. G.; Thess, A.; Nikolaev, P.; Colbert, D. T.; Smalley, R. E. Single-wall nanotubes produced by metal-catalyzed disproportionation of carbon monoxide. *Chem. Phys. Lett.* 1996, **260**, 471–475.

[17]   Kong, J.; Soh, H.; Cassell, A.; Quate, C. F.; Dai, H. Synthesis of individual single-walled carbon nanotubes on patterned silicon wafers. *Nature* 1998, **395**, 878–881.

[18]   Hafner, J.; Bronikowski, M.; Azamian, B.; Nikolaev, P.; Colbert, D.; Smalley, R. Catalytic growth of single-wall carbon nanotubes from metal particles. *Chem. Phys. Lett.* 1998, **296**, 195–202.

[19]   Kong, J.; Cassell, A. M.; Dai, H. Chemical vapor deposition of methane for single-walled carbon nanotubes. *Chem. Phys. Lett.* 1998, **292**, 567–574.

[20]   Cassell, A.; Raymakers, J.; Kong, J.; Dai, H. Large scale CVD synthesis of single-walled carbon nanotubes. *J. Phys. Chem.* 1999, **103**, 6484–6492.

[21]   Su, M.; Zheng, B.; Liu, J. A scalable CVD method for the synthesis of single walled carbon nanotubes with high catalyst productivity. *Chem. Phys. Lett.* 2000, **322**, 321–326.

[22]   Zhang, G. Y.; Qi, P. F.; Wang, X. R.; Lu, Y. R.; Li, X. L.; Tu, R.; Bangsaruntip, S.; Mann, D.; Zhang, L.; Dai, H. J. Selective etching of metallic carbon nanotubes by gas-phase reaction. *Science* 2006, **314**, 974–977.

[23]   Bachilo, S. M.; Balzano, L.; Herrera, J. E.; Pompeo, F.; Resasco, D. E.; Weisman, R. B. Narrow (n,m)-distribution of single-walled carbon nanotubes grown using a solid supported catalyst. *J. Am. Chem. Soc.* 2003, **125**, 11186–11187.

[24]   Lolli, G.; Zhang, L. A.; Balzano, L.; Sakulchaicharoen, N.; Tan, Y. Q.; Resasco, D. E. Tailoring (n,m) structure of single-walled carbon nanotubes by modifying reaction conditions and the nature of the support of CoMo Catalysts. *J. Phys. Chem. B* 2006, **110**, 2108–2115.

[25]   Kitiyanan, B.; Alvarez, W. E.; Harwell, J. H.; Resasco, D. E. Controlled production of single-wall carbon nanotubes by catalytic decomposition of CO on bimetallic Co–Mo catalysts. *Chem. Phys. Lett.* 2000, **317**, 497–503.

[26]   Nikolaev, P.; Bronikowski, M. J.; Bradley, R. K.; Rohmund, F.; Colbert, D. T.; Smith, K. A.; Smalley, R. E. Gas-phase catalytic growth of single-walled carbon nanotubes from carbon monoxide. *Chem. Phys. Lett.* 1999, **313**, 91–97.

[27]   Bronikowski, M. J.; Willis, P. A.; Colbert, D. T.; Smith, K. A.; Smalley, R. E. Gas-phase production of carbon single-walled nanotubes from carbon monoxide via the HiPCO® process: A parametric study. *J. Vac. Sci. Technol. A* 2001, **19**, 1800–1805.

[28] Li, Y. M.; Mann, D.; Rolandi, M.; Kim, W.; Ural, A.; Hung, S.; Javey, A.; Cao, J.; Wang, D. W.; Yenilmez, E.; Wang, Q.; Gibbons, J. F.; Nishi, Y.; Dai, H. J. Preferential growth of semiconducting single-walled carbon nanotubes by a plasma enhanced CVD method. *Nano Lett.* 2004, **4**, 317–321.

[29] Zhong, G. F.; Iwasaki, T.; Honda, K.; Furukawa, Y.; Ohdomari, I.; Kawarada, H. Low temperature synthesis of extremely dense and vertically aligned single-walled carbon nanotubes. *Jpn. J. App. Phys.* 2005, 1 44, 1558–1561.

[30] Zhang, G. Y.; Mann, D.; Zhang, L.; Javey, A.; Li, Y. M.; Yenilmez, E.; Wang, Q.; McVittie, J.; Nishi, Y.; Gibbons, J.; Dai, H. Ultra-high-yield growth of vertical single-walled carbon nanotubes: Hidden roles of hydrogen and oxygen. *Proc. Natl. Acad. Sci. U. S. A.* 2005, **102**, 16141–16145.

[31] Zhong, G. F.; Iwasaki, T.; Kawarada, H. Semi-quantitative study on the fabrication of densely packed and vertically aligned single-walled carbon nanotubes. *Carbon* 2006, **44**, 2009–2014.

[32] Wang, W. L.; Bai, X. D.; Xu, Z.; Liu, S.; Wang, E. G. Low temperature growth of single-walled carbon nanotubes: Small diameters with narrow distribution. *Chem. Phys. Lett.* 2006, **419**, 81–85.

[33] Kato, T.; Hatakeyama, R.; Tohji, K. Diffusion plasma chemical vapour deposition yielding freestanding individual single-walled carbon nanotubes on a silicon-based flat substrate. *Nanotechnology* 2006, **17**, 2223–2226.

[34] Zhong G.; Iwasaki T.; Robertson J.; Kawarada H. Growth kinetics of 0.5 cm vertically aligned single-walled carbon nanotubes. *J. Phys. Chem. B* 2007, **111**, 1907–1910.

[35] Chhowalla, M.; Teo, K. B. K.; Ducati, C.; Rupesinghe, N. L.; Amaratunga, G. A. J.; Ferrari, A. C.; Roy, D.; Robertson, J.; Milne, W. I. Growth process conditions of vertically aligned carbon nanotubes using plasma enhanced chemical vapor deposition. *J. Appl. Phys.* 2001, **90**, 5308–5317.

[36] Bell, M. S.; Teo, K. B. K.; Lacerda, R. G.; Milne, W. I.; Hash, D. B.; Meyyappan, M. Carbon nanotubes by plasma-enhanced chemical vapor deposition. *Pure Appl. Chem.* 2006, **78**, 1117–1125.

[37] Tam, E.; Levchenko, I.; Ostrikov, K. Deterministic shape control in plasma-aided nanotip assembly. *J. Appl. Phys.* 2006, **100**, 036104–036106.

[38] Levchenko, I.; Ostrikov, K.; Tam, E. Uniformity of postprocessing of dense nanotube arrays by neutral and ion fluxes. *Appl. Phys. Lett.* 2006, **89**, 223108-1–223108-3.

[39] Esconjauregui, S.; Bayer, B. C.; Fouquet, M.; Wirth, C. T.; Ducati, C.; Hofmann, S.; Robertson, J. Growth of high-density vertically-aligned arrays of carbon nanotubes by plasma-assisted catalyst pre-treatment. *Appl. Phys. Lett.* 2009, **95**, 173115-1–173115-3.

[40] Ostrikov, K.; Mehdipour, H. Thin single-walled carbon nanotubes with narrow chirality distribution: Constructive interplay of plasma and Gibbs–Thomson effects. *ACS Nano* 2011, **5**, 8372–8382.

[41] Ohashi T.; Kato R.; Ochiai T.; Tokune T.; Kawarada H. High quality single-walled carbon nanotube synthesis using remote plasma CVD. *Diam. Relat. Mater.* 2012, **24**, 184–187.

[42] Ostrikov, K.; Mehdipour, H. Nanoscale plasma chemistry enables fast, size-selective nanotube nucleation. *J. Am. Chem. Soc.* 2012, **134**, 4303–4312.

[43] Neyts, E. C.; van Duin, A. C. T.; Bogaerts, A. Insights in the plasma-assisted growth of carbon nanotubes through atomic scale simulations: Effect of electric field. *J. Am. Chem. Soc.* 2012, **134**, 1256–1260.

[44] Li, W. Z.; Xie, S. S.; Qian, L. X.; Chang, B. H.; Zou, B. S.; Zhou, W. Y.; Zhao, R. A.; Wang, G. Large-scale synthesis of aligned carbon nanotubes. *Science* 1996, **274**, 1701–1703.

[45] Iwasaki, T.; Motoi, T.; Den, T. Multiwalled carbon nanotubes growth in anodic alumina nanoholes. *Appl. Phys. Lett.* 1999, **75**, 2044–2046.

[46] Li, J.; Papadopoulos, C.; Xu, J. M.; Moskovits, M. Highly-ordered carbon nanotube arrays for electronic applications. *Appl. Phys. Lett.* 1999, **75**, 367–369.

[47] Rao, C. N. R.; Sen, R.; Satishkumar, B. C.; Govindaraj, A. Large aligned-nanotube bundles from ferrocene pyrolysis. *Chem. Commun.* 1998, **15**, 1525–1526.

[48] Ren, Z. F.; Huang, Z. P.; Xu, J. H.; Wang, P. B.; Siegal, M. P.; Provencio, P. N. Synthesis of large arrays of well-aligned carbon nanotubes on glass. *Science* 1998, **282**, 1105–1107.

[49] Huang, S.; Dai, L.; Mau, A. W. H. Patterned growth and contact transfer of well-aligned carbon nanotube films. *J. Phys. Chem. B* 1999, **103**, 4223–4227.

[50] Avigal, Y.; Kalish, R. Growth of aligned carbon nanotubes by biasing during growth. *Appl. Phys. Lett.* 2001, **78**, 2291–2293.

[51] Tanemura, M.; Iwata, K.; Takahashi, K.; Fujimoto, Y.; Okuyama, F.; Sugie, H.; Filip, V. Growth of aligned carbon nanotubes by plasma-enhanced chemical vapor deposition: Optimization of growth parameters. *J. Appl. Phys.* 2001, **90**, 1529–1533.

[52] Rao, A. M.; Jacques, D.; Haddon, R. C.; Zhu, W.; Bower, C.; Jin, S. In situ-grown carbon nanotube arrays with excellent field emission characteristics. *Appl. Phys. Lett.* 2000, **76**, 3813–3815.

[53] Young, L. J.; Lee, B. S. Nitrogen induced structure control of vertically aligned carbon nanotubes synthesized by microwave plasma enhanced chemical vapor deposition. *Thin Solid Films* 2002, **418**, 85–88.

[54] Tsai, S. H.; Shiu, C. T.; Lati, S. H.; Chan, L. H.; Hsieh, W. J.; Shih, H. C. In situ growing and etching of carbon nanotubes on silicon under microwave plasma. *J. Mater. Sci. Lett.* 2002, **21**, 1709–1711.

[55] Murakami, Y.; Chiashi, S.; Miyauchi, Y.; Hu, M.; Ogura, M.; Okubo, T.; Maruyama, S. Growth of vertically aligned single-wall carbon nanotube films on quartz substrates and their optical anisotropy. *Chem. Phys. Lett.* 2004, **385**, 298–303.

[56] Hata, K.; Futaba, D.; Mizuno, K.; Namai, T.; Yumura, M.; Iijima, S. Water-assisted highly efficient synthesis of impurity-free single-wall carbon nanotubes. *Science* 2004, **306**, 1362–1364.

[57] Xiong, G.; Wang, D. Z.; Ren, Z. F. Aligned millimeter-long carbon nanotube arrays grown on single crystal magnesia. *Carbon* 2006, **44**, 969–973.

[58] Li, Q. W.; Zhang, X. F.; DePaula, R. F.; Zheng, L. X.; Zhao, Y. H.; Stan, L.; Holesinger, T. G.; Arendt, P. N.; Peterson, D. E.; Zhu, Y. T. Sustained growth of ultralong carbon nanotube arrays for fiber spinning. *Adv. Mater.* 2006, **18**, 3160–3163.

[59] Yun, Y. H.; Shanov, V.; Tu, Y.; Subramaniam, S.; Schulz, M. J. Growth mechanism of long aligned multiwall carbon nanotube arrays by water-assisted chemical vapor deposition. *J. Phys. Chem. B* 2006, **110**, 23920–23925.

[60] Chakrabarti, S.; Kume, H.; Pan, L.; Nagasaka, T.; Nakayama, Y. Number of walls controlled synthesis of millimeter-long vertically aligned brushlike carbon nanotubes. *J. Phys. Chem. C* 2007, **111**, 1929–1934.

[61] Noda, S.; Hasegawa, K.; Sugime, H.; Kakehi, K.; Zhang, Z.; Maruyama, S.; Yamaguchi, Y. Millimeter-thick single-wall carbon nanotube forests: Hidden role of catalyst support. *Jpn. J. Appl. Phys.* 2007, **46**, L399–L401.

[62] Chakrabarti, S.; Gong, K.; Dai, L. Structural evaluation along the nanotube length for super-long vertically-aligned double-walled carbon nanotube arrays. *J. Phys. Chem. C* 2008, **112**, 8136–8139.

[63] Xu, Y.; Flor, E.; Schmidt, H.; Smalley, R. E.; Hauge, R. H. Effects of atomic hydrogen and active carbon species in 1 mm vertically aligned single-wall carbon nanotube growth. *Appl. Phys. Lett.* 2006, **89**, 123116–123118.

[64] Christen, H. M.; Puretzky, A. A.; Cui, H.; Belay, K.; Fleming, P. H.; Geohegan, D. B.; Lowndes, D. H. Rapid growth of long, vertically aligned carbon nanotubes through efficient catalyst optimization using metal film gradients. *Nano Lett.* 2004, **4**, 1939–1942.

[65] Yasuda, S.; Hiraoka, T.; Futaba, D.; Yamada, T.; Yumura, M.; Hata, K. Existence and kinetics of graphitic carbonaceous impurities in carbon nanotube forests to assess the absolute purity. *Nano Lett.* 2009, **9**, 769–773.

[66] Li, Y. M.; Kim, W.; Zhang, Y. G.; Rolandi, M.; Wang, D. W.; Dai, H. J. Growth of single-walled carbon nanotubes from discrete catalytic nanoparticles of various sizes. *J. Phys. Chem. B* 2001, **105**, 11424–11431.

[67] Hiraoka, T.; Bandow, S.; Shinohara, H.; Iijima, S. Control on the diameter of single-walled carbon nanotubes by changing the pressure in floating catalyst CVD. *Carbon* 2006, **44**, 1853–1859.

[68] Ohashi, T.; Ochiai, T.; Tokune, T.; Kawarada, H. Increasing the length of a single-wall carbon nanotube forest by adding titanium to a catalytic substrate. *Carbon* 2013, **57**, 79–87.

[69] Jorio, A.; Saito, R.; Hafner, J. H.; Lieber, C. M.; Hunter, M.; McClure, T.; Dresselhaus, G.; Dresselhaus, M. S. Structural (n,m) determination of isolated single-wall carbon nanotubes by resonant Raman scattering. *Phys. Rev. Lett.* 2001, **86**, 1118–1121.

[70] Ohashi, T.; Kato, R.; Tokune, T.; Kawarada, H. Understanding the stability of a sputtered Al buffer layer for single-walled carbon nanotube forest synthesis. *Carbon* 2013, **57** 401–409.

[71] Zheng, L. X.; O'Connell, M. J.; Doorn, S. K.; Liao, X. Z.; Zhao, Y. H.; Akhadov, E. A.; Hoffbauer, M. A.; Roop, B. J.; Jia, Q. X.; Dye, R. C.; Peterson, D. E.; Huang, S. M.; Liu, J.; Zhu, Y. T. Ultralong single-wall carbon nanotubes. *Nat. Mater.* 2004, **3**, 673–676.

[72] Croci, M.; Arfaoui, I.; Stöckli, T.; Chatelain, A.; Bonard, J. M. A fully sealed luminescent tube based on carbon nanotube field emission. *Microelectron J.* 2004, **35**, 329–336.

[73] Wen, Q.; Zhang, R.; Qian, W.; Wang, Y.; Tan, P.; Nie, J.; Wei, F. Growing 20 cm long DWNTs/TWNTs at a rapid growth rate of 80–90 μm/s. *Chem. Mater.* 2010, **22**, 1294–1296.

[74] Wen, Q.; Qian, W. Z.; Nie, J. Q.; Cao, A. Y.; Ning, G. Q.; Wang, Y. Hu, L.; Zhang, Q.; Huang, J.; Wei, F. 100 mm long, semiconducting triple-walled carbon nanotubes. *Adv. Mater.* 2010, **22**, 1867–1871.

[75] Huang, S. M.; Woodson, M.; Smalley, R.; Liu, J. Growth mechanism of oriented long single walled carbon nanotubes using "fast-heating" chemical vapor deposition process. *Nano Lett.* 2004, **4**, 1025–1028.

[76] Jin, Z.; Chu, H. B.; Wang, J. Y.; Hong, J. X.; Tan, W. C.; Li, Y. Ultralow feeding gas flow guiding growth of large-scale horizontally aligned single-walled carbon nanotube arrays. *Nano Lett.* 2007, **7**, 2073–2079.

[77] Su, M.; Li, Y.; Maynor, B.; Buldum, A.; Lu, J. P.; Liu, J. Lattice-oriented growth of single-walled carbon nanotubes. *J. Phys. Chem. B* 2000, **104**, 6505–6508.

[78] Ismach, A.; Segev, L.; Wachtel, E.; Joselevich, E. Atomic-step-templated formation of single wall carbon nanotube patterns. *Angew. Chem. Int. Ed.* 2004, **43**, 6140–6143.

[79] Kocabas, C.; Hur, S. H.; Gaur, A.; Meitl, M. A.; Shim, M.; Rogers, J. A. Guided growth of large-scale, horizontally aligned arrays of single-walled carbon nanotubes and their use in thin-film transistors. *Small* 2005, **1**, 1110–1116.

[80] Han, S.; Liu, X. L.; Zhou, C. W. Template-free directional growth of single-walled carbon nanotubes on a- and r-plane sapphire. *J. Am. Chem. Soc.* 2005, **127**, 5294–5295.

[81] Ismach, A.; Kantorovich, D.; Joselevich, E. Carbon nanotube graphoepitaxy: Highly oriented growth by faceted nanosteps. *J. Am. Chem. Soc.* 2005, **127**, 11554–11555.

[82] Zhang, B.; Hong, G.; Peng, B.; Zhang, J.; Choi, W.; Kim, J.; Choi, J.; Liu, Z. Grow single-walled carbon nanotubes cross-bar in one batch. *J. Phys. Chem. C* 2009, **113**, 5341–5344.

[83] Wagner, R. S.; Ellis, W. C. Vapor-liquid-solid mechanism of single crystal growth. *Appl. Phys. Lett.* 1964, **4**, 89–90.

[84] Kukovitsky, E. F.; L'vov, S. G.; Sainov, N. A. VLS-growth of carbon nanotubes from the vapour. *Chem. Phys. Lett.* 2000, **317**, 65–70.

[85] Page, A. J.; Ohta, Y.; Irle, S.; Morokuma, K. Mechanisms of single-walled carbon nanotube nucleation, growth and healing determined using QM/MD methods. *Acc. Chem. Res.* 2010, **43**, 1375–1385.

[86] Sharma, R. An environmental transmission electron microscope for in situ synthesis and characterization of nanomaterials. *J. Mater. Res.* 2005, **20**, 1695–1707.

[87] Rodriguez-Manzo, J. A.; Terrones, M.; Terrones, H.; Kroto, H. W.; Sun, L.; Banhart, F. In situ nucleation of carbon nanotubes by the injection of carbon atoms into metal particles. *Nat. Nanotechnol.* 2007, **2**, 307–311.

[88] Hofmann, S.; Sharma, R.; Ducati, C.; Du, G.; Mattevi, C.; Cepek, C.; Cantoro, M.; Pisana, S.; Parvez, A.; Cervantes-Sodi, F.; Ferrari, A. C.; Dunin-Borkowski, R.; Lizzit, S.; Petaccia, L.; Goldoni, A.; Robertson, J. In-situ observations of catalyst dynamics during surface-bound carbon nanotube nucleation. *Nano Lett.* 2007, **7**, 602–608.

[89] Lin, M.; Tan, J. P. Y.; Boothroyd, C.; Loh, K. P.; Tok, E. S.; Foo, Y. L. Dynamical observation of bamboo-like carbon nanotube growth. *Nano Lett.* 2007, **7**, 2234–2238.

[90] Yoshida, H.; Takeda, S.; Uchiyama, T.; Kohno, H.; Homma, Y. Atomic-scale in-situ observation of carbon nanotube growth from solid state iron carbide nanoparticles. *Nano Lett.* 2008, **8**, 2082–2086.

[91] Begtrup, G. E.; Gannett, W.; Meyer, J. C.; Yuzvinsky, T. D.; Ertekin, E.; Grossman, J. C.; Zettl, A. Facets of nanotube synthesis: High–resolution electron microscopy study and density functional theory calculations. *Phys. Rev. B* 2009, **79**, 205409.

[92] Moseler, M.; Cervantes-Sodi, F.; Hofmann, S.; Csanyi, G.; Ferrari, A. C. Dynamic catalyst restructuring during carbon nanotube growth. *ACS Nano* 2010, **4**, 7587–7595.

[93] Yoshida, H.; Kohno, H.; Takeda, S. In situ structural analysis of fluctuating crystalline Fe-Mo-C nanoparticle catalysts during the growth of carbon nanotubes. *Micron* 2012, **43**, 1176–1180.

[94] He, M. S.; Chernov, A. I.; Fedotov, P. V.; Obraztsova, E. D.; Sainio, J.; Rikkinen, E.; Jiang, H.; Zhu, Z.; Tian, Y.; Kauppinen, E. I.; Niemela, M.; Krause, A. O. I. Predominant growth of (6,5) single-walled carbon nanotubes on a copper promoted iron catalyst. *J. Am. Chem. Soc.* 2010, **132**, 13994–13996.

[95] Wang, B.; Wei, L.; Yao, L.; Li, L. J. Yang, Y. H.; Chen, Y. Pressure-induced single-walled carbon nanotube (n,m) selectivity on Co–Mo catalysts. *J. Phys. Chem. C* 2007, **111**, 14612–14616.

[96] An, K. H.; Park, J. S.; Yang, C. M.; Jeong, S. Y.; Lim, S. C.; Kang, C.; Son, J.; Jeong, M. J.; Lee, Y. H. A diameter-selective attack of metallic carbon nanotubes by nitronium ions. *J. Am. Chem. Soc.* 2005, **127**, 5196–5203.

[97] Yu, B.; Hou, P. X.; Li, F.; Liu, B. L.; Liu, C. Cheng, H. M. Selective removal of metallic single-walled carbon nanotubes by combined in situ and post-synthesis oxidation. *Carbon* 2010, **48**, 2941–2947.

[98] Zhang, Y. Y.; Zhang, Y.; Xian, X. J.; Zhang, J.; Liu, Z. F. Sorting out semiconducting single-walled carbon nanotube arrays by preferential destruction of metallic tubes using xenon-lamp irradiation. *J. Phys. Chem. C* 2008, **112**, 3849–3856.

[99] Hong, G.; Zhang, B.; Peng, B. H.; Zhang, J. Choi, W. M.; Choi, J. Y.; Kim, J. M.; Liu, Z. Direct growth of semiconducting single-walled carbon nanotube array. *J. Am. Chem. Soc.* 2009, **131**, 14642–14643.

[100] Miyata, Y.; Maniwa, Y.; Kataura, H. Selective oxidation of semiconducting single-wall carbon nanotubes by hydrogen peroxide. *J. Phys. Chem. B* 2006, **110**, 25–29.

[101] Li, P.; Zhang, J. Sorting out semiconducting single-walled carbon nanotube arrays by preferential destruction of metallic tubes using water. *J. Mater. Chem.* 2011, **21**, 11815–11821.

[102] Qu, L.; Du, F.; Dai, L. Preferential syntheses of semiconducting vertically aligned single-walled carbon nanotubes for direct use in FETs. *Nano Lett.* 2008, **8**, 2682–2687.

[103] Wang, Y.; Liu, Y. Q.; Li, X. L.; Cao, L. C.; Wei, D. C.; Zhang, H. L.; Shi, D.; Yu, G.; Kajiura, H.; Li, Y. Direct enrichment of metallic single-walled carbon nanotubes induced by the different molecular composition of monohydroxy alcohol homologues. *Small* 2007, **3**, 1486–1490.

[104] Wang, B.; Poa, C.; Wei, L.; Li, L. J.; Yang, Y. H.; Chen, Y. (n,m) selectivity of single-walled carbon nanotubes by different carbon precursors on Co Mo catalysts. *J. Am. Chem. Soc.* 2007, **129**, 9014–9019.

[105] Ding, L.; Tselev, A.; Wang, J. Y.; Yuan, D. N.; Chu, H. B.; McNicholas, T. P.; Li, Y.; Liu, J. Selective growth of well-aligned semiconducting single-walled carbon nanotubes. *Nano Lett.* 2009, **9**, 800–805.

[106] Chiang, W. H.; Sankaran, R. M. Linking catalyst composition to chirality distributions of as-grown single-walled carbon nanotubes by tuning NixFe1-x nanoparticles. *Nat. Mater.* 2009, **8**, 882–886.

[107] Harutyunyan, A. R.; Chen, G. G.; Paronyan, T. M.; Pigos, E. M.; Kuznetsov, O. A.; Hewaparakrama, K. Kim, S. M.; Zakharov, D.; Stach, E. A.; Sumanasekera, G. U. Preferential growth of single-walled carbon nanotubes with metallic conductivity. *Science* 2009, **326**, 116–120.

[108] Sundaram, R. M.; Koziol, K. K.; Windle, A. H. Continuous direct spinning of fibers of single-walled carbon nanotubes with metallic chirality. *Adv. Mater.* 2011, **23**, 5064–5068.

[109] Li, X. L.; Tu, X. M.; Zaric, S.; Welsher, K.; Seo, W. S.; Zhao, W.; Dai, H. Selective synthesis combined with chemical separation of single-walled carbon nanotubes for chirality selection. *J. Am. Chem. Soc.* 2007, **129**, 15770–15771.

[110]   Miyauchi, Y.; Chiashi, S.; Murakami, Y.; Hayashida, Y.; Maruyama, S. Fluorescence spectroscopy of single-walled carbon nanotubes synthesized from alcohol. *Chem. Phys. Lett.* 2004, **387**, 198–203.

[111]   Kato, T.; Hatakeyama, R. Direct growth of short single-walled carbon nanotubes with narrow-chirality distribution by time-programmed plasma chemical vapor deposition. *ACS Nano* 2010, **4**, 7395–7400.

[112]   Wang, Q.; Ng, M. F.; Yang, S. W.; Yang, Y. H.; Chen, Y. A. The mechanism of single-walled carbon nanotube growth and chirality selection induced by carbon atom and dimer addition. *ACS Nano* 2010, **4**, 939–946.

[113]   Ding, F.; Harutyunyan, A. R.; Yakobson, B. I. Dislocation theory of chirality-controlled nanotube growth. *Proc. Natl. Acad. Sci. U. S. A.* 2009, **106**, 2506–2509.

[114]   Rao, R.; Liptak, D.; Cherukuri, T.; Yakobson, B. I.; Maruyama, B. In situ evidence for chirality-dependent growth rates of individual carbon nanotubes. *Nat. Mater.* 2012, **11**, 213–216.

[115]   Botti, S.; Ciardi, R.; Asilyan, L.; Dominicis, L. D.; Fabbri, F.; Orlanducci, S.; Fiori, A. Carbon nanotubes grown by laser-annealing of SIC nano-particles. *Chem. Phys. Lett.* 2004, **400**, 264–267.

[116]   Takagi, D.; Hibino, H.; Suzuki, S.; Kobayashi, Y.; Homma, Y. Carbon nanotube growth from semiconductor nanoparticles. *Nano Lett.* 2007, **7**, 2272–2275.

[117]   Page, A. J.; Chandrakumar, K. R. S.; Irle, S.; Morokuma, K. SWNT Nucleation from Carbon-Coated SiO$_2$ Nanoparticles via a Vapor-Solid-Solid Mechanism. *J. Am. Chem. Soc.* 2011, **133**, 621–628.

[118]   Uchino, T.; Bourdakos, K. N.; de Groot, C. H.; Ashburn, P.; Kiziroglou, M. E.; Dilliway, G. D.; Smith, D. C. Metal catalyst-free low-temperature carbon nanotube growth on SiGe islands. *Appl. Phys. Lett.* 2005, **86**, 233110–233113.

[119]   Uchino, T.; Ayre, G. N.; Smith, D. C.; Hutchison, J. L.; de Groot, C. H.; Ashburn, P. Growth of single-walled carbon nanotubes using germanium nanocrystals formed by implantation. *J. Electrochem. Soc.* 2009, **156**, K144–K148.

[120]   Takagi, D.; Homma, Y.; Hibino, H.; Suzuki, S.; Kobayashi, Y. Single-walled carbon nano-tube growth from highly activated metal nanoparticles. *Nano Lett.* 2006, **6**, 2642–2645.

[121]   Ritschel, M.; Leonhardt, A.; Elefant, D.; Oswald, S.; Büchner, B. Rhenium-catalyzed growth carbon nanotubes. *J. Phys. Chem. C* 2007, **111**, 8414–8417.

[122]   Liu, H.; Takagi, D.; Chiashi, S.; Homma, Y. The growth of single-walled carbon nano-tubes on a silica substrate without using a metal catalyst. *Carbon* 2010, **48**, 114–122.

[123]   Chen, Y. B.; Zhang, J. Diameter controlled growth of single-walled carbon nanotubes from SiO$_2$ nanoparticles. *Carbon* 2011, **49**, 3316–3324.

[124]   Takagi, D.; Kobayashi, Y.; Homma, Y. Carbon nanotube growth from diamond. *J. Am. Chem. Soc.* 2009, **131**, 6922–6923.

[125]   Derycke, V.; Martel, R.; Radosvljevic, M.; Ross, F. M. R.; Avouris, P. Catalyst-free growth of ordered single-walled carbon nanotube networks. *Nano Lett.* 2002, **2**, 1043–1046.

[126]   Huang, S. M.; Cai, Q. R.; Chen, J. Y.; Qian, Y.; Zhang, L. J. Metal-catalyst-free growth of single-walled carbon nanotubes on substrates. *J. Am. Chem. Soc.* 2009, **131**, 2094–2095.

[127]   Kusunoki, M.; Rokkaku, M.; Suzuki, T. Epitaxial carbon nanotube film self-organized by sublimation decomposition of silicon carbide. *Appl. Phys. Lett.* 1997, **71**, 2620–2622.

[128] Kusunoki, M.; Suzuki, T.; Honjo, C.; Hirayama, T.; Shibata, N. Selective synthesis of zigzag-type aligned carbon nanotubes on SiC (000-1) wafers. *Chem. Phys. Lett.* 2002, **366**, 458–462.

[129] Tang, D.; Zhang, I.; Liu, C.; Yin, L.; Hou, P.; Jiang, H.; Zhu, Z.; Li, F.; Liu, B.; Kauppinen, E. I.; Cheng, H. Heteroepitaxial growth of single-walled carbon nanotubes from boron nitride. *Sci. Report* 2012, **2**, 971.

[130] Liu, J.; Wang, C.; Tu, X.; Liu, B.; Chen, L.; Zheng, M.; Zhou, C. Chirality-controlled synthesis of single-wall carbon nanotubes using vapor phase epitaxy. *Nat. Commun.* 2012, **3**, 1199.

[131] Rao, A.; Richter, R.; Bandow, S.; Chase, B.; Eklund, P.; Williams, K. A.; Fang, S.; Subbaswamy, K. R.; Menon, M.; Thess, A.; Smalley, R.; Dresselhaus, G.; Dresselhaus, M. Diameter-selective Raman scattering from vibrational modes in carbon nanotubes. *Science* 1997, **275**, 187–191.

[132] Arepalli, S.; Nikolaev, P.; Gorelik, O.; Hadjiev, V.; Holmes, W.; Files, B.; Yowell, L. Protocol for the characterization of single-wall carbon nanotube material quality. *Carbon* 2004, **42**, 1783–1791.

[133] Dresselhaus, M. S.; Dresselhaus, G.; Saito, R.; Jorio, A. Raman spectroscopy of carbon nanotubes. *Phys. Rep.* 2005, **409**, 47–99.

[134] Jorio, A.; Souza Filho, A. G.; Dresselhaus, G.; Dresselhaus, M. S.; Swan, A. K.; Ünlü, M. S.; Goldberg, B.; Pimenta, M. A.; Hafner, J. H.; Lieber, C. M.; Saito, R. G-band resonant Raman study of 62 isolated single-wall carbon nanotubes. *Phys. Rev. B* 2002, **65**, 155412-1–155419-9.

[135] Brown, S. D. M.; Jorio, A.; Corio, P.; Dresselhaus, M. S.; Dresselhaus, G.; Saito, R.; Kneipp, K. Origin of the Breit-Wigner-Fano lineshape of the tangential G-band feature of metallic carbon nanotubes. *Phys. Rev. B* 2001, **63**, 155414-1–155414-4.

[136] Dubay, O.; Kresse, G.; Kuzmany, H. Phonon softening in metallic nanotubes by a peierls-like mechanism. *Phys. Rev. Lett.* 2002, **88**, 235506-1–235506-4.

[137] Hároz, E. H.; Duque, J. G.; Rice, W. D.; Densmore, C. G.; Kono, J.; Doorn, S. K. Resonant Raman spectroscopy of armchair carbon nanotubes: Absence of broad G-feature. *Phys. Rev. B* 2011, **84**, 121403(R)-1–121403(R)-4.

[138] Colomer, J. F.; Benoit, J. M.; Stephan, C.; Lefrant, S.; Van Tendeloo, G.; Nagy, J. B. Characterization of single-wall carbon nanotubes produced by CCVD method. *Chem. Phys. Lett.* 2001, **345**, 11–17.

[139] Jishi, R.; Venkataraman, L.; Dresselhaus, M.; Dresselhaus, G. Phonon modes in carbon nanotubules. *Chem. Phys. Lett.* 1993, **209**, 77–82.

[140] Kürti, J.; Kresse, G.; Kuzmany, H. First-principles calculations of the radial breathing mode of single-wall carbon nanotubes. *Phys. Rev. B* 1998, **58**, R8869–R8872.

[141] Sanchez-Portal, D.; Artacho, E.; Soler, J.; Rubio, A.; Ordejón, P. Ab initio structural, elastic, and vibrational properties of carbon nanotubes. *Phys. Rev. B* 1999, **59**, 12678–12688.

[142] Dresselhaus, M.; Dresselhaus, G.; Jorio, A.; Souza-Filho, A. G.; Saito, R. Raman spectroscopy on isolated single wall carbon nanotubes. *Carbon* 2002, **40**, 2043–2061.

[143] Bandow, S.; Asaka, S.; Saito, Y.; Rao, A.; Grigorian, L.; Richter, E.; Eklund, P. Effect of the growth temperature on the diameter distribution and chirality of single-wall carbon nanotubes. *Phys. Rev. Lett.* 1998, **80**, 3779–3782.

[144]   Milnera, M.; Kurti, J.; Hulman, M.; Kuzmany, H. Periodic resonance excitation and intertube interaction from quasicontinuous distributed helicities in single-wall carbon nanotubes. *Phys. Rev. Lett.* 2000, **84**, 1324–1327.

[145]   Alvarez, L.; Righi, A.; Guillard, T.; Rols, S.; Anglaret, E.; Laplaze, D.; Sauvajol, J. Resonant Raman study of the structure and electronic properties of single-wall carbon nanotubes. *Chem. Phys. Lett.* 2000, **316**, 186–190.

[146]   Bandow, S.; Chen, G.; Sumanasekera, G.; Gupta, R.; Yudasaka, M.; Iijima, S.; Eklund, P. C. Diameter-selective resonant Raman scattering in double-wall carbon nanotubes. *Phys. Rev. B* 2002, **66**, 075416–075424.

[147]   Mahan, G. D. Oscillations of a thin hollow cylinder: Carbon nanotubes. *Phys. Rev. B* 2002, **65**, 235402–235409.

[148]   Araujo, P. T.; Maciel, I. O.; Pesce, P. B. C.; Pimenta, M. A.; Doorn, S. K.; Qian, H.; Hartschuh, A.; Steiner, M.; Grigorian, L.; Hata, K.; Jorio, A. Nature of the constant factor in the relation between radial breathing mode frequency and tube diameter for single-wall carbon nanotubes. *Phys. Rev. B* 2008, **77**, 241403(R)-1–241403(R)-4.

[149]   Araujo, P. T.; Fantini, C.; Lucchese, M. M.; Dresselhaus, M. S.; Jorio, A. The effect of environment on the radial breathing mode of supergrowth single wall carbon nanotubes. *Appl. Phys. Lett.* 2009, **95**, 261902-1–261902-3.

[150]   Kataura, H.; Kumazawa, Y.; Maniwa, Y.; Uemezu, I.; Suzuki, S.; Ohtsuka, Y.; Achiba, Y. Optical properties of single-wall carbon nanotubes. *Synth. Met.* 1999, **103**, 2555–2558.

[151]   Chen, J.; Hamon, M. A.; Hu, H.; Chen, Y.; Rao, A. M.; Eklund, P. C.; Haddon, R. C. Solution properties of single-walled carbon nanotubes. *Science* 1998, **282**, 95–98.

[152]   O'Connell, M. J.; Bachilo, S. M.; Huffman, C. B.; Moore, V. C.; Strano, M. S.; Haroz, E. H.; Rialon, K. L.; Boul, P. J.; Noon, W. H.; Kittrell, C.; Ma, J.; Hauge, R. H.; Weisman, R. B.; Smalley, R. E. Band gap fluorescence from individual single-walled carbon nanotubes. *Science* 2002, **297**, 593–596.

[153]   Itkis, M. E.; Perea, D. E.; Jung, R.; Niyogi, S.; Haddon, R. C. Comparison of analytical techniques for purity evaluation of single-walled carbon nanotubes. *J. Am. Chem. Soc.* 2005, **127**, 3439–3448.

[154]   Sato, Y.; Jeyadevan, B.; Hatakeyama, R.; Kasuya, A.; Tohji, K. Electronic properties of radial single-walled carbon nanotubes. *Chem. Phys. Lett.* 2004, **385**, 323–328.

[155]   Nair, N.; Usrey, M.; Kim, W.-J.; Braatz, R. D.; Strano, M. S. Deconvolution of the photo-absorption spectrum of single-walled carbon nanotubes with (n,m) resolution. *Anal. Chem.* 2006, **78**, 7689–7696.

[156]   Bachilo, S. M.; Strano, M. S.; Kittrell, C.; Hauge, R. H.; Smalley, R. E.; Weisman, R. B. Structure-assigned optical spectra of single-walled carbon nanotubes. *Science* 2002, **298**, 2361–2366.

[157]   Yang, J.; Zhang, Z.; Zhang, D.; Li, Y. Quantitative analysis of the (n,m) abundance of single-walled carbon nanotubes dispersed in ionic liquids by optical absorption spectra. *Mater. Chem. Phys.* 2013, **139**, 233–240.

[158]   Zheng, M.; Jagota, A.; Strano, M. S.; Santos, A. P.; Barone, P.; Chou, S. G.; Diner, B. A.; Dresselhaus, M. S.; Mclean, R. S.; Onoa, G. B.; Samsonidze, Ge. G.; Semke, E. D.; Usrey, M.; Walls, D. J. Structure-based carbon nanotube sorting by sequence-dependent DNA assembly. *Science* 2003, **302**, 1545–1548.

[159] Matarredona, O.; Rhoads, H.; Li, Z. R.; Harwell, J. H.; Balzano, L.; Resasco, D. E. Dispersion of single-walled carbon nanotubes in aqueous solutions of the anionic surfactant NaDDBS. *J. Phys. Chem. B* 2003, **107**, 13357–13367.

[160] Arnold, M. S.; Stupp, S. I.; Hersam, M. C. Enrichment of single-walled carbon nanotubes by diameter in density gradients. *Nano Lett.* 2005, **5**, 713–718.

[161] Arnold, M. S.; Green, A. A.; Hulvat, J. F.; Stupp, S. I.; Hersam, M. C. Sorting carbon nanotubes by electronic structure using density differentiation. *Nat. Nanotechnol.* 2006, **1**, 60–65.

[162] Nish, A.; Nicholas, R. J. Temperature induced restoration of fluorescence from oxidised single-walled carbon nanotubes in aqueous sodium dodecylsulfate solution. *Phys. Chem. Chem. Phys.* 2006, **8**, 3547–3551.

[163] Nish, A.; Hwang, J. Y.; Doig, J.; Nicholas, R. J. Highly selective dispersion of single-walled carbon nanotubes using aromatic polymers. *Nat. Nanotechnol.* 2007, **2**, 640–646.

[164] Ju, S. Y.; Doll, J.; Sharma, I.; Papadimitrakopoulos, F. Selection of carbon nanotubes with specific chiralities using helical assemblies of flavin mononucleotide. *Nat. Nanotechnol.* 2008, **3**, 356–362.

[165] Hersam, M. C. Progress towards monodisperse single-walled carbon nanotubes. *Nat. Nanotechnol.* 2008, **3**, 387–394.

[166] Green, A. A.; Hersam M. C. Colored semitransparent conductive coatings consisting of monodisperse metallic single-walled carbon nanotubes. *Nano Lett.* 2008, **8**, 1417–1422.

[167] Yanagi, K.; Miyata, Y.; Kataura, H. Optical and conductive characteristics of metallic single-wall carbon nanotubes with three basic colors; cyan, magenta, and yellow. *Appl. Phys. Express* 2008, **1**, 034003–034006.

[168] Arnold, M. S.; Suntivich, J.; Stupp, S. I.; Hersam, M. C. Hydrodynamic characterization of surfactant encapsulated carbon nanotubes using an analytical ultracentrifuge. *ACS Nano* 2008, **2**, 2291–2300.

[169] Nair, N.; Kim, W. J.; Braatz, R. D.; Strano, M. S. Dynamics of surfactant-suspended single walled carbon nanotubes in a centrifugal field. *Langmuir* 2008, **24**, 1790–1795.

[170] Tu, X.; Zheng, M. A DNA-based approach to the carbon nanotube sorting problem. *Nano Res.* 2008, **1**, 185–194.

[171] Tu, X.; Manohar, S.; Jagota, A.; Zheng, M. DNA sequence motifs for structure-specific recognition and separation of carbon nanotubes. *Nature* 2009, **460**, 250–253.

[172] Green, A. A.; Hersam, M. C. Processing and properties of highly enriched double-wall carbon nanotubes. *Nat. Nanotechnol.* 2009, **4**, 64–70.

[173] Niyogi, S.; Densmore C. G.; Doorn, S. K. Electrolyte tuning of surfactant interfacial behavior for enhanced density-based separations of single-walled carbon nanotubes. *J. Am. Chem. Soc.* 2009, **131**, 1144–1153.

[174] Odom, T. W.; Huang, J. L.; Kim, P.; Lieber, C. M. Atomic structure and electronic properties of single-walled carbon nanotubes. *Nature* 1998, **391**, 62–64.

[175] Kim, P.; Odom, T. W.; Huang, J.; Lieber, C. M. STM Study of single-walled carbon nanotubes. *Carbon* 2000, **38**, 1741–1744.

[176] Hamada, N.; Sawada, S.; Oshiyama, S. New one-dimensional conductors – graphitic microtubules. *Phys. Rev. Lett.* 1992, **68**, 1579–1581.

[177] Ouyang, M.; Huang, J.-L.; Cheung, C. L.; Lieber, C. M. Energy gaps in "metallic" single-walled carbon nanotubes. *Science* 2001, **292**, 702–705.

[178] Hashimoto, A.; Suenaga, K.; Gloter, A.; Urita, K.; Iijima, S. Direct evidence for atomic defects in graphene layers. *Nature* 2004, **430**, 870–873.

[179] Zhu, H. W.; Suenaga, K.; Mizuno, K.; Hashimoto, A.; Urita, K.; Hata, K.; Iijima, S. Atomic-resolution imaging of the nucleation points of single-walled carbon nanotubes. *Small* 2005, **1**, 1180–1183.

[180] Sato, Y.; Yanagi, K.; Miyata, Y.; Suenaga, K.; Kataura, H.; Iijima, S. Chiral-angle distribution for separated single-walled carbon nanotubes. *Nano Lett.* 2008, **8**, 3151–3154.

[181] Zhu, H.; Suenaga, K.; Wei, J.; Wang, K.; Wu, D. A strategy to control the chirality of single-walled carbon nanotubes. *J. Cryst. Growth* 2008, **310**, 5473–5476.

[182] Zoberbier, T.; Chamberlain, T. W.; Biskupek, J.; Kuganathan, N.; Eyhusen, S.; Bichoutskaia, E.; Kaiser, U.; Khlobystov, A. N. Interactions and reactions of transition metal clusters with the interior of single-walled carbon nanotubes imaged at the atomic scale. *J. Am. Chem. Soc.* 2012, **134**, 3073–3079.

[183] Zhu, H.; Suenaga, K.; Hashimoto, A.; Urita, K.; Iijima, S. Structural identification of single and double-walled carbon nanotubes by high resolution transmission electron microscopy. *Chem. Phys. Lett.* 2005, **412**, 116–120.

3

SYNTHESIS AND CHARACTERIZATION OF GRAPHENE

Santanu Das[1], P. Sudhagar[2], Yong Soo Kang[2] and Wonbong Choi[1]

[1]Department of Materials Science and Engineering, University of North Texas, Denton, TX, USA
[2]Department of Energy Engineering, Hanyang University, Seoul, South Korea

3.1   INTRODUCTION

In the contemporary research field of nanotechnology, graphene is one of the extensively studied carbon nanomaterials to date [1–5]. The graphene created unprecedented attention in the horizon of materials science and condensed-matter physics owing to its potential applications for next-generation electronics, sensors, transparent electrodes, and photonics. Graphene is a covalently functionalized, sp^2-bonded, single-atom-thick carbon allotrope that exists in a hexagonal honeycomb lattice, which forms 3D bulk graphite, when the layers of single honeycomb graphitic lattices are stacked and bounded by a weak van der Waals force (right-hand side of Fig. 3.1). Further, the graphene can be either wrapped up to form a sphere that is known as 0-dimensional fullerene or rolled up with respect to its axis to form 1-dimensional cylindrical structure, called carbon nanotube (CNT). Single graphitic layer is commonly known as monolayer graphene, and two and three graphitic layers are known as bilayer and trilayer graphenes, respectively. More than 5 layers up to 30 layers of graphene are generally called as multilayer graphene/thick graphene/nanocrystalline thin graphite.

Carbon Nanomaterials for Advanced Energy Systems: Advances in Materials Synthesis and Device Applications, First Edition. Edited by Wen Lu, Jong-Beom Baek and Liming Dai.
© 2015 John Wiley & Sons, Inc. Published 2015 by John Wiley & Sons, Inc.

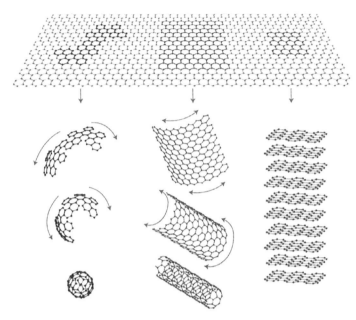

FIGURE 3.1   The 2D hexagonal nanosheets of graphene as a building block of other forms of carbon nanomaterials. Reprinted with permission from Ref. 2.

Graphene exhibits extraordinary electrical and thermal conductivity, high carrier density, high mobility, optical conductivity, and mechanical properties [2, 4–7]. Furthermore, exceptional electrical properties of graphene also include room temperature quantum Hall effect [8, 9], ballistic charge transport [1, 10], high charge carrier density [1, 2], and tunable band gap [11]. High-crystalline pristine monoatomic graphene shows semimetallic behavior as π and π^* bands touch in a single point at the Fermi level (E_f) at the corners of the Brillouin zone [8]. In this context, graphene exhibits charge carrier density in the order of $10^{13}\,cm^{-2}$ [2], whereas it shows charge carrier mobility and resistivity of ~15,000 $cm^2\,V^{-1}\,S^{-1}$ [2] and ~$10^{-6}\,\Omega$–cm, respectively; thus, it is ideally fitted for ultrafast field-effect transistor (FET) and/or THz devices [12]. Additionally, monolayer graphene shows the resemblance of the massless Dirac fermions [8] and Landau-level quantization under vertically applied magnetic field to the graphene basal plane [10]. Furthermore, the properties of graphene are varied with its number of stacked layers, thus opening up possibilities to fabricate a wide range of devices. For example, bilayer graphene exhibits gapless semiconductor behavior that occurs when parabolic bands of k and k' touch at the single points at the Brillouin zone. Thus, these phenomena bring into a negligible band overlap (about 0.0016 eV) at higher energies, which make bilayer graphene a gapless semiconductor [10]. In this context, large-scale monolayer graphene exhibits ~97% optical transmittance [13] with ~2.2 $K\Omega\,sq^{-1}$ sheet resistance, while the transmittance and sheet resistance decrease with increasing number of layers. Recent report shows that bilayer, trilayer, and 4-layer graphenes possess ~95, ~92, and ~89% transmittance with corresponding sheet resistance of 1 KΩ sq^{-1}, ~700 Ω sq^{-1}, and ~400 Ω sq^{-1}, respectively [13].

Exceptional electrical properties of graphene have attracted applications for future electronics such as ballistic transistors, flexible displays, photonic devices, integrated circuit components, transparent conducting electrodes, and sensors. Graphene has a high electron (or hole) mobility as well as low Johnson noise (electronic noise generated by the thermal agitation of the charge carriers in an electrical conductor at equilibrium, which happens regardless of any applied voltage), allowing it to be utilized as the channel in a FET [14]. The high electrical conductivity and high optical transparency promote graphene as a candidate for transparent conducting electrodes, required for applications in touch screens, liquid crystal displays, organic photovoltaic cells, and organic light-emitting diodes (OLEDs) [2]. Numerous graphene-based devices have already been demonstrated in the potential applications in transistors [1, 15], photonic devices, Li-ion battery [16], supercapacitor [17], and solar cells [18–21]. Many reports are available for graphene synthesis, and most of them are based on mechanical exfoliation [1] from graphite, chemical exfoliation [22], chemical process [23], epitaxy on SiC surface [24], and chemical vapor deposition (CVD) [25].

Thus, graphene created a tremendous impact in the scientific/industrial community as demonstrated by the year-wise number of publication trends in Figure 3.2a. Figure 3.2b shows the scalability versus cost and graphene quality trends that vary with different synthesis techniques. This chapter summarizes the graphene synthesis processes, including mechanical, chemical, and epitaxial growth process together with the detailed discussion of process parameters and their feasibility. In addition, graphene characterization techniques are also reviewed in later section.

3.2   OVERVIEW OF GRAPHENE SYNTHESIS METHODOLOGIES

Till date, several methodologies have been demonstrated for the synthesis of graphene and graphene-derived materials. To name a few are mechanical cleaving (exfoliation) [1], chemical exfoliation [22], chemical synthesis [23], thermal CVD synthesis [25, 27], and epitaxial growth [24] methods. Besides these, several other processes are also demonstrated such as unzipping of CNT [28, 29], electrochemical exfoliation [30], laser ablation process, and several others [31]. The graphene synthesis methodologies can be classified in two types: top-down and bottom-up approach (Fig. 3.3). A top-down process involves the synthesis of nanoscale materials by reducing its sizes from the bulk, whereas a bottom-up approach is associated with the strategy to build up a structure by atomic or molecular arrangements. For example, all kinds of exfoliation processes (e.g., mechanical and chemical, as shown in Fig. 3.3) are top-down processes, whereas CVD, pyrolysis, and epitaxial processes are the so-called bottom-up approaches (Fig. 3.3).

In 1999, the mechanical cleaving of highly ordered pyrolytic graphite (HOPG) by atomic force microscopy (AFM) tips was first invented in order to fabricate few

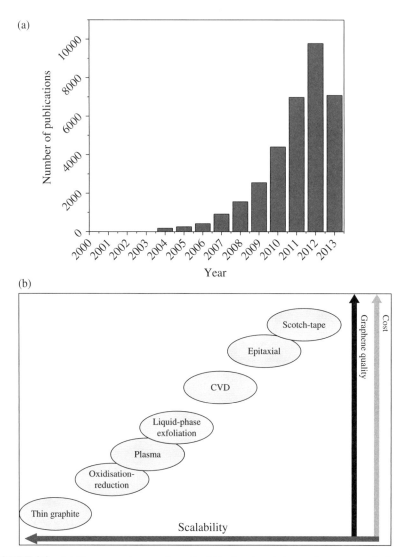

FIGURE 3.2   (a) The year-wise number of publication trends in graphene research showing exponentially increasing behavior [using SciFinder, American Chemical Society database, Search Date: June 19, 2013]. (b) Scalability versus cost and graphene quality trends for different synthesis techniques of graphene. Reprinted with permission from Ref. 26.

layers or single layer of graphene from bulk graphite [32]. The mechanical cleaving method by using AFM cantilever was capable of fabricating ~10 nm thick graphene, which is comparable to 30-monolayer graphene. The concept of fabricating single atomic thick layer graphene was first reported in the year 2004 where simple scotch tape was used for the exfoliation of graphene layers to form a bulk graphite [1]. This technique is the first reported technique to produce a single layer of graphene with

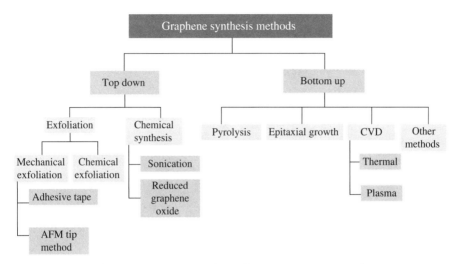

FIGURE 3.3   The classifications of the different graphene synthesis processes.

relatively easy fabrication. Similarly, chemical exfoliation is a method where solution-dispersed graphite is exfoliated by inserting large alkali ions between the graphite layers. Similarly, the chemical synthesis process involves fabrication of graphite oxide (GO) dispersed in a solution followed by reduction process. Like CNT synthesis, catalytic thermal CVD is proved to be the best method for large-scale graphene fabrication where thermally dissociated carbon is deposited onto a catalytically active transition metal surface and forms a honeycomb graphite lattice at elevated temperature under ambient or low pressure.

To sum it all, all the techniques stand popular in the individual field of experimentalists. However, all synthesis methods have their own advantages as well as limitations based on the property requirements as well as final applications of graphene. For example, the mechanical exfoliation method is capable of fabricating different layers of graphene (from monolayer to few layers); however, the reliability and repeatability of obtaining the same quality of graphene are a real challenge. Moreover, the fabrication of large-area graphene by mechanical exfoliation process is a real challenge at present. On the other hand, chemical synthesis processes (which involve the synthesis of GO and reduction back to graphene in a liquid medium) are low-temperature processes that make it easier to fabricate graphene on any types of substrates at ambient temperature, particularly on polymeric substrates. However, the process is susceptible to the incomplete reduction process and often produced GO instead of pure graphene, thus resulting in successive degradation of electrical and optical properties of the final product based upon its degree of reduction. The thermal CVD process is a quite fascinating method for a large-area graphene fabrication and is easily transferred to any flexible substrates via roll-to-roll process, which is also favorable for adoption in the post-Si-based flexible CMOS technology [33]. However, the defect-free and large-scale homogeneous graphene formation is another serious challenge. It has been reported that the grain boundary (GB) and ripple formation during the CVD process also successively depreciate

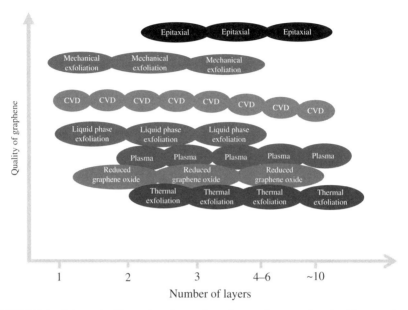

FIGURE 3.4 Schematic illustrating the quality of a graphene produced with a number of layers for the different synthesis processes.

graphene's properties [34]. Graphene obtained from the thermal graphitization of a SiC surface is one of the best quality graphenes, but the high process temperature, high vacuum, and inability to separate SiC from graphene put together make this process ineffective to the industrial use. Figure 3.4 presents the overview of the quality of graphene versus number of layers obtained using different synthesis processes. In the upcoming sections, several of these graphene synthesis methodologies and their scientific and technological importance are elaborately described.

3.2.1   Mechanical Exfoliation

Mechanical exfoliation is a top-down technique of graphene synthesis, by which a longitudinal/transverse stress is applied on the surface of the layered graphite using simple scotch tape or AFM tip. Basically, the graphite structure is composed of stacked monoatomic graphene layers bonded with the weak van der Waals force. The interlayer distance and the interlayer bond energy between two graphene honeycomb lattices are $3.34\,\text{Å}$ and $2\,\text{eV}\,\text{nm}^{-2}$, respectively. Monolayer to few-layer graphene sheets can be produced via mechanical cleaving method by slicing down the layers from graphitic materials such as HOPG, single-crystal graphite, or natural graphite. For mechanical cleaving, $\sim\!300\,\text{nN}\,\mu\text{m}^{-2}$ external force is required in order to separate one graphene layer from graphite, which is subsequently low [35]. In 1999, Ruoff et al. [32] first proposed the mechanical exfoliation technique of plasma-etched pillared HOPG using AFM tip to fabricate graphene. As seen in Figure 3.5a and b, the thin multilayered graphite was fabricated with a thickness of $\sim\!200\,\text{nm}$, which consists of 500–600 layers of monolayer graphene. The first result on the production of single-layer graphene and

(a)   (b)

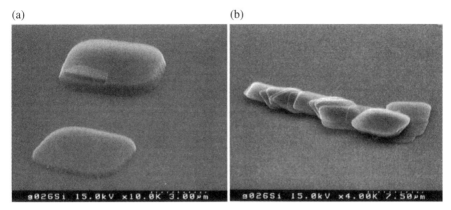

FIGURE 3.5   (a and b) Scanning electron micrographs of mechanically exfoliated thin graphite layers from highly oriented pyrolytic graphite (HOPG) by AFM tip. Reprinted with permission from Ref. 32.

its unusual electronic properties was made by Novoselov and Geim. The invention of novel approach for graphene synthesis and the extraordinary electrical properties of 2D graphene brought them Nobel Prize in Physics in the year 2010.

Novoselov et al. [1] used adhesive tape to produce single graphene layer by mechanically cleaving technique from 1 mm thick HOPG. The process involves the dry etching by oxygen plasma to prepare graphite mesas of few mm deep on the top of the graphite platelets. The graphite mesa surface was then compressed against a 1 mm thick layer of wet photoresist over a glass substrate followed by baking in order to attach the HOPG mesas with the photoresist layer. Then, the graphite flakes were gradually exfoliated by using scotch tape and released in acetone. Finally, the dispersed graphene was transferred onto a SiO_2/Si (n-doped Si with a SiO_2 top layer) wafer after cleaning it with water and propanol. Finally, thin flakes of graphene (thickness <10 nm) were found to adhere on the surface by van der Waals and/or capillary forces as reported in the reference no. 1. Figure 3.6a shows optical micro-graph of multiple numbers of graphene flakes produced on top of oxide grown Si substrate using mechanical exfoliation. A single graphene flake with different number of layers can be easily identified on the top of the 300 nm SiO_2 coated Si substrate as shown in Figure 3.6b. Figure 3.6c shows the optical micro-graph of a transistor fabricated using the mechanically exfoliated graphene and Figure 3.6d demonstrates the bright-field transmission electron microscopy (TEM) image of an exfoliated suspended graphene.

Kim et al. [35] further tried to improve the graphene production method in large scale by cleaving the HOPG using a tipless AFM cantilever as shown in Figure 3.7a and b. The exfoliation process was performed using an AFM tip with predetermined cantilever spring constant (~300 nN), which propagates the required shear stresses onto graphene flakes in order to slice down the single layer. A thinnest graphene flake of ~10 nm thickness was produced by this method; nonetheless, the technique was quite unable to produce single to few layers of graphene.

A FET device based on the exfoliated graphene was fabricated and tested for its electrical characterization (Figs. 3.6c and 3.7d). Most importantly, the exfoliation

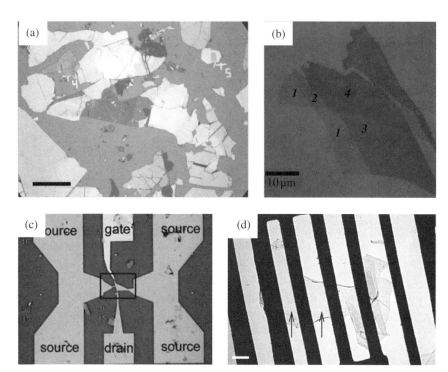

FIGURE 3.6   (a) Optical micrograph of a few layer graphene flake produced by scotch tape methods on top of a 300 nm thick thermal oxide coated Si substrate (Scale bar 50 μm) [Reprinted with permission from: Geim, A. K.; MacDonald, A. H., Graphene: Exploring carbon flatland. Physics Today 2007, 60 (8), 35–41.]; (b) A graphene layer of various thicknesses on SiO_2/Si substrate [Reprinted with permission from Ref: Z. H. Ni, H. M. Wang, J. Kasim, H. M. Fan, T. Yu, Y. H. Wu, Y. P. Feng, Z. X. Shen, Graphene thickness determination using reflection and contrast spectroscopy. Nano Letters 2007, 7, 2758. *existing ref no 119*]; (c) Optical micrograph shows the field effect transistor (FET) device fabricated using nano-patterning on mechanically exfoliated graphene on 300 nm thick SiO_2/Si [Reprinted with permission from Ref Lin, Y.-M.; Jenkins, K. A.; Valdes-Garcia, A.; Small, J. P.; Farmer, D. B.; Avouris, P., Operation of Graphene Transistors at Gigahertz Frequencies. Nano Letters 2008, 9 (1), 422–426.]; and (d) bright-field TEM image of suspended graphene (scale bar 500 nm) [Reprinted with permission from Ref Meyer, J. C.; Geim, A. K.; Katsnelson, M. I.; Novoselov, K. S.; Booth, T. J.; Roth, S., The structure of suspended graphene sheets. Nature 2007, 446 (7131), 60–63.].

process has also been applied for fabricating other two-dimensional (2D) planar materials like boron nitride (BN), molybdenum disilicide (MoS_2), $NbSe_2$, and $Bi_2Sr_2CaCu_2O$ out of their bulk materials [37, 38].

However, for its feasible applica tions in nanoelectronics, the scalability and reproducibility of mechanically exfoliating graphene need to be improved further. In this regard, Liang et al. [39] proposed an interesting method for wafer-scale graphene fabrication by cut-and-choose transfer printing method for integrated circuit; however, still uniform large-scale graphene fabrication with controlled layer is yet to be achieved.

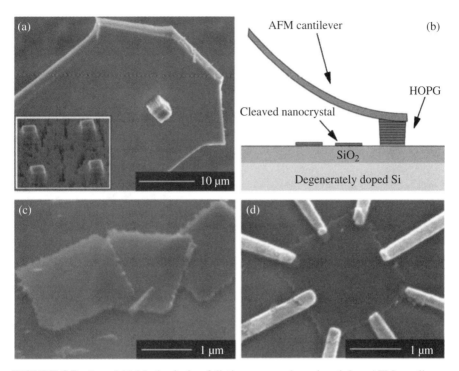

FIGURE 3.7 (a and b) Mechanical exfoliation process by using tipless AFM cantilever. (c) Scanning electron micrograph (SEM) image showing the exfoliated graphene layers. (d) A mesoscopic graphite-based devise demonstrated out of the peeled-off graphene sheet. Reprinted with permission from Ref. 35.

3.2.2 Chemical Exfoliation

In chemical exfoliation process, alkali metals are introduced in a solution of graphite in order to intercalate the single-layer graphene from the bulk graphite structure followed by dispersion in a liquid medium. Alkali metals are the groups of materials in the periodic table that can easily form graphite intercalation compounds (GICs) with various stoichiometric ratio of graphite to alkali metals. The reaction occurs between alkali metals and graphene owing to the difference of the ionization potential between alkali metals and the graphite. For example, potassium (K) ionization potential (4.34 eV) is less than graphite's electron affinity (4.6 eV); thus, K reacts moderately with graphite and forms GICs. Similarly, cesium (Cs) (3.894 eV) possesses much lower ionization potential than K (4.34 eV); therefore, Cs reacts with graphite more vigorously than potassium, which creates a significant improvement in the intercalation of graphite at significantly low temperature and ambient pressure. Additionally, sodium–potassium alloy (Na–K_2) forms eutectic melting at −12.62°C; thus, an exfoliation reaction is expected to occur at room temperature and ambient pressure. In particular, the graphene produced by using Na–K_2 alloy exhibits a wide range of thicknesses from 2 to 150 nm. The major advantage of alkali metals is their atomic radius, which is smaller than the graphite interlayer spacing and hence fits easily in the interlayer spacing as shown in the schematic in Figure 3.8a.

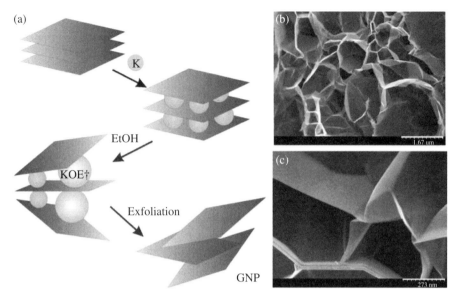

FIGURE 3.8 (a) Schematic illustrating the mechanism of chemical exfoliation process by Viculis et al. [31]. (b) and (c) The SEM pictures of chemically exfoliated graphite nanoplatelets, which show ~10 nm thickness of ~30 layers of single graphite sheet. Reprinted with permission from Ref. 31.

As basic methodologies, graphite, alkali metals, and GICs have been used as starting materials for reaction in order to obtain colloidal dispersions of single-layer graphene flakes. For example, Viculis et al. reported [31] chemically exfoliated graphite nanoplatelets using potassium (K) as intercalating compound forming alkali metal. Potassium (K) forms KC_8 intercalated compound when reacting with graphite at 200°C under inert helium atmosphere (<1 ppm H_2O and O_2). The GIC KC_8 undergoes an exothermic reaction when it reacts with the aqueous solution of ethanol (CH_3CH_2OH) as per Equation 3.1:

$$KC_8 + CH_3CH_2OH \rightarrow 8C + KOCH_2CH_3 + 1/2H_2 \qquad (3.1)$$

Hence, potassium ions dissolve into the solution forming potassium ethoxide followed by the hydrogen gas formation, which further facilitates to separate the graphite layers. Precaution must be taken for this type of reactions as alkali metals react vigorously with water and alcohol. For scalable production, the reaction chamber needs to be kept in an ice bath in order to dissipate the abruptly generated heat during the reaction. The formation of graphitic nanoplatelet structures consists of few-layer graphene (FLG) as shown in Figure 3.8b and c, which consists of 40 ± 15 layers of monoatomic layer of graphene. In this regard, the same researchers also explored the exfoliation process via GIC formation using other alkali metals such as Cs and NaK_2 alloy [40]. Vallés et al. [41] show an exfoliation process of graphene synthesis by reacting a separately prepared alkali metal-GICs and *N*-methylpyrrolidone (NMP) solution as shown in a reaction schematic in Figure 3.9. In this study, potassium ternary salt $K(THF)_x$ C_{24} tetrahydrofurane (THF) is used as alkali metal-GICs material for the reaction process. The process spontaneously exfoliates graphite in *N*-methylpyrrolidone (NMP) solution, yielding stable solutions of negatively charged monolayer graphene sheets and graphene nanoribbons (GNR).

Furthermore, in a recent report, one interesting graphene synthesis process has been delineated via chemical exfoliation followed by the separation of the flakes using density differentiation. Green et al. demonstrated the graphene flake synthesis

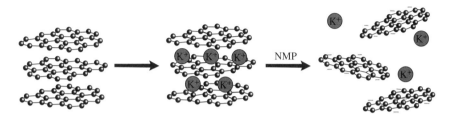

FIGURE 3.9 Flow diagram of GIC-assisted exfoliation process of graphene. Reprinted with permission from Ref. 41.

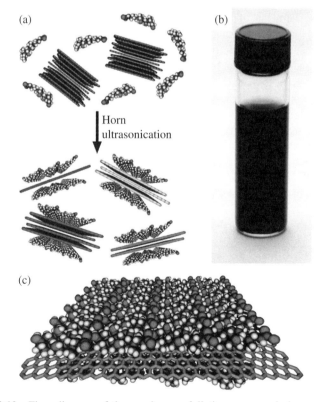

FIGURE 3.10 Flow diagram of the graphene exfoliation process via horn sonication followed by ultracentrifugation. (b) Photograph showing 90 μg ml^{-1} graphene dispersion in SC 6 weeks after it was prepared. (c) Schematic illustrating an ordered SC monolayer on graphene. Reprinted with permission from Ref. 42. (*See insert for color representation of the figure.*)

with controlled thickness by combining the ultrasonication and the density gradient ultracentrifugation (DGU) [42] (Fig. 3.10). Here, they used naturally occurring graphite and sodium cholate as starting materials for this process, which yield a stably dispersed graphene solution and sodium cholate-encapsulated graphene.

Chemical exfoliation process is highly important as it could produce large amount of exfoliated graphene at low temperature, which makes the process highly distinct among reported graphene synthesis process. Furthermore, the process is scalable and can be extended to produce a wide range of functionalized graphenes in a solution process, hence exhibiting high technological significance.

A novel approach was proposed separately regarding the dispersion and exfoliation of pure graphite in organic solvents such as NMP. Hernandez et al. [22] reported the exfoliation of pure graphite in NMP by a simple sonication process. The report showed high-quality, unoxidized monolayer graphene synthesis at a yield of ~1%, and further improvement of the process could potentially improve the yield up to 7–12% of the starting graphite mass by employing sediment recycling (the detailed process is given in Ref. [22]). The morphology of graphite and graphene by the sonication process is shown in Figure 3.11a and b, respectively. The proposed mechanism states that the exfoliation of a layered structure is possible upon the addition of mechanical energy, if the solute and solvent surface energy is the same. In this context, the energy required to exfoliate graphene should be equivalent to the solvent–graphene interaction for the solvents whose surface energies are analogous to that of the suspended graphene. The process is versatile as it is a low-cost solution-phase method and it can be capable of depositing graphene in a wide range of substrates in large scale. Furthermore, the method can be extended to produce graphene-based composites/films, which are the key requirement for special applications, such as transparent conductive electrodes, protective coatings and reinforcements of composite, etc.

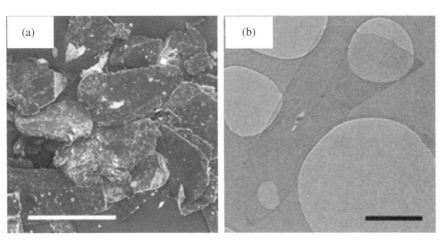

FIGURE 3.11 (a) SEM image of pristine graphite before sonication and (b) transmission electron microscopy of graphene flake prepared in *N*-methylpyrrolidone after the sonication process. Reprinted with permission from Ref. 22.

3.2.3 Chemical Synthesis: Graphene from Reduced Graphene Oxide

Chemical synthesis of graphene is a top-down indirect synthesis method of graphene and, more preciously, first ever method that demonstrated the graphene synthesis by chemical route. In the year 1962, Bochm ct al. first demonstrated monolayer flakes of reduced graphene oxide (RGO), which is recently acknowledged by the Nobel Prize awardee Geim [43]. The method involves oxidation of graphite, dispersion of the flakes by sonication, and reduction of graphene oxide. There are three popular methods available for GO synthesis, which are the Brodie method [36], Staudenmaier method [44], and Hummers and Offeman method [45]. These three methods consist of oxidation of graphite using strong acids and oxidants. The degree of the oxidation can be varied by the reaction conditions (temperature, pressure, etc.), stoichiometry, and the type of graphite used as a starting material. GO was first prepared by Brodie et al. [36], by mixing graphite with potassium chlorate and nitric acid. However, the process contains several steps that are time consuming and hazardous. In order to overcome those problems, Hummers [45] developed the oxidation method for graphite by mixing graphite with sodium nitrite, sulfuric acid, and potassium permanganate, which is well known as the Hummers method. In this process, sulfuric acid and potassium permanganate react to form dimanganese heptoxide, which is a powerful oxidizing agent shown in the reactions as follows [46]:

$$KMnO_4 + 3H_2SO_4 \rightarrow K^+ + MnO_3^+ + H_3O^+ + 3HSO_4^- \tag{3.2}$$

$$MnO_3^+ + MnO_4^- \rightarrow Mn_2O_7 \tag{3.3}$$

When graphite turns into GO, the interlayer spacing is increased two or three times larger than the pristine graphite. For pristine graphite, the interlayer distance is 3.34Å, which expanded up to 5.62Å after 1 h oxidative reaction, and then further interlayer expansion has taken place to $7.0 \pm 0.35\text{Å}$ upon prolonged oxidation to 24 h. As reported by Boehm et al. [43], the interlayer distances can be further increased by inserting polar liquids, for example, sodium hydroxide. As a result, the interlayer distance was expanded, which resulted in separation of single layer from the GO bulk materials. Upon treatment with hydrazine hydrate, GO reduces back to graphene. The chemical reduction process is carried out using dimethylhydrazine or hydrazine in the presence of either polymer or surfactant in order to produce homogeneous colloidal suspensions of graphene [47]. In this regard, reduction of GO is also performed using sodium borohydrate [48, 49] ($NaBH_4$) and hydroquinone [48] as reported in few reports. The process flow chart of the chemical synthesis of graphene is shown in the schematic in Figure 3.12.

FIGURE 3.12 The process flow chart of graphene synthesis derived from graphene oxide.

In 2006, Ruoff and his coworkers produced monoatomic graphene by chemical synthesis process [50, 51]. They prepared GO by Hummers method and chemically modified GO in order to produce a water-dispersible GO. GO is a stacked layer of squeezed sheets with AB stacking, which exhibits oxygen-containing functional group like hydroxyl and epoxide to their basal plane when it is highly oxidized [52]. The attached functional groups (carbonyl and carboxyl) are hydrophilic in nature, which facilitates the exfoliation of GO upon ultrasonication in an aqueous medium. Thus, the hydrophilic functional groups accelerate the intercalation of water molecules between the GO layers. In this process, functionalized GO is used as the precursor material for graphene production, which forms graphene upon reduction of graphene oxide with dimethylhydrazine [47]. Stankovich et al. [51] reported that chemical functionalization of GO flakes by organic molecules leads to the homogeneous suspension of GO flakes in organic solvents. They reported that the reaction of graphite oxide with isocyanate results in isocyanate-modified graphene oxide, which can be dispersed uniformly in polar aprotic solvents like N,N-dimethylformamide (DMF), NMP, dimethyl sulfoxide (DMSO), and hexamethylphosphoramide (HMPA). The proposed mechanism states that the reaction of isocyanate with hydroxyl and carboxyl groups generates the carbamate and amide functional groups, which get attached to the GO flakes (as shown in Fig. 3.13).

Xu et al. [53] further reported the colloidal suspensions of chemically modified graphene (CMG) decorated with small organic molecules or nanoparticles. They demonstrated noncovalent functionalization of graphene oxide sheets using 1-pyrenebutyrate (PB⁻), known as organic molecule with a strong adsorption affinity toward the graphite basal plane via π stacking. PB⁻-functionalized graphene was prepared by dispersing GO in pyrenebutyric acid followed by reducing it with hydrazine monohydrate at 80°C for 24 h. The resultant product was a homogeneous black colloidal suspension, which is the PB⁻-functionalized graphene dispersed in water. However, dispersed graphene solution needs stabilizers or surfactants, which introduce additional functional groups attached to the graphene sheets during device fabrication. Moreover, removal of stabilizers or surfactants readily agglomerates the dispersed graphene; hence, obtaining pristine monolayer graphene is challenging. Therefore, fabrication of stabilizer- or surfactant-free dispersed graphene via chemical synthesis became an important issue. Few reports have been found relating to the synthesis methods of stabilizer- or surfactant-free and unagglomerated graphene sheets. Li et al. demonstrated the surfactant- and stabilizer-free aqueous suspension ($0.5 \, mg \, ml^{-1}$) of RGO sheets under basic conditions (pH 10) [54]. They found that electrostatically stabilized graphene dispersion is strongly dependent on pH. This can be further explained by the fact that, the as-prepared GO sheets, containing carboxylic acid and phenolic hydroxyl group and thus having a highly negative surface charge (zeta potential), can form a stable suspension upon their reduction by hydrazine in the presence of ammonia at a high pH (~10). As illustrated in Figure 3.14a, at pH 10, neutral carboxylic group converts into negatively charged carboxylate during the reduction reaction, which results in the retardation of further agglomeration of the suspended graphene. The reduced graphene exhibited substantial amount of surface negative charge, which was confirmed by zeta potential

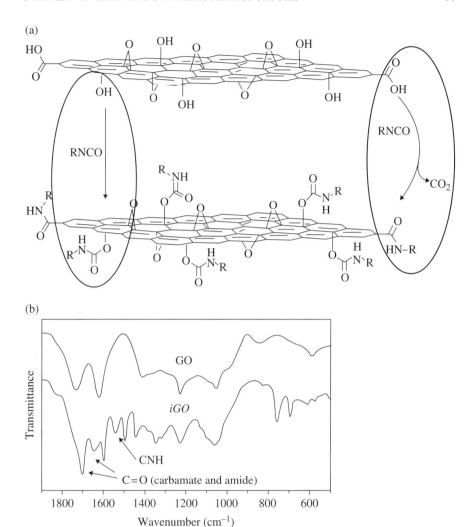

FIGURE 3.13 (a) Mechanism proposed by Stankovich et al. on isocyanate-treated GO where organic isocyanates react with the hydroxyl (left oval) and carboxyl groups (right oval) of graphene oxide sheets to form carbamate and amide functionalities, respectively. (b) Representative FTIR spectra of GO and phenyl isocyanate-functionalized GO. Reprinted with permission from Ref. 51.

measurement (Fig. 3.14b). The thickness of this dispersed chemically converted graphene (CCG) on SiO₂/Si wafer was reported as ~1 nm using tapping mode AFM as shown in Figure 3.14c. It was concluded that the transformation of negatively charged stable GO colloids was due to the electrostatic repulsion, not due to just the hydrophilicity of GO as per the earlier report [55]. Later on, Tung et al. [56] reported the synthesis of large-scale (~20×40 mm) single sheets of graphene using graphite oxide paper, where they tried to remove oxygen functionalities from GO in order to

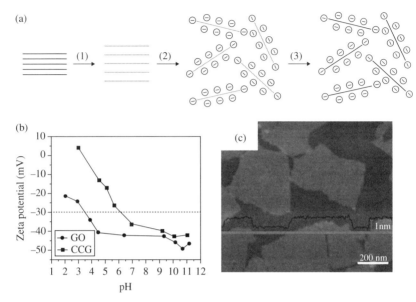

FIGURE 3.14 (a) Schematic showing the aqueous suspension of graphene fabrication mechanism via chemical technique. The process step consists of (1) graphite oxide production with greater interlayer distance, (2) sonication of GO in order to prepare mechanically exfoliated colloidal suspension of GO in water, and (3) conversion of GO to graphene using hydrazine reduction. (b) Representative data of zeta potential of GO and chemically converted graphene (CCG) as a function of pH. (c) Tapping mode atomic force micrograph of drop-casted CCG flakes on silicon wafer. Reprinted with permission from Ref. 55.

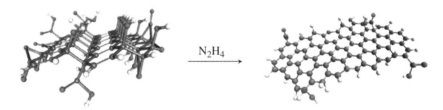

FIGURE 3.15 Schematic showing the 3D GO (carbon in gray, oxygen in red, and hydrogen in white) restoring its planar structure when reduced and dispersed with N_2H_4. Reprinted with permission from Ref. 56. (*See insert for color representation of the figure.*)

restore the planar geometry of the single sheets of CCG. In this approach, reduction as well as the dispersion of the GO film was done directly in hydrazine, which creates hydrazinium graphene (HG) through the formation of counter ions. In this regard, HG is composed of a negatively charged, reduced graphene sheet surrounded by $N_2H_4^+$ counterions as shown in the schematic in Figure 3.15. Finally, single-layer stable graphene sheet of ~0.6 nm thickness was obtained by this process.

Few reports have also revealed that the reduction of GO occurs at significantly high temperature with faster heating rate during the chemical route of process [57]. One interesting chemical technique, Langmuir–Blodgett assembly of GO single layers, was successfully demonstrated by Cote et al. [58]. The results are summarized as follows: (i) water-supported monolayers of GO single layers were suspended without any surfactant or stabilizing agent, (ii) the single layers formed stable dispersion when bound at the 2D air–water interface, (iii) the edge-to-edge repulsion between the single layers prevented them from overlapping during monolayer compression, and (iv) the process is novel as one could obtain the GO sheet of ~1 nm thickness.

The aforementioned processes are basically composed of the chemical approach for synthesizing graphene via transforming through the GO method. However, direct graphene synthesis using electrochemical methods was reported by Liu et al. [30]. The method is environment friendly and leads to the production of colloidal suspension of imidazolium ion-functionalized graphene sheets by direct electro-chemical treatment of graphite. The imidazolium ion is covalently attached to the graphene nanosheets by electrochemically through the breaking of C–C π bond. As shown in Figure 3.16, 10–20 V potential was applied in order to originate graphene nanosheets from the graphite anode. Further, dispersion of the modified graphene nanosheet was done in DMF, and the thickness of the GNS was found to be ~1.1 nm.

Several other reports are also found based on graphene functionalization with poly(m-phenylenevinylene-co-2,5-dioctoxy-p-phenylenevinylene) (PmPV) [59], 1,2-distearoyl-sn-glycero-3-phosphoethanolamine-N[methoxy(polyethylenegly col)-5000] (DSPE-mPEG) [60], poly(tert-butyl acrylate), and so on. In view of technological applications on the device level, several recent reports are also found based on the poly(N-vinylpyrrolidone) graphene nanocomposite for humidity sens-ing [61], GO/polymer for organic solar cells [62], dye-sensitized solar cell [63], organic memory devices [64], Li-ion battery [65], and so on. Recent reports are also

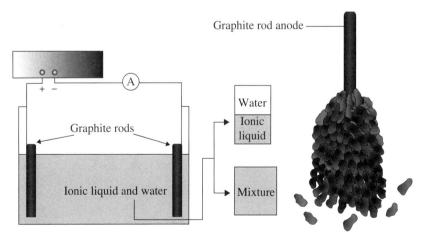

FIGURE 3.16 Schematic diagram illustrating the electrochemical process for graphene synthesis. Reprinted with permission from Ref. 30.

available based on the modification of graphene with inorganic nanoparticles like Au [66], TiO_2 [67], Fe_3O_4 [68], WO_3 [69], CuO [70], and ZnO [71, 72].

In conclusion, the chemical method for the production of graphene, exhibiting an advantage of low-temperature processibility, could be readily used for processing graphene on those substrates with high flexibility. *In situ* functionalized graphene with different functional groups can be easily synthesized via this route toward chemical and bio applications. Further, the process is low in cost as graphite is abundant (supplied natural graphite worldwide has been estimated at 800 M tons) in nature. In contrast, the chemical methods have several demerits including, small yield of final products, yield of defective graphene, and partially reduced GO (i.e., incomplete reduction of graphene), which degrade the electronic properties of graphene and its usefulness for ultra-fast electronic devices. Moreover, the process involves too many tedious steps along with use of less safe, hazardous explosive chemicals like hydrazine. During chemical reduction of GO, incomplete reduction dictates the possible deterioration of the conductivity, charge carrier concentration, carrier mobility, etc. Finally, graphene produced by chemical methods is not superior in purity grade compared to that by other available methods; therefore, it needs further development for finding their applications in real products.

3.2.4 Direct Chemical Synthesis

All chemical synthesis processes described earlier (Sections 3.2.2 and 3.2.3) are top-down approaches as the processes involve the oxidation of bulk graphite, exfoliation of GO, and then reduction back to graphene. In this section, a bottom-up approach of chemical synthesis of graphene, named as solvothermal method, is introduced [73]. In this method, laboratory-grade ethanol and sodium were used as starting materials to synthesize sodium ethoxide followed by the pyrolyzation, which yields a fused array of graphene sheets that can be easily dispersed using mild sonication. The solvothermal reaction involved a reaction of 1:1 molar ratio of sodium (2 g) and ethanol (5 ml) in a sealed reactor vessel at 220°C for 72 h, resulting in a yield of sodium ethoxide, which was used as a graphene precursor for further reaction. The resultant solid (sodium ethoxide) was rapidly pyrolyzed, vacuum filtered, and dried in a vacuum oven at 100°C for 24 h. The process yield was 0.1 g/1 ml of ethanol, typically yielding ~0.5 g per reaction. Raman spectroscopy of the resultant sheet showed a broad D-band at 1353 cm^{-1}, a G-band at 1590 cm^{-1}, and the intensity ratio of $I_G/I_D \sim 1.16$, which demonstrates highly defective graphene. Finally, 4 ± 1 Å thickness of graphene was obtained by this process. The advantage of this process is a low-temperature, low-cost, and bottom-up process that can be further extended to more controlled fabrication of high-purity and functionalized graphene. However, the quality of graphene is still not satisfactory as it contains a large number of defects.

3.2.5 CVD Process

Thermal CVD is a chemical process by which a substrate is exposed to thermally decomposed precursors and the desired product deposited onto the substrate surface at high temperature. Because high temperature is not desired in many cases,

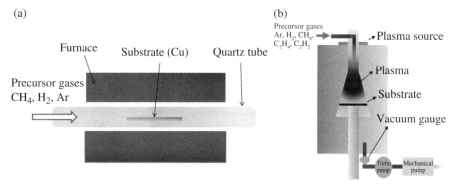

FIGURE 3.17   Schematic of (a) thermal CVD and (b) plasma-enhanced CVD (PECVD).

plasma-assisted decomposition and reaction have been applied to lower the process temperature. Figure 3.17a and b demonstrates the schematic of a thermal and plasma-enhanced chemical vapor deposition (PECVD), respectively.

The advantages of the CVD process include, high quality, high purity, large scale synthesis of graphene. In addition, by controlling the CVD process parameters, control over the morphology and crystallinity of the desired product is possible. However, tailoring high-precision atomic-level synthesis is still under investigation.

3.2.5.1 Graphene Synthesis by CVD Process

In 1975, Lang et al. first demonstrated the monolayer graphitic structure growth on Pt by thermal CVD method [74]. It was found that the graphitic structure formation took place on platinum due to the decomposition of ethylene. Later, Eizenberg et al. reported [75] the graphite layer formation on Ni (111) using the CVD process. Here, the process includes first carbon doping in single-crystal Ni (111) at 1200–1300 K for a long period of time (~7 days), followed by a quenching process. In this context, rapid quenching process causes the carbon phase condensation on Ni (111) surface. Therefore, it was concluded based on the detailed thermodynamic analysis that the carbon phase segregation on Ni (111) was solely dependent on the rate of quenching.

Since then, almost for two decades, the single graphitic layer deposition had not been explored further due to the inadequacy in finding applications of thin graphite film. Immediately after the discovery of graphene in the beginning of twenty-first century along with the potential search for the high-speed electronic materials, graphene constitutes emerging research in the field of nanotechnology. Graphene attracted tremendous attention to the scientific and industrial communities, owing to its unusual electronic and optoelectronic properties, high mechanical strength, and good thermal conductivity. In this context, the physics and chemistry of graphene structure were also scrutinized considerably in order to open up the possibilities of several applications of graphene [1, 4, 76, 77].

In 2006, the first attempt at graphene synthesis on Ni foil using CVD was found using camphor (terpenoid, a white transparent solid of chemical formula $C_{10}H_{16}O$) as the precursor material [78]. Here, the graphene synthesis was carried out in a two-step process: camphor deposition on Ni foil at ~180°C and subsequent pyrolyzation

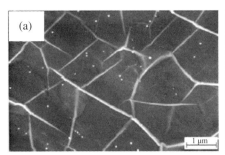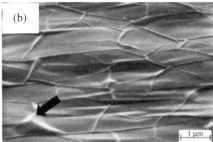

FIGURE 3.18   (a and b) Scanning electron micrograph of graphene syntheses on Ni(111) by DC discharge method. Reprinted with permission from Ref. 79.

at 700–850°C in Ar atmosphere. Upon the investigation using TEM, the obtained product was found to have a hexagonal planar few-layer graphite-like structure, which consisted of almost ~35 layers of stacked single graphene layer with interlayer distance of 0.34 nm. This investigation created a novel pathway of large-scale graphene growth using thermal CVD. Nevertheless, large-scale mono- or bilayer graphene growth using thermal CVD was still in demand until Obraztsov et al. [79, 80] reported the deposition of FLG on Ni. A thin layer (1–2 nm) of graphene was produced at 40–80 mT pressure and 950°C under a DC discharge of hydrocarbons. They used a gas mixture of H_2:CH_4 = 92:8 as a precursor under the DC discharge current of ~0.5 A cm^{-2}. Figure 3.18 shows the as-deposited structure of graphene consisting of anomalous surface ridges. The ridge formation in the graphene was due to the thermal expansion coefficient mismatch between graphene and Ni substrate. Indeed, the CVD graphene in this report was ~1–2 nm thick FLG as confirmed by Raman spectroscopy and scanning tunneling microscopy (STM). Although well-ordered few graphene layers were found on the Ni surface, the same experimental process did not yield well-ordered graphene on Si except amorphous carbon.

Similarly, Pei et al. [81] also demonstrated the synthesis of high-quality graphene on polycrystalline Ni surface using a thermal CVD of methane (CH_4). The FLG was synthesized at 1000°C using CH_4:H_2:Ar = 0.15:1:2 with a total gas flow rate of 315 standard cubic centimeters per minute (sccm) under normal atmospheric pressure. Figure 3.19a shows the high-resolution transmission electron microscopy (HRTEM) image of CVD graphene on Ni with different numbers of layers (1–4 layers of graphene on Ni), which suggested that the graphene deposition took place due to the segregation of carbon on Ni. The report also emphasized the effects of cooling rates (fast ~20°C s^{-1}, medium ~10°C s^{-1}, and slow ~0.1°C s^{-1}), which significantly control the formation of different numbers of graphene layers (as shown in Fig. 3.19b). In the year 2009 [82], the large-scale growth of graphene films as stretchable transparent electrodes was reported. Kim et al. [82] demonstrated the wafer-scale graphene growth over an e-beam evaporated Ni, followed by thermal CVD of CH_4:H_2: Ar ~550:65:200 at 1000°C. Another graphene transfer process on a flexible PDMS substrate by chemical wet process was also demonstrated [82]. The transferred graphene on PDMS exhibited a sheet resistance of ~280 Ω sq^{-1} with more than

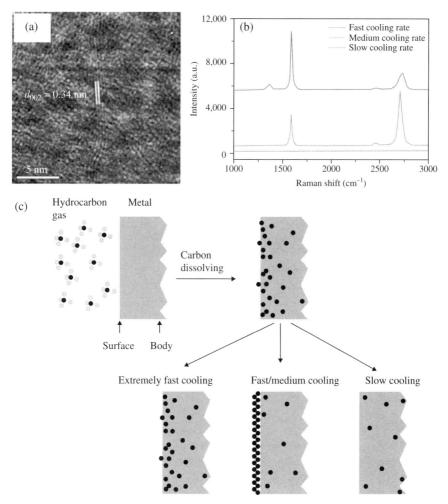

FIGURE 3.19 (a) HRTEM image of graphene precipitated on Ni. (b) Raman spectra confirming the effect of cooling rate on graphene formation. (c) Schematic representing the mechanism of carbon segregation on Ni. Reprinted with permission from Ref. 81.

80% transmittance under visible light wavelength. The film on SiO_2/Si substrate showed electrical conductivity and carrier density of ~3750 $cm^2 V^{-1} s^{-1}$ and $5 \times 10^{12} cm^{-2}$, respectively.

Successively, several other reports also demonstrated single- to few-layer graphene synthesis on the surface of different metals. Reina et al. [83] used e-beam evaporated Ni on SiO_2/Si substrate, followed by annealing of the substrate at 900–1000°C under a reducing atmosphere of Ar and H_2. Then, the graphene synthesis was performed by the decomposition of diluted hydrocarbon gas at elevated temperature and ambient pressure. They found 1–10 layers of graphene on the Ni surface as shown in the HRTEM images in Figure 3.20a, b, and c and Raman spectroscopy in Figure 3.20d.

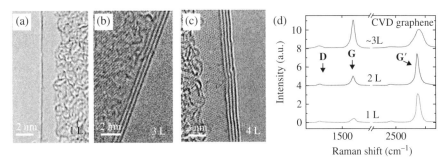

FIGURE 3.20   (a–c) Pictures illustrate HRTEM images of one to few layers of graphene grown on Ni using thermal CVD process. (d) Representative Raman plot for successive layers of graphene. Reprinted with permission from Ref. 83.

A wet chemical method was adopted in order to transfer the CVD-grown graphene on any substrates. Similarly, Wang et al. [84] proposed a new approach for the production of gram-scale graphene by using thermal CVD at 1000°C. In this process, Co-supported MgO was used as a catalyst, and a precursor gas mixture of CH_4:Ar (1:4 volume ratio, total 375 ml min^{-1} flow rate) was used as carbon source. Finally, graphene was obtained after washing away the catalyst particles using concentrated hydrochloric acid solution (HCl), and the process yield was almost 0.05 gm graphene from a batch of 500 mg catalyst powder. Although the process was claimed as a unique process due to its low cost and gram-scale graphene production, single-layer graphene was not obtained by this process. The graphene synthesized by this process was randomly aggregated few layers (5–6 layers) of graphene sheet.

Later, Li et al. demonstrated large-scale graphene growth on Cu surfaces by thermal CVD at 1000°C temperature [25]. High-purity Cu foil and hydrocarbon gas mixture were used as the substrate and the source of carbon, respectively. The Cu foil was annealed at 1000°C under the 40 mTorr of reduced atmosphere, followed by deposition of graphene by flowing 35 sccm of methane gas (CH_4) at ~500 mTorr pressure. Furthermore, the synthesized graphene on Cu foil was transferred on other substrates by solution etching of Cu and transfer process. The process yields different layers of graphene, which was further confirmed by HRTEM and Raman spectroscopy. The mechanism of graphene deposition on Cu surface is explained as a high-temperature surface-catalyzed process associated with the limited solubility of carbon in Cu [25]. In addition, it is also found that the graphene growth on Cu occurs by the segregation of supersaturated carbon and surface adsorbed carbon on Cu surface during the rapid cooling process in CVD [25, 85].

In this regard, Choi's group developed a large-scale graphene synthesis on Cu foil by using a CVD process. In this process, a large Cu foil was rolled up and placed inside of a quartz tube furnace, and then the graphene was transferred on different substrates including flexible polymer substrates using a hot press lamination process. Verma et al. [27] reported large-area graphene growth as large as 15×5 cm using thermal CVD technique as shown in Figure 3.21a and b. The graphene deposition was performed at 1000°C with a mixture of H_2:CH_4 (1:4) at ambient atmospheric

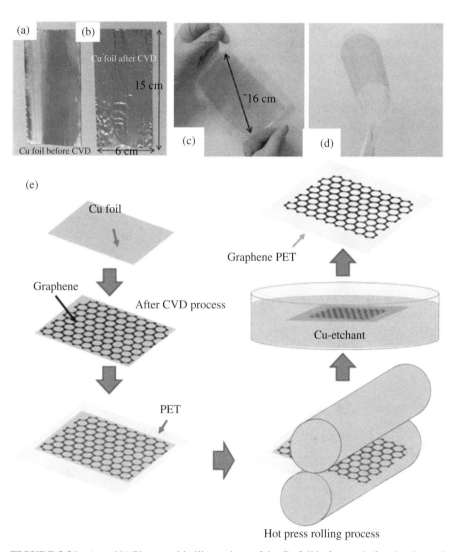

FIGURE 3.21 (a and b) Pictographic illustrations of the Cu foil before and after the thermal CVD process of graphene growth, respectively; (c and d) demonstrate the large scale (~16 cm diagonal length) and flexible graphene on PET, respectively; (e) the hot press lamination method of fabricating graphene-PET film. Reprinted with permission from Ref. 27.

pressure. The as-grown graphene was further transferred onto a polyethylene terephthalate (PET) using a unique hot press lamination process (as shown in Fig. 3.21e). In addition, the transferred large-area graphene on flexible film (as shown in Fig. 3.21c and d) was assembled in a flexible field emission (FE) device.

On the other hand, Bae et al. [86] reported the roll-to-roll production of 30 in. flexible graphene as shown in Figure 3.22a. The graphene was synthesized on Cu foil by using a catalytic decomposition of methane in a CVD and transferred onto a flexible PET substrate by roll-to-roll transfer process. The formation of different numbers of graphene layers was confirmed by HRTEM and Raman spectroscopy analysis (Fig. 3.22b). The transferred graphene on flexible substrate was demonstrated to be useful as a touch screen panel for future electronic devices.

Several other reports were found demonstrating the graphene synthesis on different metal foils and metal thin films [87]. However, Ni and Cu have been known as the most common substrates for graphene growth via thermal CVD methods. Similarly, a few more novel approaches have been reported on graphene synthesis by CVD [21, 88]. GNR synthesis is also reported recently using a CVD process [89].

Although CVD graphene synthesis on different transition metal surfaces has been demonstrated extensively, direct graphene synthesis on a dielectric surface is quite challenging since the transition metal surface plays a role as catalyst. In this perspective, Ismach et al. [90] reported graphene growth directly onto insulator substrates. A very thin Cu film deposition was carried out on dielectric substrates by e-beam evaporation, followed by the graphene growth using thermal CVD at 1000°C under 100–500 mT pressure (Fig. 3.23). The graphene precipitation on the dielectric surfaces occurs due to the concurrent process of the surface-catalyzed process of Cu and the copper film dewetting, which leads to direct graphene deposition on the bare dielectric substrates as shown in Figure 3.23. Further, the CVD process leads to the formation of well-crystallized graphene on the dielectric surfaces containing low

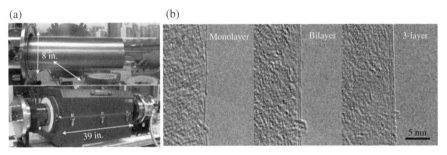

FIGURE 3.22 (a) Large-scale CVD of graphene on Cu foil. (b) HRTEM images demonstrating the growth of 1-layer, 2-layer, and 3-layer graphene on Cu. Reprinted with permission from Ref. 86.

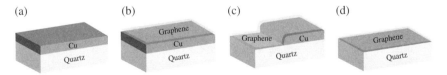

FIGURE 3.23 Schematic illustration of the mechanism of graphene synthesis via CVD on a dielectric substrate (a) deposition of very thin Cu film on dielectric substrate, (b) graphene growth using thermal CVD, (c) de-wetting of copper films, and (d) deposition of graphene on the dielectric substrate. Reprinted with permission from Ref. 90.

defects and 1–2 layers of graphene. Later, Rümmeli et al. [91] reported another technique to deposit graphene on insulator substrates using thermal CVD at significantly lower temperatures. The low-temperature CVD process produced graphene that is several nanometers to a few hundred nanometers thick directly on magnesium oxide (MgO) nanocrystal powder using cyclohexane, acetylene, and argon as a feedstock gas mixture at temperatures between 325 and 875°C. The domain size of graphene was found from few hundred nanometers to several micrometers having FLG.

Therefore, direct synthesis of graphene on dielectric substrates bypasses the post-synthesis graphene transfer process, which leads to the inclusion of defects and contamination in graphene. However, the graphene deposition on dielectric surface needs to be explored further in order to achieve well-ordered, large-scale, defect-free graphene for electronic applications.

3.2.5.2 Graphene Synthesis by Plasma CVD Process

Plasma CVD process involves chemical reactions of the reacting gases in presence of plasma inside a vacuum chamber leading to the deposition of thin film on substrates, which is known as PECVD process (Fig. 3.17b). The plasma can be generated inside a PECVD system using plasma sources such as RF (AC frequency), microwave, and inductive coupling (electrical currents produced by an electromagnetic induction). By PECVD technique, a graphene synthesis process can be performed at relatively low temperature compared to other CVD processes; hence, the process is more feasible for industrial applications. Moreover, catalyst-free growth [92] can be carried out by controlling the process parameters, which can easily influence the physical properties of the final products.

The synthesis of thin graphitic layer using PECVD process was first demonstrated by Obraztsov et al. [93] by DC discharge CVD of a gas mixture containing CH_4 and H_2 (0–25% CH_4) at 10–150 Torr. In this report, Si wafer and different metal sheets of Ni, W, and Mo were used as substrates for nanocrystalline graphite (NG) growth. Using optical emission spectra embedded with the PECVD system, it was observed that the presence of C_2 dimers in CH_4 plasma played an important role in nanocrystalline graphitic layer formation. However, the graphene produced by this process was a thick layer graphene. Wang et al. [94] attempted to deposit graphene using PECVD on different substrates with process parameters of 900 W RF power, 10 sccm CH_4, and chamber pressure of ~12 Pa. In addition, they tried to control graphene deposition by varying methane concentration (5–100%) at different temperatures from 600 to 900°C. The typical deposition time was kept 5–40 min. It was also observed that the increase of CH_4 concentration and substrate temperature resulted in the increase of graphene growth rate. Finally, FLG nanosheets with a thickness of ~1–2 nm were grown. Similarly, Zhu et al. also reported the synthesis of vertically aligned free-standing graphene (thickness ~1 nm) by inductively coupled RF-PECVD system on catalyst-free substrates by using hydrocarbon and hydrogen as precursor gas mixture. Several other reports are available on graphene synthesis using different other synthesis processes, such as microwave plasma [95], atmospheric pressure process [96], petal-like graphene formation [97], nitrogen (N_2)-doped graphene/CNT hybrid structure [98, 99], and several others. PECVD process can produce only the vertically oriented graphene, which is not yet demonstrated using any other graphene synthesis processes. The PECVD method produced high-purity and high-crystalline graphene; however, uniform large-area and single-layer graphene production using this process is still under vigorous investigation.

3.2.5.3 Grain and GBs in CVD Graphene

Graphene growth on Cu by CVD process is one of the most promising and scalable methods of graphene synthesis. Hence, producing continuous layer of high-quality large-area single-crystalline graphene is becoming a challenge as there is a formation of grain and GBs, which is considered as defects of graphene. The structure and morphology of the GBs have significant effect on the electrical, thermal transport [100], and mechanical properties [101] of the final graphene film. Primarily, graphene deposition process on Cu occurs due to the formation of nucleation sites, followed by the growth process. Therefore, based on the nucleation process, two major types of GBs are found in the structure: (i) intragranular GBs and (ii) intergranular GBs. Intragranular GBs are formed when a graphene grain formed from a single nucleation site, whereas intergranular GBs are formed when two individual graphene grains are originated from two different nucleation sites and merged together at a single point. Therefore, orientation of single-crystalline graphene grain and GBs strictly depends on the crystallinity, purity, and morphology of the substrate.

It has been shown that graphene nucleation sites were varied with the Cu crystal planes and their orientations. Wood et al. [102] showed that polycrystalline Cu exhibits different low- and high-index crystal facets, GBs, annealing twins, etc. as shown in Figure 3.24a and b and during the CVD process low-index Cu facets produce more monolayer graphene with fewer defects as compared to the high-index surfaces. It is also stated that monolayer graphene formed on Cu (110) with high area coverage within short growth time compared to the other high-index Cu crystal facets [102]. On the other hand, substrate surface and surface morphologies of Cu play an important role in nucleation seeds and graphene domain sizes during the CVD process [106]. Han et al. showed that the number of graphene nucleation seed formation on polished Cu is much lower than that of the unpolished one. Hence, the domain sizes of graphene flakes are larger on the top of the polished Cu surface. Furthermore, reducing the density of nucleation sites and controlling gas flow rates help in growing millimeter-scale large single-crystal grain (0.5 mm) as shown in Figure 3.24c. In this context, the edge of the large single-crystal graphene exhibited corrugated edge with dendritic morphologies, which indicates that the graphene growth mechanism is based on the restructuring of the individual carbon atoms rather than the atom cluster. Similarly, two different grains, which originate from different nucleation sites, create intergranular GBs as shown in Figure 3.24c. The large-area single-crystal graphene growth on Cu is not only varied with the controlled nucleation sites on the substrates but also varied with the other growth parameters such as temperature, pressure, gas flow rates, and the ratio of the gas mixtures. High pressure and high temperature could also help in growing large single-crystal graphene up to wafer-scale size of $4.5\,mm^2$ (Fig. 3.24d, e) on commercial Cu foils [104]. It is also demonstrated that hexagonally shaped single-crystal

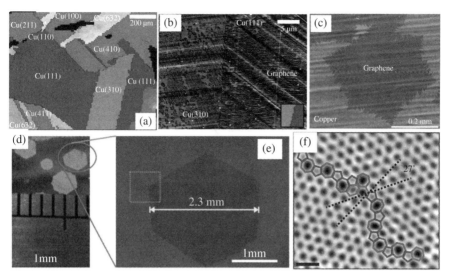

FIGURE 3.24 (a) Electron beam-scattered diffraction (EBSD) of a polycrystalline Cu foil showing different crystal planes of Cu. (b) Graphene on two Cu grains, Cu(310) and Cu(111). Reprinted with permission from Ref. 102. (c) Large-scale deposition of graphene on Cu using CVD process. Reprinted with permission from Ref. 103. (d and e) The 2.3 mm size of a single-crystal graphene grown on Cu foil using a CVD method. Reprinted with permission from Ref. 104. (f) The clear formation of grain boundaries at the junction between the two graphene grains, producing tilt boundaries. Reprinted with permission from Ref. 105.

graphene grains can be grown continuously across the GBs of the Cu crystal [104, 107]; thus, there was no such relation between the formations of the graphene GBs based on the polycrystalline Cu GBs. Additionally, in Ref. [107], it has also been demonstrated that there was no epitaxial correlation between the crystal orientation of the graphene and the underlayer polycrystalline Cu owing to the weak inter-action between the carbon and Cu atoms, which has been shown in several other reports as well [108, 109]. In this regard, characteristics of graphene GBs are more significant in representing the transport properties of graphene. Using atomic res-olution imaging, one recent report demonstrated a technique to locate and identify the graphene GBs [107]. Interestingly, it was found that graphene GBs are formed when different grains stitch together predominantly through pentagon–heptagon pairs as shown in Figure 3.24f. The electrical characterizations of graphene GBs show the sharp voltage drops due to the high resistance of the GBs. In particular, the grain size and the formation of the number graphene layers are varied with different growth conditions.

3.2.6 Epitaxial Growth of Graphene on SiC Surface

Epitaxial thermal synthesis process of graphene on single-crystalline silicon carbide (SiC) surface is one of the most renowned synthesis processes, which has been explored vigorously for the last 7–8 years. The term "epitaxy" can be defined as a method that allows depositing a single-crystalline film on a single-crystal substrate. The deposited film is referred as "epitaxial film" and the process is known as epitaxial growth process. The "epitaxial graphene growth" is a process to fabricate highly crystalline graphene on single-crystalline SiC substrates. When the deposited film on a substrate is of same material, it is known as homoepitaxial layer, and if the films are different materials than the substrates, then they are called heteroepitaxial films. For example, single-layer graphite or graphene formation on the SiC is known as heteroepitaxial layer, which will be discussed in this section.

The investigations on the electronic properties of graphene consist of two successive directions. One is based on the exfoliated graphene, and the other one is related to the wafer-scale synthesis of epitaxial graphene, which stipulates the most feasible and scalable approach toward graphene electronics. Extensive conscientious research areas have been explored such as (i) epitaxial graphene-based electronics, (ii) band gap opening, (iii) epitaxial growth mechanism, (iv) heterointerfaces of graphene/SiC, and several others, which focused on research toward a goal of improvements in large-scale graphene-based electronics.

Bommel et al. [110] first reported the graphite formation on both of the 6H-SiC(0001) and (000$\bar{1}$) surfaces in the year 1975. Their work showed the graphite synthesis on both of the SiC polar planes (0001) and (000$\bar{1}$) at 1000–1500°C under ultrahigh vacuum (UHV) (~10^{-10} mbar). However, the crystal structure, crystal orientations, and stacking of SiC are described in details elsewhere [111]. In the year 2004, de Heer and coworkers reported the fabrication of ultrathin (few layers, 1–3 monoatomic graphene layers) graphitic layers on Si-terminated (0001) face of single-crystal 6H-SiC and its electronic properties [112]. The detailed experimental procedure includes the following: (i) surface preparation using oxidation or H$_2$ etching, (ii) surface cleaning by electron bombardment at 1000°C at ~10^{-10} Torr pressure, and (iii) the heat treatment of the samples at 1250–1450°C for 1–20 min. Furthermore, they demonstrated the fabrication of 1–2 layers of epitaxial graphene on the (0001) face of a 6H-SiC wafer using the same thermal decomposition method at high temperature [113]. The number of epitaxial graphene layers formed on the SiC substrate was characterized using X-ray photoelectron spectroscopy (XPS) and angle-resolved photoemission spectroscopy. Similarly, few recent studies also demonstrated the exceptionally high-quality graphene production on the 4H-SiC C-face in an RF furnace under pressure ~10^{-4}–10^{-3} Torr. After the removal of surface oxides by preheat treatment at 1200°C, the epitaxial graphene formation occurred at 1420°C, which was comparatively higher than the normal UHV graphitization on SiC. However, the growth rate on 4H-SiC was relatively higher than that of the other SiC surface; therefore, it is quite difficult to produce very thin layer of graphene as well as to control the number of graphene layer by changing time and temperatures. In general, 4–5 layers of graphene film can be obtained easily at 1420°C within ~6 min by the thermal decomposition of SiC. Epitaxial graphene on SiC exhibits high quality (Fig. 3.4), which can be utilized for high-performance electronics; however, the transfer of graphene from SiC to other substrates is very difficult. Thus, this causes a serious disadvantage of epitaxial graphene synthesis process for versatile applications.

In a recent approach, Juang et al. [114] reported the epitaxial graphene growth on a catalyst thin film-coated SiC substrate at a comparatively low temperature. The authors illustrated the low-temperature epitaxial graphene growth process on SiC substrate coated with Ni thin film. The process involves (i) the deposition of 200 nm Ni thin film on 6H-SiC(0001) and 3C-SiC substrates and (ii) the growth process at ~750°C under ~10^{-7} Torr pressure. The graphene film was synthesized over the continuous Ni thin film surface. The mechanism of graphene formation at such a low temperature was elucidated as follows: (i) there is a Ni dissolution taking place at the Ni/SiC interface due to the rapid heat treatment; (ii) a formation of nickel silicide/carbon mixed phase occurred; (iii) diffusion of carbon atoms into the Ni thin film matrix; and finally (iv) the carbon atoms segregated onto the surface of the Ni during the cooling process similar to thermal CVD explained in the thermal CVD section. The process was delineated as the versatile large-scale facile method as it is a comparatively lower-temperature process than others. Further, the graphene obtained by this process is easily transferrable to any substrates for other applications. The feasibility of the graphene-based 2D electronics is based on fabricating wafer-scale graphene with controlled thickness, width, and specified crystallographic orientations in order to achieve precious control over the graphene electronic properties. The "epitaxial graphene on SiC" research attracted huge attention both academically and industrially due to its scalability, best electronic properties, and above all the high-quality graphene obtained from the process. The major advantage of this process is large-scale fabrication of graphene on an insulator or semiconductor surface, which can be used for post-CMOS-based electronics [33, 115]. Furthermore, epitaxial multilayered graphene over the SiC substrate behaves as an isolated graphene, which would be an added advantage of its applications in graphene-based nanoelectronics. Nevertheless, the high growth temperature and very low process pressure are the major disadvantages of this process. More specifically, the final epitaxial graphene exhibits smaller grain size. One novel procedure for atmospheric pressure graphene synthesized on 6H-SiC(0001) with much larger domain size has been reported recently [116]. Although high-quality superior-grade epitaxial graphene formation was reported, the graphene transfer to any other substrates is difficult, which seriously limits its versatility in a wide range of electronic applications.

3.3 GRAPHENE CHARACTERIZATIONS

Characterizations of 2D graphene involve different types of microscopic and spectroscopic techniques in order to obtain the structural, morphological, and chemical information of as-synthesized or as-transferred graphene. Thus, the characterization of graphene covers an important part of graphene research. Mostly, characteristics information regarding graphene layers are important for end applications as graphene properties are strictly dependent on its number of layers. Furthermore, the visualization of graphene is also important as it offers information regarding the shape, size, and morphology of graphene. Similarly, the characterization process is also associated with the measure of the purity and defects of graphene. Basically, graphene's purity varies with different synthesis processes and/or processing parameters. Therefore, the information regarding purity and defects will further help in controlling the synthesis parameters. Major structural characterizations of 2D graphene belong to the characterization of graphene number of layers. In this context, optical microscopy is one of the straightforward and versatile methods by which one can easily detect the number of layers in graphene. Graphene layers could be measured more accurately using HRTEM and AFM. On the other hand, Raman spectroscopy is a nondestructive technique by which fingerprints of almost all of the carbon allotropes can be detected by their various structural and bonding information. Similarly, graphene's number of layers, its crystallinity, phonon transitions, and bonding information could be procured by using Raman spectroscopy. Furthermore, XPS and Raman spectroscopy are the prevailing methods for the measurement of graphene's chemical purity and detection of the functional groups attached to the graphene. This section summarizes the characterization methods based on various microscopic and spectroscopic techniques, which are basically used to evaluate graphene fingerprints.

3.3.1 Optical Microscopy

Optical microscopy is one of the straightforward and effective methods for nondestructive characterization of large-area graphene sample. In particular, optical microscopy is used for characterizing the number of layers in graphene by using the contrast difference between the graphene layers and the underneath dielectric substrate. Additionally, optical microscopy is useful in visualizing the size and shape of the graphene flakes, which makes any device fabrication process rapid and faster. In this context, the characterization of graphene using optical microscopy involves the design of the substrate, and this is significant in order to visualize the graphene crystallites and distinguish its layers. For the past few decades, low-contrast transparent thin film samples have been characterized using the optical interference technique. Specifically, the incorporation of a thin dielectric layer in between the transparent materials and a reflective substrate enhance the contrast difference for the fluorescent layers. This is well known as Fabry–Pérot interference in the dielectric material, which further modulates the incident light intensity between the graphene layer and the substrate. As SiO_2 and Si_3N_4 are commonly deposited dielectric layers on Si, they act as good contrast-enhancing layers for thin graphene flakes. Following similar contrast difference method, identification of mono- to few-layer graphene sheets has been demonstrated using an optical microscope under white light illumination condition on 300 nm SiO_2/Si substrate [117]. As graphene is one atom thick material of ~98% transparency, the contrast difference is very weak on the substrate surface. In Ref. [117], using the Fresnel theory, it has been demonstrated that the contrast of graphene flakes on SiO_2/Si surface varies with the thickness of the SiO_2 layer and wavelengths of incident light [117, 118]. Thus, the contrast changes between the different numbers of graphene flakes are clearly quantified as shown in

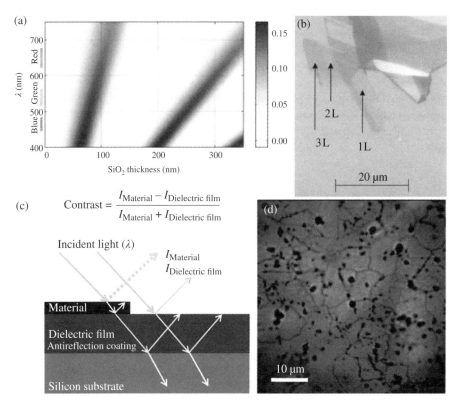

FIGURE 3.25 (a) Color plot for the expected contrast as a function of SiO$_2$ thickness and wavelength. Reprinted with permission from Ref. 117. (b) Optical micrograph showing single- (1L), bi- (2L), and trilayer (3L) graphene on SiO$_2$/Si substrate. Reprinted with permission from Ref. 119. (c) Schematic illustrating the mechanism of the Michelson contrast as per the equation. Reprinted with permission from Ref. 120. (d) Optical micrograph of CVD graphene on Cu showing graphene grain boundaries. Reprinted with permission from Ref. 121.

Figure 3.25a [119]. Apart from SiO$_2$, contrast of graphene layers can also be visualized on the 50 nm Si$_3$N$_4$ and on the 90 nm PMMA using blue light and white light, respectively. Figure 3.25b shows the characterization of different numbers of exfoliated graphene layers over the SiO$_2$/Si substrate using optical microscopy [119]. On the other hand, the visibility of the graphene oxide flakes was characterized by applying the Michelson contrast (C) relationship (Fig. 3.25c) as depicted in the equation [120]

$$C = R_{materials} - R_{dielectrics} / R_{materials} + R_{dielectrics} \qquad (3.4)$$

where R$_{material}$ is the reflected light intensity with the material and R$_{dielectric}$ is the reflected light intensity without the material. Jung et al. demonstrated the visualization of single-layer graphene oxide on SiO$_2$/Si substrate, whereas the optical contrast of the graphene film strictly depends on the thickness and optical properties of the intermediate dielectric layer [120].

Large-scale CVD graphene and its grain and GBs can also be characterized using optical microscope. Duong et al. [121] demonstrated that grains and GBs of large-scale graphene on Cu can be visualized using selective oxidation of Cu substrate via graphene GBs functionalized with O$^-$ and OH$^-$ radicals as shown in Figure 3.25d. They demonstrated the selective etching of GBs under the exposure of UV in a moist environment and visualization of GBs under the optical microscopy. Due to its easy accessibility and convenience, optical microscopy attracted huge attention for demonstrating graphene and its number of layers. Moreover, the technique is low cost and nondestructive; hence, it became one of the essential parts of the graphene research. However, exact quantification/measurement of the graphene layers and its properties using the optical microscopy is not so easy. In addition, substrate preparation and its dielectric layer's contrast are equally important for observing the exact number of graphene layers.

3.3.2 Raman Spectroscopy

For half a century, Raman spectroscopy is one of the most reliable characterization techniques for carbon allotropes. Starting from fullerene to graphite, individual carbon nanomaterials (including bulk carbon materials) show their unique fingerprints in Raman spectra. Basically, Raman spectroscopy is a nondestructive characterization technique, which reveals the information related to the bonding information of carbon nanomaterials, the signature of sp^2–sp^3 hybridization, the introduction of chemical impurities, the optical energy gap, elastic constants, doping, and other crystal disorders [122–123]. For graphene and graphene-derived materials, Raman spectroscopy provides useful information correlating its defects, edge constructions, functionalizations, induced strain, lattice mismatch, and the number of graphene layers [124, 125]. Raman spectra of monolayer graphene show its characteristics D-band, G-band, and the 2D band at its respective positions of ~1350, ~1580, and ~2690 cm^{-1}, respectively. The appearance of the three characteristics of Raman peaks is due to the electronic bands of graphene and its interband phonon transitions. The D-band appears in a Raman spectrum due to the induced defects in graphene. Primarily, this is a double resonance scattering, which includes the intervalley scattering of optical (iTO) phonon near the K point of the Brillouin zone. The presence of a symmetric G-band and 2D band in a Raman spectrum of graphene manifests the key signature of high-crystalline single-layer graphene. The origin of G-band is from a single resonance doubly degenerate E$_{2g}$ phonon mode at the Brillouin zone, while 2D peak generates due to the double resonance scattering with iTO phonon near the K point of the Brillouin zone. The D-band and G-band in the Raman spectrum of graphite symbolize the disorder bands and tangential bands, respectively, whereas the symmetric 2D band illustrates the stacking order of graphene layers, particularly the electronic interactions between the interlayer and substrates [123, 126]. A comparative Raman spectrum of graphene and graphite is shown in Figure 3.26a, which distinctly demonstrates the symmetric and nonsymmetric peak of graphene and graphite, respectively. The position and

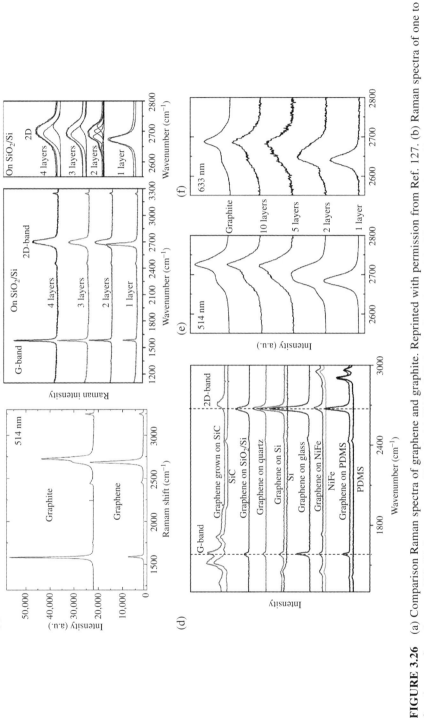

FIGURE 3.26 (a) Comparison Raman spectra of graphene and graphite. Reprinted with permission from Ref. 127. (b) Raman spectra of one to four layers of graphene on SiO₂/Si (laser power ~532 nm). (c) Shift of 2D band of graphene with different numbers of layers (laser power ~532 nm). (d) Effect of substrates on the Raman spectrum of single-layer graphene (laser power ~532 nm). Reprinted with permission from Ref. 128. (e and f) Effect of laser power on the 2D peak position of single-layer graphene. Reprinted with permission from Ref. 127.

intensity of characteristic bands of graphene change with its number of layers as shown in Figure 3.26b and c. Similarly, the intensity ratio of the D peak and G peak (i.e., I_D/I_G) signifies the induced defect concentrations in graphene, while the intensity ratio of the G peak and 2D peak (i.e., I_G/I_{2D}) gives the number of layers. Monolayer graphene exhibits an I_G/I_{2D} ratio of ~0.2–0.5, which further increases with increasing number of graphene layers [129]. Traces of the surface-functionalized molecules in graphene could be identified using this technique as well. Since functionalization is a broad subject in "graphene chemistry," a discussion on the effect of chemical functionalizations of graphene's Raman spectra is beyond the scope of this article and could be found elsewhere in details [130]. As graphene is a single to few atomic layered thick material, the appearance of the graphene's interatomic bond vibrations is highly influenced by the substrate surface and the dangling bond between graphene and substrate [108, 131]. Therefore, the appearance of characteristic peak positions and the relative intensity of graphene's Raman spectra are varied with different substrates as demonstrated in Figure 3.26d [128, 132]. In this context, the wavelength of the Raman laser also plays an important role in shifting the band position and intensity as shown in Figure 3.26e and f [127]. Furthermore, graphene characteristic peak positions also vary with some external influences like pressure [133], temperature [134], doping [20], and induced strain [135]. Hence, the experimentation technique and its setup for the Raman spectroscopy of graphene are significantly important.

3.3.3 High Resolution Transmission Electron Microscopy

Despite the presence of noninvasive techniques such as optical and Raman spectroscopy, HRTEM is one of the very powerful, frequently used, reliable characterization techniques for graphene's structural characterizations. In this technique, an electron beam is transmitted through the ultrathin sample and reaches to the imaging lenses and detector. Using HRTEM, atomic resolution images of any materials can be explored; thus, this is one of the powerful tools for characterizing the atomic arrangements of structure and interfaces of graphene (Fig. 3.27). More specifically, HRTEM is constructed with high-resolution imaging modes and several other advanced features compared to the conventional TEM. Since traditional TEMs were used over few decades to characterize the structural features of materials, visualizing atomic resolution of graphene is highly challenging. In contrast, TEM is quite advantageous to observe the micron-sized graphene flakes as shown in Figure 3.27a [136]. Recent advances in the HRTEM technology allow characterizing the atomic arrangements up to 1Å resolutions of materials under low operational voltages. Low-voltage operation is appropriate for HRTEM characterization of graphene as high-voltage electron beam induces defect in the graphene. Meyer et al. showed direct imaging of hexagonal close-packed honeycomb carbon atoms in a suspended single-layer graphene lattice using HRTEM as shown in Figure 3.27b, c, and d [137]. They also demonstrated the imperfections in single-layer graphene membranes such as defects, vacancies, adsorbates, and edges in graphene. They combined aberration correction with a monochromator to achieve 1Å resolution in HRTEM under an acceleration voltage of only 80 kV [137]. Similarly, in a recent report, the dynamics of carbon atoms along the edges and holes of a free-standing graphene is demonstrated using

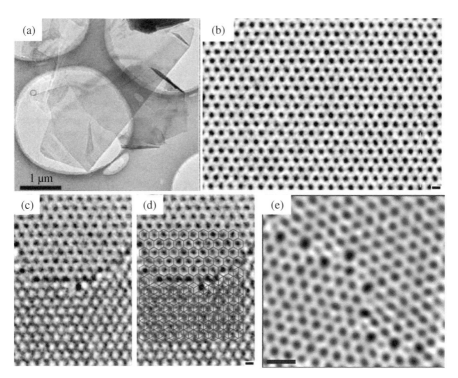

FIGURE 3.27 (a) TEM image showing a graphene flake. Reprinted with permission from Ref. 136. (b) The hexagonal honeycomb lattice of single-layer suspended graphene under HRTEM (scale bar 2 Å). (c) The clear step between single- and bilayer graphene. (d) The same HRTEM image of (c) marked with the two overlay layers (red line, bottom layer; blue line, top layer) of graphene (scale bar 2 Å). Reprinted with permission from Ref. 137. (e) HRTEM image of grain boundary of CVD graphene (scale bar 5 Å). Reprinted with permission from Ref. 105. (*See insert for color representation of the figure.*)

the same aberration-corrected HRTEM [138]. Graphene GBs and its defect structures are also visualized using HRTEM as shown in Figure 3.27e. Figure 3.27e shows the joining of two grains at the GBs associated with a relative misorientation between the grains, which forms a tilt boundary. HRTEM is one of the powerful and reliable techniques that allow us to see the atoms, atomic arrangements, stacking orders, and defects in graphene. However, the technique has serious limitations in its sample preparation, which needs a high level of expertise in order to obtain atomic resolution images. In addition, HRTEM is an expensive tool, which needs a high level of instrumentation and precision control to obtain an ideal atomic resolution image of graphene.

3.3.4 Scanning Probe Microscopy

Scanning probe microscopy (SPM) is used to characterize the topography of nanomaterials by scanning a surface with the help of a nanometer resolution probe. For the characterizations of graphene, SPM is the most popularly used technique using two different modes, for example, AFM and STM. The fundamental difference between AFM and STM is their principle of sensing mechanisms. In AFM, the cantilever probe deflects based on the topography of graphene and forms a 2D and 3D image resembling the surface topography. The AFM also exhibits two different modes including contact mode and noncontact tapping mode. In the former case, the probe touches the sample surface, whereas in the latter case, the probe doesn't touch the surface and small attractive forces act in between the graphene and probe. Figure 3.28a shows the AFM image of a mechanically exfoliated graphene flake with atomic-level thickness. Similarly, an AFM image of GO flakes of ~3 nm thickness (i.e., few layers of GO) is shown in Figure 3.28b. In this context, AFM is also used to measure the thickness of graphene, its surface roughness, topological disorders, ripples, etc. In some recent reports, it has been demonstrated that AFM can also be used for several other useful purposes despite surface characterizations of graphene. Giesbers et al. demonstrate AFM-based nanolithography of graphene, which can create a 30 nm trench on graphene [142]. Likewise, AFM can also be used as an effective tool for repairing the defect in GO in order to restore its electrical properties [143]. In this context, patterning and cutting of graphene in large scale along various lattice

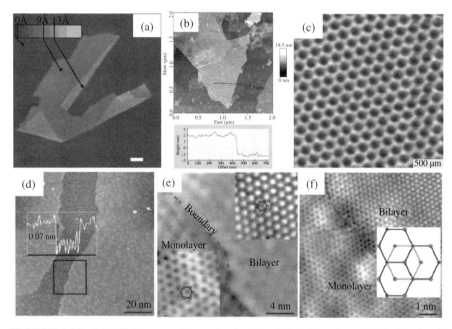

FIGURE 3.28 (a) AFM image of a mechanically exfoliated graphene (scale bar 1 μm). Reprinted with permission from Ref. 37. (b) AFM image illustrating the surface topography of graphene oxide flakes and the step height measurement profile of graphene flakes using AFM. Reprinted with permission from Ref. 139. (c) Atomic resolution scanning tunneling microscopy of graphene showing hexagonal close-packed lattice structure. Reprinted with permission from Ref. 140. (d–f) High-resolution STM images of epitaxial graphene on SiC and the coexistence of single-layer and bilayer domain with distinct domain boundaries. Reprinted with permission from Ref. 141.

orientations are demonstrated using an AFM-based nanorobot technique [144]. However, large-area precision scanning of graphene samples is quite inconvenient and needs extreme expertise. On the other hand, STM senses the tunneling current between the graphene and SPM nanoprobe and forms a pattern similar to the surface topography of the sample. Since STM senses the tunneling current between the sample and probe, the sample surface must be electrically conducting. The major advantage of graphene characterizations using STM is to produce high-contrast atomic resolution images as shown in Figure 3.28c. Several other information such as the atomic arrangements, surface defects, adsorbates, and graphene edge structure can also be obtained using the STM as well. Figure 3.28d, e, and f shows the atomic resolution hexagonal graphene lattice characterized via STM. Wong et al. [141] reported the illustration of the number of layers, substrate mismatch, and the rotational disorder of epitaxial graphene using STM as observed in Figure 3.28d, e, and f. Figure 3.28e and f shows the manifestation of the atomic arrangements of monolayer and bilayer graphene with clear distinction with domain boundary in epitaxial graphene [141]. Like other high-precision instruments, STM also requires extreme level of precision and calibration in experimental setup including massive instrumentation, which seriously demerits the easy access of it for graphene characterization.

3.4 SUMMARY AND OUTLOOK

In summary, to date, several graphene synthesis techniques have been well demonstrated using both top-down and bottom-up approaches of nanotechnology. Among all the graphene synthesis routes, each process has its own merits and demerits depending upon the final applications of graphene. Mechanical exfoliation of graphene using scotch tape is the ever reported simple technique for graphene fabrication. The process has capability of fabricating single-crystal, micron-range flakes of graphene with different numbers of layers. Although the technique is simple, the control over the wafer-scale synthesis and reproducibility is yet to be demonstrated. Chemical synthesis methods are useful for gram-scale production of graphene at low temperature and the process can produce graphene film/graphene coatings on wide variety of substrates. Also, the process is highly advantageous for the synthesis of functionalized graphene. However, the chemical synthesis process produce defective graphene, partially reduced GO, which seriously deteriorates graphene's electronic properties and limits its applications for electronic devices. As of now, few reports have been claimed the direct synthesis of graphene using solution process, which is impressive and the process need to improve further for obtaining better quality graphene. Furthermore, graphene synthesized using chemical process exhibits high surface area, edge-plane like surface defects, and easy functionalization. These graphene show excellent catalytic properties toward chemical reactions for bio-sensors, energy generations, and energy storage. Therefore, chemically synthesized graphene have enormous potential applications in future bio-sensor and energy related devices. High-temperature chemical process such as thermal CVD has been proven to be a more feasible and scalable approach for wafer-scale graphene synthesis. Also, these processes have enormous possibilities for controlling various morphologies of graphene and its grain and GBs. However, CVD processes are incapable of producing graphene on a wide range of substrates. Graphene produced by CVD process needs to be transferred on other substrates, which causes its structural and electrical deterioration owing to defect formations and contaminations. Epitaxial graphene synthesized on the single crystal SiC surface produce high-purity graphene with excellent electronic properties. However, very high processing cost with limited fabrication on multiple substrates restrict the wide acceptance of the process. To sum it all, bottom-up approaches were found to have better control over the scalable synthesis of graphene than the top-down approaches; however, synthesizing high purity, high crystalline graphene via bottom-up methods is truly challenging. A numerous techniques are available to evaluate graphene's structural characterizations associated with atomic arrangements, defects, and number of layers. Among them, optical microscopy, Raman spectroscopy, HRTEM, and SPM are widely used graphene characterization methods, which are prompt, powerful, and reliable. Most importantly, graphene paved a pathway that also contributed in inventing, synthesizing, and fabricating devices based on other 2D materials and their hybrid structures. As a result, a vast areas of research and development has been opened up with further expectation toward the inventions and applications of new devices in the field of electronics and energy.

REFERENCES

[1] K. S. Novoselov, A. K. Geim, S. V. Morozov, D. Jiang, Y. Zhang, S. V. Dubonos, I. V. Grigorieva, A. A. Firsov, Electric field effect in atomically thin carbon films. *Science* 2004, **306**, 666.

[2] A. K. Geim, K. S. Novoselov, The rise of graphene. *Nature Materials* 2007, **6**, 183.

[3] A. K. Geim, Graphene: status and prospects. *Science* 2009, **324**, 1530.

[4] A. K. Geim, P. Kim, Carbon wonderland. *Scientific American* 2008, **298**, 90.

[5] W. Choi, I. Lahiri, R. Seelaboyina, Y. S. Kang, Synthesis of graphene and its applications: a review. *Critical Reviews in Solid State and Materials Sciences* 2010, **35**, 52.

[6] D. Lahiri, S. Das, W. Choi, A. Agarwal, Unfolding the damping behavior of multilayer graphene membrane in the low-frequency regime. *ACS Nano* 2012, **6**, 3992.

[7] (a) R. R. Nair, P. Blake, A. N. Grigorenko, K. S. Novoselov, T. J. Booth, T. Stauber, N. M. R. Peres, A. K. Geim, Fine structure constant defines visual transparency of graphene. *Science* 2008, **320**, 1308; (b) C. Lee, X. D. Wei, J. W. Kysar, J. Hone, Measurement of the elastic properties and intrinsic strength of monolayer graphene. *Science* 2008, **321**, 385.

[8] K. S. Novoselov, A. K. Geim, S. V. Morozov, D. Jiang, M. I. Katsnelson, I. V. Grigorieva, S. V. Dubonos, A. A. Firsov, Two-dimensional gas of massless Dirac fermions in graphene. *Nature* 2005, **438**, 197.

[9] K. S. Novoselov, Z. Jiang, Y. Zhang, S. V. Morozov, H. L. Stormer, U. Zeitler, J. C. Maan, G. S. Boebinger, P. Kim, A. K. Geim, Room-temperature quantum hall effect in graphene. *Science* 2007, **315**, 1379.

[10] A. H. Castro Neto, F. Guinea, N. M. R. Peres, K. S. Novoselov, A. K. Geim, The electronic properties of graphene. *Reviews of Modern Physics* 2009, **81**, 109.

[11] S. M. Kozlov, F. Viñes, A. Görling, Bandgap engineering of graphene by physisorbed adsorbates. *Advanced Materials* 2011, **23**, 2638.

[12] M. Karabiyik, C. Al-Amin, S. Das, N. Pala, W. B. Choi, Subwavelength, multimode, tunable plasmonic terahertz lenses and detectors. *SPIE Defense, Security, and Sensing* 2012, 83630L.

[13] X. S. Li, Y. W. Zhu, W. W. Cai, M. Borysiak, B. Y. Han, D. Chen, R. D. Piner, L. Colombo, R. S. Ruoff, Transfer of large-area graphene films for high-performance transparent conductive electrodes. *Nano Letters* 2009, **9**, 4359.

[14] D. Choi, C. Kuru, C. Choi, K. Noh, S.-K. Hong, S. Das, W. Choi, S. Jin, Nanopatterned graphene field effect transistor fabricated using block co-polymer lithography. *Materials Research Letters* 2014, **2**, 131.

[15] F. Schwierz, Graphene transistors. *Nature Nanotechnology* 2010, **5**, 487.

[16] G. Kucinskis, G. Bajars, J. Kleperis, Graphene in lithium ion battery cathode materials: a review. *Journal of Power Sources* 2013, **240**, 66.

[17] Y. Huang, J. Liang, Y. Chen, An overview of the applications of graphene-based materials in supercapacitors. *Small* 2012, **8**, 1805.

[18] (a) L. Kavan, J. H. Yum, M. Grätzel, Optically transparent cathode for dye-sensitized solar cells based on graphene nanoplatelets. *ACS Nano* 2010, **5**, 165; (b) S. Das, P. Sudhagar, V. Verma, D. Song, E. Ito, S. Y. Lee, Y. S. Kang, W. Choi, Amplifying charge-transfer characteristics of graphene for tri-iodide reduction in dye-sensitized solar cells. *Advanced Functional Materials* 2011, **21**, 3729; (c) S. Das, P. Sudhagar, S. Nagarajan, E. Ito, S. Y. Lee, Y. S. Kang, W. Choi, Synthesis of graphene-CoS electro-catalytic electrodes for dye sensitized solar cells. *Carbon* 2012, **50**, 4815; (d) X. Li, H. Zhu, K. Wang, A. Cao, J. Wei, C. Li, Y. Jia, Z. Li, X. Li, D. Wu, Graphene-on-silicon Schottky junction solar cells. *Advanced Materials* 2010, **22**, 2743.

[19] S. Das, P. Sudhagar, Y. S. Kang, W. Choi, Graphene synthesis and application for solar cells. *Journal of Materials Research* 2014, **29**, 299.

[20] S. Das, P. Sudhagar, E. Ito, D.-y. Lee, S. Nagarajan, S. Y. Lee, Y. S. Kang, W. Choi, Effect of HNO_3 functionalization on large scale graphene for enhanced tri-iodide reduction in dye-sensitized solar cells. *Journal of Materials Chemistry* 2012, **22**, 20490.

[21] L. G. De Arco, Y. Zhang, C. W. Schlenker, K. Ryu, M. E. Thompson, C. W. Zhou, Highly flexible, and transparent graphene films by chemical vapor deposition for organic photovoltaics. *ACS Nano* 2010, **4**, 2865.

[22] Y. Hernandez, V. Nicolosi, M. Lotya, F. M. Blighe, Z. Y. Sun, S. De, I. T. McGovern, B. Holland, M. Byrne, Y. K. Gun'ko, J. J. Boland, P. Niraj, G. Duesberg, S. Krishnamurthy, R. Goodhue, J. Hutchison, V. Scardaci, A. C. Ferrari, J. N. Coleman, High-yield production of graphene by liquid-phase exfoliation of graphite. *Nature Nanotechnology* 2008, **3**, 563.

[23] S. Park, R. S. Ruoff, Chemical methods for the production of graphenes. *Nature Nanotechnology* 2009, **4**, 217.

[24] E. Rollings, G. H. Gweon, S. Y. Zhou, B. S. Mun, J. L. McChesney, B. S. Hussain, A. Fedorov, P. N. First, W. A. de Heer, A. Lanzara, Synthesis and characterization of atomically thin graphite films on a silicon carbide substrate. *Journal of Physics and Chemistry of Solids* 2006, **67**, 2172.

[25]   X. S. Li, W. W. Cai, J. H. An, S. Kim, J. Nah, D. X. Yang, R. Piner, A. Velamakanni, I. Jung, E. Tutuc, S. K. Banerjee, L. Colombo, R. S. Ruoff, Large-area synthesis of high-quality and uniform graphene films on copper foils. *Science* 2009, **324**, 1312.

[26]   K. Ghaffarzadeh, *IDTechEx Forecasts a $100 Million Graphene Market in 2018*, Vol. **2012**, IDTechEx Ltd, Cambridge, 2012, https://www.cambridgenetwork.co.uk/news/idtechex-forecasts-a-100-million-graphene-market-in-2018/

[27]   V. P. Verma, S. Das, I. Lahiri, W. Choi, Large-area graphene on polymer film for flexible and transparent anode in field emission device. *Applied Physics Letters* 2010, **96**, 203108.

[28]   D. V. Kosynkin, A. L. Higginbotham, A. Sinitskii, J. R. Lomeda, A. Dimiev, B. K. Price, J. M. Tour, Longitudinal unzipping of carbon nanotubes to form graphene nanoribbons. *Nature* 2009, **458**, 872.

[29]   (a) L. Y. Jiao, X. R. Wang, G. Diankov, H. L. Wang, H. J. Dai, Facile synthesis of high-quality graphene nanoribbons. *Nature Nanotechnology* 2010, **5**, 321; (b) D. V. Kosynkin, A. L. Higginbotham, A. Sinitskii, J. R. Lomeda, A. Dimiev, B. K. Price, J. M. Tour, Longitudinal unzipping of carbon nanotubes to form graphene nanoribbons. *Nature* 2009, **458**, 872.

[30]   N. Liu, F. Luo, H. X. Wu, Y. H. Liu, C. Zhang, J. Chen, One-step ionic-liquid-assisted electrochemical synthesis of ionic-liquid-functionalized graphene sheets directly from graphite. *Advanced Functional Materials* 2008, **18**, 1518.

[31]   L. M. Viculis, J. J. Mack, O. M. Mayer, H. T. Hahn, R. B. Kaner, Intercalation and exfoliation routes to graphite nanoplatelets. *Journal of Materials Chemistry* 2005, **15**, 974.

[32]   X. K. Lu, M. F. Yu, H. Huang, R. S. Ruoff, Tailoring graphite with the goal of achieving single sheets. *Nanotechnology* 1999, **10**, 269.

[33]   P. Sutter, Epitaxial graphene: how silicon leaves the scene. *Nature Materials* 2009, **8**, 171.

[34]   M. Luisier, T. B. Boykin, Z. Ye, A. Martini, G. Klimeck, N. Kharche, X. Jiang, S. Nayak, Investigation of ripple-limited low-field mobility in large-scale graphene nanoribbons. *Applied Physics Letters* 2013, **102**, 253506.

[35]   Y. B. Zhang, J. P. Small, W. V. Pontius, P. Kim, Fabrication and electric-field-dependent transport measurements of mesoscopic graphite devices. *Applied Physics Letters* 2005, **86**, 073104.

[36]   B. C. Brodie, Sur le poids atomique du graphite. *Annales de Chimie Physique* 1860, **59**, 466.

[37]   K. S. Novoselov, D. Jiang, F. Schedin, T. J. Booth, V. V. Khotkevich, S. V. Morozov, A. K. Geim, Two-dimensional atomic crystals. *Proceedings of the National Academy of Sciences of the United States of America* 2005, **102**, 10451.

[38]   S. Das, M. Kim, J.-W. Lee, W. Choi, Synthesis, Properties, and Applications of 2-D Materials: A Comprehensive Review. *Critical Reviews in Solid State and Materials Sciences* 2014, **39**, 231.

[39]   X. Liang, Z. Fu, S. Y. Chou, Graphene transistors fabricated via transfer-printing in device active-areas on large wafer. *Nano Letters* 2007, **7**, 3840.

[40]   L. M. Viculis, J. J. Mack, R. B. Kaner, Chemical route to carbon nanoscrolls. *Science* 2003, **299**, 1361.

[41]   C. Vallés, C. Drummond, H. Saadaoui, C. A. Furtado, M. He, O. Roubeau, L. Ortolani, M. Monthioux, A. Pénicaud, Solutions of negatively charged graphene sheets and ribbons. *Journal of the American Chemical Society* 2008, **130**, 15802.

[42] A. A. Green, M. C. Hersam, Solution phase production of graphene with controlled thickness via density differentiation. *Nano Letters* 2009, **9**, 4031.

[43] (a) B. Hanns-Peter, Graphene-how a laboratory curiosity suddenly became extremely interesting. *Angewandte Chemie International Edition* 2010, **49**, 9332–9335; (b) H. P. Boehm, A. Clauss, G. O. Fischer, U. D. Hofmann, Das adsorptionsverhalten sehr dünner kohlenstoff-folien. *Anorganische und Allgemeine Chemie* 1962, **316**, 119.

[44] L. Staudenmaier, Verfahren zur Darstellung der Graphitsäure. *Berichte Der Deutschen Botanischen Gesellschaft* 1898, **31**, 1481.

[45] W. S. Hummers, R. E. Offeman, Preparation of graphitic oxide. *Journal of the American Chemical Society* 1958, **80**, 1339.

[46] K. R. Koch, An impressive demonstration of the powerful oxidizing property of dimanganeseheptoxide. *Journal of Chemical Education* 1982, **59**, 973.

[47] (a) H. A. Becerril, J. Mao, Z. Liu, R. M. Stoltenberg, Z. Bao, Y. Chen, Evaluation of solution-processed reduced graphene oxide films as transparent conductors. *ACS Nano* 2008, **2**, 463; (b) G. Eda, Y.-Y. Lin, S. Miller, C.-W. Chen, W.-F. Su, M. Chhowalla, Transparent and conducting electrodes for organic electronics from reduced graphene oxide. *Applied Physics Letters* 2008, **92**, 233305.

[48] A. B. Bourlinos, D. Gournis, D. Petridis, T. Szabó, A. Szeri, I. Dékány, Chemical reduction to graphite and surface modification with primary aliphatic amines and amino acids. *Langmuir* 2003, **19**, 6050.

[49] H.-J. Shin, K. K. Kim, A. Benayad, S.-M. Yoon, H. K. Park, I.-S. Jung, M. H. Jin, H.-K. Jeong, J. M. Kim, J.-Y. Choi, Y. H. Lee, Efficient reduction of graphite oxide by sodium borohydride and its effect on electrical conductance. *Advanced Functional Materials* 2009, **19**, 1987.

[50] S. Stankovich, D. A. Dikin, G. H. B. Dommett, K. M. Kohlhaas, E. J. Zimney, E. A. Stach, R. D. Piner, S. T. Nguyen, R. S. Ruoff, Graphene-based composite materials. *Nature* 2006, **442**, 282.

[51] S. Stankovich, R. D. Piner, S. T. Nguyen, R. S. Ruoff, Synthesis and exfoliation of isocyanate-treated graphene oxide nanoplatelets. *Carbon* 2006, **44**, 3342.

[52] H. K. Jeong, Y. P. Lee, R. Lahaye, M. H. Park, K. H. An, I. J. Kim, C. W. Yang, C. Y. Park, R. S. Ruoff, Y. H. Lee, Evidence of graphitic AB stacking order of graphite oxides. *Journal of the American Chemical Society* 2008, **130**, 1362.

[53] Y. X. Xu, H. Bai, G. W. Lu, C. Li, G. Q. Shi, Flexible graphene films via the filtration of water-soluble noncovalent functionalized graphene sheets. *Journal of the American Chemical Society* 2008, **130**, 5856.

[54] D. Li, M. B. Muller, S. Gilje, R. B. Kaner, G. G. Wallace, Processable aqueous dispersions of graphene nanosheets. *Nature Nanotechnology* 2008, **3**, 101.

[55] S. Stankovich, D. A. Dikin, R. D. Piner, K. A. Kohlhaas, A. Kleinhammes, Y. Jia, Y. Wu, S. T. Nguyen, R. S. Ruoff, Synthesis of graphene-based nanosheets via chemical reduction of exfoliated graphite oxide. *Carbon* 2007, **45**, 1558.

[56] V. C. Tung, M. J. Allen, Y. Yang, R. B. Kaner, High-throughput solution processing of large-scale graphene. *Nature Nanotechnology* 2009, **4**, 25.

[57] (a) H. C. Schniepp, J. L. Li, M. J. McAllister, H. Sai, M. Herrera-Alonso, D. H. Adamson, R. K. Prud'homme, R. Car, D. A. Saville, I. A. Aksay, Functionalized single graphene sheets derived from splitting graphite oxide. *Journal of Physical Chemistry B* 2006, **110**, 8535; (b) M. J. McAllister, J. L. Li, D. H. Adamson, H. C. Schniepp, A. A. Abdala, J. Liu, M. Herrera-Alonso, D. L. Milius, R. Car, R. K. Prud'homme,

I. A. Aksay, Single sheet functionalized graphene by oxidation and thermal expansion of graphite. *Chemistry of Materials* 2007, **19**, 4396.

[58] L. J. Cote, F. Kim, J. Huang, Langmuir−Blodgett assembly of graphite oxide single layers. *Journal of the American Chemical Society* 2008, **131**, 1043.

[59] X. L. Li, X. R. Wang, L. Zhang, S. W. Lee, H. J. Dai, Highly conducting graphene sheets and Langmuir–Blodgett films. *Science* 2008, **319**, 1229.

[60] X. L. Li, G. Y. Zhang, X. D. Bai, X. M. Sun, X. R. Wang, E. Wang, H. J. Dai, Highly conducting graphene sheets and Langmuir–Blodgett films. *Nature Nanotechnology* 2008, **3**, 538.

[61] J. L. Zhang, G. X. Shen, W. J. Wang, X. J. Zhou, S. W. Guo, Individual nanocomposite sheets of chemically reduced graphene oxide and poly(N-vinyl pyrrolidone). *Journal of Materials Chemistry* 2010, **20**, 10824.

[62] S. S. Li, K. H. Tu, C. C. Lin, C. W. Chen, M. Chhowalla, Solution-processable graphene oxide as an efficient hole transport layer in polymer solar cells. *ACS Nano* 2010, **4**, 3169.

[63] (a) P. Hasin, M. A. Alpuche-Aviles, Y. Y. Wu, Electrocatalytic activity of graphene multi layers toward I-/I-3(-): effect of preparation conditions and polyelectrolyte modification. *Journal of Physical Chemistry C* 2010, **114**, 15857; (b) J. D. Roy-Mayhew, D. J. Bozym, C. Punckt, I. A. Aksay, Functionalized graphene as a catalytic counter electrode in dye-sensitized solar cells. *ACS Nano* 2010, **4**, 6203.

[64] G. L. Li, G. Liu, M. Li, D. Wan, K. G. Neoh, E. T. Kang, Organo- and water-dispersible graphene oxide-polymer nanosheets for organic electronic memory and gold nanocomposites. *Journal of Physical Chemistry C* 2010, **114**, 12742.

[65] S. Q. Chen, Y. Wang, Microwave-assisted synthesis of a Co_3O_4-graphene sheet-on-sheet nanocomposite as a superior anode material for Li-ion batteries. *Journal of Materials Chemistry* 2010, **20**, 9735.

[66] R. Muszynski, B. Seger, P. V. Kamat, Decorating graphene sheets with gold nanoparticles. *Journal of Physical Chemistry C* 2008, **112**, 5263.

[67] (a) J. Qian, P. Liu, Y. Xiao, Y. Jiang, Y. Cao, X. Ai, H. Yang, TiO_2-coated multilayered SnO_2 hollow microspheres for dye-sensitized solar cells. *Advanced Materials* 2009, **21**, 3663; (b) N. L. Yang, J. Zhai, D. Wang, Y. S. Chen, L. Jiang, Two-dimensional graphene bridges enhanced photoinduced charge transport in dye-sensitized solar cells. *ACS Nano* 2010, **4**, 887.

[68] K. F. Zhou, Y. H. Zhu, X. L. Yang, C. Z. Li, One-pot preparation of graphene/ Fe_3O_4 composites by a solvothermal reaction. *New Journal of Chemistry* 2010, **34**, 2950.

[69] A. Devadoss, P. Sudhagar, S. Das, S. Y. Lee, C. Terashima, K. Nakata, A. Fujishima, W. Choi, Y. S. Kang, U. Paik, Synergistic Metal–metal oxide nanoparticles supported electrocatalytic graphene for improved photoelectrochemical glucose oxidation. *ACS Applied Materials and Interfaces* 2014, **6**, 4864.

[70] J. W. Zhu, G. Y. Zeng, F. D. Nie, X. M. Xu, S. Chen, Q. F. Han, X. Wang, Decorating graphene oxide with CuO nanoparticles in a water-isopropanol system. *Nanoscale* 2010, **2**, 988.

[71] P. K. Vabbina, M. Karabiyik, C. Al-Amin, N. Pala, S. Das, W. Choi, T. Saxena, M. Shur, Controlled synthesis of single-crystalline ZnO nanoflakes on arbitrary substrates at ambient conditions. *Particle & Particle Systems Characterization* 2014, **31**, 190.

[72] P. K. Vabbina, S. Das, N. Pala, W. Choi, Synthesis of crystalline ZnO nanosheets on graphene and other substrates at ambient conditions. *MRS Online Proceedings Library* 2012, **1449**, 121.

[73] M. Choucair, P. Thordarson, J. A. Stride, Gram-scale production of graphene based on solvothermal synthesis and sonication. *Nature Nanotechnology* 2009, **4**, 30.

[74] B. Lang, A LEED study of the deposition of carbon on platinum crystal surfaces. *Surface Science* 1975, **53**, 317.

[75] M. Eizenberg, J. M. Blakely, Carbon monolayer phase condensation on Ni(111). *Surface Science* 1979, **82**, 228.

[76] (a) M. I. Katsnelson, Graphene: carbon in two dimensions. *Materials Today* 2007, **10**, 20; (b) D. R. Dreyer, S. Park, C. W. Bielawski, R. S. Ruoff, The chemistry of graphene oxide. *Chemical Society Reviews* 2010, **39**, 228.

[77] S. Das, *Mechanical and Materials Engineering*, Vol. PhD, Florida International University, Miami, Florida, 2012, 199.

[78] P. R. Somani, S. P. Somani, M. Umeno, Planer nano-graphenes from camphor by CVD. *Chemical Physics Letters* 2006, **430**, 56.

[79] A. N. Obraztsov, E. A. Obraztsova, A. V. Tyurnina, A. A. Zolotukhin, Chemical vapor deposition of thin graphite films of nanometer thickness. *Carbon* 2007, **45**, 2017.

[80] A. N. Obraztsov, Chemical vapour deposition: making graphene on a large scale. *Nature Nanotechnology* 2009, **4**, 212.

[81] Q. K. Yu, J. Lian, S. Siriponglert, H. Li, Y. P. Chen, S. S. Pei, Graphene segregated on Ni surfaces and transferred to insulators. *Applied Physics Letters* 2008, **93**, 113103.

[82] K. S. Kim, Y. Zhao, H. Jang, S. Y. Lee, J. M. Kim, J. H. Ahn, P. Kim, J. Y. Choi, B. H. Hong, Large-scale pattern growth of graphene films for stretchable transparent electrodes. *Nature* 2009, **457**, 706.

[83] A. Reina, X. T. Jia, J. Ho, D. Nezich, H. B. Son, V. Bulovic, M. S. Dresselhaus, J. Kong, Large area, few-layer graphene films on arbitrary substrates by chemical vapor deposition. *Nano Letters* 2009, **9**, 30.

[84] X. B. Wang, H. J. You, F. M. Liu, M. J. Li, L. Wan, S. Q. Li, Q. Li, Y. Xu, R. Tian, Z. Y. Yu, D. Xiang, J. Cheng, Large-scale synthesis of few-layered graphene using CVD. *Chemical Vapor Deposition* 2009, **15**, 53.

[85] X. S. Li, W. W. Cai, L. Colombo, R. S. Ruoff, Evolution of graphene growth on Ni and Cu by carbon isotope labeling. *Nano Letters* 2009, **9**, 4268.

[86] S. Bae, H. Kim, Y. Lee, X. F. Xu, J. S. Park, Y. Zheng, J. Balakrishnan, T. Lei, H. R. Kim, Y. I. Song, Y. J. Kim, K. S. Kim, B. Ozyilmaz, J. H. Ahn, B. H. Hong, S. Iijima, Roll-to-roll production of 30-inch graphene films for transparent electrodes. *Nature Nanotechnology* 2010, **5**, 574.

[87] A. Reina, S. Thiele, X. T. Jia, S. Bhaviripudi, M. S. Dresselhaus, J. A. Schaefer, J. Kong, Growth of large-area single- and Bi-layer graphene by controlled carbon precipitation on polycrystalline Ni surfaces. *Nano Research* 2009, **2**, 509.

[88] (a) L. B. Gao, W. C. Ren, J. P. Zhao, L. P. Ma, Z. P. Chen, H. M. Cheng, Efficient growth of high-quality graphene films on Cu foils by ambient pressure chemical vapor deposition. *Applied Physics Letters* 2010, **97**, 3; (b) Z. Z. Sun, Z. Yan, J. Yao, E. Beitler, Y. Zhu, J. M. Tour, Growth of graphene from solid carbon sources. *Nature* 2010, **468**, 549.

[89] (a) J. Campos-Delgado, Y. A. Kim, T. Hayashi, A. Morelos-Gomez, M. Hofmann, H. Muramatsu, M. Endo, H. Terrones, R. D. Shull, M. S. Dresselhaus, M. Terrones,

Thermal stability studies of CVD-grown graphene nanoribbons: defect annealing and loop formation. *Chemical Physics Letters* 2009, **469**, 177; (b) J. Campos-Delgado, J. M. Romo-Herrera, X. T. Jia, D. A. Cullen, H. Muramatsu, Y. A. Kim, T. Hayashi, Z. F. Ren, D. J. Smith, Y. Okuno, T. Ohba, H. Kanoh, K. Kaneko, M. Endo, H. Terrones, M. S. Dresselhaus, M. Terrones, Bulk production of a new form of sp(2) carbon: crystalline graphene nanoribbons. *Nano Letters* 2008, **8**, 2773.

[90] A. Ismach, C. Druzgalski, S. Penwell, A. Schwartzberg, M. Zheng, A. Javey, J. Bokor, Y. G. Zhang, Direct chemical vapor deposition of graphene on dielectric surfaces. *Nano Letters* 2010, **10**, 1542.

[91] M. H. Rummeli, A. Bachmatiuk, A. Scott, F. Borrnert, J. H. Warner, V. Hoffman, J. H. Lin, G. Cuniberti, B. Buchner, Direct low-temperature nanographene CVD synthesis over a dielectric insulator. *ACS Nano* 2010, **4**, 4206.

[92] N. G. Shang, P. Papakonstantinou, M. McMullan, M. Chu, A. Stamboulis, A. Potenza, S. S. Dhesi, H. Marchetto, Catalyst-free efficient growth, orientation and biosensing properties of multilayer graphene nanoflake films with sharp edge planes. *Advanced Functional Materials* 2008, **18**, 3506.

[93] A. N. Obraztsov, A. A. Zolotukhin, A. O. Ustinov, A. P. Volkov, Y. Svirko, K. Jefimovs, DC discharge plasma studies for nanostructured carbon CVD. *Diamond and Related Materials* 2003, **12**, 917.

[94] (a) J. J. Wang, M. Y. Zhu, R. A. Outlaw, X. Zhao, D. M. Manos, B. C. Holloway, Synthesis of carbon nanosheets by inductively coupled radio-frequency plasma enhanced chemical vapor deposition. *Carbon* 2004, **42**, 2867; (b) J. J. Wang, M. Y. Zhu, R. A. Outlaw, X. Zhao, D. M. Manos, B. C. Holloway, V. P. Mammana, Free-standing subnanometer graphite sheets. *Applied Physics Letters* 2004, **85**, 1265.

[95] (a) G. D. Yuan, W. J. Zhang, Y. Yang, Y. B. Tang, Y. Q. Li, J. X. Wang, X. M. Meng, Z. B. He, C. M. L. Wu, I. Bello, C. S. Lee, S. T. Lee, Graphene sheets via microwave chemical vapor deposition. *Chemical Physics Letters* 2009, **467**, 361; (b) R. Vitchev, A. Malesevic, R. H. Petrov, R. Kemps, M. Mertens, A. Vanhulsel, C. Van Haesendonck, Initial stages of few-layer graphene growth by microwave plasma-enhanced chemical vapour deposition. *Nanotechnology* 2010, **21**, 7.

[96] O. Jasek, P. Synek, L. Zajickova, M. Elias, V. Kudrle, Synthesis of carbon nanostructures by plasma enhanced chemical vapour deposition at atmospheric pressure. *Journal of Electrical Engineering-Elektrotechnicky Casopis* 2010, **61**, 311.

[97] (a) J. L. Qi, X. Wang, W. T. Zheng, H. W. Tian, C. Liu, Y. L. Lu, Y. S. Peng, G. Cheng, Effects of total CH_4/Ar gas pressure on the structures and field electron emission properties of carbon nanomaterials grown by plasma-enhanced chemical vapor deposition. *Applied Surface Science* 2009, **256**, 1542; (b) T. Bhuvana, A. Kumar, A. Sood, R. H. Gerzeski, J. J. Hu, V. S. Bhadram, C. Narayana, T. S. Fisher, Contiguous petal-like carbon nanosheet outgrowths from graphite fibers by plasma CVD. *ACS Applied Materials and Interfaces* 2010, **2**, 644.

[98] D. H. Lee, J. A. Lee, W. J. Lee, D. S. Choi, S. O. Kim, Facile fabrication and field emission of metal-particle-decorated vertical N-doped carbon nanotube/graphene hybrid films. *Journal of Physical Chemistry C* 2010, **114**, 21184.

[99] S. Das, R. Seelaboyina, V. Verma, I. Lahiri, J. Y. Hwang, R. Banerjee, W. Choi, Synthesis and characterization of self-organized multilayered graphene-carbon nanotube hybrid films. *Journal of Materials Chemistry* 2011, **21**, 7289.

[100] (a) J. C. Koepke, J. D. Wood, D. Estrada, Z.-Y. Ong, K. T. He, E. Pop, J. W. Lyding, Atomic-scale evidence for potential barriers and strong carrier scattering at graphene grain boundaries: a scanning tunneling microscopy study. *ACS Nano* 2012, **7**, 75; (b) A. Y. Serov, Z.-Y. Ong, E. Pop, Effect of grain boundaries on thermal. *Applied Physics Letters* 2013, **102**, 033104.

[101] J. Kotakoski, J. C. Meyer, Mechanical properties of polycrystalline graphene based on a realistic atomistic model. *Physical Review B* 2012, **85**, 195447.

[102] J. D. Wood, S. W. Schmucker, A. S. Lyons, E. Pop, J. W. Lyding, Effects of polycrystalline Cu substrate on graphene growth by chemical vapor deposition. *Nano Letters* 2011, **11**, 4547.

[103] X. Li, C. W. Magnuson, A. Venugopal, R. M. Tromp, J. B. Hannon, E. M. Vogel, L. Colombo, R. S. Ruoff, Large-area graphene single crystals grown by low-pressure chemical vapor deposition of methane on copper. *Journal of the American Chemical Society* 2011, **133**, 2816.

[104] Z. Yan, J. Lin, Z. Peng, Z. Sun, Y. Zhu, L. Li, C. Xiang, E. L. Samuel, C. Kittrell, J. M. Tour, Toward the synthesis of wafer-scale single-crystal graphene on copper foils. *ACS Nano* 2012, **6**, 9110.

[105] P. Y. Huang, C. S. Ruiz-Vargas, A. M. van der Zande, W. S. Whitney, M. P. Levendorf, J. W. Kevek, S. Garg, J. S. Alden, C. J. Hustedt, Y. Zhu, J. Park, P. L. McEuen, D. A. Muller, Grains and grain boundaries in single-layer graphene atomic patchwork quilts. *Nature* 2011, **469**, 389.

[106] G. H. Han, F. Güneş, J. J. Bae, E. S. Kim, S. J. Chae, H.-J. Shin, J.-Y. Choi, D. Pribat, Y. H. Lee, Influence of copper morphology in forming nucleation seeds for graphene growth. *Nano Letters* 2011, **11**, 4144.

[107] Q. K. Yu, L. A. Jauregui, W. Wu, R. Colby, J. F. Tian, Z. H. Su, H. L. Cao, Z. H. Liu, D. Pandey, D. G. Wei, T. F. Chung, P. Peng, N. P. Guisinger, E. A. Stach, J. M. Bao, S. S. Pei, Y. P. Chen, Control and characterization of individual grains and grain boundaries in graphene grown by chemical vapour deposition. *Nature Materials* 2011, **10**, 443.

[108] S. Das, D. Lahiri, D.-Y. Lee, A. Agarwal, W. Choi, Measurements of the adhesion energy of graphene to metallic substrates. *Carbon* 2013, **59**, 121.

[109] N. Wilson, A. Marsden, M. Saghir, C. Bromley, R. Schaub, G. Costantini, T. White, C. Partridge, A. Barinov, P. Dudin, A. Sanchez, J. Mudd, M. Walker, G. Bell, Weak mismatch epitaxy and structural Feedback in graphene growth on copper foil. *Nano Research* 2013, **6**, 99.

[110] A. J. Van Bommel, J. E. Crombeen, A. Van Tooren, LEED and Auger electron observations of the SiC(0001) surface. *Surface Science* 1975, **48**, 463.

[111] J. Hass, W. A. de Heer, E. H. Conrad, The growth and morphology of epitaxial multilayer graphene. *Journal of Physics-Condensed Matter* 2008, **20**, 27.

[112] C. Berger, Z. M. Song, T. B. Li, X. B. Li, A. Y. Ogbazghi, R. Feng, Z. T. Dai, A. N. Marchenkov, E. H. Conrad, P. N. First, W. A. de Heer, Ultrathin epitaxial graphite: 2D electron gas properties and a route toward graphene-based nanoelectronics. *Journal of Physical Chemistry B* 2004, **108**, 19912.

[113] W. A. de Heer, C. Berger, X. S. Wu, P. N. First, E. H. Conrad, X. B. Li, T. B. Li, M. Sprinkle, J. Hass, M. L. Sadowski, M. Potemski, G. Martinez, Epitaxial graphene. *Solid State Communications* 2007, **143**, 92.

[114] Z. Y. Juang, C. Y. Wu, C. W. Lo, W. Y. Chen, C. F. Huang, J. C. Hwang, F. R. Chen, K. C. Leou, C. H. Tsai, Synthesis of graphene on silicon carbide substrates at low temperature. *Carbon* 2009, **47**, 2026.

[115] B. Dlubak, M.-B. Martin, C. Deranlot, B. Servet, S. Xavier, R. Mattana, M. Sprinkle, C. Berger, W. A. De Heer, F. Petroff, A. Anane, P. Seneor, A. Fert, Highly efficient spin transport in epitaxial graphene on SiC. *Nature Physics* 2012, **8**, 557.

[116] K. V. Emtsev, A. Bostwick, K. Horn, J. Jobst, G. L. Kellogg, L. Ley, J. L. McChesney, T. Ohta, S. A. Reshanov, J. Rohrl, E. Rotenberg, A. K. Schmid, D. Waldmann, H. B. Weber, T. Seyller, Towards wafer-size graphene layers by atmospheric pressure graphitization of silicon carbide. *Nature Materials* 2009, **8**, 203.

[117] P. Blake, E. W. Hill, A. H. C. Neto, K. S. Novoselov, D. Jiang, R. Yang, T. J. Booth, A. K. Geim, Making graphene visible. *Applied Physics Letters* 2007, **91**, 063124.

[118] Z. H. Ni, H. M. Wang, J. Kasim, H. M. Fan, T. Yu, Y. H. Wu, Y. P. Feng, Z. X. Shen, Graphene thickness determination using reflection and contrast spectroscopy. *Nano Letters* 2007, **7**, 2758.

[119] J. S. Park, A. Reina, R. Saito, J. Kong, G. Dresselhaus, M. S. Dresselhaus, G′ band Raman spectra of single, double and triple layer graphene. *Carbon* 2009, **47**, 1303.

[120] I. Jung, M. Pelton, R. Piner, D. A. Dikin, S. Stankovich, S. Watcharotone, M. Hausner, R. S. Ruoff, Simple approach for high-contrast optical imaging and characterization of graphene-based sheets. *Nano Letters* 2007, **7**, 3569.

[121] D. L. Duong, G. H. Han, S. M. Lee, F. Gunes, E. S. Kim, S. T. Kim, H. Kim, T. Quang Huy, K. P. So, S. J. Yoon, S. J. Chae, Y. W. Jo, M. H. Park, S. H. Chae, S. C. Lim, J. Y. Choi, Y. H. Lee, Probing graphene grain boundaries with optical microscopy. *Nature* 2012, **490**, 235.

[122] M. S. Dresselhaus, A. Jorio, R. Saito, Characterizing graphene, graphite, and carbon nanotubes by Raman spectroscopy. *Annual Review of Condensed Matter Physics* 2010, **1**, 89.

[123] M. S. Dresselhaus, A. Jorio, M. Hofmann, G. Dresselhaus, R. Saito, Perspectives on carbon nanotubes and graphene Raman spectroscopy. *Nano Letters* 2010, **10**, 751.

[124] C. Casiraghi, A. Hartschuh, H. Qian, S. Piscanec, C. Georgi, A. Fasoli, K. S. Novoselov, D. M. Basko, A. C. Ferrari, Raman spectroscopy of graphene edges. *Nano Letters* 2009, **9**, 1433.

[125] L. M. Malard, M. A. Pimenta, G. Dresselhaus, M. S. Dresselhaus, Raman spectroscopy in graphene. *Physics Reports* 2009, **473**, 51.

[126] Z. H. Ni, Y. Y. Wang, T. Yu, Z. X. Shen, Raman spectroscopy and imaging of graphene. *Nano Research* 2008, **1**, 273.

[127] A. C. Ferrari, J. C. Meyer, V. Scardaci, C. Casiraghi, M. Lazzeri, F. Mauri, S. Piscanec, D. Jiang, K. S. Novoselov, S. Roth, A. K. Geim, Raman spectrum of graphene and graphene layers. *Physical Review Letters* 2006, **97**, 4.

[128] Y. Y. Wang, Z. H. Ni, T. Yu, Z. X. Shen, H. M. Wang, Y. H. Wu, W. Chen, A. T. S. Wee, Raman studies of monolayer graphene: the substrate effect. *Journal of Physical Chemistry C* 2008, **112**, 10637–10640.

[129] A. Das, B. Chakraborty, A. K. Sood, Raman spectroscopy of graphene on different substrates and influence of defects. *Bulletin of Materials Science* 2008, **31**, 579.

[130]   V. Georgakilas, M. Otyepka, A. B. Bourlinos, V. Chandra, N. Kim, K. C. Kemp, P. Hobza, R. Zboril, K. S. Kim, Functionalization of graphene: covalent and non-covalent approaches, derivatives and applications. *Chemical Reviews* 2012, **112**, 6156.

[131]   S. Das, D. Lahiri, A. Agarwal, W. Choi, Interfacial bonding characteristics between graphene and dielectric substrates. *Nanotechnology* 2014, **25**, 045707.

[132]   I. Calizo, W. Z. Bao, F. Miao, C. N. Lau, A. A. Balandin, The effect of substrates on the Raman spectrum of graphene: graphene-on-sapphire and graphene-on-glass. *Applied Physics Letters* 2007, **91**, 201904.

[133]   A. Hadjikhani, J. Chen, S. Das, W. Choi, Raman spectroscopy of graphene and plasma treated graphene under high pressure, in *Supplemental Proceedings: Materials Properties, Characterization, and Modeling*, Vol. **2**, John Wiley & Sons, Inc., Hoboken, NJ, 2012, 75.

[134]   I. Calizo, A. A. Balandin, W. Bao, F. Miao, C. N. Lau, Temperature dependence of the Raman spectra of graphene and graphene multilayers. *Nano Letters* 2007, **7**, 2645.

[135]   M. Huang, H. Yan, T. F. Heinz, J. Hone, Probing strain-induced electronic structure change in graphene by Raman spectroscopy. *Nano Letters* 2010, **10**, 4074.

[136]   M. Yi, Z. Shen, X. Zhang, S. Ma, Achieving concentrated graphene dispersions in water/acetone mixtures by the strategy of tailoring Hansen solubility parameters. *Journal of Physics D-Applied Physics* 2013, **46**, 025301.

[137]   J. C. C. Kisielowski, R. Erni, M. D. Rossell, M. F. Crommie, A. Zettl, Direct imaging of lattice atoms and topological defects in graphene membranes. *Nano Letters* 2008, **8**, 3582.

[138]   Ç. Ö. Girit, J. C. Meyer, R. Erni, M. D. Rossell, C. Kisielowski, L. Yang, C.-H. Park, M. F. Crommie, M. L. Cohen, S. G. Louie, A. Zettl, Graphene at the edge: stability and dynamics. *Science* 2009, **323**, 1705.

[139]   H.-C. Hsu, I. Shown, H.-Y. Wei, Y.-C. Chang, H.-Y. Du, Y.-G. Lin, C.-A. Tseng, C.-H. Wang, L.-C. Chen, Y.-C. Lin, K.-H. Chen, Graphene oxide as a promising photocatalyst for CO_2 to methanol conversion. *Nanoscale* 2013, **5**, 262.

[140]   A. Luican, G. Li, E. Y. Andrei, Scanning tunneling microscopy and spectroscopy of graphene layers on graphite. *Solid State Communications* 2009, **149**, 1151.

[141]   S. L. Wong, H. Huang, W. Chen, A. T. S. Wee, STM studies of epitaxial graphene. *MRS Bulletin* 2012, **37**, 1195.

[142]   A. J. M. Giesbers, U. Zeitler, S. Neubeck, F. Freitag, K. S. Novoselov, J. C. Maan, Nanolithography and manipulation of graphene using an atomic force microscope. *Solid State Communications* 2008, **147**, 366.

[143]   M. Cheng, R. Yang, L. Zhang, Z. Shi, W. Yang, D. Wang, G. Xie, D. Shi, G. Zhang, Restoration of graphene from graphene oxide by defect repair. *Carbon* 2012, **50**, 2581.

[144]   Y. Zhang, Y. Gao, L. Liu, N. Xi, Y. Wang, L. Ma, Z. Dong, U. C. Wejinya, Cutting forces related with lattice orientations of graphene using an atomic force microscopy based nanorobot. *Applied Physics Letters* 2012, **101**, 213101.

4

DOPING CARBON NANOMATERIALS WITH HETEROATOMS

TOMA SUSI[1] AND PAOLA AYALA[1,2]

[1]*Faculty of Physics, University of Vienna, Vienna, Austria*
[2]*Yachay Tech University, School of Physical Sciences and Nanotechnology, Urcuquí, Ecuador*

4.1 INTRODUCTION

Carbon nanotubes (CNTs) and graphene are both allotropes of carbon, which have exceptional and unique properties. The electronic nature (i.e., semiconducting or metallic) of single-walled CNTs (SWCNTs) depends on their precise geometric structure [1, 2]. This can represent a problem for applications to a certain extent, since nanotubes of a specific type are required in some cases. Great effort has been directed to controlling the electronic properties of an ensemble of CNTs, with significant progress [3–5]. Specially in the last years, the advances toward the purification and separation of nanotube samples have propelled the understanding of their physical properties compared to theoretical predictions. In the case of graphene, efforts have been directed into opening its band gap and to tuning the carrier concentration.

There are several possible ways to control the electronic and optical properties of graphene and nanotubes. Although it is not fully understood if a perfect graphitic lattice is inert, which would limit its usefulness for applications, one certain way to gain this control is the structural functionalization with different methods [6, 7]. Substitutional heteroatoms are ideal in this context. It should be noted from the outset that substitutional doping is distinct from other methods of donating charge to the nanotubes, such as from adsorbed species (chemical doping or intercalation), by

Carbon Nanomaterials for Advanced Energy Systems: Advances in Materials Synthesis and Device Applications, First Edition. Edited by Wen Lu, Jong-Beom Baek and Liming Dai.
© 2015 John Wiley & Sons, Inc. Published 2015 by John Wiley & Sons, Inc.

electrochemical charging, or by simple electrical gating. The two natural substitutional candidates are the nearest neighbors to carbon in the periodic table, boron (B) and nitrogen (N) [7–11]. Nitrogen in particular has attracted much attention due to its suitable atomic size and additional electron compared to carbon. Although nitrogen-doped multiwalled CNTs (N-MWCNTs) are widely available, the synthesis of nitrogen-doped single-walled CNTs (N-SWCNTs) presented in the beginning more difficulties. However, the synthesis of these single-walled structures is nowadays performed with lesser difficulties. Several research questions remain largely unanswered, chief among them, the local atomic structure of the dopant sites. Nitrogen-doped graphene (N-graphene) [12–14], on the other hand, has seen surge of high research activity in recent years (see Fig. 4.1). Phosphorus (P) is also a possible substitutional dopant [15, 16], which like nitrogen has five valence electrons, but in the third electron shell instead of the second. Although P has been theoretically proposed long time ago as ideal dopant for such carbon systems, the first experimental reports on phosphorus doping of CNTs and graphene have been recently published [17–20]. Other heteroatoms such as S, Si, Al, and Ni, among others, have been

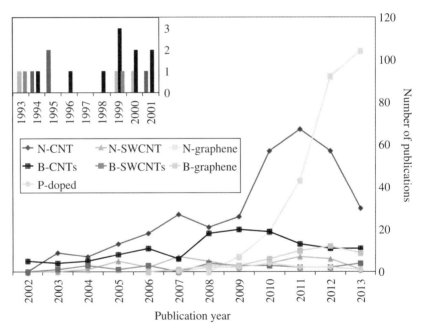

FIGURE 4.1 The number of publications yearly from 2002 to 2013 on nitrogen-doped carbon nanotubes (N-CNTs) in general, nitrogen-doped single-walled carbon nanotubes (N-SWCNTs) specifically, nitrogen-doped graphene (N-graphene), boron-doped CNTs (B-CNTs), boron-doped SWCNTs (B-SWCNTs), boron-doped graphene (B-graphene), and both phosphorus-doped nanotubes and graphene (P doped). The inset shows the few publications from the years 1993 to 2001. Data from the ISI Web of Science; note that the data for 2013 is incomplete and that works discussing several types of systems are duplicated in the counts.

suggested, but the work in this direction is still limited, and we will omit further discussion here for brevity. More information can be found elsewhere [13, 16, 21–25].

In this chapter, we summarize the structure, synthesis, characterization, and applications of doped single-walled CNTs and graphene doped with heteroatoms of N, B, and P. Note that analogue heterostructures regarding their overall morphology such as boron nitride (BN) nanotubes and sheets will be left outside this short review chapter. These are not doped structures but completely different materials, and the interested reader can find more related information in other recent articles [7, 26, 27].

4.2 LOCAL BONDING OF THE DOPANTS

Achieving the desired functionality or modification of the properties of the carbon materials here discussed depends on the local bonding of the dopants. For different heteroatoms, different challenges arise. For example, particularly in nanotubes, nitrogen tends to bond in structurally and electronically distinct configurations, whereas boron mainly stays in a direct-substitutional configuration. Since nitrogen has an extra electron compared to carbon, introducing N atoms into the graphitic lattice could be expected to contribute to n-type doping of the host. Indeed, theoretical works have shown that a direct substitution of a C atom by N results in localized states above the Fermi level [28–30] (Fig. 4.2a). This is because N uses three electrons in σ bonds and one in π, with the remaining fifth valence electron forced to occupy the π^* donor state. Since the size and electronic structure of nitrogen are different than those of carbon, N can also create a defect in the lattice. The most widely discussed of such defects is the so-called triple pyridinic vacancy configuration, where 3 N atoms each form sp^2 bonds to two carbon atoms around a single vacancy [34] (Fig. 4.2b). This is energetically favorable since the site presents no dangling bonds, although other possibilities have also been considered (Fig. 4.2). However, electronically, these sites are p type [30, 34, 35]. Thus, the actual effect N doping has on the electronic structure of SWCNTs or graphene will also depend on the relative amounts of the different doping configurations incorporated into the lattice.

A simple substitution of B is the most favorable bonding configuration for this type of atom. However, attempts to synthesize B-doped material have reported the formation of B nanodomains along the nanotube structure [36, 37]. Only low concentrations of B atoms are expected to lead to the formation of a shallow acceptor state in the band gap. Early calculations [8] found an acceptor state located at 0.16 eV above the Fermi energy for an (8,0) nanotube with a B/C ratio of 1/80. Further studies discussed the evolution of this acceptor-like level with increasing levels of B, pointing out that high doping entail critical differences in the resulting physical properties [38].

Phosphorus has five valence electrons, but since they are on the third electron shell, P has a significantly larger atomic radius (107 pm) that is expected to cause it to protrude from the graphitic lattice [17, 39] (Fig. 4.3a). A higher curvature in small diameter SWCNTs is thought to mitigate the resulting stress [18]. As for the effect a P substitution will have on the electronic structure of SWCNTs, no consensus seems

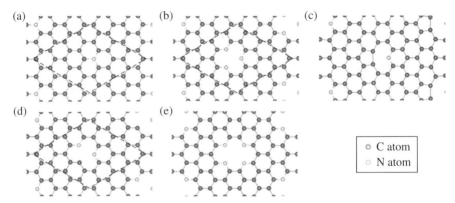

FIGURE 4.2 Nitrogen dopant configurations in a model graphene unit cell marked by the dashed line: (a) substitutional (also called graphitic or quaternary), (b) triple pyridinic vacancy, (c) single pyridinic vacancy, (d) double pyridinic vacancy, and (e) quadruple pyridinic divacancy. Note that nitrogen substitutions on both 2nd [31] and 3rd [32] nearest neighboring sites have also been observed in graphene and that the boron simple substitution corresponds geometrically to (a). All structures are very closely planar. Adapted from Ref. 33, reproduced with permission. © 2011, The American Physical Society.

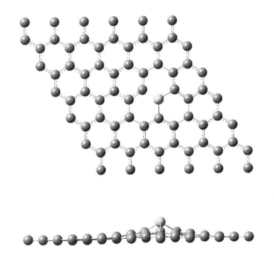

FIGURE 4.3 Substitutions with atoms significantly larger than carbon, such as phosphorus, are expected to protrude from the graphene plane based on density functional theory calculations. Adapted from Ref. [16], reproduced with permission. © 2010, Elsevier. Note that other more complex configurations, such as the P analogue of the triple pyridinic vacancy, have been proposed, again with expected significant out-of-plane character [39].

to have been reached in the available literature yet. It has been suggested that P will act as a net donor both in the substitutional configuration [19] and in a P analogue of the N pyridinic site (3 P atoms around a vacancy) [39]; however, Krstic et al. proposed that the substitutional site will be oxidized in ambient, converting it into a net acceptor

[19]. Alternatively, Cruz-Silva et al. have suggested that the P atom will bond in sp^3 hybridization creating a nondispersive localized state that does not act as a net donor [17, 18]. However, direct evidence for the incorporation of P into the SWCNT lattice or for its electronic nature is still lacking.

Although we will not discuss them in detail here, worth mentioning are also heterostructures of BC_3, BC_2N, and C_3N_4, which among others represent stable stoichiometries that contain C, B, and N atoms forming a sheet or a tubular structure [40]. These kinds of materials have often been reported as "doped" graphene or nanotubes with very high heteroatom contents. Indeed, their completely new stable stoichiometry makes of them totally different structures, where their novel properties are not obtained anymore by variations induced by the incorporation of a dopant. Experimental results on BC_3 nanotubes have showed the formation of a uniform energy gap of 0.4 eV [41], with an acceptor-like band approximately 0.1 eV above the Fermi level. Miyamoto et al. [42] performed first-principles and tight-binding calculations analyzing the stability of nanotubes formed from hexagonal stable BC_2N sheets and suggested a preferential chiral nanotube formation compared to graphite and BN, which has also been observed experimentally [43]. Graphene-like C_3N_4 arrangements are also expected to be stable in planar forms [44, 45] and are expected to form nanotubes of the same stoichiometry. Although the synthesis of C_3N_4-MWCNTs has been reported [46], no experimental evidence for C_3N_4-SWCNTs has been published to our knowledge.

4.3 SYNTHESIS OF HETERODOPED NANOCARBONS

There are several review articles in the literature related to the synthesis of pristine carbon nanomaterials, and guide to this topic can also be found in this book. These synthesis methods can be divided into high- and low-temperature methods, depending on the mechanism for the decomposition or evaporation of the carbon sources. Chemical vapor deposition (CVD), which depends on the pyrolysis of precursor gases, has proven to be the most versatile and attractive technique, and it has been most commonly reported for the synthesis of doped nanotubes and graphene [7]. High-temperature methods, often called physical methods, involve physical means to produce the energy for breaking the precursor molecules such as by arc discharge or laser ablation of a graphite target.

Most CVD-based methods for nanocarbons have used recipes created for making the corresponding pristine material, with the addition of a precursor containing the desired dopant. Alternatively and less frequently reported, doping can be effected by postsynthesis treatments of pristine materials, either by thermochemical substitution reactions [47, 48] or by ion bombardment [49, 50].

Until recent years, it was common practice in the literature to report only the relative amount of the dopant atoms to carbon atoms in the precursors, without a detailed analysis of the concentration of dopants in the actual product. This was partly due to the challenge of characterizing low doping concentrations in the very limited amounts of material that could be produced. This is still the case in studies on P-SWCNTs, which have reported only indirect measurements such as Raman or transport to infer the actual presence of dopants in the nanotube walls. To ensure that correct conclusions on the effects of doping are made, it is vital to have clear and direct proof of the incorporation of heteroatoms into the structure, preferably with several complementary techniques.

Since publications on both N or B-doped MWCNT and graphene are so numerous (Fig. 4.1), we will not attempt to discuss them here in detail. Instead, we will focus the following discussion on B- and N-doped single-walled nanotubes and phosphorus-doped nanocarbons. Recent reviews on nitrogen-doped CNTs [10] and graphene [14], and of doped graphene sheets and nanoribbons [13, 51], are available for the interested reader.

Nitrogen and boron doping of CNTs was proposed theoretically already in 1993 by Yi and Bernholc [8]. After pioneering work by Stephan et al. on BN codoping using arc discharge [52] (Table 4.1), first reports on the synthesis of N-MWCNTs are from 1997 by Yudasaka et al. [59] and Sen et al. [67] and from 1999 by Terrones et al. [68]. The first experimental work on N-SWCNT synthesis was by Glerup et al. in 2004 using arc discharge [57]. Successful synthesis was later achieved also by laser ablation [54]. In addition, there are already several reports on N-SWCNTs synthesized using variations of CVD methods [61, 69–80].

TABLE 4.1   First Reported Instances of Doped Nanocarbon Syntheses Found in the Literature

Method	Material	Precursor	Reference
Laser ablation	BCN-MWCNT	B	[52]
	B-SWCNT	B	[53]
	N-SWCNT	N_2	[54]
Arc discharge	BCN-MWCNT	BN	[55]
	B-MWCNT	BN	[56]
	B-SWCNT	B	[34]
	N-SWCNT	Melamine	[57]
	B-graphene	B_2H_6,B	[58]
	N-graphene	Pyridine, NH_3	[58]
CVD	N-MWCNT	Ni phthalocyanine	[59]
	B-SWCNT	Triisopropyl borate	[60]
	N-SWCNT	Acetonitrile	[61]
	B-graphene	Phenylboronic acid	[62]
	N-graphene	NH_3	[12]
Thermochemical	BCN-SWCNT	B_2O_3, N_2	[63]
	B-graphene	B_2O_3	[64]
Solvothermal	B-graphene	BBr_3	[65]
	N-graphene	Li_3N	[31]
	N-MWCNTs	Cyanuric chloride	[46]
Ion bombardment	N-MWCNT	N_2^+	[49]
	N-SWCNT	N_2^+	[66]

Laser ablation was the first technique to produce B-SWCNTs by the introduction of boron into graphite–Co–Ni targets [53, 81]. Gai et al. [53] noted the difficulty of incorporating B in the graphitic network when characterizing the samples by transmission electron microscopy (TEM) and electron energy loss spectroscopy (EELS), only finding SWCNTs in the products when the B content in the target material was less than 3.5 at.%. Using CVD, boron-doped double-walled CNTs were first synthesized on bimetallic Fe/Mo catalysts supported in MgO in 2007 by Panchakarla et al. [82] by decomposing CH_4 and Ar mixtures in presence of diborane. The first reported synthesis of B-SWCNTs involved the use of nondiluted precursors (triisopropyl borate) in a high vacuum CVD system [60]. In such method, high-quality SWCNTs could be formed in a wide temperature range, but it was again found that it is difficult to get B incorporated into the CNT network. More than 80 at.% of the B content in the samples was found to be boron carbide and boron oxide.

The production of phosphorus-doped SWCNTs has been successfully accomplished only recently. CVD material has been produced using triphenylphosphine as the P precursor [18, 83, 84]. Krstic et al. reported on the synthesis of P-SWCNTs using arc discharge, with red phosphorus mixed into the anode rod acting as P precursor [19]. On the other hand, the synthesis of P-doped graphene has also been successful to a certain extent. Its synthesis using ionic liquid 1-butyl-3-methlyimidazolium hexafluorophosphate [85] has been reported. In addition, P/N-heterodoped graphene using triphenylphosphine and triphenylamine [20] and by chemical treatment of N-graphene using phosphoric acid [86] is available.

4.4 CHARACTERIZATION OF HETERODOPED NANOTUBES AND GRAPHENE

The characterization of a doped nanotube sample depends closely on the aim of the study and application purpose of the material. If morphology and collective behavior—regardless of the presence of impurities—are sufficient, the use of cutting-edge techniques is important but not decisive. Optical absorption and Raman spectroscopy are particularly interesting to have an overall picture of the diameter of the tubes and the changes in the optical response upon doping (depending on the doping level). Among the most commonly used techniques for studying both pristine and doped carbon nanostructures, we will give particular attention to three techniques that are of interest to analyze the bonding and concentration of dopants. These techniques are X-ray photoelectron spectroscopy (XPS), scanning tunneling microscopy/spectroscopy (STM/STS), and TEM/EELS (Table 4.2). Raman spectroscopy has also been employed to indirectly probe signatures that have been correlated with doping, and measurable effects on optical absorption spectra have also been discussed. More details on these and other techniques such as other X-ray absorption, bulk-sensitive EELS, X-ray diffraction, and others can be found in a recent review [7].

XPS is a technique that probes the elemental composition of a material. XPS spectra are obtained by irradiating a material with a beam of energetic photons while measuring the kinetic energy and number of electrons that escape. The binding

TABLE 4.2   First Reported Instances in the Literature of Characterizing Doped Nanocarbons with the Most Important Analytical Techniques

Technique	Material	Reported Maximum Dopant Concentration (at.%)	Reference
TEM/EELS	BCN-MWCNT	10	[52]
	B-MWCNT	—	[55]
	N-MWCNT	2	[68]
	BCN-SWCNT	10 (B), 2 (N)	[63]
	B-SWCNT	3	[87]
	N-SWCNT	1	[57]
	B-graphene	13.85	[88]
	N-graphene	0.9	[58]
XPS	N-MWCNT	1	[59]
	B-SWCNT	1	[60]
	N-SWCNT	4	[69]
	B-graphene	3.2	[58]
	N-graphene	8.9	[12]
STM	BCN-MWCNT	—	[36]
	N-MWCNT	—	[34]
	B-SWCNT	—	[89]
	N-SWCNT	—	[54]
	B-graphene	—	[90]
	N-graphene	—	[31]

energy of a core electron depends not only upon the level from which photoemission is occurring but also upon the local chemical and physical environment. These cause small shifts in the peak positions in the XPS spectrum, which are interpreted in association to the type of system studied and its properties. The advantages of XPS are its superior energy resolution and long integration times, allowing the detection of small amounts of dopants. Because XPS is a more bulk-sensitive technique compared to local probes, it is very important to take into account the sample quality and preparation. Detecting doping levels is generally not compromised by the energy resolution of the spectrometer, unless signals from the different elements in a sample overlap as in the case shown in Figure 4.4a, where the N $1s$ overlaps with the crystal splitting of $Mo3p_{3/2}$ present from a remaining of catalytic material in the samples.

Because N-doped CNTs are the most abundantly synthesized materials among the ones described here, several XPS studies have been reported. It is generally well understood how to identify N bonding environments in such systems, but in many cases reported in the literature, the bonding configuration assignment has not been done taking into account the type of carbon system and its structure [92–94]. The N $1s$ core-level signal in XPS has allowed identifying several components while measuring nanotube samples. A peak around 398.5 eV is commonly attributed to pyridinic nitrogen, usually implicitly referring to the triple pyridinic configuration (Fig. 4.2). Because the binding energy of a substitutional atom must be closely related to the system in question, the binding energy values reported for a N substitution has can be found variously assigned from around 400.5 eV up to 401.5 eV,

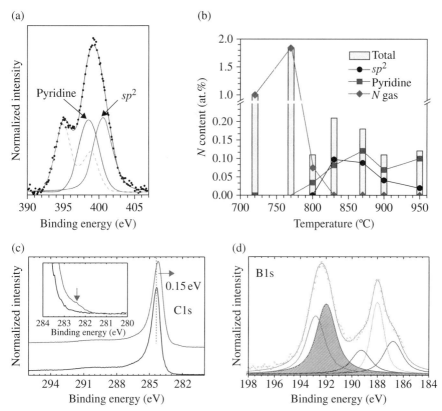

FIGURE 4.4 XPS spectra corresponding to N- and B-doped SWCNTs. The N $1s$ signal recorded from a sample of N-SWCNTs produced with benzylalmine is shown in (a). The line-shape analysis takes into account the presence of pyridinic and sp^2-like bonding environments. The dotted line represents the contribution to the spectral shape arising from the presence of catalytic material, which must be carefully considered in order to use XPS effectively for the characterization of doping levels and doping configurations. The nitrogen incorporation profile according to synthesis temperatures using the same methods is shown in (b). The sp^2, pyridine, and N volatile species that can be found in that specific kind of material are shown. The lower panels show the C $1s$ core signal corresponding to B-doped SWCNTs synthesized by arc discharge, compared to a scan corresponding to pristine material of similar characteristics. The incorporation of B in the lattice induces the formation of a shoulder at lower binding energy in the B-SWCNT sample. The spectrum shown in (d) corresponds to the B $1s$ signal recorded from a similar sample, where it is clear that in the best synthesis conditions, the stronger signals correspond to substitutional doping or elemental boron remaining from the synthesis process. (a, b) Adapted from Ref. 72, reproduced with permission. © 2007, American Chemical Society. (c, d) Adapted from Ref. 91, reproduced with permission. © 2010, AIP Publishing LLC.

depending on the system (i.e., graphene or nanotubes, with higher values typical for N-graphene). Signals observed near 400.0 eV have been attributed to pyrrolic nitrogen (N in a five-membered ring), which is thought to be responsible for the inner compartments typical for N-MWCNTs (Fig. 4.5). Obviously, there is no consensus about a specific assignment for the core-level binding energies of N atoms in

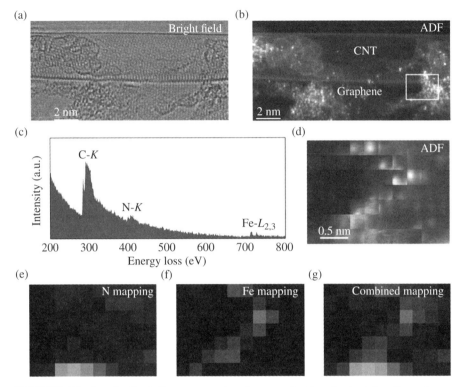

FIGURE 4.5 Longitudinal slices extracted at the same depth and orientation from shape-sensitive reconstructions of EELS energy-filtered images of an N-MWCNT, measured at the C–K edge for the carbon map and the N–K edge for the nitrogen map. The panels show the mean density (left), C and N 3D elemental maps (middle), and C-to-N 3D relative map (right). The relative map was obtained by superimposing the two elemental 3D maps with different colors, nitrogen in green and carbon in red. The presence of two types of arches (i.e., transverses and rounded ones), which is typical for the highly doped MWCNTs, can be observed. Reproduced with permission from Ref. 95. © 2012, American Chemical Society. (*See insert for color representation of the figure.*)

nanotubes, in particular. This is because each type of material is measured in specific conditions and surrounded by a molecular environment proper to the synthesis conditions. For this reason, only using ultrapure material and experiments done in controlled conditions can be used for establishing footprints in XPS [92] or even angle-resolved photoemission for the case of graphene [96]. The assignment of the N 1s components into specific atomic structures is thus still a contentious question, which can be solved for every specific system correlating XPS with atomically resolved techniques.

In boron-doped systems, the identification of the components of the B 1s response is relatively easier than in the case of nitrogen. This is because the possibilities of boron to form gaseous compounds that could attach to the outer walls of nanotubes or adsorb onto the surface of either graphene or nanotubes are rather limited. In SWCNTs, a signal around 192 eV has been assigned to substitutional boron (Fig. 4.4) [60, 82, 91]. Components corresponding to boric acid, boron oxide, or elemental boron (specially in materials synthesized using targets or rods containing elemental B) can be found. In boron-doped graphene, binding energy values from 188 to 191 eV have been reported for a possible B substitutional configuration [62, 88, 97, 98], but as reported for other atoms onto graphene, even the C $1s$ is very sensitive to the substrate on which the measurements are done. For this reason, determining a fix value for the binding energy for this configuration is not straightforward in the case of graphene.

As for phosphorus doping, very limited data is currently available. A P $2p$ signature around 130.5 eV has been attributed to substitutional P in doped and P/N-codoped MWCNTs [99, 100] and graphene [20]. An additional peak around 132 eV has commonly been attributed to P–O bonds; however, some authors have assigned this peak to the P substitution in purely P-doped graphene [85]. No XPS data has been reported for P-SWCNTs.

STM is based on the quantum tunneling of electrons between a probe tip and a sample. The tunneling current is a function of tip position, applied voltage, and the local density of states (LDOS) of the sample. In scanning tunneling spectroscopy (STS), the LDOS as a function of energy at a specific location in the sample can be obtained by recording the current as a function of bias voltage. STM and STS enable atomically resolved images of SWCNTs along with local spectroscopy [101, 102]. Visualizing the topography of a single dopant atom in the carbon network is thus possible by STM, the change in LDOS induced by the dopant can be investigated by STS, and both measurements compared to simulations [30].

In N-SWCNTs, several signatures attributed to N doping have been detected by atomic resolution STM/STS [34, 35, 103]. Unfortunately, such measurements are limited to very small areas, and the interpretation of the results remains complicated. Only very recently was a simple N substitution identified for the first time [35] (Fig. 4.6). A tentative conclusion from the researchers was that most dopant configurations found in N-SWCNTs do not correspond to the widely considered simple cases (i.e., the substitution or the simple pyridinic configurations), but rather are more complicated nitrogen-vacancy complexes [35, 105], as yet unidentified.

In graphene, STM measurements are simplified by the lack of curvature. This makes not only the interpretation of the images simpler, but likely contributes to the preference for the simple N substitution configuration [31, 32, 106–108] that has been observed for N-graphene (Fig. 4.6) in contrast to the nanotube case. However, it should be noted that pyridinic sites are also possible depending on the synthesis process [14, 104, 109]. Very recently, the first STM/STS study on the B substitution in graphene was published [90]. Interestingly, B dopants are reported to interact strongly with the underlying Cu substrate, unlike N substitutions. The authors also note that this result in a random distribution of B dopants between the graphene sublattices, whereas N dopants tend to favor the same sublattice. Some authors have noted the same sublattice preference in N-graphene [107, 108], whereas others have observed no clear effect [32, 104, 106, 109]. No STM measurements of P-doped nanocarbons have been reported.

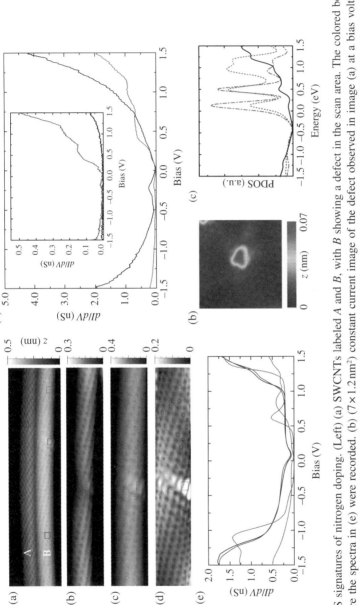

FIGURE 4.6 STM/STS signatures of nitrogen doping. (Left) (a) SWCNTs labeled A and B, with B showing a defect in the scan area. The colored boxes in image (a) represent the points where the spectra in (e) were recorded. (b) ($7 \times 1.2\,nm^2$) constant current image of the defect observed in image (a) at a bias voltage $V_s = -1.00\,V$, (c) ($7 \times 1.2\,nm^2$) constant current image of the defect observed in image (a) at $V_s = +1.00\,V$. (d) STM image shown in panel (c) corrected by a line-by-line flattening; the red dots indicate the periodicity of a superstructure "donut" pattern. (e) Tunneling spectra measured on the two tubes, with the colors of the lines corresponding to the positions of the colored boxes in (a). Note that the peak around $0.9\,eV$ in the red curve corresponds to the donor state measured at the defect, while the peak at $0.6\,eV$ in the blue curve corresponds to the first Van Hove singularity of the pristine tube A assigned with a (12,4) chirality. Reproduced with permission from Ref. 35. © 2013, American Chemical Society. (Right) (a) STM image of a single substitutional N dopant in graphene on copper foil. Inset: Line profile across the dopant shows atomic corrugation and the apparent height of the site. (b) STM image of N-graphene showing 14 substitutional dopants and extended electronic perturbations due to intervalley scattering. Inset: Fast Fourier transform of the topography shows both atomic peaks (outer hexagon) and intervalley scattering peaks (inner hexagon). (c) dI/dV curves taken on a N atom (bottom) and on the bright topographic features near the N atom, offset vertically for clarity. The top curve was taken approximately 2 nm away from the dopant. Inset: positions where

In a transmission electron microscope (TEM), a parallel beam of electrons is focused onto a sample. The electron beam travels through the specimen, some of the electrons are scattered, and inelastically scattered electrons experience energy loss. This energy loss can be measured in TEM/EELS. The transmitted portion is focused by an objective lens to project a magnified image onto a screen. In basic TEM, the whole image area is illuminated with electrons. Alternatively, in scanning TEM (STEM), a very narrow electron beam is scanned in a raster over the sample. By collecting the image from electrons scattered by the sample instead of those transmitted through it (annular dark-field (ADF) imaging), STEM is very sensitive to differences in the atomic number in the sample (so-called Z-contrast imaging) [110] or even local variations in the charge density [106].

STEM/EELS has been particularly useful for studying nanotubes [52, 57, 68, 95, 103, 111]. In particular, these measurements have probed the content and spatial distribution of nitrogen in the inner compartments characteristic of the so-called bamboo-like N-MWCNTs (see, e.g., Ref. 95 and Fig. 4.5 for a recent example). However, spatial and energy resolution limitations have not yet allowed an unambiguous identification of the dopant configurations [57, 103, 112], and the possibility of knock-on damage (i.e., the ejection of atoms from the material under study by inelastic collisions of the highly energetic probe electrons) has been an issue. More recently, cutting-edge developments in instrumentation have enabled atom-by-atom analysis of graphene [110, 113] and even direct imaging of nitrogen sites [32, 106, 114]. A recent study looking in detail at knock-on damage in N-doped nanocarbons found that acceleration voltages below 80 kV are needed to avoid electron beam damage of the studied structures [32]. Thus, aberration-corrected STEM/EELS machines operating at reduced voltages (Fig. 4.7) look promising for further progress in conclusively identifying the dopant sites also in doped SWCNTs in the near future.

Although extremely convenient and useful, Raman and optical absorption spectroscopies cannot directly demonstrate the presence of dopants in carbon nanomaterial systems. Therefore, while electronic structure changes caused by doping can lead to optically measurable signatures, it is very important to note that the interpretation of optical measurements should be made with great care, particularly for nanotubes. This is because the introduction of a dopant precursors almost invariably also causes changes in the diameter distribution of the nanotubes in the sample, which alters optical resonance conditions and causes peak shifts regardless of doping.

Gerber et al. [116] predicted theoretically that nitrogen substitution at a 1.6% level (1 N in a 64 atom unit cell) would increase the E_{11} transition energy by 0.05 eV while decreasing the E_{22} energy by 0.09 eV. They also found that doping would induce a downshift of the Raman radial breathing mode frequency in the order of 5 cm^{-1} at a 3.1% doping level. However, the nanotube they considered had a very small diameter (<0.7 nm), and both these effects would be much weaker in typical larger-diameter tubes. An increase in the Raman D to G intensity ratio when SWCNTs are doped with N has been often reported [61, 69, 71, 75, 117, 118], which can be attributed to increased disorder in the walls caused by the nitrogen dopants. A decrease in the diameters of the N-SWCNTs produced with some methods while increasing the amount of N precursor has also often [32, 61, 69, 80] been observed,

Mean density   Carbon map   Nitrogen map   3D relative map

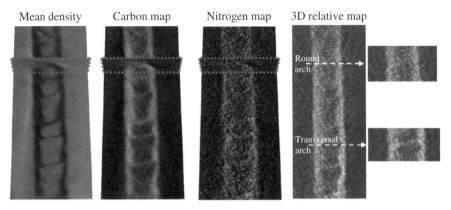

FIGURE 4.7 Scanning transmission electron microscopy (STEM) imaging and electron energy loss spectroscopy (EELS) mapping of iron and nitrogen atoms in a doped carbon nanotube–graphene hybrid structure. (a) Bright-field and (b) annular dark-field (ADF) STEM images of a carbon nanotube partially covered with nanosized graphene pieces. The area marked by the white square in (b) was characterized by (c) EELS, (d) ADF intensity mapping, (e) nitrogen EELS mapping, (f) iron EELS mapping, and (g) an overlaid iron and nitrogen EELS map. In this system, iron atoms were frequently observed on the edges of graphene sheets close to nitrogen. Adapted from Ref. 115, reproduced with permission. © 2012, rights managed by Nature Publishing Group.

but this is not an established rule [78]. A softening (downshift) of the G band phonon has been reported for N-doped nanotubes with more than one wall [119, 120], but not for N-SWCNTs. As for the G′ (2D) band, Maciel et al. [83] proposed that it is sensitive to the presence of charged dopants, leading to a hardening of the K-point phonon and a change in the 2D line shape and position (Fig. 4.8). They find a splitting of the 2D peak according to the nanotubes charge state. These are very useful tools, but unless changes in the diameter distribution of the samples are very carefully taken into account, Raman shifts alone should not be used to infer the presence of dopants.

4.5 POTENTIAL APPLICATIONS

Understanding and controlling the properties of doped nanotubes is not only of fundamental physical interest, but required for their use in applications. Multiwalled species have been investigated much more exhaustively, mainly in the fields of composite materials, energy, and biocompatibility tests of carbon materials [10, 122]. Until recently, most reported application studies utilizing doped SWCNTs were constrained to theoretical predictions, because of the scarce availability of clean samples of doped nanotubes or nanotube material where dopants had been clearly identified.

Regarding quantum transport, in SWCNTs conduction occurs via the delocalized electron system. At room temperature, both metallic and semiconducting SWCNTs

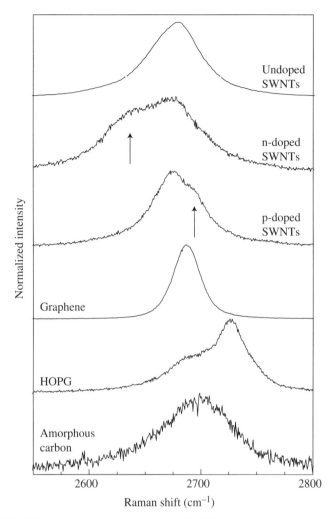

FIGURE 4.8 Shifts and splitting in the Raman G' (2D) response have been proposed as a convenient signature of doping. Reproduced with permission from Ref. 121. © 2008, rights managed by Nature Publishing Group.

are ballistic conductors along the tube axis. To achieve significant charge transfer in a bundled nanotube system or a network, one requirement would be that the charge transfer is of a uniform electrical nature. This has proven challenging particularly for N-CNTs. Even if only one type of doping was achieved, a decrease in conductance could still result from increased electron backscattering from the dopant states. A single dopant site in an SWCNT mainly induces a dip in the conductance at the resonant energy [28] (Fig. 4.9a). For a realistic random distribution of such sites, however, the effect on transport is severe. It has been theoretically proposed that the mean free path decreases linearly with substitutional dopant concentration

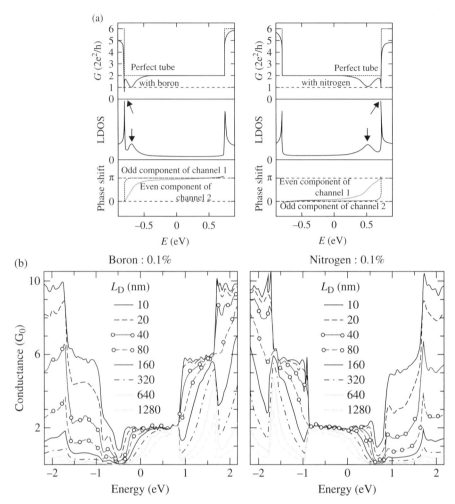

FIGURE 4.9 The effect of B or N doping on the transport properties of SWCNTs. (a) Conductance of a (10,10) SWCNT doped with a boron (left) or a nitrogen (right) substitution, as calculated using an ab initio pseudopontential method. Boron or nitrogen dopants produce quasibound impurity states of a definite parity, reducing the tube conductance by one quantum ($2e^2h$) via resonant backscattering. Adapted from Ref. 28, reproduced with permission. © 2000 by The American Physical Society. (b) Quantum conductance of a single (10,10) nanotube containing 0.1% of randomly positioned boron (left) or nitrogen (right) impurities, calculated using a tight-binding model. The conductance is plotted as a function of energy for different device lengths, and a decrease is observed near the energies of dopant donor/acceptor states and near the Van Hove singularities. Adapted from Ref. 123, reproduced with permission. © 2004 by The American Physical Society.

at low (0.5%) doping levels (Fig. 4.9b) [123]. Only few studies have tried to experimentally address the issue of the conductivity of N-SWCNTs. Villalpando-Paez [70] showed that N-SWCNTs exhibited improved conductivity at reduced temperatures with small amounts of N in the precursors; however, no direct measurements were shown to indicate the level of wall doping. Similarly, Krstic et al. reported n-type conduction in individual metallic N-SWCNTs [124] without directly characterizing the doping. More recently, it was shown that at a fixed temperature, the conductivity of N-SWCNTs films with both substitutional doping and pyridinic nitrogen doping is decreased [75, 112], even when the influence of shorter lengths—and the correspondingly greater number of resistive tube–tube contacts—of the doped tubes is taken into account [112] (Fig. 4.10). On the contrary, for boron, Liu et al. [87] have reported that the sheet resistances of B-SWCNT thin films at fixed

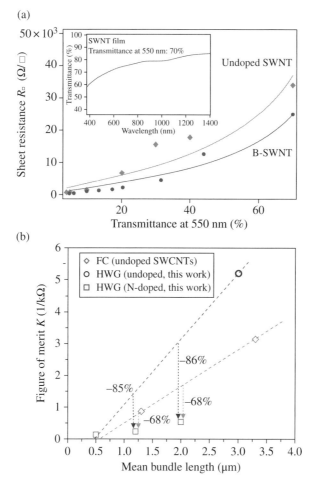

FIGURE 4.10 The effect of B or N doping on the resistances of SWCNT films. (a) The sheet resistance and transmittance measured at 550 nm of undoped and B-doped SWCNT films, showing reduced resistance of the films upon doping. Adapted from Ref. 87, reproduced with permission. © 2008, American Chemical Society. (b) The optoelectronic figures of merit (related to the inverse of sheet resistance) used to compare films with different transmittance for undoped and N-doped SWCNT films, showing a significant increase in resistance (68–86% decrease in the figure of merit) upon doping, even when the effect of decreased bundle lengths is taken into account. Adapted from Ref. 112, reproduced with permission. © 2011, American Chemical Society.

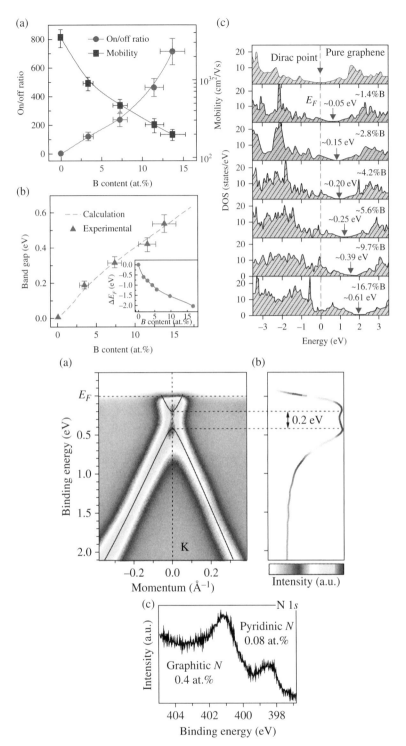

FIGURE 4.11 The effect of B or N doping on the electronic structure of graphene. (Top) (a) On/off current and carrier mobility of B-graphene synthesized with various doping levels using a reactive trimethylboron microwave plasma treatment, as determined from measurements of 20 field-effect transistor (FET) devices. (b) Experimental band gaps and corresponding results from density functional theory calculation. Inset: The dependence of doping level (the

transmittance are decreased by 30% compared to pristine SWCNT films from the same batch (Fig. 4.10). In N-MWCNT films, Wiggins-Camacho et al. observed a slight enhancement of conductivity at moderate nitrogen doping levels [125], which they attributed to an increased density of states at the Fermi level and an increased number of charge carriers. Nevertheless, it is important to keep in mind that most of such experimental studies have not been done on purified doped nanotubes, which still limits the possibility to provide conclusive answers regarding these arguments.

In N- or B-doped graphene, the opening of a tunable band gap has been recently reported [126, 127] (Fig. 4.11). As for the effect of phosphorus incorporation on transport, it has only been calculated that a substitutional P atom induces a 50% reduction in conduction at the Fermi level for metallic tubes [18]. Limited transport data has reported by Krstic et al. [19], who found p-type doping, which they attributed to the oxidation of the substitutional P sites. However, direct evidence for P incorporation into the lattice was not reported. Since pristine graphene is a zero-band gap semiconductor with a very high electron mobility, there has been a lot of interest in engineering a gap using various approaches. Using heteroatom doping, a band gap opening and significant shifting of the Fermi level have been recently observed both with boron [88] and nitrogen [126] doping, with potential implications in using doped graphenes in field-effect transistor devices (Fig. 4.11).

Another motivation for doping is the introduction of reactive sites on the otherwise inert walls of CNTs. This is expected to enhance their bonding in composites [128, 129], enhance field emission [56, 130–132], increase hydrogen or lithium storage [133, 134], and increase their reactivity with adsorbed molecules for, for example, molecular [135, 136] or biosensing [137, 138] applications. In a very intensively studied recent development [139], N-MWCNTs have also shown promise as metal-free electrocatalysts. Particularly, the topic of using N-MWCNTs for the oxygen reduction reaction has garnered a great number of works in the last few years. Recent reviews cover this topic in detail [140, 141]. Another recent review also covers the potential of graphene doped with heteroatoms other than N for energy applications [142]. Additionally, CNTs rank among potential candidates for a new family of nanoscale devices for molecular sensing applications. The sensing mechanism in the best available sensors in the market today is based on the binding of the molecule one wishes to detect to the surface of the sensor, which causes a change in the electric current through the device. Since the surface to volume ratio of SWCNTs is significantly higher than traditional semiconductor devices and their electronic transport properties are exceptional, small changes to

---

FIGURE 4.11 (*Continued*) position of the Fermi level from the Dirac point) on substitutional B content. (c) Calculated DOS of graphenes doped with different B content, with the Fermi level labeled by the dashed line. From Ref. 88, reproduced with permission. © 2012, American Chemical Society. (Bottom) (a) Angle-resolved photoemission spectrum of N-graphene synthesized by CVD of molecular precursors (s-triazine) on a tungsten-supported 10 nm Ni film and annealed in high temperature after gold intercalation. The spectrum was measured at a photon energy of 35 eV, through the K-point and a few degrees off the direction perpendicular to ΓK direction in reciprocal space. (b) Photoemission spectrum at the K-point. (c) N 1s XPS spectrum of the sample, showing the predominance of graphitic nitrogen. From Ref. 126, reproduced with permission. © 2011, American Chemical Society.

the surface of the system can cause measurable changes in the current, possibly reaching single-molecule detection. Based on numerical simulations, it is expected that nitrogen dopant sites would enhance selectivity, sensitivity, or both for various gases [143, 144]. For the case of phosphorus, preliminary theoretical work suggests slightly different but quite strong selectivity for several typical analyte gases [136].

4.6 SUMMARY AND OUTLOOK

This short chapter has provided an introduction to structural properties of CNTs doped with substitutional heteroatoms. An overview of the field of hybrid nanotubes containing B, C, and N atoms has also been given. Great progress has been made in the recent years particularly in the fields of graphene and N-MWCNTs. Further developments in the next few years should provide conclusive answers to many open questions, especially regarding the atomic configuration of the heteroatoms and the influence of their incorporation in the physical properties of nanotubes and graphene. Different and novel materials are expected to find potential applications particularly suited to their specific properties. In this respect, electrocatalysis for N-MWCNTs and sensing using doped single-walled species represent very promising fields. Whether fine-tuned control over the electrical properties of semiconducting nanotubes for transistor applications can be realized is still uncertain. On the other hand, doped graphenes are receiving a lot of attention in this direction.

REFERENCES

[1] Charlier, J.-C.; Blase, X.; Roche, S. Electronic and transport properties of nanotubes. *Rev Mod Phys* 2007, **79**, 677.

[2] Saito, R.; Dresselhaus, G.; Dresselhaus, M. *Physical Properties of Carbon Nanotubes*; Imperial College Press, London, 1998.

[3] Arnold, M.; Green, A.; Hulvat, J.; Stupp, S.; Hersam, M. Sorting carbon nanotubes by electronic structure using density differentiation. *Nat Nanotechnol* 2006, **1**, 60.

[4] Miyata, Y.; Yanagi, K.; Maniwa, Y.; Tanaka, T.; Kataura, H. Diameter analysis of rebundled single-wall carbon nanotubes using X-ray diffraction: verification of chirality assignment based on optical spectra. *J Phys Chem C* 2008, **112**, 15997–16001.

[5] Krupke, R.; Hennrich, F.; Lohneysen, H.; Kappes, M. Separation of metallic from semiconducting single-walled carbon nanotubes. *Int S Techn Pol Inn* 2003, **301**, 344.

[6] Pichler, T.; Kukovecz, A.; Kuzmany, H.; Kataura, H. Charge transfer in doped single wall carbon nanotubes. *Synthetic Met* 2003, **135**, 717–719.

[7] Ayala, P.; Arenal, R.; Loiseau, A.; Rubio, A.; Pichler, T. The physical and chemical properties of heteronanotubes. *Rev Mod Phys* 2010, **82**, 1843–1885.

[8] Yi, J.-Y.; Bernholc, J. Atomic structure and doping of microtubules. *Phys Rev B* 1993, **47**, 1708–1711.

[9] Ewels, C.; Glerup, M. Nitrogen doping in carbon nanotubes. *J Nanosci Nanotechnol* 2005, **5**, 1345–1363.

[10] Ayala, P.; Arenal, R.; Rmmeli, M. H.; Rubio, A.; Pichler, T. The doping of carbon nanotubes with nitrogen and their potential applications. *Carbon* 2010, **48**, 585–586.

[11] Terrones, M.; Ajayan, P.; Blase, X.; Blau, W.; Carroll, D.; Charlier, J.; Czerw, R.; Foley, B.; Grobert, N.; Kamalakaran, R.; Kohler-Redlich, P.; Ruhle, M.; Seeger, T.; Terrones, H. Advances on the growth and properties of N- and B-doped carbon nanotubes. *Aip Conf Proc* 2001, **591**, 212–216.

[12] Wei, D.; Liu, Y.; Wang, Y.; Zhang, H.; Huang, L.; Yu, G. Synthesis of N-doped graphene by chemical vapor deposition and its electrical properties. *Nano Lett* 2009, **9**, 1752–1758.

[13] Terrones, H.; Lv, R.; Terrones, M.; Dresselhaus, M. S. The role of defects and doping in 2D graphene sheets and 1D nanoribbons. *Rep Prog Phys* 2012, **75**, 062501.

[14] Wang, H.; Maiyalagan, T.; Wang, X. Review on recent progress in nitrogen-doped graphene: synthesis, characterization, and its potential applications. *ACS Catal* 2012, **2**, 781–794.

[15] Strelko, V.; Kuts, V.; Thrower, P. On the mechanism of possible influence of heteroatoms of nitrogen, boron and phosphorus in a carbon matrix on the catalytic activity of carbons in electron transfer reactions. *Carbon* 2000, **38**, 1499–1503.

[16] Denis, P. A. Bandgap opening of monolayer and bilayer graphene doped with aluminium, silicon, phosphorus, and sulfur. *Chem Phys Lett* 2010, **492**, 251–257.

[17] Cruz-Silva, E.; Cullen, D. A.; Gu, L.; Romo-Herrera, J. M.; Munoz-Sandoval, E.; Lopez-Urias, F.; Sumpter, B. G.; Meunier, V.; Charlier, J.-C.; Smith, D. J.; Terrones, H.; Terrones, M. Heterodoped nanotubes: theory, synthesis, and characterization of phosphorus-nitrogen doped multiwalled carbon nanotubes. *ACS Nano* 2008, **2**, 441–448.

[18] Cruz-Silva, E.; Lopez-Urias, F.; Munoz-Sandoval, E.; Sumpter, B. G.; Terrones, H.; Charlier, J.-C.; Meunier, V.; Terrones, M. Electronic transport and mechanical properties of phosphorus and phosphorus-nitrogen-doped carbon nanotubes. *ACS Nano* 2009, **3**, 1913–1921.

[19] Krsti, V.; Ewels, C. P.; Wgberg, T.; Ferreira, M. S.; Janssens, A. M.; Stephan, O.; Glerup, M. Indirect magnetic coupling in light-element-doped single-walled carbon nanotubes. *ACS Nano* 2010, **4**, 5081–5086.

[20] Some, S.; Kim, J.; Lee, K.; Kulkarni, A.; Yoon, Y.; Lee, S.; Kim, T.; Lee, H. Highly air-stable phosphorus-doped n-type graphene field-effect transistors. *Adv Mater* 2012, **24**, 5481–5486.

[21] Sumpter, B. G.; Huang, J.; Meunier, V.; Romo-Herrera, J. M.; Cruz-Silva, E.; Terrones, H.; Terrones, M. A theoretical and experimental study on manipulating the structure and properties of carbon nanotubes using substitutional dopants. *Int J Quantum Chem* 2009, **109**, 97–118.

[22] Jiang, H.; Zhang, D.; Wang, R. Silicon-doped carbon nanotubes: a potential resource for the detection of chlorophenols/chlorophenoxy radicals. *Nanotechnology* 2009, **20**, 145501.

[23] Yang, S.; Zhi, L.; Tang, K.; Feng, X.; Maier, J.; Müllen, K. Efficient synthesis of heteroatom (N or S)-doped graphene based on ultrathin graphene oxide-porous silica sheets for oxygen reduction reactions. *Adv Funct Mater* 2012, **22**, 3634–3640.

[24] Audiffred, M.; Elías, A. L.; Gutiérrez, H. R.; López-Urías, F.; Terrones, H.; Merino, G.; Terrones, M. Nitrogen–silicon heterodoping of carbon nanotubes. *J Phys Chem C* 2013, **117**, 8481–8490.

[25]   Zhou, W.; Kapetanakis, M.; Prange, M.; Pantelides, S.; Pennycook, S.; Idrobo, J.-C. Direct determination of the chemical bonding of individual impurities in graphene. *Phys Rev Lett* 2012, **109**, 206803.

[26]   Golberg, D.; Bando, Y.; Huang, Y.; Terao, T.; Mitome, M. Boron nitride nanotubes and nanosheets. *ACS Nano* 2010, **4**, 2979–2993.

[27]   Arenal, R.; Blase, X.; Loiseau, A. Boron-nitride and boron-carbonitride nanotubes: synthesis, characterization and theory. *Adv Physiol* 2010, **59**, 101–179.

[28]   Choi, H.; Ihm, J.; Louie, S.; Cohen, M. Defects, quasibound states, and quantum conductance in metallic carbon nanotubes. *Phys Rev Lett* 2000, **84**, 2917–2920.

[29]   Nevidomskyy, A.; Csanyi, G.; Payne, M. Chemically active substitutional nitrogen impurity in carbon nanotubes. *Phys Rev Lett* 2003, **91**, 105502.

[30]   Zheng, B.; Hermet, P.; Henrard, L. Scanning tunneling microscopy simulations of nitrogen-and boron-doped graphene and single-walled carbon nanotubes. *ACS Nano* 2010, **4**, 4165.

[31]   Deng, D.; Pan, X.; Yu, L.; Cui, Y.; Jiang, Y.; Qi, J.; Li, W.-X.; Fu, Q.; Ma, X.; Xue, Q.; Sun, G.; Bao, X. Toward N-doped graphene via solvothermal synthesis. *Chem Mater* 2011, **23**, 1188–1193.

[32]   Susi, T.; Kotakoski, J.; Arenal, R.; Kurasch, S.; Jiang, H.; Skakalova, V.; Stephan, O.; Krasheninnikov, A. V.; Kauppinen, E. I.; Kaiser, U.; Meyer, J. C. Atomistic description of electron beam damage in nitrogen-doped graphene and single-walled carbon nanotubes. *ACS Nano* 2012, **6**, 8837–8846.

[33]   Fujimoto, Y.; Saito, S. Formation, stabilities, and electronic properties of nitrogen defects in graphene. *Phys Rev B* 2011, **84**, 245446.

[34]   Czerw, R.; Terrones, M.; Charlier, J.; Blase, X.; Foley, B.; Kamalakaran, R.; Grobert, N.; Terrones, H.; Tekleab, D.; Ajayan, P.; Blau, W.; Ruhle, M.; Carroll, D. Identification of electron donor states in N-doped carbon nanotubes. *Nano Lett* 2001, **1**, 457–460.

[35]   Tison, Y.; Lin, H.; Lagoute, J.; Repain, V.; Chacon, C.; Girard, Y.; Rousset, S.; Henrard, L.; Zheng, B.; Susi, T.; Kauppinen, E. I.; Ducastelle, F.; Loiseau, A. Identification of nitrogen dopants in single-walled carbon nanotubes by scanning tunneling microscopy. *ACS Nano* 2013, **7**, 7219–7226.

[36]   Carroll, D.; Redlich, P.; Blase, X.; Charlier, J.; Curran, S.; Ajayan, P.; Roth, S.; Ruhle, M. Effects of nanodomain formation on the electronic structure of doped carbon nanotubes. *Phys Rev Lett* 1998, **81**, 2332–2335.

[37]   Quandt, A.; Özdogbreve, C.; Kunstmann, J.; Fehske, H. Boron doped graphene nanostructures. *Phys Status Solidi(B)* 2008, **245**, 2077–2081.

[38]   Wirtz, L.; Rubio, A. Band structure of boron doped carbon nanotubes. *Aip Conf Proc* 2003, **685**, 402.

[39]   Garcia, A. G.; Baltazar, S. E.; Castro, A. H. R.; Robles, J. F. P.; Rubio, A. Influence of S and P doping in a graphene sheet. *J Comput Theor Nanosci* 2008, **5**, 2221–2229.

[40]   Mazzoni, M. S. C.; Nunes, R. W.; Azevedo, S.; Chacham, H. Electronic structure and energetics of $B_xC_yN_z$ layered structures. *Phys Rev B* 2006, **73**, 073108.

[41]   Fuentes, G.; Borowiak-Palen, E.; Knupfer, M.; Pichler, T.; Fink, J.; Wirtz, L.; Rubio, A. Formation and electronic properties of BC_3 single-wall nanotubes upon boron substitution of carbon nanotubes. *Phys Rev B* 2004, **69**, 245403.

[42]   Miyamoto, Y.; Rubio, A.; Cohen, M.; Louie, S. Chiral tubules of hexagonal BC_2N. *Phys Rev B* 1994, **50**, 4976–4979.

[43] Liu, A. Y.; Wentzcovitch, R. M.; Cohen, M. L. Atomic arrangement and electronic structure of BC_2N. *Phys Rev B* 1989, **39**, 1760–1765.

[44] Teter, D. M.; Hemley, R. J. Low-compressibility carbon nitrides. *Int S Techn Pol Inn* 1996, **271**, 53–55.

[45] Komatsu, T. Attempted chemical synthesis of graphitic-like carbon nitride. *J Mater Chem* 2000, **11**, 799.

[46] Cao, C.; Huang, F.; Cao, C.; Li, J.; Zhu, H. Synthesis of carbon nitride nanotubes via a catalytic-assembly solvothermal route. *Chem Mater* 2004, **16**, 5213–5215.

[47] Borowiak-Palen, E.; Pichler, T.; Fuentes, G.; Graff, A.; Kalenczuk, R.; Knupfer, M.; Fink, J. Efficient production of B-substituted single-wall carbon nanotubes. *Chem Phys Lett* 2003, **378**, 516–520.

[48] Golberg, D.; Bando, Y.; Kurashima, K.; Sato, T. MoO_3-promoted synthesis of multi-walled BN nanotubes from C nanotube templates. *Chem Phys Lett* 2000, **323**, 185–191.

[49] Morant, C.; Andrey, J.; Prieto, P.; Mendiola, D.; Sanz, J.; Elizalde, E. XPS characterization of nitrogen-doped carbon nanotubes. *Phys Status Solidi A* 2006, **203**, 1069–1075.

[50] Lin, C. H.; Chang, H. L.; Hsu, C. M.; Lo, A. Y.; Kuo, C. T. The role of nitrogen in carbon nanotube formation. *Diam Relat Mater* 2003, **12**, 1851–1857.

[51] Lv, R.; Li, Q.; Botello-Méndez, A. R.; Hayashi, T.; Wang, B.; Berkdemir, A.; Hao, Q.; Elas, A. L.; Cruz-Silva, R.; Gutirrez, H. R.; Kim, Y. A.; Muramatsu, H.; Zhu, J.; Endo, M.; Terrones, H.; Charlier, J. C.; Pan, M.; Terrones, M. Nitrogen-doped graphene: beyond single substitution and enhanced molecular sensing. *Sci Rep* 2012, **2**, 586.

[52] Stephan, O.; Ajayan, P. Doping graphitic and carbon nanotube structures with boron and nitrogen. *Int S Techn Pol Inn* 1994, **266**, 1863–1865.

[53] Gai, P.; Stephan, O.; McGuire, K.; Rao, A.; Dresselhaus, M.; Dresselhaus, G.; Colliex, C. Structural systematics in boron-doped single wall carbon nanotubes. *J Mater Chem* 2004, **14**, 669–675.

[54] Lin, H.; Lagoute, J.; Chacon, C.; Arenal, R.; Stephan, O.; Repain, V.; Girard, Y.; Enouz, S.; Bresson, L.; Rousset, S.; Loiseau, A. Combined STM/STS, TEM/EELS investigation of CNx-SWNTs. *Phys Status Solidi B* 2008, **245**, 1986–1989.

[55] Terrones, M.; Hsu, W. K.; Ramos, S.; Castillo, R.; Terrenes, H. The role of boron nitride in graphite plasma arcs. *Fullerene Sci Technol* 1998, **6**, 787–800.

[56] Charlier, J.; Terrones, M.; Baxendale, M.; Meunier, V.; Zacharia, T.; Rupesinghe, N.; Hsu, W.; Grobert, N.; Terrones, H.; Amaratunga, G. Enhanced electron field emission in B-doped carbon nanotubes. *Nano Lett* 2002, **2**, 1191–1195.

[57] Glerup, M.; Steinmetz, J.; Samaille, D.; Stephan, O.; Enouz, S.; Loiseau, A.; Roth, S.; Bernier, P. Synthesis of N-doped SWNT using the arc-discharge procedure. *Chem Phys Lett* 2004, **387**, 193–197.

[58] Panchakarla, L.; Subrahmanyam, K.; Saha, S.; Govindaraj, A.; Krishnamurthy, H.; Waghmare, U.; Rao, C. Synthesis, structure, and properties of boron- and nitrogen-doped graphene. *Adv Mater* 2009, **21**, 4726–4730.

[59] Yudasaka, M.; Kikuchi, R.; Ohki, Y.; Yoshimura, S. Nitrogen-containing carbon nanotube growth from Ni phthalocyanine by chemical vapor deposition. *Carbon* 1997, **35**, 195–201.

[60] Ayala, P.; Plank, W.; Gruneis, A.; Kauppinen, E. I.; Rummeli, M. H.; Kuzmany, H.; Pichler, T. A one step approach to B-doped single-walled carbon nanotubes. *J Mater Chem* 2008, **18**, 5676–5681.

[61] Keskar, G.; Rao, R.; Luo, J.; Hudson, J.; Chen, J.; Rao, A. Growth, nitrogen dping and characterization of isolated single-wall carbon nanotubes using liquid precursors. *Chem Phys Lett* 2005, **412**, 269.

[62] Wang, H.; Zhou, Y.; Wu, D.; Liao, L.; Zhao, S.; Peng, H.; Liu, Z. Synthesis of boron-doped graphene monolayers using the sole solid feedstock by chemical vapor deposition. *Small* 2013, **9**, 1316–1320.

[63] Golberg, D.; Bando, Y.; Burgeois, L.; Kurashima, K.; Sato, T. Large-scale synthesis and HRTEM analysis of single-walled B- and N-doped carbon nanotube bundles. *Carbon* 2000, **38**, 2017–2027.

[64] Sheng, Z.-H.; Shao, L.; Chen, J.-J.; Bao, W.-J.; Wang, F.-B.; Xia, X.-H. Catalyst-free synthesis of nitrogen-doped graphene via thermal annealing graphite oxide with melamine and its excellent electrocatalysis. *ACS Nano* 2011, **5**, 4350–4358.

[65] Lin, H.; Lagoute, J.; Repain, V.; Chacon, C.; Girard, Y.; Lauret, J. S.; Ducastelle, F.; Loiseau, A.; Rousset, S. Many-body effects in electronic bandgaps of carbon nanotubes measured by scanning tunnelling spectroscopy. *Nat Mater* 2010, **9**, 235–238.

[66] Xu, F.; Minniti, M.; Giallombardo, C.; Cupolillo, A.; Barone, P.; Oliva, A.; Papagno, L. Nitrogen ion implantation in single wall carbon nanotubes. *Surf Sci* 2007, **601**, 2819–2822.

[67] Sen, R.; Satishkumar, C. B.; Govindaraj, A.; Harikumar, K. R.; Renganathan, M. K.; Rao, C. N. R. Nitrogen-containing carbon nanotubes. *J Mater Chem* 1997, **7**, 2335–2337.

[68] Terrones, M.; Terrones, H.; Grobert, N.; Hsu, W. K.; Zhu, Y. Q.; Hare, J. P.; Kroto, H. W.; Walton, D. R. M.; Kohler-Redlich, P.; Rühle, M.; Zhang, J. P.; Cheetham, A. K. Efficient route to large arrays of CN_x nanofibers by pyrolysis of ferrocene/melamine mixtures. *Appl Phys Lett* 1999, **75**, 3932–3934.

[69] Min, Y.-S.; Bae, E. J.; Asanov, I. P.; Kim, U. J.; Park, W. Growth and characterization of nitrogen-doped single-walled carbon nanotubes by water-plasma chemical vapour deposition. *Nanotechnology* 2007, **18**, 285601.

[70] Villalpando-Paez, F.; Zamudio, A.; Elias, A. L.; Son, H.; Barros, E. B.; Chou, S. G.; Kim, Y. A.; Muramatsu, H.; Hayashi, T.; Kong, J.; Terrones, H.; Dresselhaus, G.; Endo, M.; Terrones, M.; Dresselhaus, M. Synthesis and characterization of long strands of nitrogen-doped single-walled carbon nanotubes. *Chem Phys Lett* 2006, **424**, 345–352.

[71] Ayala, P.; Grueneis, A.; Gemming, T.; Buechner, B.; Rümmeli, M. H.; Grimm, D.; Schumann, J.; Kaltofen, R.; Freire, F., Jr.; Fonseca-Filho, H.; Pichler, T. Influence of the catalyst hydrogen pretreatment on the growth of vertically aligned nitrogen-doped carbon nanotubes. *Chem Mater* 2007, **19**, 6131–6137.

[72] Ayala, P.; Grüneis, A.; Gemming, T.; Grimm, D.; Kramberger, C.; Rümmeli, M.; Freire, F., Jr.; Kuzmany, H.; Pfeiffer, R.; Barreiro, A.; Büchner, B.; Pichler, T. Tailoring N-doped single and double wall carbon nanotubes from a non-diluted carbon/nitrogen feedstock. *J Phys Chem C* 2007, **111**, 2879–2884.

[73] Ayala, P.; Freire, F. L., Jr.; Ruemmeli, M. H.; Grueneis, A.; Pichler, T. Chemical vapor deposition of functionalized single-walled carbon nanotubes with defined nitrogen doping. *Phys Status Solidi B* 2007, **244**, 4051–4055.

[74] Elias, A. L.; Ayala, P.; Zamudio, A.; Grobosch, M.; Cruz-Silva, E.; Romo-Herrera, J. M.; Campos-Delgado, J.; Terrones, H.; Pichler, T.; Terrones, M. Spectroscopic characterization of N-doped single-walled carbon nanotube strands: an X-ray photoelectron spectroscopy and Raman study. *J Nanosci Nanotechnol* 2010, **10**, 3959–3964.

[75] Ibrahim, E.; Khavrus, V.; Leonhardt, A. Synthesis, characterization, and electrical properties of nitrogen-doped single-walled carbon nanotubes with different nitrogen content. *Diam Relat Mater* 2010, **19**, 1199–1206.

[76] Liu, Y.; Jin, Z.; Wang, J.; Cui, R.; Sun, H.; Peng, F.; Wei, L.; Wang, Z.; Liang, X.; Peng, L. Nitrogen-doped single-walled carbon nanotubes grown on substrates: evidence for framework doping and their enhanced properties. *Adv Funct Mater* 2011, **21**, 986–992.

[77] Susi, T.; Nasibulin, A. G.; Ayala, P.; Tian, Y.; Zhu, Z.; Jiang, H.; Roquelet, C.; Garrot, D.; Lauret, J. S.; Kauppinen, E. I. High quality SWCNT synthesis in the presence of NH_3 using a vertical flow aerosol reactor. *Phys Status Solidi B* 2009, **246**, 2507–2510.

[78] Pint, C. L.; Sun, Z.; Moghazy, S.; Xu, Y.-Q.; Tour, J. M.; Hauge, R. H. Supergrowth of nitrogen-doped single-walled carbon nanotube arrays: active species, dopant characterization, and doped/undoped heterojunctions. *ACS Nano* 2011, **5**, 6925–6934.

[79] Koos, A. A.; Dillon, F.; Nicholls, R. J.; Bulusheva, L.; Grobert, N. N-SWCNTs production by aerosol-assisted CVD method. *Chem Phys Lett* 2012, **538**, 108–111.

[80] Thurakitseree, T.; Kramberger, C.; Zhao, P.; Aikawa, S.; Harish, S.; Chiashi, S.; Einarsson, E.; Maruyama, S. Diameter-controlled and nitrogen-doped vertically aligned single-walled carbon nanotubes. *Carbon* 2012, **50**, 2635–2640.

[81] McGuire, K.; Gothard, N.; Gai, P.; Dresselhaus, M.; Sumanasekera, G.; Rao, A. Synthesis and Raman characterization of boron-doped single-walled carbon nanotubes. *Carbon* 2005, **43**, 219.

[82] Panchakarla, L. S.; Govindaraj, A.; Rao, C. N. R. Nitrogen- and boron-doped double-walled carbon nanotubes. *ACS Nano* 2007, **1**, 494–500.

[83] Maciel, I.; Campos-Delgado, J.; Cruz-Silva, E.; Pimenta, M. A.; Sumpter, B. G.; Meunier, V.; Lopez-Urias, F.; Munoz-Sandoval, E.; Terrones, H.; Terrones, M.; Jorio, A. Synthesis, electronic structure, and Raman scattering of phosphorus-doped single-wall carbon nanotubes. *Nano Lett* 2009, **9**, 2267–2272.

[84] Campos-Delgado, J.; Maciel, I. O.; Cullen, D. A.; Smith, D. J.; Jorio, A.; Pimenta, M. A.; Terrones, H.; Terrones, M. Chemical vapor deposition synthesis of N-, P-, and Si-doped single-walled carbon nanotubes. *ACS Nano* 2010, **4**, 1696–1702.

[85] Li, R.; Wei, Z.; Gou, X.; Xu, W. Phosphorus-doped graphene nanosheets as efficient metal-free oxygen reduction electrocatalysts. *RSC Adv* 2013, **3**, 9978–9984.

[86] Choi, C. H.; Chung, M. W.; Kwon, H. C.; Park, S. H.; Woo, S. I. B, N- and P, N-doped graphene as highly active catalysts for oxygen reduction reactions in acidic media. *J Mater Chem A* 2013, **1**, 3694–3699.

[87] Liu, X. M.; Romero, H. E.; Gutierrez, H. R.; Adu, K.; Eklund, P. C. Transparent boron-doped carbon nanotube films. *Nano Lett* 2008, **8**, 2613–2619.

[88] Tang, Y.-B.; Yin, L.-C.; Yang, Y.; Bo, X.-H.; Cao, Y.-L.; Wang, H.-E.; Zhang, W.-J.; Bello, I.; Lee, S.-T.; Cheng, H.-M.; Lee, C.-S. Tunable band gaps and p-type transport properties of boron-doped graphenes by controllable ion doping using reactive microwave plasma. *ACS Nano* 2012, **6**, 1970–1978.

[89] Czerw, R.; Chiu, P.-W.; Choi, Y.-M.; Lee, D.-S.; Carroll, D.; Roth, S.; Park, Y.-W. Substitutional boron-doping of carbon nanotubes. *Curr Appl Phys* 2002, **2**, 473–477.

[90] Zhao, L.; Levendorf, M. P.; Goncher, S. J.; Schiros, T.; Palova, L.; Zabet-Khosousi, A.; Rim, K. T.; Gutierrez, C.; Nordlund, D.; Jaye, C.; Hybertsen, M. S.; Reichman, D. R.; Flynn, G. W.; Park, J.; Pasupathy, A. N. Local atomic and electronic structure of boron chemical doping in monolayer graphene. *Nano Lett* 2013, **13**, 4659–4665.

[91]   Ayala, P.; Reppert, J.; Grobosch, M.; Knupfer, M.; Pichler, T.; Rao, A. M. Evidence for substitutional boron in doped single-walled carbon nanotubes. *Appl Phys Lett* 2010, **96**, 183110–183113.

[92]   Ayala, P.; Miyata, Y.; De Blauwe, K.; Shiozawa, H.; Feng, Y.; Yanagi, K.; Kramberger, C.; Silva, S.; Follath, R.; Kataura, H. Disentanglement of the electronic properties of metallicity-selected single-walled carbon nanotubes. *Phys Rev B* 2009, **80**, 205427.

[93]   Ayala, P.; Kitaura, R.; Kramberger, C.; Shiozawa, H.; Imazu, N.; Kobayashi, K.; Mowbray, D. J.; Hoffmann, P.; Shinohara, H.; Pichler, T. A resonant photoemission insight to the electronic structure of Gd nanowires templated in the hollow core of SWCNTs. *Mat Express* 2011, **1**, 30–35.

[94]   Mowbray, D. J.; Ayala, P.; Pichler, T.; Rubio, A. Computing C1s X-ray absorption for single-walled carbon nanotubes with distinct electronic type. *Mat Express* 2011, **1**, 225–230.

[95]   Florea, I.; Ersen, O.; Arenal, R.; Ihiawakrim, D.; Messaoudi, C.; Chizari, K.; Janowska, I.; Pham-Huu, C. 3D analysis of the morphology and spatial distribution of nitrogen in nitrogen-doped carbon nanotubes by energy-filtered transmission electron microscopy tomography. *J Am Chem Soc* 2012, **134**, 9672–9680.

[96]   Haberer, D.; Vyalikh, D. V.; Taioli, S.; Dora, B.; Farjam, M.; Fink, J.; Marchenko, D.; Pichler, T.; Ziegler, K.; Simonucci, S.; Dresselhaus, M. S.; Knupfer, M.; Buchner, B.; Gruneis, A. Tunable band gap in hydrogenated quasi-free-standing graphene. *Nano Lett* 2010, **10**, 3360–3366.

[97]   Sheng, Z.-H.; Gao, H.-L.; Bao, W.-J.; Wang, F.-B.; Xia, X.-H. Synthesis of boron doped graphene for oxygen reduction reaction in fuel cells. *J Mater Chem* 2011, **22**, 390.

[98]   Kim, Y. A.; Fujisawa, K.; Muramatsu, H.; Hayashi, T.; Endo, M.; Fujimori, T.; Kaneko, K.; Terrones, M.; Behrends, J.; Eckmann, A.; Casiraghi, C.; Novoselov, K. S.; Saito, R.; Dresselhaus, M. S. Raman spectroscopy of boron-doped single-layer graphene. *ACS Nano* 2012, **6**, 6293–6300.

[99]   Yu, D.; Xue, Y.; Dai, L. Vertically aligned carbon nanotube arrays co-doped with phosphorus and nitrogen as efficient metal-free electrocatalysts for oxygen reduction. *J Phys Chem Lett* 2012, **3**, 2863–2870.

[100] Liu, J.; Liu, H.; Zhang, Y.; Li, R.; Liang, G.; Gauthier, M.; Sun, X. Synthesis and characterization of phosphorus-nitrogen doped multiwalled carbon nanotubes. *Carbon* 2011, **49**, 5014–5021.

[101] Odom, T. W.; Huang, J.-L.; Kim, P.; Lieber, C. M. Atomic structure and electronic properties of single-walled carbon nanotubes. *Nature* 1998, **391**, 62–64.

[102] Wilder, J. W. G.; Venema, L. C.; Rinzler, A. G.; Smalley, R. E.; Dekker, C. Electronic structure of atomically resolved carbon nanotubes. *Nature* 1998, **391**, 59–62.

[103] Lin, H.; Arenal, R.; Enouz-Vedrenne, S.; Stephan, O.; Loiseau, A. Nitrogen configuration in individual CNx-SWNTs synthesized by laser vaporization technique. *J Phys Chem C* 2009, **113**, 9509–9511.

[104] Joucken, F.; Tison, Y.; Lagoute, J.; Dumont, J.; Cabosart, D.; Zheng, B.; Repain, V.; Chacon, C.; Girard, Y.; Botello-Méndez, A. R.; Rousset, S.; Sporken, R.; Charlier, J.-C.; Henrard, L. Localized state and charge transfer in nitrogen-doped graphene. *Phys Rev B* 2012, **85**, 161408.

[105] Lin, H.; Lagoute, J.; Repain, V.; Chacon, C.; Girard, Y.; Lauret, J.-S.; Arenal, R.; Ducastelle, F.; Rousset, S.; Loiseau, A. Coupled study by TEM/EELS and STM/STS of electronic properties of C- and CNx-nanotubes. *C R Phys* 2011, **12**, 909–920.

[106] Meyer, J. C.; Kurasch, S.; Park, H. J.; Skakalova, V.; Knzel, D.; Gro, A.; Chuvilin, A.; Algara-Siller, G.; Roth, S.; Iwasaki, T.; Starke, U.; Smet, J. H.; Kaiser, U. Experimental analysis of charge redistribution due to chemical bonding by high-resolution transmission electron microscopy. *Nat Mater* 2011, **10**, 209–215.

[107] Zhao, L.; He, R.; Rim, K. T.; Schiros, T.; Kim, K. S.; Zhou, H.; Gutirrez, C.; Chockalingam, S. P.; Arguello, C. J.; Plov, L.; Nordlund, D.; Hybertsen, M. S.; Reichman, D. R.; Heinz, T. F.; Kim, P.; Pinczuk, A.; Flynn, G. W.; Pasupathy, A. N. Visualizing individual nitrogen dopants in monolayer graphene. *Int S Techn Pol Inn* 2011, **333**, 999–1003.

[108] Lv, R.; Terrones, M. Towards new graphene materials: doped graphene sheets and nanoribbons. *Mater Lett* 2012, **78**, 209–218.

[109] Schiros, T.; Nordlund, D.; Plov, L.; Prezzi, D.; Zhao, L.; Kim, K. S.; Wurstbauer, U.; Gutirrez, C.; Delongchamp, D.; Jaye, C.; Fischer, D.; Ogasawara, H.; Pettersson, L. G. M.; Reichman, D. R.; Kim, P.; Hybertsen, M. S.; Pasupathy, A. N. Connecting dopant bond type with electronic structure in N-doped graphene. *Nano Lett* 2012, **12**, 4025–4031.

[110] Krivanek, O. L.; Chisholm, M. F.; Nicolosi, V.; Pennycook, T. J.; Corbin, G. J.; Dellby, N.; Murfitt, M. F.; Own, C. S.; Szilagyi, Z. S.; Oxley, M. P.; Pantelides, S. T.; Pennycook, S. J. Atom-by-atom structural and chemical analysis by annular dark-field electron microscopy. *Nature* 2010, **464**, 571–574.

[111] Trasobares, S.; Stephan, O.; Colliex, C.; Hsu, W. K.; Kroto, H. W.; Walton, D. R. M. Compartmentalized CNx nanotubes: chemistry, morphology, and growth. *J Chem Phys* 2002, **116**, 8966–8972.

[112] Susi, T.; Kaskela, A.; Zhu, Z.; Ayala, P.; Arenal, R.; Tian, Y.; Laiho, P.; Mali, J.; Nasibulin, A. G.; Jiang, H.; Lanzani, G.; Stephan, O.; Laasonen, K.; Pichler, T.; Loiseau, A.; Kauppinen, E. I. Nitrogen-doped single-walled carbon nanotube thin films exhibiting anomalous sheet resistances. *Chem Mater* 2011, **23**, 2201–2208.

[113] Suenaga, K.; Koshino, M. Atom-by-atom spectroscopy at graphene edge. *Nature* 2010, **468**, 1088–1090.

[114] Nicholls, R. J.; Murdock, A. T.; Tsang, J.; Britton, J.; Pennycook, T. J.; Kos, A.; Nellist, P. D.; Grobert, N.; Yates, J. R. Probing the bonding in nitrogen-doped graphene using electron energy loss spectroscopy. *ACS Nano* 2013, **7**, 7145–7150.

[115] Li, Y.; Zhou, W.; Wang, H.; Xie, L.; Liang, Y.; Wei, F.; Idrobo, J.-C.; Pennycook, S. J.; Dai, H. An oxygen reduction electrocatalyst based on carbon nanotube–graphene complexes. *Nat Nanotechnol* 2012, **7**, 394–400.

[116] Gerber, I. C.; Puech, P.; Gannouni, A.; Bacsa, W. Influence of nitrogen doping on the radial breathing mode in carbon nanotubes. *Phys Rev B* 2009, **79**, 075423.

[117] Susi, T.; Zhu, Z.; Ruiz-Soria, G.; Arenal, R.; Ayala, P.; Nasibulin, A. G.; Lin, H.; Jiang, H.; Stephan, O.; Pichler, T.; Loiseau, A.; Kauppinen, E. I. Nitrogen-doped SWCNTs synthesized using ammonia and carbon monoxide. *Phys. Status Solidi B* 2010, **247**, 2726–2729.

[118] Ayala, P.; Grüneis, A.; Kramberger, C.; Rümmeli, M.; Solórzano, M.; Freire, F., Jr.; Pichler, T. Effects of the reaction atmosphere composition on the synthesis of single and multiwall nitrogen doped nanotubes. *J Chem Phys* 2007, **127**, 184709.

[119] Liu, H.; Zhang, Y.; Li, R.; Sun, X.; Dsilets, S.; Abou-Rachid, H.; Jaidann, M.; Lussier, L.-S. Structural and morphological control of aligned nitrogen-doped carbon nanotubes. *Carbon* 2009, **48**, 1498–1507.

[120] Yang, Q.; Hou, P.; Unno, M.; Yamauchi, S.; Saito, R.; Kyotani, T. Dual Raman features of double coaxial carbon nanotubes with N-doped and B-doped multiwalls. *Nano Lett* 2005, **5**, 2465–2469.

[121] Maciel, I.; Anderson, N.; Pimenta, M.; Hartschuh, A.; Quian, H.; Terrones, M.; Terrones, H.; Campos-Delgado, J.; Rao, A.; Novotny, L.; Jorio, A. Electron and phonon renormalization near charged defects in carbon nanotubes. *Nat Mater* 2008, **7**, 878–883.

[122] Terrones, M.; Grobert, N.; Terrones, H.; Ajayan, P.; Banhart, F.; Blase, X.; Czerw, D. C. R.; Foley, B.; Charlier, J.; Kamalakaran, R.; Kohler-Redlich, P.; Ruhle, M.; Seeger, T. Doping and connecting nanotubes. *Mol Cryst Liq Cryst* 2002, **387**, 51.

[123] Latil, S.; Roche, S.; Mayou, D.; Charlier, J. Mesoscopic transport in chemically doped carbon nanotubes. *Phys Rev Lett* 2004, **92**, 256805.

[124] Krstic, V.; Rikken, G. L. J. A.; Bernier, P.; Roth, S.; Glerup, M. Nitrogen doping of metallic single-walled carbon nanotubes: n-type conduction and dipole scattering. *Eur Phys Lett* 2007, **77**, 107–110.

[125] Wiggins-Camacho, J.; Stevenson, K. Effect of nitrogen concentration on capacitance, density of states, electronic conductivity, and morphology of N-doped carbon nanotube electrodes. *J Phys Chem C* 2009, **113**, 9.

[126] Usachov, D.; Vilkov, O.; Grneis, A.; Haberer, D.; Fedorov, A.; Adamchuk, V. K.; Preobrajenski, A. B.; Dudin, P.; Barinov, A.; Oehzelt, M.; Laubschat, C.; Vyalikh, D. V. Nitrogen-doped graphene: efficient growth, structure, and electronic properties. *Nano Lett* 2011, **11**, 5401–5407.

[127] Tsang, J.; Freitag, M.; Perebeinos, V.; Liu, J.; Avouris, P. Doping and phonon renormalization in carbon nanotubes. *Nat Nanotechnol* 2007, **2**, 725–730.

[128] Fragneaud, B.; Masenelli-Varlot, K.; Gonzalez-Montiel, A.; Terrones, M.; Cavaille, J. Efficient coating of N-doped carbon nanotubes with polystyrene using atomic transfer radical polymerization. *Chem Phys Lett* 2006, **419**, 567–573.

[129] Carrero-Sanchez, J. C.; Elias, A. L.; Mancilla, R.; Arrellin, G.; Terrones, H.; Laclette, J. P.; Terrones, M. Biocompatibility and toxicological studies of carbon nanotubes doped ith nitrogen. *Nano Lett* 2006, **6**, 1609–1616.

[130] Zhang, G.; Duan, W.; Gu, B. Effect of substitutional atoms in the tip on field-emission properties f capped carbon nanotubes. *Appl Phys Lett* 2002, **80**, 2589–2591.

[131] Ahn, H.; Lee, K.; Kim, D.; Han, S. Field emission of doped carbon nanotubes. *Appl Phys Lett* 2006, **88**, 093122.

[132] Srivastava, S. K.; Vankar, V. D.; Rao, D. V. S.; Kumar, V. Enhanced field emission characteristics of nitrogen-doped carbon nanotube films grown by microwave plasma enhanced chemical vapor deposition process. *Thin Solid Films* 2006, **515**, 1851–1856.

[133] Zhou, Z.; Ci, L.; Song, L.; Yan, X.; Liu, D.; Yuan, H.; Gao, Y.; Wang, J.; Liu, L.; Zhou, W.; Wang, G.; Xie, S. Random networks of single-walled carbon nanotubes. *J Phys Chem B* 2004, **108**, 10751–10753.

[134] Zhou, Z.; Gao, X. P.; Yan, J.; Song, D. Y. Doping effects of B and N on hydrogen adsorption in single-walled carbon nanotubes through density functional calculations. *Carbon* 2006, **44**, 939–947.

[135] Rocha, A. R.; Rossi, M.; Fazzio, A.; Da Silva, A. J. R. Designing real nanotube-based gas sensors. *Phys Rev Lett* 2008, **100**, 1–4.

[136] Cruz-Silva, E.; Lopez-Urias, F.; Munoz-Sandoval, E.; Sumpter, B. G.; Terrones, H.; Charlier, J.-C.; Meunier, V.; Terrones, M. Phosphorus and phosphorus–nitrogen doped carbon nanotubes for ultrasensitive and selective molecular detection. *Nanoscale* 2011, **3**, 1008.

[137] Deng, C.; Chen, J.; Chen, X.; Mao, C.; Nie, L.; Yao, S. Direct electrochemistry of glucose oxidase and biosensing for glucose based on boron-doped carbon nanotubes modified electrode. *Biosens Bioelectron* 2008, **23**, 1272–1277.

[138] Deng, C.; Chen, J.; Chen, X.; Mao, C.; Nie, Z.; Yao, S. Boron-doped carbon nanotubes modified electrode for electroanalysis of NADH. *Electrochem Commun* 2008, **10**, 907–909.

[139] Gong, K.; Du, F.; Xia, Z.; Durstock, M.; Dai, L. Nitrogen-doped carbon nanotube arrays with high electrocatalytic activity for oxygen reduction. *Int S Techn Pol Inn* 2009, **323**, 760–764.

[140] Yu, D.; Nagelli, E.; Du, F.; Dai, L. Metal-free carbon nanomaterials become more active than metal catalysts and last longer. *J Phys Chem Lett* 2010, **1**, 2165–2173.

[141] Machado, B. F.; Serp, P. Graphene-based materials for catalysis. *Catal Sci Technol* 2012, **2**, 54–75.

[142] Paraknowitsch, J. P.; Thomas, A. Doping carbons beyond nitrogen: an overview of advanced heteroatom doped carbons with boron, sulphur and phosphorus for energy applications. *Energy Environ Sci* 2013, **6**, 2839–2855.

[143] Rocha, A. R.; Padilha, J. E.; Fazzio, A.; da Silva, A. J. R. Transport properties of single vacancies in nanotubes. *Phys Rev B* 2008, **77**, 153406.

[144] Rocha, A.; Rossi, M.; Silva, A.; Fazzio, A. Realistic calculations of carbon-based disordered systems. *J Phys D Appl Phys* 2010, **43**, 374002.

PART II

CARBON NANOMATERIALS FOR ENERGY CONVERSION

5

HIGH-PERFORMANCE POLYMER SOLAR CELLS CONTAINING CARBON NANOMATERIALS

JUN LIU[1] AND LIMING DAI[2]

[1] State Key Laboratory of Polymer Physics and Chemistry, Changchun Institute of Applied Chemistry, Chinese Academy of Sciences, Changchun, China
[2] Center of Advanced Science and Engineering for Carbon (Case4Carbon), Department of Macromolecular Science and Engineering, Case Western Reserve University, Cleveland, OH, USA

5.1 INTRODUCTION

Polymer solar cells (PSCs) have recently received great attention due to its low-cost solution-based processing and flexibility. In the past decade, the power conversion efficiency (PCE) of PSCs based on conjugated polymers has dramatically increased from about 3% to over 10% thanks to the development of novel polymeric/organic materials, optimization of device structure, and improvement of fabrication techniques [1–7]. To be competitive with conventional photovoltaic technologies based on silicon or other inorganic semiconductors, however, the efficiency, lifetime, and cost of PSCs still need to be further improved.

Figure 5.1a shows a typical device structure for a PSC, while Figure 5.1b illustrates molecular structures for a typical polymer electron donor (i.e., poly(3-hexylthiophene) (P3HT)) and a typical organic electron acceptor (i.e., [6,6]-phenyl-C61-butyric acid methyl ester ($PC_{61}BM$)) that are blended into a bulk heterojunction configuration as the active layer. As can be seen in Figure 5.1a, the active layer is sandwiched between

Carbon Nanomaterials for Advanced Energy Systems: Advances in Materials Synthesis and Device Applications, First Edition. Edited by Wen Lu, Jong-Beom Baek and Liming Dai.
© 2015 John Wiley & Sons, Inc. Published 2015 by John Wiley & Sons, Inc.

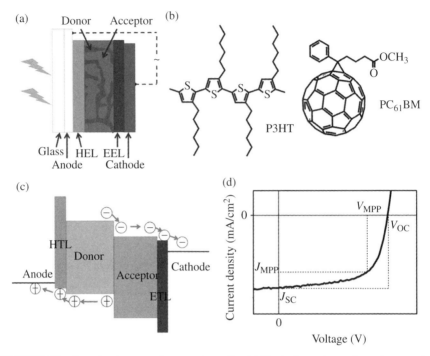

FIGURE 5.1   Introduction to polymer solar cells. (a) Typical structure of a polymer solar cell device. (b) Chemical structures of a typical polymer electron donor (P3HT) and a typical organic electron acceptor ($PC_{61}BM$). (c) Energy level diagram of a typical polymer solar cell device. (d) A typical current density–voltage curve of a polymer solar cell device under illumination.

a transparent anode (e.g., tin-doped indium oxide (ITO)) and a metal cathode (e.g., Al). In addition, a hole extraction layer (HEL) (e.g., poly(3,4-ethylenedioxythiophene) doped with poly(styrenesulfonate), PEDOT:PSS) is inserted between the anode and the active layer, while an electron extraction layer (EEL) (e.g., LiF) is used between the cathode and the active layer [1]. Figure 5.1c depicts the energy level diagram and operation principle for a typical PSC device. Upon illumination of a PSC device, the polymer electron donor in the active layer absorbs photons to generate excitons (hole/electron pairs), followed by the exciton diffusion to the donor/acceptor interface for charge separation into free holes and electrons to be collected at the anode and the cathode, respectively [3].

Several parameters including the open-circuit voltage (V_{OC}), short-circuit current density (J_{SC}), fill factor (FF), and PCE have been developed to evaluate the performance of a PSC device. As shown in Figure 5.1d, V_{OC} is the (maximum) voltage at the current density of zero, and J_{SC} is the (maximum) current density at the voltage of zero. $FF = J_{MPP} \times V_{MPP} / J_{SC} \times V_{OC}$, where J_{MPP} and V_{MPP} are the current density and voltage, respectively, at the maximum power output (Fig. 5.1d). PCE is defined as the maximum power output per incident light power density (P_{in}) and is given by $PCE = V_{OC} \times J_{SC} \times FF / P_{in}$. An excellent PSC device with P3HT:PCBM blend as the

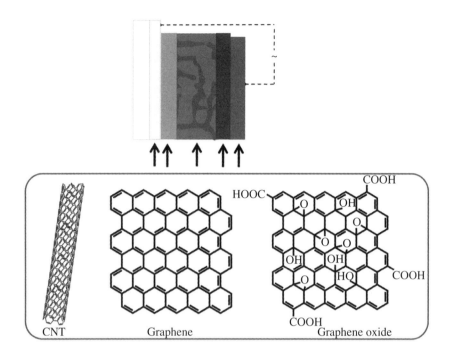

FIGURE 5.2 Carbon nanomaterials used in every layer in polymer solar cell devices. Top: typical polymer solar cell device structure. Bottom: schematic of the structures of carbon nanomaterials, including CNT, graphene, and graphene oxide.

active layer often has a V_{OC} of 0.55–0.65 V, J_{SC} of 8–11.5 mA/cm², FF of 0.6–0.7, and PCE of 3–5% [1–7].

The performance of a PSC device strongly relies on the property of the materials employed. Carbon nanomaterials, such as fullerene, carbon nanotubes (CNTs), and graphene, have been demonstrated to be very promising to improve the PSC performance due to their fully conjugated molecular structures and unique material geometries (Fig. 5.2) [8–17]. Particularly, $PC_{61}BM$, a fullerene derivative, has been the state-of-the-art electron acceptor since the discovery of PSCs in 1995 [1, 17]. As shown in Figure 5.2, CNTs and graphene materials have indeed been used in every component of high-performance PSCs. In this chapter, we will give an overview on the recent progress in the development of CNTs and graphene materials for high-performance PSCs by summarizing recent important studies in this exciting field, including our own ones.

5.2 CARBON NANOMATERIALS AS TRANSPARENT ELECTRODES

Conventional PSC devices always use ITO as the transparent electrode due to its low sheet resistance and high transparency. However, the large-scale application of ITO is limited by its high cost and scarcity. Indium composes nearly 75% of the mass of an ITO film, but indium is a fairly rare metal with limited supply on the earth.

Besides, ITO is normally fabricated through a costly and slow vacuum-based process. While these factors contribute to the increasing price of ITO in the past decade, ITO is too brittle to be used for flexible devices. Therefore, alternative transparent electrodes based on other conducting materials, including conducting polymers, metal nanowires, metal grids, CNTs, and graphene, have been intensively investigated recently [18–20]. Among them, CNTs and graphene are very promising for the transparent electrode of high-performance PSC devices.

5.2.1　CNT Electrode

Percolative networks consisting of randomly distributed CNTs can transport electricity via the CNT framework and allow light transmittance through the void. The flexible, yet mechanically strong characteristics intrinsically associated with the CNT materials offer additional advantages. Thus, CNTs are regarded as promising candidate as transparent electrode materials for PSCs, particularly in a flexible form. Transparent CNT electrodes can be prepared by solution processing (e.g., vacuum filtration, spin coating, spray coating, roll-to-roll coating) of CNT "inks," in which CNTs are dispersed in solvents (e.g., water with the aid of surfactants). These transparent CNT films often have a thickness of 5–50 nm, optical transmittance of 60–90%, and sheet resistance of several hundreds to several tens of Ω/sq. CNT films can be deposited on plastic substrates to make flexible electrodes, which are desired for flexible electronics.

In spite of the recent considerable efforts devoted to the research and development of CNT transparent electrodes, the large-scale application of CNT transparent electrodes in high-performance PSCs is still limited by two critical factors: (i) the low conductivity of CNT electrodes due to the large contact resistance between tubes and (ii) the surface roughness of CNT electrodes, which may result in poor uniformity of the active layer and hence degrade device efficiency or even short the device [21–24]. Some novel approaches have been developed to make smooth CNT transparent films [23], while the CNT film conductivity can be improved by lowering the contact resistance and using high-quality CNTs, such as long, pure, debundled, single- and/or double-walled CNTs.

Lagemaat et al. [21] prepared a highly conductive transparent electrode with spray-coated single-walled CNTs (SWCNTs) showing a sheet resistance of 50 Ω/sq at a transmittance of 70%. They used a very thick P3HT:PCBM (500–1000 nm) active layer to alleviate the electrode roughness effect. As a result, the PSC device exhibited a V_{OC} of 0.56 V, J_{SC} of 9.2 mA/cm^2, FF of 0.29, and PCE of 1.5%. In a separated work, the same authors further optimized the device and obtained a high PCE of 2.65% with a V_{OC} of 0.52 V, J_{SC} of 11.2 mA/cm^2, and FF of 0.46 [22]. Their results had been verified by the National Renewable Energy Laboratory (NREL) and indicated that SWCNT network could be used to replace ITO in efficient PSCs. On the other hand, Rowell et al. [23] made smooth SWCNT films by a transfer printing technique. In this case, a SWCNT film prepared by vacuum filtration of SWCNT dispersion over a porous alumina membrane was lifted off with a poly(dimethylsiloxane) (PDMS) stamp and transferred to a flexible poly(ethylene terephthalate) (PET) substrate by contact printing. The electrode thus prepared showed a sheet resistance of

200 Ω/sq at a transmittance of 85%. The resulting PSC device with a thin P3HT:PCBM active layer showed a PCE of 2.5%, which was a little lower than 3% of the ITO-based device due to the relatively large sheet resistance of the SWCNT electrode. Highly conductive and ultrasmooth SWCNT electrodes with a sheet resistance of 60 Ω/sq and a surface roughness of less than 3 nm on a 10 μm scale had also been developed by ultrasonic spraying of surfactant-dispersed SWCNTs, followed by nitric acid treatment to remove the surfactant residue [24]. The resulting PSC device with the P3HT:PCBM active layer exhibited a PCE of 3.1%, which was fairly comparable to 3.6% for an ITO-based counterpart.

5.2.2 Graphene Electrode

The high electrical conductivity and low optical absorption (2.3% absorption for one-layer graphene) make graphene an excellent transparent electrode. Other advantages of graphene as electrode include good surface smoothness, chemical stability, and flexibility. Peumans et al. [25] have predicted that the sheet resistance (Rs) of graphene will vary with the number of layers as Rs~62.4/$N\Omega$/sq for highly doped graphene, where N is the number of layers. Kim et al. [26] have demonstrated the preparation of large-area graphene films with a sheet resistance of 30 Ω/sq and a transmittance of 90% at 550 nm. This performance is fairly comparable to that of ITO and is sufficient for the transparent electrode application in small-area PSCs. Among various methods for preparing graphene, including micromechanical exfoliation, epitaxial growth, chemical vapor deposition (CVD), and reduction of graphene oxide (GO), the CVD approach can produce high-quality graphene sheets with a large size, low defect content, and high conductivity [27]. The GO-reducing approach suffers from high defect content and low conductivity but has advantages of high-throughput preparation, low cost, and simple film/device fabrication [28].

Although significant progress has been achieved, the development of high-performance PSCs based on graphene electrodes is still hampered by two factors: (i) the low conductivity reported for many graphene films due to the nonoptimal fabrication technique of graphene electrode and (ii) the difficulty of coating HEL. The hydrophobicity of graphene prevents the uniform coating of hydrophilic HEL on the graphene electrode.

Geng et al. [29] had used reduced graphene oxide (rGO) as the transparent electrode to fabricate PSC devices. The rGO was carefully thermal annealed and gave a sheet resistance of $6 \times 10^3 \Omega$/sq at a transmittance of 78%. Owing to the high sheet resistance, the resulting PSC device showed a PCE of 1.01 ± 0.05%, which equaled to half of the corresponding value (2.01 ± 0.1%) for a reference device based on an ITO electrode. Compared to PSCs with an ITO electrode, the rGO-based PSC always shows inferior photovoltaic performance due to the very high sheet resistance of rGO. Highly conductive graphene electrodes can be obtained by synthesis of graphene via CVD approach, layer-by-layer stacking of graphene sheets, and acid doping, as reported by Kim et al. [26]. Wang et al. [30] used CVD graphene with a sheet resistance of 210 Ω/sq at a transmittance of 72% as the transparent electrode for PSCs. The resulting device with a P3HT:PCBM active layer exhibited a PCE as low as 0.21% due to the hydrophobic property of graphene, which prevented the uniform coating of hydrophilic HEL PEDOT:PSS. After the graphene electrode was modified by pyrenebutanoic acid succinimidyl ester to improve the surface wettability for spin coating PEDOT:PSS, the PCE significantly increased up to 1.71% with a V_{OC} of 0.55 V, J_{SC} of 6.05 mA/cm^2, and FF of 0.51. This work highlighted the importance of interface engineering of graphene electrode for PSC device applications. In a separated but closely related work, Wang et al. [31] made highly conductive graphene electrode by layer-by-layer transfer method (Fig. 5.3a–c), followed by acid doping.

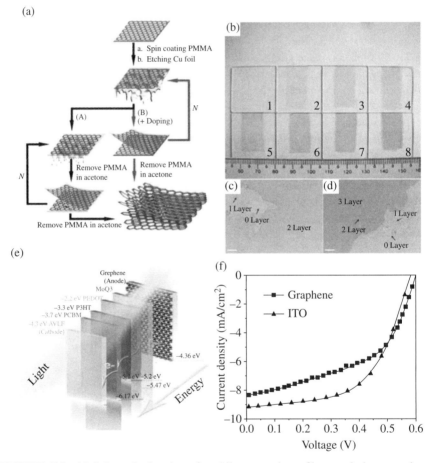

FIGURE 5.3 (a) Schematic drawing of multilayer graphene films made by normal wet transfer (A) and by LBL assembly (B) (N=0,1,2,3…). (b) Optical images of multilayer graphene films (from 1 to 8 layers) on quartz substrates. (c, d) Typical optical microscope images of 2- and 3-layer graphene films on SiO$_2$/Si substrates. (e) Schematic of photovoltaic device structure. (f) Current density–voltage curves of the devices with the anode of ITO or MoO$_3$-coated graphene (device structure: anode/PEDOT:PSS/P3HT:PCBM/LiF/Al). Adapted from Ref. 31 with permission. © 2011, John Wiley & Sons.

The four-layer graphene film exhibited a sheet resistance of $80\,\Omega$/sq with a transmittance of 90% at 550 nm and outperformed the ITO electrode on a flexible PET substrate (sheet resistance, $100\,\Omega$/sq; transmittance, 80%). During the PSC device fabrication, a thin layer of MoO_3 was evaporated on the graphene electrode to improve its hydrophobicity for spin coating PEDOT:PSS (Fig. 5.3d). The resultant PSC device exhibited a V_{OC} of 0.59 V, J_{SC} of 8.5 mA/cm^2, and FF of 0.51, corresponding to a PCE of 2.5%. In comparison, the ITO counterpart had a V_{OC} of 0.58 V, J_{SC} of 9.2 mA/cm^2, and FF of 0.57, corresponding to a PCE of 3% (Fig. 5.3e). For interface engineering of the graphene electrodes, Park et al. [32] developed poly(3,4-ethylenedioxythiophene)-block-poly(ethylene glycol) doped with perchlorate (PEDOT:PEG(PC)) (Fig. 5.4a). PEDOT:PEG(PC) was deposited on the hydrophobic graphene electrode to facilitate the uniform deposition of PEDOT:PSS (Fig. 5.4a–d). With the interface improvement, the graphene electrode could give a device performance fairly comparable to that of its ITO counterpart in either bilayer organic solar cells or single-layer bulk heterojunction PSCs (Fig. 5.4e).

5.2.3 Graphene/CNT Hybrid Electrode

Just like composite materials often show properties characteristic of each of the components with a synergistic effect, the combination of CNTs and graphene in transparent electrodes is expected to improve the conductivity. In this context, Tung et al. [33] developed a hybrid nanocomposite film comprised of graphene and CNT (G-CNT) by codissolving GO and CNT in anhydrous hydrazine. The resultant G-CNT film gave a sheet resistance of $240\,\Omega$/sq and a transmittance of 86% (Fig. 5.5a, b). PSC devices with P3HT:PCBM active layer and G-CNT electrode exhibited a V_{OC} of 0.58 V, J_{SC} of 3.47 mA/cm^2, FF of 0.42, and PCE of 0.85% (Fig. 5.5c). The low J_{SC} and FF were likely due to the poor contact at the G-CNT/P3HT:PCBM interface. Huang et al. [34] reported a strategy to tune the work function of rGO–SWCNT hybrid composite electrode through doping with alkali carbonates. The work function can be readily tuned within the range of -5.1 eV to -3.4 eV (Fig. 5.6a). Inverted PSC devices employing Cs_2CO_3-doped rGO–SWCNT as cathode gave a PCE of 1.27% (Fig. 5.6b).

5.3 CARBON NANOMATERIALS AS CHARGE EXTRACTION LAYERS

The HEL between the anode and the active layer and the EEL between the cathode and the active layer (see Fig. 5.1) play critical roles in regulating the overall device performance of PSCs [35]. The function of these charge extraction layers includes (i) to minimize the energy barrier for charge extraction, (ii) to selectively extract one sort of charge carriers and block the opposite charge carriers, (iii) to modify the interface between the electrode and the active layer, and (iv) to act as an optical spacer. An HEL should have a relatively high work function to allow for the built-in electrical field across the active layer and for holes to transport toward the anode. Similarly, an EEL needs to have a low work function for electrons to efficiently

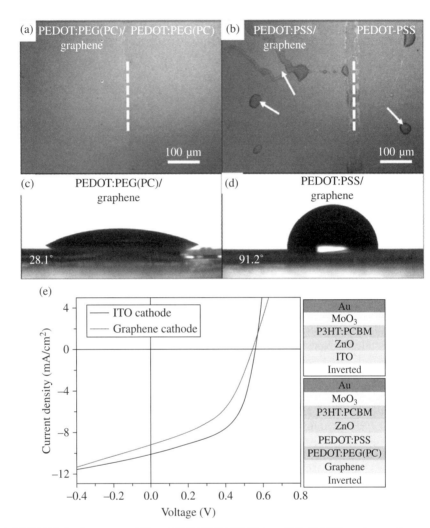

FIGURE 5.4 Better wettability of PEDOT:PEG(PC) than PEDOT:PSS on the graphene surface. Optical microscopy images of PEDOT:PEG(PC) (a) and PEDOT:PSS (b) on graphene and bare quartz substrates. The white dotted lines indicate the edge of the graphene, and the arrows denote dewetted PEDOT:PSS. Contact angle images of graphene/PEDOT:PEG(PC) (c) and graphene/PEDOT:PSS (d). (e) Current density–voltage curves for the graphene-based PSC device and the ITO-based PSC device. Adapted from Ref. 32 with permission. © 2013, Nature Publishing Group. (*See insert for color representation of the figure.*)

transport to the cathode. Moreover, the hole/electron extraction materials should be solution processable.

The most widely used HEL in PSCs is PEDOT:PSS. However, PEDOT:PSS suffers from its hygroscopicity, which degrades the device efficiency and lifetime. Moreover, PEDOT:PSS itches ITO electrode during long-term device operation due

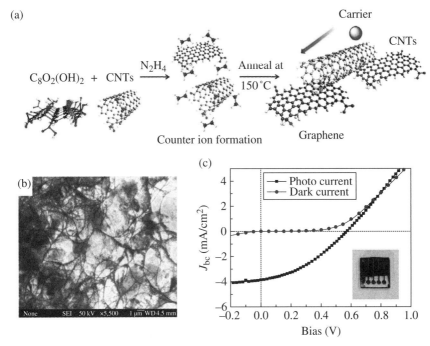

FIGURE 5.5 (a) Schematic illustration of G-CNT electrode. (b) A representative SEM image of a G-CNT film. (c) Current density–voltage curves of the device with a G-CNT film as the electrode in dark and under AM 1.5G illumination. Adapted from Ref. 33 with permission. © 2009, American Chemical Society.

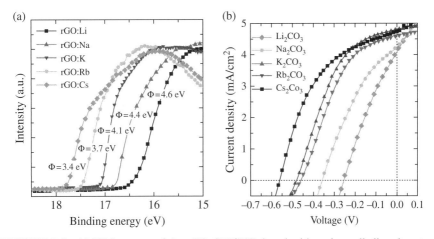

FIGURE 5.6 (a) UPS spectra of the rGO–SWCNT doped with various alkali carbonates. The work functions were determined from the UPS secondary electron cutoff. (b) Current density–voltage curves of the inverted P3HT:PCBM PSCs incorporating rGO–SWCNT doped with various alkali carbonates as the cathode. Adapted from Ref. 34 with permission. © 2011, American Chemical Society.

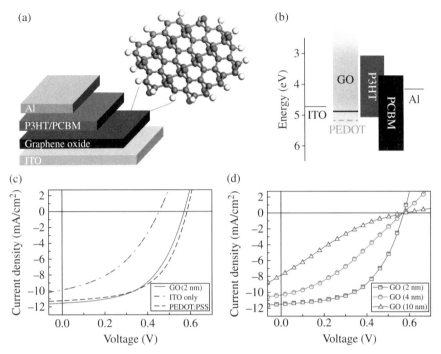

FIGURE 5.7 (a) Schematic of the PSC device structure with GO as the hole extraction layer (ITO/GO/P3HT:PCBM/Al). (b) Energy level alignment in the PSC devices. (c) Current density–voltage characteristics of PSC devices with no hole extraction layer and with 30 nm PEDOT:PSS layer and 2 nm thick GO film. (d) Current density–voltage characteristics of ITO/GO/P3HT:PCBM/Al devices with different GO layer thickness under simulated AM 1.5 illumination. Adapted from Ref. 36 with permission. © 2010, American Chemical Society.

to its strong acidity (pH = 1–2). Therefore, certain inorganic oxide semiconductors, such as NiO, MoO_3, and V_2O_5, have been recently introduced as HEL for PSCs. The commonly used EEL in PSCs includes, but is not limit to, inorganic salts (e.g., LiF), low work function metals (e.g., Ca), inorganic semiconductors (e.g., TiO_2 and ZnO), and conjugated polymer electrolyte. As we shall see later, GO derivative is the only family that can be used as either HEL or EEL in PSCs.

Li et al. and Gao et al. [36, 37] demonstrated that a thin layer of GO (about 2 nm) could act as an excellent HEL for PSCs to show a fairly comparable device efficiency to that of PEDOT:PSS (Fig. 5.7a–c). GO has a work function of −4.8 eV and band gap of about 3.6 eV and can be spin cast from its aqueous solution into a uniform film. It was found that the *FF* of the resulting device decreased from 0.53 to 0.19 and the PCE decreased from 3.5% to 0.9% with the GO layer thickness increasing from 2 to 10 nm (Fig. 5.7d). This was due to the insulating property of GO that led to a high series resistance of the resulting device. Murray et al. [38] reported a highly efficient and stable PSC with GO as the HEL and PTB7:PC_{71}BM as the active layer (Fig. 5.8). The GO-based device showed a PCE of 7.39%, which was fairly

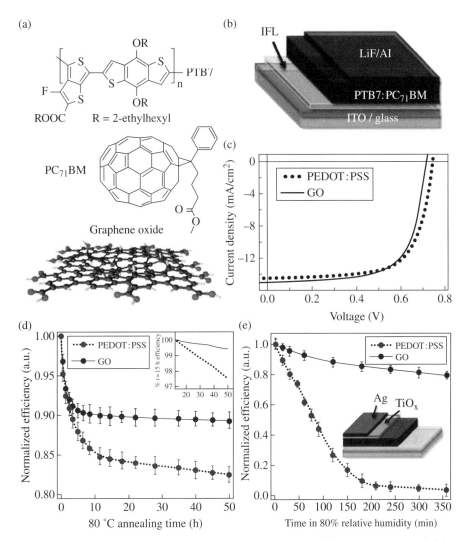

FIGURE 5.8 (a) Chemical structures of the PTB7 donor, PC$_{71}$BM acceptor, and GO. (b) Schematic of the PSC device indicating the location of the GO. (c) Representative current density–voltage plots under AM 1.5G solar simulated light for PSCs with PEDOT:PSS and GO as the hole extraction layer. (d) Thermal degradation of encapsulated devices at 80°C under a N$_2$ atmosphere. (e) Environmental degradation of unencapsulated devices fabricated with air-stable electrodes at 25°C under 80% relative humidity. Adapted from Ref. 38 with permission. © 2011, American Chemical Society.

comparable to the corresponding value of 7.46% for a PEDOT:PSS-based device (Fig. 5.8c). More importantly, the GO-based device provided a 5 times enhancement in thermal aging lifetime and a 20 times enhancement in humid ambient lifetime compared with the PEDOT:PSS-based device (Fig. 5.8d, e). The aforementioned results indicated that GO was a promising HEL for efficient and stable PSCs.

Although excellent device performance has been achieved, the performance of PSC device with GO as an HEL is highly sensitive to the GO layer thickness due to its insulating property. To address this problem, Liu et al. [39] treated GO with oleum and developed sulfated graphene oxide (GO–OSO$_3$H), in which −OSO$_3$H groups were introduced to the basal plane of reduced GO (Fig. 5.9a). GO–OSO$_3$H had the advantages of good solubility for solution processing due to the presence of −OSO$_3$H groups and improved conductivity due to the reduced basal plane. The much improved conductivity of GO–OSO$_3$H (1.3 S/m vs. 0.004 S/m) led to greatly improved *FF* (0.71 vs. 0.58) and PCE (4.37% vs. 3.34%) for a PSC device based on GO–OSO$_3$H HEL with respect to the GO-based counterpart (Fig. 5.9c). On the other hand, Yun et al. [40] reported the use of graphene oxide (pr-GO) reduced by *p*-toluenesulfonyl hydrazide (*p*-TosNHNH$_2$) as an HEL. pr-GO could be dispersed with high concentration (0.6 mg/mL) and could give uniform film by spin coating. PSC devices based on the pr-GO HEL exhibited a fairly comparable photovoltaic efficiency and much improved lifetime compared with those of the PEDOT:PSS device. Various studies reported by many other groups have also proven that photovoltaic performance of GO-based PSC devices could be significantly improved by increasing the conductivity of the GO layer. Examples include the blending of highly conductive SWCNTs

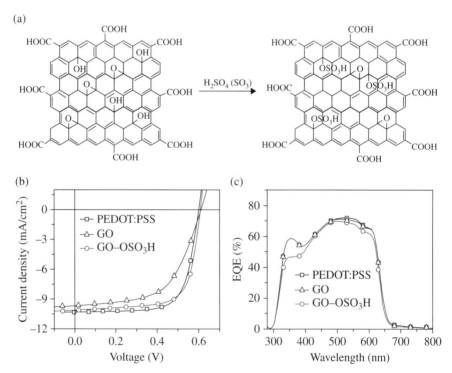

FIGURE 5.9 (a) Synthetic route to GO–OSO$_3$H. Current density–voltage curves (b) and external quantum efficiency curves (c) of the PSC devices with PEDOT:PSS (25 nm), GO (2 nm), or GO–OSO$_3$H (2 nm) as the hole extraction layer. Adapted from Ref. 39 with permission. © 2012, American Chemical Society.

in the GO layer to increase the GO layer lateral conductivity [41] and the use of thermal annealing [42] and plasma treatment [43] to reduce GO during device fabrication for increasing its conductivity.

Liu et al. [44] reported that cesium-neutralized graphene oxide (GO–Cs), in which the COOH groups at the periphery of GO were replaced by COOCs by simple neutralization (Fig. 5.10a), could act as an EEL for PSCs. GO itself had a work function

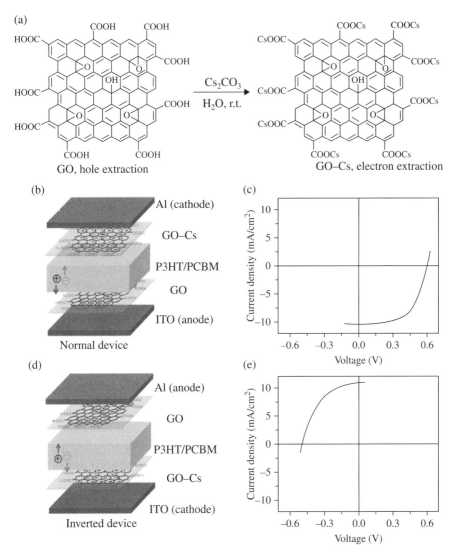

FIGURE 5.10 (a) Synthetic route from GO to GO–Cs. Device structure (b) and current density–voltage curve (c) of the normal device with GO as the hole extraction layer and GO–Cs as the electron extraction layer. Device structure (d) and current density–voltage curve (e) of the inverted device with GO as the hole extraction layer and GO–Cs as the electron extraction layer. Adapted from Ref. 44 with permission. © 2012, John Wiley & Sons.

of $-4.7\,eV$, but GO–Cs-modified electrode had a work function of $-3.8\,eV$ to match well with the LUMO level of $PC_{61}BM$ acceptor for electron extraction. Indeed, it was found that the electron extraction performance of GO–Cs was fairly comparable to that of the state-of-the-art electron extraction material, LiF. These authors also fabricated normal device and inverted device both with GO as the HEL and GO–Cs as the EEL. The normal device exhibited a V_{OC} of 0.61 V, J_{SC} of 10.30 mA/cm^2, FF of 0.59, and PCE of 3.67%, while the inverted device exhibited a V_{OC} of 0.51 V, J_{SC} of 10.69 mA/cm^2, FF of 0.54, and PCE of 2.97%. These results were fairly comparable to those of the standard PSC devices, indicating that the hole extraction capability of GO and electron extraction capability of GO–Cs were independent of the electrode materials. This was the first single system reported to be both the electron and hole extraction materials for PSCs.

5.4   CARBON NANOMATERIALS IN THE ACTIVE LAYER

5.4.1   Carbon Nanomaterials as an Electron Acceptor

The most widely used electron acceptor in PSCs is fullerene derivatives, such as $PC_{61}BM$. Although much efforts have been devoted to the development of new acceptor materials, only a little improvement has been achieved. Therefore, other carbon nanomaterials, such as CNTs, graphene, and graphene quantum dots (GQD), have recently been investigated as acceptor materials.

Geng et al. [45] reported a P3HT/SWCNT photovoltaic device using SWCNTs as the acceptor and P3HT as the donor. They found that SWCNTs had significant interaction with P3HT, which was helpful to form continuous active film with an interpenetrating structure and improved crystallinity. Kymakis et al. [46] prepared a PSC device with P3OT:SWCNT (1 wt%) as the active layer, which exhibited a V_{OC} of 0.75 V, J_{SC} of 0.5 mA/cm^2, FF of 0.60, and PCE of 0.22% after thermal annealing. Generally speaking, PSC devices with CNTs as acceptors show low PCEs due to several unfavorable factors, including the incomplete exciton dissociation with low percentage of CNTs, nonuniformity of CNTs in the active layer, and mixture of semiconducting and conductive SWCNTs in commercially available SWCNT materials. The semiconducting SWCNTs are the desirable acceptor materials, while the conductive SWCNTs often lead to hole/electron recombination for a deteriorate device efficiency.

Graphene derivatives can be easily functionalized and modified for solution processing in PSC devices [47]. For example, Liu et al. [48] reported PSC devices employing solution-processable functionalized graphene (SPFGraphene) as the acceptor and poly(3-octylthiophene) (P3OT) as the donor. A PCE of 0.32% was obtained for the PSC device with 5% SPFGraphene in the active layer. After annealing at 160°C for 20 min, the PCE increased to 1.4% with a V_{OC} of 0.92 V, J_{SC} of 4.2 mA/cm^2, and FF of 0.37. Li et al. [49] synthesized GQD with a uniform size of 3–5 nm through an electrochemical approach. Owing to its high specific surface area for large interface, high mobility, and tunable band gap, GQD showed great potential as an electron acceptor. PSC devices with GQD as the acceptor and P3HT as the donor exhibited a PCE of 1.28%

with a J_{SC} of 6.33 mA/cm², V_{OC} of 0.67 V, and FF of 0.3. The overall performance of the GQD-based PSC was comparable to many PSCs with acceptors other than fullerene. Gupta et al. [50] also used GQD synthesized by hydrothermal approach as acceptors in PSCs. Just like those PSCs with CNT electron acceptors, PSCs with graphene as the acceptor show poorer photovoltaic performance than that with fullerene. As graphene has emerged for only several years, much more effort is required to further improve the performance of PSCs with graphene electron acceptors (Fig. 5.11).

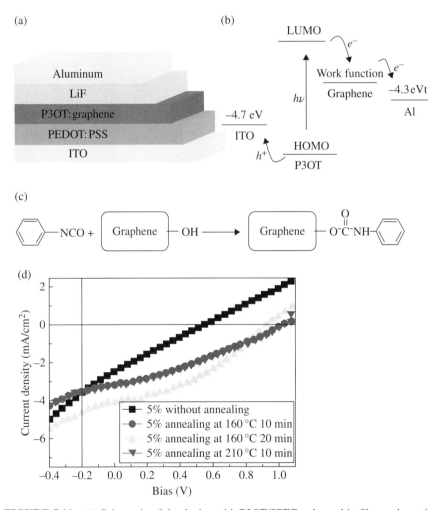

FIGURE 5.11 (a) Schematic of the device with P3OT/SPFGraphene thin film as the active layer. (b) Energy level diagram of P3OT and SPFGraphene. (c) Schematic representation of the reaction of phenyl isocyanate with graphene oxide to form SPFGraphene. (d) Current density–voltage curves of PSC devices based on P3OT/SPFGraphene composite with an SPFGraphene content of 5 wt% without or with thermal annealing at different conditions. Adapted from Ref. 48 with permission. © 2008, John Wiley & Sons.

5.4.2   Carbon Nanomaterials as Additives

The electron donor and electron acceptor in the active layer of PSC devices are blended in a bicontinuous network configuration, which maximizes the donor/acceptor interface for charge separation but are less optimal for charge transport. One straightforward approach to improve the charge transport is to incorporate CNTs with high conductivity into the active layer. For this purpose, Lee et al. [51] discovered that the charge selectivity of CNTs was crucial to prevent hole/electron recombination on CNTs and hence to improve PSC device efficiency. Boron-doped CNTs (B-CNTs) have a work function of 5.2 eV, which matches well with the HOMO level of P3HT donor, and could selectively transport holes (Fig. 5.12a, b). Nitrogen-doped CNTs (N-CNTs) have a work function of 4.4 eV, which matches well with the LUMO level of $PC_{61}BM$ acceptor, and could selectively transport electrons (Fig. 5.12a, b). As a result, PSCs with 1 wt% B-CNTs in the $P3HT/PC_{61}BM$ active layer exhibited a PCE of 4.1%, while the corresponding devices with 1 wt% N-CNTs in the active layer showed a PCE of 3.7%. Both the two PCE values are much higher than 3.0% of the

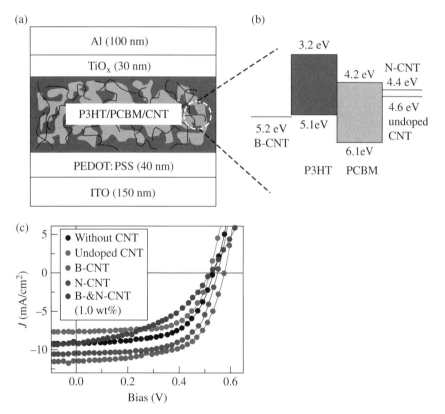

FIGURE 5.12   Schematic device structure (a) and energy diagram (b) of a PSC device with CNTs blended in the active layer. (c) Current density–voltage characteristics of the PSC devices without or with different CNTs. Adapted from Ref. 51 with permission. © 2011, John Wiley & Sons. (*See insert for color representation of the figure.*)

control device without any CNT additive (Fig. 5.12c). In contrast, the corresponding device with 1 wt% undoped CNTs exhibited a decreased PCE of 2.6% due to the lack of charge selectivity (Fig. 5.12c). Lu et al. [52] also incorporated N-doped multi-walled carbon nanotubes (N-MWCNTs) into the active layer of a PSC based on PTB7 and $PC_{71}BM$ and demonstrated a PCE as high as 8.6%. The incorporation of N-MWCNTs resulted in not only an increased nanocrystallite size but also a reduced phase-separated domain size of both PTB7 and $PC_{71}BM$. Hence, N-MWCNTs facilitated charge separation and transport in the active layer to improve the device efficiency. Lee et al. [53] further incorporated a hybrid nanomaterial, which consisted of indium phosphide quantum dots (InP QDs) and N-CNTs, into the active layer of PSC devices to improve exciton dissociation and electron transport (Fig. 5.13a, b).

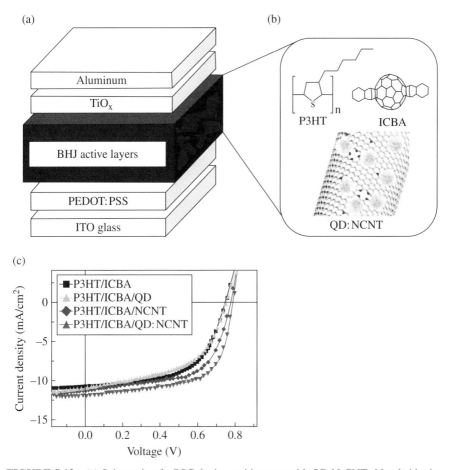

FIGURE 5.13 (a) Schematic of a PSC device architecture with QD:N-CNTs blended in the active layer. (b) Chemical structure of P3HT donor and ICBA acceptor as well as schematic of QD:N-CNTs. (c) Current density–voltage characteristics of the devices with QD, N-CNT, and QD:N-CNT. Adapted from Ref. 53 with permission. © 2013, John Wiley & Sons. (*See insert for color representation of the figure.*)

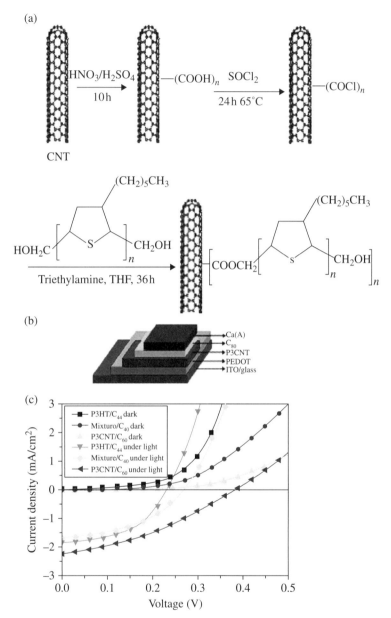

FIGURE 5.14   (a) Reaction scheme for grafting P3HT chains to carbon nanotubes. Device structure (b) and current density–voltage curves (c) of photovoltaic cells based on the P3HT, mixture of P3HT and MWCNTs (1%), and P3CNT under AM 1.5G illumination. Adapted from Ref. 56 with permission. © 2010, American Chemical Society.

The energy level of InP QDs was well balanced within the donor/acceptor materials and promoted charge separation of excitons. Moreover, N-CNTs effectively transported the separated electrons through their high aspect ratio and one-dimensional nanostructure. This synergistic effect successfully improved the PCE of the PSC devices from 4.68% to 6.11%, corresponding to 31% improvement (Fig. 5.13c).

5.4.3 Donor/Acceptor Functionalized with Carbon Nanomaterials

Along with the aforementioned physical blending of CNTs into the active layer, an alternative approach to use carbon nanomaterials to improve charge transport is to covalently attach carbon nanomaterials with the conventional donor or acceptor. The covalent bonding is expected to not only enhance charge transport but also ensure a uniform dispersion of carbon nanomaterials in the active layer. In this regard, Li et al. [54, 55] had prepared MWCNTs–C_{60} complex, in which C_{60} served as an electron acceptor and MWCNTs served as an efficient charge transporter. They found that carboxylated MWCNTs gave better PSC device performance than that of octadecylamine-modified MWCNTs because the alkyl spacer in the latter case blocked electron transfer from C_{60} to MWCNTs. Kuila et al. [56] developed P3HT-grafted CNTs (P3CNTs) through ester reaction of CH_2OH-terminated P3HT and acid-oxidized CNTs (Fig. 5.14a). The P3CNTs showed a blueshifted absorption spectrum compared

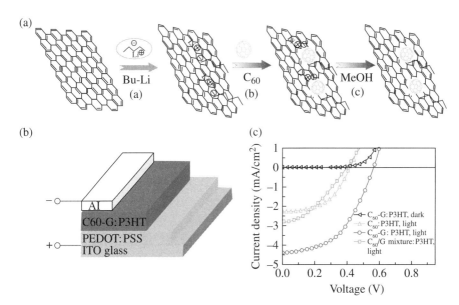

FIGURE 5.15 (a) Schematic representation of grafting C_{60} onto graphene through lithiation reaction with *n*-butyllithium. (b) Schematic of a device with the C_{60}-G:P3HT blend as the active layer. (c) Current density–voltage curves of the photovoltaic devices with the C_{60}-G:P3HT, C_{60}:P3HT, or C_{60}/G mixture (12 wt% G):P3HT as the active layer. Adapted from Ref. 57 with permission. © 2011, American Chemical Society.

with P3HT. However, cyclic voltammetry indicated that P3CNTs had a lower band gap than that of P3HT because of electron delocalization between the covalently linked P3HT and CNTs. Bilayer PSC devices based on the P3CNTs exhibited an increase in PCE by about 40% with respect to its counterpart based on pure P3HT (Fig. 5.14b, c).

As shown in Figure 5.15a, Yu et al. [57] covalently attached C_{60} to graphene sheet by lithiation reaction to afford a graphene–C_{60} hybrid. The resultant C_{60}-grafted graphene was used as an electron acceptor in PSCs to significantly improve electron transport and hence the overall device performance. As a result, the PSC device with C_{60}-grafted graphene/P3HT blend as the active layer showed 2.5 times higher PCE compared to the control device with C_{60}:P3HT blend (Fig. 5.15b, c). In a separated work, these authors also developed P3HT-grafted graphene (G-P3HT) via esterification reaction between CH_2OH-terminated P3HT and graphene oxide (Fig. 5.16a) [58]. The covalent bonding of P3HT and graphene sheets was confirmed by spectroscopic analyses and electrochemical measurements. As shown in Figure 5.16b and c,

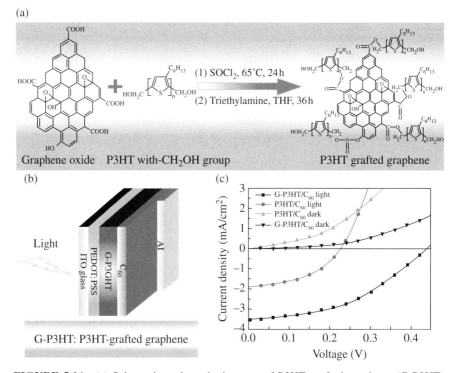

FIGURE 5.16 (a) Schematic and synthetic route of P3HT-grafted graphene (G-P3HT). (b) Schematic of the bilayer device with the structure of ITO/PEDOT/C_{60}:G-P3HT/Al. (c) Current density–voltage curves of the PSC devices with C_{60}:P3HT or C_{60}:G-P3HT as the active layer in dark and under AM 1.5G illumination. Adapted from Ref. 58 with permission. © 2010, American Chemical Society.

bilayer photovoltaic devices based on C_{60}:G-P3HT showed two times increase in PCE (0.61%) with respect to the C_{60}:P3HT counterpart.

5.5 CONCLUDING REMARKS

As we can see from the aforementioned discussions, carbon nanomaterials (e.g., CNTs, graphene) are versatile for applications in PSCs as the electrode, as the charge extraction layer, or as the active layer. Excellent device performance (e.g., high efficiency and long lifetime) has been achieved for various PSCs based on CNT and graphene electrode, though their certain properties (e.g., conductivity of CNTs and graphene electrodes) need to be further improved. The unique mechanical flexibility and optical property have made CNTs and graphene very promising for flexible PSCs. Furthermore, the good solution processability and low production cost of graphene oxide offer additional advantages for their use in practical PSCs. Given that current PSC devices with a high content of CNTs/graphene in the active layer still show inferior device performance than expected, there should be considerable room for further improvement of the PSC performance. Continued research and development efforts in this embryonic field could give birth to a flourishing area of photovoltaic technologies.

ACKNOWLEDGMENTS

The authors are very grateful for the financial support from AFOSR, AFOSR-MURI, AFRL, NSF, and NSF-NSFC WMN in the United States and the "Thousand Talents Program" in PR China.

REFERENCES

[1] G. Yu, J. Gao, J. C. Hummelen, F. Wudl, A. J. Heeger, Polymer Photovoltaic Cells: Enhanced Efficiencies via a Network of Internal Donor-Acceptor Heterojunctions. *Science* 1995, **270**, 1789–1791.

[2] Y. Y. Liang, Z. Xu, J. B. Xia, S. T. Tsai, Y. Wu, G. Li, C. Ray, L. P. Yu, For the Bright Future-Bulk Heterojunction Polymer Solar Cells with Power Conversion Efficiency of 7.4%. *Adv. Mater.* 2010, **22**, E135–E138.

[3] T. Y. Chu, J. P. Lu, S. Beaupre, Y. G. Zhang, J. R. Pouliot, S. Wakim, J. Y. Zhou, M. Leclerc, Z. Li, J. F. Ding, Y. Tao, Bulk Heterojunction Solar Cells Using Thieno[3,4-*c*]pyrrole-4,6-dione and Dithieno[3,2-*b*:2',3'-*d*]silole Copolymer with a Power Conversion Efficiency of 7.3%. *J. Am. Chem. Soc.* 2011, **133**, 4250–4253.

[4] H. Y. Chen, J. H. Hou, S. Q. Zhang, Y. Y. Liang, G. W. Yang, Y. Yang, L. P. Yu, Y. Wu, G. Li, Polymer Solar Cells with Enhanced Open-Circuit Voltage and Efficiency. *Nat. Photonics* 2009, **3**, 649–653.

[5]   J. B. You, L. T. Dou, K. Yoshimura, T. Kato, K. Ohya, T. Moriarty, K. Emery, C.-C. Chen, J. Gao, G. Li, Y. Yang, A Polymer Tandem Solar Cell with 10.6% Power Conversion Efficiency. *Nat. Commun.* 2013, **4**, 1446.

[6]   B. C. Thompson, J. M. J. Fréchet, Polymer-Fullerene Composite Solar Cells. *Angew. Chem. Int. Ed.* 2007, **47**, 58–77.

[7]   Z. C. He, C. M. Zhong, S. J. Su, M. Xu, H. B. Wu, Y. Cao, Enhanced Power-Conversion Efficiency in Polymer Solar Cells Using an Inverted Device Structure. *Nat. Photonics* 2012, **6**, 591–595.

[8]   A. K. Geim, K. S. Novoselov, The Rise of Graphene. *Nat. Mater.* 2007, **6**, 183–191.

[9]   A. K. Geim, Graphene: Status and Prospects. *Science* 2009, **324**, 1530–1534.

[10] D. R. Dreyer, R. S. Ruoff, C. W. Bielawski, From Conception to Realization: An Historial Account of Graphene and Some Perspectives for Its Future. *Angew. Chem. Int. Ed.* 2010, **49**, 9336–9344.

[11] L. M. Dai, D. W. Chang, J.-B. Baek, W. Lu, Carbon Nanomaterials for Advanced Energy Conversion and Storage. *Small* 2012, **8**, 1130–1166.

[12] J. Liu, Y. H. Xue, M. Zhang, L. M. Dai, Graphene-Based Materials for Energy Applications. *MRS Bull.* 2012, **37**, 1265–1272.

[13] S. Iijima, Helical Microtubules of Graphitic Carbon. *Nature* 1991, **354**, 56–58.

[14] X. J. Wan, G. K. Long, L. Huang, Y. S. Chen, Graphene-a Promising Material for Organic Photovoltaic Cells. *Adv. Mater.* 2011, **23**, 5342–5348.

[15] S. Cataldo, P. Salice, E. Menna, B. Pignataro, Carbon Nanotubes and Organic Solar Cells. *Energy Environ. Sci.* 2012, **5**, 5919–5940.

[16] A. Iwan, A. Chuchmała, Perspectives of Applied Graphene: Polymer Solar Cells. *Prog. Polym. Sci.* 2012, **37**, 1805–1828.

[17] Y. J. He, Y. F. Li, Fullerene Derivative Acceptors for High Performance Polymer Solar Cells. *Phys. Chem. Chem. Phys.* 2011,**13**, 1970–1983.

[18] D. S. Hecht, L. B. Hu, G. Irvin, Emerging Transparent Electrodes Based on Thin Films of Carbon Nanotubes, Graphene, and Metallic Nanostructures. *Adv. Mater.* 2011, **23**, 1482–1513.

[19] M. G. Kang, L. J. Guo, Nanoimprinted Semitransparent Metal Electrodes and Their Application in Organic Light-Emitting Diodes. *Adv. Mater.* 2007, **19**, 1391–1396.

[20] Y. J. Xia, K. Sun, J. Y. Ouyang, Solution-Processed Metallic Conducting Polymer Films as Transparent Electrode of Optoelectronic Devices. *Adv. Mater.* 2012, **24**, 2436–2440.

[21] J. vande Langemaat, T. M. Barnes, G. Rumbles, S. E. Shaheen, T. J. Coutts, C. Weeks, I. Levitsky, J. Peltola, P. Glatkowski, Organic Solar Cells with Carbon Nanotubes Replacing In2O3:Sn as the Transparent Electrode. *Appl. Phys. Lett.* 2006, **88**, 233503.

[22] T. M. Barnes, J. D. Bergeson, R. C. Tenent, B. A. Larsen, G. Teeter, L. M. Jones, J. L. Blackburn, J. vande Langemaat, Carbon Nanotube Network Electrodes Enabling Efficient Organic Solar Cells without a Hole Transport Layer. *Appl. Phys. Lett.* 2010, **96**, 243309.

[23] M. W. Rowell, M. A. Topinka, M. D. McGehee, H. J. Prall, G. Dennler, N. S. Sariciftci, L. Hu, G. Grüner, Organic Solar Cells with Carbon Nanotube Network Electrodes. *Appl. Phys. Lett.* 2006, **88**, 233506.

[24] R. C. Tenent, T. M. Barnes, J. D. Bergson, A. J. Ferguson, B. To, L. M. Gedvilas, M. J. Heben, J. L. Blackburn, Ultrasmooth, Large-Area, High-Uniformity, Conductive

Transparent Single-Walled-Carbon-Nanotube Films for Photovoltaics Produced by Ultrasonic Spraying. *Adv. Mater.* 2009, **21**, 3210–3216.

[25] J. B. Wu, M. Agrawal, H. A. Becerril, Z. N. Bao, Z. F. Liu, Y. S. Chen, P. Peumans, Organic Light-Emitting Diodes on Solution-Processed Graphene Transparent Electrodes. *ACS Nano* 2010, **4**, 43–48.

[26] S. Bae, H. Kim, Y. Lee, X. F. Xu, J.-S. Park, Y. Zheng, J. Balakrishnan, Y. Lei, H. R. Kim, Y. I. Song, Y. J. Kim, B. Öyilmaz, J. H. Ahn, B. H. Hong, S. Lijima, Roll-to-Roll Production of 30-inch Graphene Films for Transparent Electrodes. *Nat. Nanotechnol.* 2010, **5**, 574–578.

[27] X. S. Li, W. W. Cai, J. H. An, S. Kim, J. Nah, D. X. Yang, R. D. Piner, A. Velamakanni, I. Jung, E. Tutuc, S. K. Banerjee, L. Colombo, R. S. Ruoff, Large-Area Synthesis of High-Quality and Uniform Graphene Films on Copper Foils. *Science* 2009, **324**, 1312–1314.

[28] G. Eda, M. Chhowalla, Chemically Derived Graphene Oxide: Towards Large-Area Thin-Film Electronics and Optoelectronics. *Adv. Mater.* 2010, **22**, 2392–2415.

[29] J. X. Geng, L. J. Liu, S. B. Yang, S. C. Youn, D. W. Kim, J. S. Lee, J. K. Choi, H. T. Jung, A Simple Approach for Preparing Transparent Conductive Graphene Films Using the Controlled Chemical Reduction of Exfoliated Graphene Oxide in an Aqueous Suspension. *J. Phys. Chem. C* 2010, **114**, 14433–14440.

[30] Y. Wang, X. H. Chen, Y. L. Zhong, F. R. Zhu, K. P. Loh, Large Area, Continuous, Few-Layered Graphene as Anodes in Organic Photovoltaic Devices. *Appl. Phys. Lett.* 2009, **95**, 063302.

[31] Y. Wang, S. W. Tong, X. F. Xu, B. Özyilmaz, K. P. Loh, Interface Engineering of Layer-by-Layer Stacked Graphene Anodes for High-Performance Organic Solar Cells. *Adv. Mater.* 2011, **23**, 1514–1518.

[32] H. Park, S. Chang, M. Smith, S. Gradečak, J. Kong, Interface Engineering of Graphene for Universal Applications as both Anode and Cathode in Organic Photovoltaics. *Sci. Rep.* 2013, **3**, 1581.

[33] V. C. Tung, L. M. Chen, M. J. Allen, J. K. Wassei, K. Nelson, R. B. Kaner, Y. Yang, Low-Temperature Solution Processing of Graphene-Carbon Nanotube Hybrid Materials for High-Performance Transparent Conductors. *Nano Lett.* 2009, **9**, 1949–1955.

[34] J. H. Huang, J. H. Fang, C. C. Liu, C. W. Chu, Effective Work Function Modulation of Graphene/Carbon Nanotube Composite Films As Transparent Cathodes for Organic Optoelectronics. *ACS Nano* 2011, **5**, 6262–6271.

[35] H.-L. Yip, A. K.-Y. Jen, Recent Advances in Solution-Processed Interfacial Materials for Efficient and Stable Polymer Solar Cells. *Energy Environ. Sci.* 2012, **5**, 5994–6011.

[36] S. S. Li, K. H. Tu, C. C. Lin, C. W. Chen, M. Chhowalla, Solution-Processable Graphene Oxide as an Efficient Hole Transport Layer in Polymer Solar Cells. *ACS Nano* 2010, **4**, 3169–3174.

[37] Y. Gao, H. L. Yip, S. K. Hau, K. M. Malley, N. C. Cho, H. Z. Chen, A. K.-Y. Jen, Anode Modification of Inverted Polymer Solar Cells using Graphene Oxide. *Appl. Phys. Lett.* 2010, **97**, 203306.

[38] I. P. Murray, S. J. Lou, L. J. Cote, S. Loser, C. J. Kadleck, T. Xu, J. M. Szarko, B. S. Rolczynski, J. E. Johns, J. X. Huang, L. P. Yu, L. X. Chen, T. J. Marks, M. C. Hersam, Graphene Oxide Interlayers for Robust, High-Efficiency Organic Photovoltaics. *J. Phys. Chem. Lett.* 2011, **2**, 3006–3012.

[39] J. Liu, Y. H. Xue, L. M. Dai, Sulfated Graphene Oxide as a Hole-Extraction Layer in High Performance Polymer Solar Cells. *J. Phys. Chem. Lett.* 2012, **3**, 1928–1933.

[40] J. M. Yun, J. S. Yeo, J. Kim, H. G. Jeong, D. Y. Kim, Y. J. Noh, S. S. Kim, B. C. Ku, S. I. Na, Solution-Processable Reduced Graphene Oxide as a Novel Alternative to PEDOT:PSS Hole Transport Layers for Highly Efficient and Stable Polymer Solar Cells. *Adv. Mater.* 2011, **23**, 4923–4928.

[41] J. Kim, V. C. Tung, J. X. Huang, Water Processable Graphene Oxide: Single Walled Carbon Nanotube Composite as Anode Modifier for Polymer Solar Cells. *Adv. Energy Mater.* 2011, **1**, 1052–1057.

[42] X. D. Liu, H. Kim, L. J. Guo, Optimization of Thermally Reduced Graphene Oxide for an Efficient Hole Transport Layer in Polymer Solar Cells. *Org. Electron.* 2013, **14**, 591–598.

[43] D. Yang, L. Y. Zhou, L. C. Chen, B. Zhao, J. Zhang, C. Li, Chemically Modified Graphene Oxides as a Hole Transport Layer in Organic Solar Cells. *Chem. Commun.* 2012, **48**, 8078–8080.

[44] J. Liu, Y. H. Xue, Y. X. Gao, D. S. Yu, M. Durstock, L. M. Dai, Hole and Electron Extraction Layers Based on Graphene Oxide Derivatives for High-Performance Bulk Heterojunction Solar Cells. *Adv. Mater.* 2012, **24**, 2228–2233.

[45] J. X. Geng, T. Y. Zeng, Influence of Single-Walled Carbon Nanotubes Induced Crystallinity Enhancement and Morphology Change on Polymer Photovoltaic Devices. *J. Am. Chem. Soc.* 2006, **128**, 16827–16833.

[46] E. Kymakis, E. Koudoumas, I. Franghiadakis, G. A. J. Amaratunga, Post-Fabrication Annealing Effects in Polymer-Nanotube Photovoltaic Cells. *J. Phys. D Appl. Phys.* 2006, **39**, 1058–1062.

[47] L. Dai, Functionalization of Graphene for Efficient Energy Conversion and Storage. *Acc. Chem. Res.* 2013, **46**, 31–42.

[48] Z. F. Liu, Q. Liu, Y. Huang, Y. F. Ma, S. G. Yin, X. Y. Zhang, W. Sun, Y. S. Chen, Organic Photovoltaic Devices Based on a Novel Acceptor Material: Graphene. *Adv. Mater.* 2008, **20**, 3924–3930.

[49] Y. Li, Y. Hu, Y. Zhao, G. Q. Shi, L. E. Deng, Y. B. Hou, L. T. Qu, An Electrochemical Avenue to Green-Luminescent Graphene Quantum Dots as Potential Electron-Acceptors for Photovoltaics. *Adv. Mater.* 2011, **23**, 776–780.

[50] V. Gupta, N. Chaudhary, R. Srivastava, G. D. Sharma, R. Bhardwaj, S. Chand, Luminscent Graphene Quantum Dots for Organic Photovoltaic Devices. *J. Am. Chem. Soc.* 2011, **133**, 9960–9963.

[51] J. M. Lee, J. S. Park, S. H. Lee, H. Kim, S. Yoo, S. O. Kim, Selective Electron- or Hole-Transport Enhancement in Bulk-Heterojunction Organic Solar Cells with N- or B-Doped Carbon Nanotubes. *Adv. Mater.* 2011, **23**, 629–633.

[52] L. Y. Lu, T. Xu, W. Chen, J. M. Lee, Z. Q. Luo, I. H. Jung, J. I. Park, S. O. Kim, L. P. Yu, The Role of N-Doped Multiwall Carbon Nanotubes in Achieving Highly Efficient Polymer Bulk Heterojunction Solar Cells. *Nano Lett.* 2013, **13**, 2365–2369.

[53] J. M. Lee, B.-H. Kwon, H. Il Park, H. Kim, M. G. Kim, J. S. Park, E. S. Kim, S. Yoo, D. Y. Jeon, S. O. Kim, Exciton Dissociation and Charge-Transport Enhancement in Organic Solar Cells with Quantum-Dot/N-doped CNT Hybrid Nanomaterials. *Adv. Mater.* 2013, **25**, 2011–2017.

[54] C. Li, Y. H. Chen, S. A. Ntim, S. Mitra, Fullerene-Multiwalled Carbon Nanotube Complexes for Bulk Heterojunction Photovoltaic Cells. *Appl. Phys. Lett.* 2010, **96**, 143303.

[55] C. Li, Y. H. Chen, Y. Wang, Z. Iqbal, M. Chhowalla, S. Mitra, A Fullerene-Single Wall Carbon Nanotube Complex for Polymer Bulk Heterojunction Photovoltaic Cells. *J. Mater. Chem.* 2007, **17**, 2406–2411.

[56] B. K. Kuila, K. Park, L. Dai, Soluble P3HT-Grafted Carbon Nanotubes: Synthesis and Photovoltaic Application. *Macromolecules* 2010, **43**, 6699–6705.

[57] D. Yu, Y. Yang, M. Durstock, J. B. Baek, L. Dai, Soluble P3HT-Grafted Graphene for Efficient Bilayer–Heterojunction Photovoltaic Devices. *ACS Nano* 2010, **4**, 5633–5640.

[58] D. Yu, K. Park, M. Durstock, L. Dai, Fullerene-Grafted Graphene for Efficient Bulk Heterojunction Polymer Photovoltaic Devices. *J. Phys. Chem. Lett.* 2011, **2**, 1113–1118.

6

GRAPHENE FOR ENERGY SOLUTIONS AND ITS PRINTABLE APPLICATIONS

Lorenzo Grande[1], Vishnu T. Chundi[2] and Di Wei[3]

[1] *Helmholtz Institute Ulm, Ulm, Germany*
[2] *Department of Materials Science and Metallurgy, University of Cambridge, Cambridge, UK*
[3] *Nokia R&D UK Ltd, Cambridge, UK*

6.1 INTRODUCTION TO GRAPHENE

Graphene is a two-dimensional material with a one atom thick planar sheet of sp^2- bonded carbon atoms that are densely packed in a honeycomb crystal lattice. It is regarded as the thinnest material in the universe with tremendous application potential [1–3]. Graphene has attracted stronger scientific and technological interest [4–10] since the recent award of Nobel Prize in Physics in 2010 [11]. It has shown great promise in many applications, such as electronics [12], energy harvesting and storage devices (supercapacitors [13], batteries [14, 15], fuel cells [16–20], solar cells [21, 22]), and bioscience/biotechnologies [23–28] because of its unique physicochemical properties such as high surface area [4, 8], excellent thermal conductivity [29], electrical conductivity (intrinsic mobility of ~200,000 cm^{2}V^{-1}s^{-1} [4, 30, 31]), and strong mechanical strength [32]. In addition, graphene is one of the few materials available that has excellent conductivity and excellent transparency (nearly 97.7% transmittance [33]).

Many methods have already been developed to produce graphene [8, 34]. In 2004, Geim and coworkers [35] first reported graphene sheets prepared by *mechanical exfoliation* (repeated peeling) of highly oriented pyrolytic graphite (HOPG).

Carbon Nanomaterials for Advanced Energy Systems: Advances in Materials Synthesis and Device Applications, First Edition. Edited by Wen Lu, Jong-Beom Baek and Liming Dai.
© 2015 John Wiley & Sons, Inc. Published 2015 by John Wiley & Sons, Inc.

This method, called *scotch-tape peeling* [7, 8] is still widely used in many laboratories to obtain pristine perfect structured graphene layers for basic scientific research and for making proof-of-concept devices. However, it is unsuitable for mass production. Graphene has also been prepared by *thermal decomposition of SiC* wafer under ultrahigh vacuum conditions [36–38] and by chemical vapor deposition (*CVD*) *growth on metal substrate* (Ru [39], Ni [40, 41], Pd [42], Ir [43], Co [44], Re [45], and Cu [46, 47]) or by *substrate-free CVD* [48]. CVD has the potential to enable large-scale graphene production for electronics applications such as thin film transistors, solar cells, touch panels, and LCDs [49]. Recently, a one-step transfer method of ultralarge graphene area to plastic substrate by a roll-to-roll lamination process without creating additional defects has been proposed. The *roll-to-roll technique* is similar to a newspaper printing press to transfer the graphene between different substrates [50]. A thin graphene layer on transparent PET film is obtained. The transferred graphene film has a transmittance of 96.7% at 550 nm and $1.96 \, k\Omega \, sq^{-1}$ of sheet resistance. In a similar process, sheet resistance of $30 \, \Omega \, sq^{-1}$ at transmittance of 90% has been achieved after p doping with HNO_3 [50, 51] This is comparable to existing state-of-the-art indium tin oxide (ITO) transparent conductors. There are also solution-based methods for graphene synthesis, such as chemical reduction of graphene oxide [52–55], electrochemical exfoliation [56–58], and liquid-phase exfoliation [59–61] that are suitable for printing applications. Different synthesis methods will impart graphene different properties.

6.2   ENERGY HARVESTING FROM SOLAR CELLS

Solar power has already proven to be a vital and sustainable option to be considered in the energy mix, due to the large, constant amount of solar radiation hitting our planet ($3*10^{24}$ joules every year [62]) and the possibility to integrate this technology at every scale (from solar farms to single houses and further down to mobile phones and MEMS devices), as well as to the great performance improvements obtained in the last few decades.

The principle of operation of solar harvesting devices relies on the conversion of energy given by photons into electrical energy (photovoltaic effect); conventionally, this is done by using mono- or polycrystalline-doped semiconductors (Si, Ge, etc.) assembled into p–n junctions, where minority charge carriers are induced by illumination. When representing the operating principle on an I–V chart (Fig. 6.1), one can observe several performance indicators, such as the open-circuit voltage (V_{oc}), the short-circuit current density (J_{sc}), and the fill factor (FF).

V_{oc} is the maximum voltage that can be reached in open-circuit conditions, that is, when the current flow is zero, and gives an indication of the amount of recombination in the device. The short-circuit current J_{sc} is the maximum current that can be drawn from a solar cell and occurs when the voltage across the cell is equal to zero. The FF is given by the ratio of the maximum power attainable to the product of V_{oc} and J_{sc}. It also depends on the diode quality factor, which limits the highest possible FF to 0.86 [64]. These values can be used, together with the power of incident light

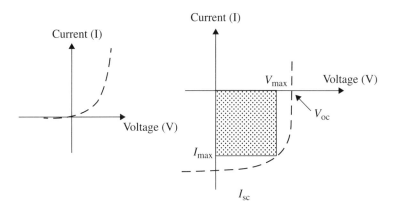

FIGURE 6.1   Schematic of an I–V chart. Reprinted with permission from Ref. 63. © Elsevier.

P_{in}, to evaluate the device's power conversion efficiency PCE (or η), given by following equation:

$$\eta = \frac{J_{sc} \times V_{oc} \times FF}{P_{in}}$$

Air mass (AM) is also one important parameter that is often indicated when describing solar cell performances, as it refers to the direct optical length that sunlight has to travel before hitting the Earth. A value of AM comprised between 1 and 1.1 is typical of tropical latitudes, where incident sunlight is nearly perpendicular throughout the year, while the value of AM 1.5, which refers to temperate latitudes, is commonly taken as a standard, since most of the populated areas in the world lie in these regions. In modern commercial solar panels, power conversion efficiencies as high as 17% at 1.5 AM are attainable [65]. However, generating energy from conventional solar panels can still cost up to 20¢ kWh^{-1}, as compared to three cents for a conventional coal-fired power plant; moreover, Shockley–Queisser efficiency limit for single-junction silicon is around 30%, due to spectrum losses, recombination, and black-body radiation [66], while stacking cells into multijunction configurations might bring this limit up to 68% [67]; finally, given the growth rate of the PV industry, we might soon face a shortage of solar-grade silicon [68]. These factors are only part of the reason why several research groups in the world are now working on different cell chemistries other than conventional semiconductors; these comprise thin film photovoltaics, quantum dots, dye-sensitized solar cells (DSSCs), and organic photovoltaics (OPVs). In this section, we will present the latter two and the improvements brought about by the introduction of graphene.

6.2.1   DSSCs

DSSCs are a relatively old technology, whose operating principle was first observed in the nineteenth century; they acquired a renewed interest during the early 1990s, thanks to the work of Grätzel and others [69], who were able to boost their power conversion efficiencies up to 10% and turn them into valuable competitors for conventional solar cells. DSSCs consist of three parts:

1. A semiconductor oxide (TiO_2, ZnO, SnO_2, Nb_2O_5) deposited onto a transparent conducting oxide (typically fluorine tin oxide, FTO, or indium tin oxide, ITO); the oxide particles are coated with a light-absorbing dye.
2. An organic electrolyte containing a redox couple (I^{3-}/I^-).
3. A counter electrode coated with platinum, where the redox shuttle can be regenerated through a catalytic reaction.

A schematic is shown in Figure 6.2. The principle of operation differs from standard inorganic PV in that no hole/electron pairs are generated by the interaction with photons, but rather excitons, an excited state where an electron and a hole are paired together by a binding energy that is higher than thermal agitation [71].

The reason why this technology had not yielded relatively high efficiencies until Grätzel's contribution is the fact that the oxides used (typically TiO_2 anatase) only absorb a small fraction of the incident light, mostly in the UV spectrum; therefore, a light-sensitive dye coating has to be cast on top of them to ensure sensitivity to a

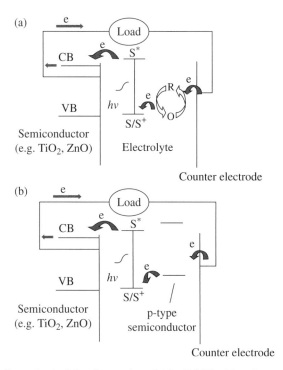

FIGURE 6.2 General principle of operation of (a) a DSSC with redox couple in the liquid electrolyte. (b) a solid state DSSC with a p-type semiconductor. Reprinted with permission from Ref. 70.

broader spectrum range. A monomolecular layer of cis-$RuL_2(SCN)_2$ dye, a ruthenium complex where L stands for 2,2'-bipyridyl-4-4'-dicarboxylate, can however only give a 0.13% incident photon-to-current efficiency (IPCE) at its 530 nm absorption peak when deposited on a flat TiO_2 surface; adding extra layers would further decrease this value by effectively creating a series of filters for light absorption. The use of nanostructured TiO_2 solved this problem by allowing the adsorption of larger amounts of dye and hence overcoming the intrinsic limitations of a flat, unstructured morphology. This resulted in a 600-fold increase in IPCE and paved the way to an organic-based solar harvesting alternative.

Despite these groundbreaking advancements, progress in increasing DSSC performances has since been slow and only incremental. Several reviews on DSSCs have been published in recent years, focusing on the state of the art, the perspectives, and the new cell chemistries that can be considered [70, 72–74]. In recent years, the ubiquitous applications found for graphene, a bidimensional allotropic form of carbon, have fueled research into the solar harvesting field too. In the following is a review of the most relevant as well as recent works in the field of DSSCs.

6.2.2   Graphene and DSSCs

6.2.2.1   Counter Electrode   One of the applications of graphene in DSSCs has mainly been devoted toward the replacement of the Pt electrocatalyst at the counter electrode. Platinum is generally added in relatively small amount (<0.1 g m^{-2}), but its high cost and possible side reaction with the iodide/triiodide redox shuttle have driven researchers toward cheaper and more reliable alternatives. Among the first who explored carbonaceous materials are Kay and Grätzel themselves, who added around 20% carbon black to a graphite dispersion and produced a counter electrode for a low-cost solar cell [75]. Later, they also tried grinding together carbon black and TiO_2 to get a 15 μm thick carbon film, which could deliver an outstanding PCE of 9.1% [76]. These promising results led to further efforts from other groups (including ours) who went on to compare the PCEs obtained with graphite, activated carbon, and single- and multiwalled carbon nanotubes (MWCNTs) [77, 78]; the highest efficiency (7.7% under AM 1.5) was obtained by Lee et al., by using defect-rich multiwalled carbon nanotubes [79]. More recently, an interesting approach on the production of large-effective-area carbonaceous materials was provided by Lee and coworkers [80]. Large-effective-surface-area polyaromatic hydrocarbons (LPAH) were produced via a hydrogen arc discharge and then assembled with a graphite film to give an all-carbon counter electrode. The average surface roughness of this material was 37.4 nm and contributed to the production of a DSSC with a remarkable PCE of 8.63% and a FF of 80%. Hung et al. [81] obtained graphene oxide (GO) pastes via freeze-drying, which increased the material's porosity, advisable for increased wettability and thus higher number of reduction sites for I^{3-}. This is confirmed by the attained PCE of 6.21%, compared to 5.62 of a blank DSSC. GO was also produced via the Hummer's and the Staudenmaier method [82, 83], but despite obtaining relatively small flakes (down to 5 and 19 nm, respectively), results from DSSC tests were not encouraging (Fig. 6.3).

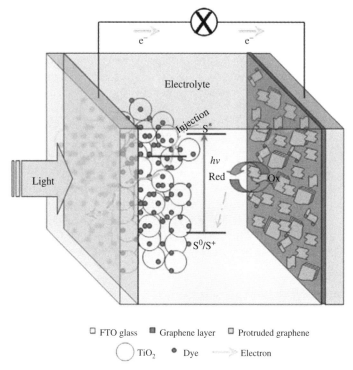

FTO glass   Graphene layer   Protruded graphene

TiO$_2$ Dye   Electron

FIGURE 6.3   Schematic diagram of GN-based DSSC. Reprinted with permission from Ref. 84. © Elsevier.

Early attempts at using graphene-like materials can be ascribed to Meng and coworkers, who carbonized sugar at 1000°C in an Ar atmosphere and then milled this powder to obtain 1 µm-sized particles. These were then screen printed onto FTO together with conductive carbon black and PVdF in NMP [85]. Through activation at 900°C in water vapor, it was possible to get a PCE of 5.7%, as compared to 4.7% with an untreated powder. An alternative, production-oriented approach is the one followed by Shi et al., who prepared a water-stable dispersion of 1-pyrenebutyrate-functionalized graphene sheets that was then spin coated onto ITO. The performance of cells built this way was however very poor (PCE: 2.2%) due to the low quality of the graphene [86]. This is what led the same group to try functionalization by means of the more conductive polystyrene sulfonate doped poly(3,4-ethylenedioxythiophene) (PEDOT:PSS), which has a superior film-forming ability [87]. A film thickness of 2.1 nm was recorded, as well as a PCE that was more than double the one previously obtained (4.5%), but still lower than that of a reference Pt cell (6.3%). In fact, despite analogous values of J_{sc} and V_{oc}, the difference is largely due to a much lower FF (0.48 vs. 0.68). What can be noted here is that even small amounts of graphene (1 wt% in this case) can actively perform as electrocatalysts. In comparison, the electrodes by Meng previously described contain a whopping 75 wt% of carbon. Wu et al. electro-deposited the PEDOT:PSS film onto graphene flakes with a weight ratio varying from 0.01 to 0.15 wt%, using a current density of 10 mA cm^{-2}, and subsequently coated the composite onto FTO [88]; the best performances were obtained with the DSSCs electrodes containing 0.05 wt% graphene: the recorded PCE was higher than the reference Pt cell (7.86 vs. 7.31%) while having similar charge transfer resistance values (2.74 vs. 2.82 Ω cm^2). Huang and coworkers produced graphene nanosheets through the conventional hydrazine reduction pathway but then created a paste with a 5 wt% of ethyl cellulose in terpineol, which after screen printing onto FTO and heating at different temperatures gave ready-to-use, efficient (6.89% PCE) DSSC counter electrodes [84]. Attempts with hot filament CVD have also been made, with modest results [89].

Traditionally sputtered Pt thin films are generally 20–40 nm thick. The large amounts of graphene needed to compensate for the absence of Pt have led to intermediate solutions to the complete replacement of Pt [90]. Ma et al. were the first ones to report a Pt–graphene composite material to be used as a counter electrode in DSSCs [91]; they added an aqueous solution of H$_2$PtCl$_6$ and ethylene glycol to a dispersion of graphene sheets in deionized water, then heated at 120°C, washed, centrifuged, and then dried the resulting black powder. The Pt nanoparticles, which had a 4–7 nm average size, made up 17 wt% of the powder and showed an astonishing 6.35% PCE, in comparison with 2.89% of Pt-free graphene electrodes and 5.27% of a reference cell with a 50 nm thick Pt layer. Misra and coworkers, instead, used thermally reduced few-layer graphene (FLG) where Pt nanoparticles were deposited via PLD. With 27% Pt loading, the DSSC exhibited a PCE of 2.9%, nearly 30% higher than that of a control reference cell with a Pt layer of 20–25 nm thickness [92]. A "greener" approach was followed by Kim et al., who synthesized graphene starting from common vitamin C (ascorbic acid); they performed a coreduction of GO and H$_2$PtCl$_6$ in ascorbic acid to give a water-dispersible Pt–graphene nanohybrid [93] (Fig. 6.4). A further Pt nanoparticle doping followed by thermal annealing of the CE gave the fabricated device a PCE of 8.9%, which is comparable to that of conventional, state-of-the-art DSSCs.

Graphene can also be used as a substrate where cheaper-than-Pt electrocatalysts can be deposited. Gao et al. embedded Ni$_{12}$P$_5$ particles, which have electrocatalytic activity, within graphene sheets [94]. The procedure used involved hydrothermal reduction of red phosphorus, nickel chloride, and GO in an ethylene/water solution at 180°C. The obtained nanocrystallites, with an average size of 20–40 nm, showed promising performances when coupled with graphene (PCE 5.7%). NiS$_2$ has been also considered as a Pt substitute in reduced graphene oxide (rGO) composite

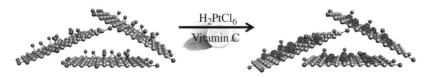

FIGURE 6.4 Schematic description of the preparation of GNS/Pt-NHB-based CEs. Reprinted with permission from Ref. 93. © ACS.

electrodes [95]; such DSSCs showed performances (8.55% PCE), nearly analogous to those of Pt-based electrodes.

Functionalization of graphene sheets can also occur by making use of oxygen-rich defective sites. Roy-Mayhew and coworkers [96] took on the work on PEDOT:PSS by showing that the C/O ratio in graphene films, determined by the amount of hydroxyl, carbonyl, and epoxy groups on the surface, as well as lattice defects, can be tuned by choosing the appropriate annealing temperature. The resulting material can be then used as an ink and cast onto nonconductive substrates, to give performances which are only 10% lower than traditional Pt electrodes. Lee et al. introduced nitrogen chemical doping into graphene sheets to lower their sheet resistance down to $60 \, \Omega \, sq^{-1}$ and then spin coated them with a PEDOT film [97]. The composite film, 2.2 nm in thickness and with 70% transmittance at 500 nm, showed comparable performances to those of a reference Pt/ITO counter electrode when assembled into a DSSC device (6.26 vs. 6.68%, respectively). Other graphene composites include those obtained by grafting MWCNTs at 900°C [98]; however, the recorded performance was relatively poor (PCE 3%). Other groups unzipped MWCNTs to produce graphene nanoribbons (GNR) for DSSCs [99].

In several of the works mentioned so far, graphene has been either produced through high temperature exfoliation (900–1100°C) or via chemical reduction. These pathways not only employ extremely dangerous and toxic compounds like hydrazine but also are neither cost effective nor easy to scale up. Ramaprabhu et al. have shown the advantages of lowering the operating temperature to 200°C while using H_2 as reducing agent [100]. The same group performed graphene thermal exfoliation at 1050°C under Argon atmosphere [101], but surprisingly it was possible to obtain graphene in gram quantities and with better quality sheets through the first method. As a matter of fact, in the case of H_2 exfoliation, a PCE of 3.61% was achieved, compared with 2.8% of thermally exfoliated graphene. Others tried obtaining graphene-functionalized counter electrodes via electrophoresis, by creating a dispersion of GO in magnesium nitrate [102]; then, a voltage of 30 V was applied for a short time (5, 10, 15 or 30 s) to deposit 5–16 layers of graphene onto FTO. Unfortunately, despite these sheets had a transparency of 55% at 550 nm and a remarkably low charge transfer resistance (between 9.3 and 12.7 Ω), the resulting DSSCs only showed modest PCEs (1.3–2.3%).

6.2.2.2 Photoanode As it can be seen, most of the efforts of incorporating graphene into DSSCs have been devoted to the replacement of expensive platinum and/or transparent conducting oxides (TCO) on the counter electrode. However, graphene's unique versatility has also shown that it can be a good candidate in photoanodes to replace conventional semiconductor oxides like TiO_2, ZnO, SnO_2, Nb_2O_5, and so on. Already in 2008, Hamann et al. had foreseen graphene's advent without directly mentioning it: "an ideal photoanode should be highly transparent, have high surface area and porosity, be easy to fabricate and exhibit fast electron transport [64]." TiO_2 is cheap, naturally abundant, and environmental friendly, but it has a low electron diffusion coefficient and limits the choice of new dyes and redox couples that can be used. On the other hand, it has been proven that when added to TiO_2, graphene can enhance the photocatalytic activity of the latter [103, 104], because of the highly porous anodes that can be fabricated and onto which larger amounts of dye can be chemisorbed. Moreover, graphene is a zero band gap material, as opposed to TiO_2 (3.2 eV), and this property can be exploited to prevent charge recombination during the electron transfer process. Nair et al. have created TiO_2–graphene composites via electrospinning, which yielded 150 nm thick fibers [105]. By adding 0.7 wt% of graphene, they were able to increase the short-circuit current of a DSSC device from 13.9 to 16.2 mA cm^{-2}, thereby obtaining a higher PCE than a TiO_2-only control cell (7.6 vs. 6.3%). The synergic contribute of graphene–TiO_2 composites with as little as 0.5 wt% graphene content has also been proven via alternative syntheses, such as ball milling [106], heterogeneous coagulation from Nafion [107], spraying [108], and simultaneous reduction hydrolysis of GO with a Ti precursor [109], good for preventing the collapse and restacking of the sheets, even with a graphene content as high as 5 wt%. In all cases, graphene was able to improve the cell performance. The size of TiO_2 particles attached to the graphene sheets was tuned by He et al. [110] who could produce nanometer-sized spherical particles as well as TiO_2 nanorods and graft them onto graphene via a solvothermal approach. The device containing ultrasmall TiO_2 particles (2 nm) showed the best performance with a PCE of 7.25%.

Among alternative methods to produce multicomposite graphene photoanodes, the work by Jung et al. shows an interesting approach [111]: graphene-wrapped alumina particles were prepared by the coreduction of CO gas and AlN and later mixed with TiO_2; this composite, which only contained 1 wt% graphene, showed an 11% increase in performance as compared to a graphene-free device. Ma and coworkers tried to reduce charge recombination at the photoanode by grafting acid-treated MWCNTs onto graphene [112]. The obtained TiO_2-nanostructured hybrid paste, with a 0.07 wt% content of MWCNT and 0.03 wt% graphene, showed a charge transfer resistance of 13 Ω cm^2 as compared to a TiO_2-only electrode (25.34 Ω cm^2); as expected, an improvement in PCE was attained, not only due to a lower charge recombination but also to a higher degree of dye adsorption.

6.2.2.3 Transparent Conducting Oxide Graphene's considerable optical transparency has also encouraged its use as an alternative to TCOs, such as in a pioneering work by Müllen [113], who fabricated a 10 nm thick graphene window electrode via thermal reduction of GO, with a transmittance of 70% between 1000 and 3000 nm. Graphene can also be used to improve the interface junction between TiO_2 and the TCO (typically FTO). The earliest and best resulted reported so far is that of Li and coworkers [114], who spin coated the FTO surface with graphene sheets of 0.3–3 µm in size. Repeating the spin-coating process for three times only lowered FTO's transmittance from 84.6% down to 82.1% at 651 nm, indicating excellent optical transparency retention. A composite 0.02 wt% graphene–TiO_2 paste with an 11 µm thickness was deposited on top of FTO, and the device thus fabricated reported excellent performances, such as a PCE of 8.13%, as compared to 5.8% for a blank sample. Others also attempted to improve the TiO_2–FTO junction by using UV-reduced GO [115].

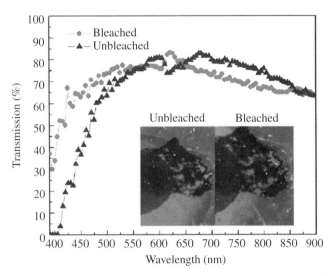

FIGURE 6.5 Enhanced transparency due to the addition of graphene nanoribbons in the electrolyte. Reprinted with permission from Ref. 99. © ACS.

6.2.2.4 Electrolyte A developing area is represented by quasisolid-state electrolytes. Gun'ko et al. incorporated graphene into the ionic liquid 1-methyl-3-propylimidazolium iodide [116]. They found out that a 30 wt% addition of graphene resulted in a quasisolid-state electrolyte that helped deliver a 2.1% PCE without causing an internal short circuit. If 12 wt% graphene and 3 wt% CNTs were incorporated to obtain a hybrid material, this value could be boosted up to 2.5%, with an increase in J_{sc} from 5.3 to 7.32 mA cm^{-2} but a slight drop in both V_{oc} and FF, down from 0.62 to 0.59 V and from 0.49 to 0.44 respectively. Others incorporated graphene into polyacrylonitrile (PAN) to make a gel polymer electrolyte in much smaller amounts (0.1–1 wt%) [117], and a PCE of 5.41% was obtained, much better than its liquid-state equivalent (3.72%). An interesting finding by Zakhidov et al. [99] (Fig. 6.5) was the bleaching effect of GNR when added to the electrolyte, that is, their ability to increase its transparency. They used GNRs both as a counter electrode and as an electrolyte additive.

What was found was that, despite a slight decrease in photocurrent, the overall efficiency was kept constant via a boost in the FF. The unzipping of MWCNTs led to flakes with a GNR morphology, which were then drop-cast onto FTO. The same GNRs were added as a suspension with an estimated concentration of 0.04 mg ml^{-1}, and a 20% boost in PCE was observed with respect to DSSCs made with an unmodified electrolyte.

6.3 OPV DEVICES

OPVs go by different names, including polymer solar cells, and represent the other big fields into which alternative solar cell chemistries are developing. In their modern aspect, OPV cells consist of an organic bulk heterojunction (BHJ) sandwiched between

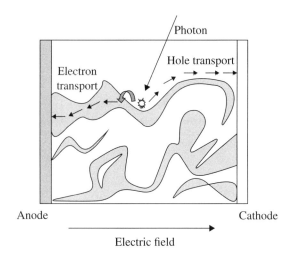

Photon

Hole transport

Electron transport

Anode

Cathode

Electric field

FIGURE 6.6   Schematic of a bulk heterojunction solar cell. Reprinted with permission from Ref. 118. © Elsevier.

a TCO and a semitransparent metal electrode; the BHJ is made of two interpenetrating block copolymers with different electronic properties, namely, with a p-type (or donor) and an n-type (or acceptor) behavior. The interface between the two polymers is where excitonic hole/electron pairs are generated, and dissociation occurs once they reach the metal interface (Fig. 6.6).

Although lower PCEs have been attained so far, as compared to DSSCs, a lot of attention has been devoted to this technology, since it will pave the way for easy-to-manufacture, cheap, and flexible electronic devices. OPV cells can be fabricated via a wet processing, which does not require the high temperatures generally needed to obtain high-quality silicon, leading to a reduction in energy use as well as energy payback time. Furthermore, it has been forecast that OPV cells in a tandem configuration (i.e., where different materials are stacked on top of each other to absorb different parts of the spectrum) might drive cell efficiencies up to 15%, thus turning polymer solar cells into a clean and sustainable alternative to inorganic PV, even for large-scale energy production [119]. As of April 2012, the world record for commercial OPV cells has already reached 12% [120].

6.3.1   Graphene and OPVs

6.3.1.1   Transparent Conducting Oxide   OPV cells have benefited too from the incorporation of graphene within the cell chemistry. Most of the work has been devoted to its use as a transparent electrode, due to its already mentioned exceptional optical transmittance across a wide range of the visible spectrum. Graphene can provide an economically attractive alternative to ITO while at the same time enabling applications in next-generation IT devices, thanks to its flexibility, which ITO lacks in [121] (Fig. 6.7).

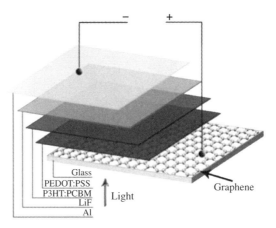

FIGURE 6.7 Incorporation of a graphene layer within an OPV cell. Reprinted with permission from Ref. 122. © AIP.

Among the first reports is the one by Müllen et al. [21], who transposed their previous work on DSSCs onto OPVs, showing that graphene films with an average thickness of 4–30 nm can be obtained from thermal fusion of LPAHs and with transparencies up to 90% at 500 nm that only dip down to 85% when the material is assembled into a functioning device. With a sheet resistance of $1.6\,k\Omega\,sq^{-1}$ and a roughness of 0.4–0.7 nm, the OPV cells fabricated had a 1.53% PCE when exposed to monochromatic light at 510 nm and 0.29% when tested under simulated solar light. Peumans et al. [22] thermally reduced GO and obtained graphene sheets with a thickness less than 20 nm, a transmittance higher than 80% at 550 nm, and a sheet resistance varying between $5\,k\Omega\,sq^{-1}$ and $1\,M\Omega\,sq^{-1}$; transmittance values of 95% and sheet resistances of $100\,k\Omega\,sq^{-1}$ were also obtained with smaller 4–7 nm thick graphene films. A proof-of-concept OPV cell was assembled and showed a promising PCE of 0.4%, nearly half as much as a reference ITO-based cell. Others have focused on cost reduction by adopting solution-processable graphene [123] or naturally occurring starting products, such as camphor [124]. Significant improvements in performance have rapidly occurred since the early developmental stage, with works like that of Loh [122], where CVD-grown FLG was shown to have similar size and transparency as in previous works (6 nm and 91% at 550 nm, respectively) while lowering the sheet resistance down to $210\,\Omega\,sq^{-1}$, thanks to the noncovalent functionalization with pyrenebutanoic acid succinimidyl ester (PBASE). An increase in graphene's work function from 4.2 to 4.7 eV as well as improved contact with PEDOT:PSS led to a PCE of 1.71%. Others have also used fluorine-functionalized graphene [125] or $AuCl^{3-}$ decorated sheets [126].

One problem often ascribed to graphene is the possibility to produce it on a large scale. The CVD approach offers several advantages, due to the possibility to grow large, continuous films with a high degree of uniformity, which are also easy to transfer to other substrates; moreover, control of the synthesis procedure allows a good tuning of the final film properties. These unique selling points were the driving force behind the work of Gomez et al. [127], who lowered sheet resistances down to 230 Ω sq^{-1} at 72% transmittance or 91% transparency with 8300 Ω sq^{-1} resistance for 1.3 nm thick sheets. The average roughness of 0.9 nm confirmed that highly regular sheets are paramount in order to reach such performances. Moreover, they proved that such a technique can be scaled up for industrial production; the efficiency output remains comparable to that of standard devices (PCE 1.18 vs. 1.27% for ITO-based OPVs), but devices made of graphene sheets can be bent up to 138% without showing a drop in performances. A layer-by-layer method developed by Loh et al. allowed a controlled deposition of up to 8 CVD-grown, acid-doped graphene monolayers on top of each other; the device yielded a PCE of 2.5%, more than 80% that of a control ITO device [128]. Yin et al. also focused on CVD-grown GO [129], preparing the active layer through inexpensive spin coating instead of thermal evaporation such as in the previous case. 4–21 nm thick graphene sheets were produced, with a transparency varying between 55 and 88% at 550 nm and a sheet resistance between 1.6 and 16 kΩ sq^{-1}. In this work, it is shown that a transmittance of 65% is the boundary above which lowering the sheet resistance will have a positive influence over the device performance. Incremental improvements are taking place in order to produce graphene sheets with a higher degree of perfection [130].

Insofar, graphene-based OPVs still lag behind the performance typically exhibited by ITO devices, which can boast PCE values up to 7% when in a tandem configuration. Lee et al. were the first ones to break the 2% PCE proof-of-concept barrier [131], by producing 15-layer stacks of CVD-grown graphene sheets, adding a TiO$_x$ hole-blocking layer between P3HT:PCBM and the Al cathode, reaching 2.6% PCE. The sheet resistance can also be lowered through doping, as shown in a work by Yan et al. [132], where the conductance of a single layer was increased by a factor of 4 using Au nanoparticles and PEDOT:PSS. Illumination from the graphene side gave a PCE of 2.7%, which decreases to 2.3% when the active area was increased from 20 to 50 mm^2, due to the diminished edge effect, often a cause for the overestimate of the real performance.

6.3.1.2 BHJ
Another prominent area of uses found for graphene in OPVs is BHJ. [6,6]-Phenyl C$_{61}$-butyric acid methyl ester (PCBM) is traditionally used as electron acceptor, while poly(3-hexylthiophene-2,5-diyl) regioregular (P3HT) and poly(3-octylthiophene) (P3OT) are generally used as electron donors. These materials provide for a good HOMO/LUMO match, but allotropic forms of carbon could be used to have a stronger light absorption, higher charge mobility, as well as stability. The first report in this area was published by Chen et al. [133] where it was shown that PCBM could be replaced with solution-processable functionalized graphene (SPFG) and mixed with P3OT to get a PCE of 1.4%. An optimum in performance could be achieved with 5 wt% SPFG, as well as by controlling the temperature and time during the annealing step. In the authors' view, adding a sufficient amount of graphene is crucial in order to form a sufficiently continuous donor/acceptor interface inside the electron donor matrix, while exceeding this value might favor the formation of aggregates, which are deleterious for charge transport and separation within the active layer. Mixing graphene with P3HT was also tested by the same group [134, 135], and

in this case, lower efficiencies (1.1%) were achieved with twice as much graphene content (10 wt%). He et al. used the same amount of SPFG but functionalized it with phenyl isocyanate in order to make it water soluble [136]. The annealing temperature was also lowered from 160 to 120°C without showing significant losses in terms of performance. In a successive work, functionalized MWCNTs were incorporated, and an improvement in carrier mobility, exciton splitting, and suppression of charge recombination increased the device's PCE from 0.88 to 1.05% and J_{sc} from 3.72 to 4.7 mA cm^{-2} [137].

6.3.1.3 Hole Transport Layer   Graphene can also be found in BHJ as a hole transport layer (HTL). This role is usually covered by poly(ethylenedioxythiophene):poly (styrene sulfonate) (PEDOT:PSS), thanks to its high electrical conductivity (up to 10 S cm^{-1}) that can enhance hole collection from ITO; however, some negative drawbacks of this polymer, such as microstructural inhomogeneities and its corrosive behavior at the electrode interface, have encouraged research toward alternative materials. A layer of 2–3 nm thick GO was initially found to increase P3HT's electrical conductivity by a factor of 6 with respect to PEDOT:PSS, due to the resulting protonic doping of the surface of P3HT [138]. When assembled into an OPV cell and sandwiched between a P3HT:PCBM layer and the TCO, graphene oxide could help deliver a PCE of 3.5–3.61% both in direct [139] and inverted configurations [140], comparable to that of control devices containing PEDOT:PSS; functionalization with molybdenum [141] and a NiO$_x$ film [142] has also been made, generally showing an improvement over the GO-only devices. Green pathways have been pursued, such as in the work by Yun et al. [143], who used *p*-toluenesulfonyl hydrazide (*p*-TosNHNH$_2$) instead of hydrazine, giving similar results. GO has also been found useful as an additive to PEDOT:PSS rather than just a substitute, with encouraging results [144].

6.4   LITHIUM-ION BATTERIES

Lithium-ion batteries are widely regarded as the current main energy storage technology for mobile devices, laptops, and hybrid electric vehicles. The reason of their success lies in several technical advantages they have over other battery chemistries, such as high operating voltage (4 V) and a specific energy in the 100–150 Wh kg^{-1} [145]. The principle of operation is the so-called "rocking chair" model: two electrodes with different standard potentials can reversibly intercalate lithium ions within their structure, and the shuttling of the lithium ions from the positive to the negative electrode allows for energy to be stored in the system. This value is often dependent on the speed at which a battery is discharged, that is, the C rate: a value of 1C corresponds to the current that has to be applied to allow for a complete discharge (or charge) in 1 h and is often taken as the reference value. A 10C implies a 10 times faster charge–discharge rate, while C/10 indicates a process which is 10 times slower than 1C (Fig. 6.8).

The growing demands from the consumers' view, as well as the utility companies in terms of energy storage capability, have raised some doubts as to whether Li-ion

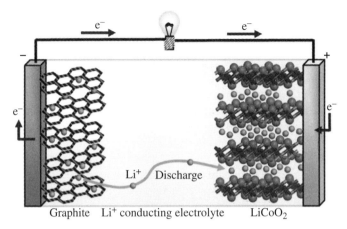

FIGURE 6.8 Schematic of a lithium-ion battery. Reprinted with permission from Ref. 146. © Elsevier.

batteries can fulfill the need for a system that is able to store large amounts of energy in a small enough, lightweight device in the future. One way of looking at this is the capability that the anode and the cathode have of storing lithium ions, mainly defined as specific capacity. A typical anode, such as graphite, has a specific capacity of $372 \, \text{mAh g}^{-1}$, while cathodes are typically in the $140–170 \, \text{mAh g}^{-1}$ range. This means that nowadays, the main challenge lies on the cathode side, since a doubling of capacity will lead to an immediate weight reduction of the battery pack, while an analogous increase on the anode side will lead to an increase in active material needed for the positive electrode. Other ways of improving the energy density include the development of high-voltage cathodes, that is, compounds that can generate a higher cell potential. Other less conventional approaches involve the development of completely different cell chemistries, implying a switch from the "rocking chair" concept to conversion materials.

6.4.1 Graphene and Lithium-Ion Batteries

6.4.1.1 Anode Material Several reviews, to this day, have been written with a specific focus on the overlap between electrochemical energy storage and graphene [147–151]. Intuitively, the immediate application of graphene in lithium-ion batteries is as a replacement to common graphite. Graphite's specific capacity is limited by the fact that it takes six carbon atoms in a ring configuration to coordinate one lithium ion (LiC_6). Graphene, as a two-dimensional crystal, can theoretically allow two lithium ions to be coordinated, by having one on each side of the same carbon ring (Li_2C_6). In this way, a doubling of the capacity is attainable, by reaching $744 \, \text{mAh g}^{-1}$, with a theoretical limit of $1176 \, \text{mAh g}^{-1}$ forecast by Sato et al., if Li_2 covalent molecules can form [152]. Our group achieved hitting the $750 \, \text{mAh g}^{-1}$ target by producing graphene via electrochemical exfoliation in different ionic liquids [153]. Others have pushed this boundary up to $1488 \, \text{mAh g}^{-1}$ by assuming a

Li_4C_6 stoichiometry [154]. Initial screenings on graphene paper gave modest results, with a specific capacity dropping to only 82 mAh g^{-1} after the first cycle [155]; however, the promising values reached during the first charge (680 mAh g^{-1}) fueled research into this field. Graphene nanosheets with an interlayer spacing of 0.365 nm and a thickness corresponding to 10–20 layers were obtained and cycled as anodes on their own or in combination with CNT and C_{60} molecules [14]. Indeed, a higher capacity than that of common graphite was observed, and despite a quick capacity fading, after 20 cycles cells made of pure graphene anodes still showed 540 mAh g^{-1}. An increase in d spacing was correlated to an enhanced lithium storage capability, but other factors such as sheet disorder, defects, surface area, and functional groups have also been suggested to add up toward higher values of specific capacity [156]. Pan et al. were among the first who tried tuning these parameters by reducing GO at different temperatures, with hydrazine or via electron beam irradiation, proving that after 16 cycles capacities around 800 mAh g^{-1} could still be retained. A control on the morphology is what led to the design of hollow carbon microspheres obtained by impregnating a silica template in hexa-peri-hexabenzocoronene (HBC), pyrolyzing the material up to 1000°C, and then etching away the silicon matrix [157] (Fig. 6.9). The final product was a hollow sphere that was functionalized by several graphene platelets, with a good electrical contact ensured by the inner graphitic walls. A good reversible capacity of around 600 mAh g^{-1} was retained still after 150 cycles at a C/5 rate, which was lowered by a factor of three when operating at 10C, but a large capacity loss of about 1000 mAh g^{-1} was however observed during the first cycle.

Wang et al. showed that after 100 cycles a value of 460 mAh g^{-1} could still be retained, which corresponds to an improvement of about 20% from conventional graphitic carbon, but again a drop from 950 to 650 mAh g^{-1} was observed after the first cycle [158]. This phenomenon, which goes by the name of solid electrolyte interphase (SEI) formation, is the creation of an organic film on top of graphite from the decomposition of the electrolyte. The formation of this organic film is necessary

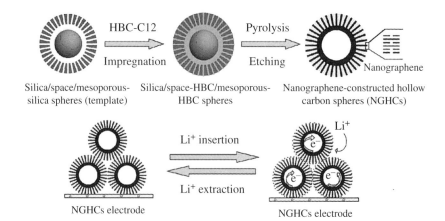

FIGURE 6.9 Hollow, graphene-functionalized carbon spheres as anode materials. Reprinted with permission from Ref. 157.

to prevent lithium from irreversibly plating on top of the anode, a fact that would stop the intercalation process within the graphene planes. SEI formation is unavoidable and is largely surface dependent, posing a real limitation to graphene, due to its extremely high surface area ($2600\,m^2\,g^{-1}$). This is particularly true for graphene nanosheets, where the irreversible capacity loss can be as high as 50%, down to 672 from $1233\,mAh\,g^{-1}$ during the first charge–discharge cycle [159]; furthermore, nanocavities and defects can greatly contribute to the performance suppression of graphene anodes. Another cause for graphene's instability lies in the decomposition of oxygen-containing functional groups present on its surface during delithiation; this can lead to a partial oxidation of the electrolyte and induce a quick drop in performance as well as potentially dangerous side reactions. Finally, a significant downside related to graphene anodes is the lack of a voltage plateau during discharge, an issue that poses serious limitations for practical uses.

To address such problems, the focus has been shifted to doped and composite anode materials. Phosphorus, nitrogen, and boron are the most common doping agents used so far for common graphite; thus, an improvement in performance could also be expected by transferring this approach on graphene. By doping graphene with nitrogen, for instance, it is possible to create more active sites as well as increase the interactions with lithium ions, enhancing the anode's electronic behavior. B doping and N doping of graphene were performed via heat treatment with either BCl_3 or NH_3 at 800 and 600°C, respectively [160]. These electrodes showed high charge rates, with a full charge occurring in 30 s, as well as a high capacity when cycled at lower currents ($1549\,mAh\,g^{-1}$ for the B-doped anode and $1043\,mAh\,g^{-1}$ for the N-doped material, after 30 cycles). Others showed that the increase of defect sites in nitrogen-doped graphene nanosheets (N-GNS) during cycling can be beneficial for an improvement in performance [161]. Despite initial values similar to graphite, the specific capacity of N-GNS constantly rose while cycling, until reaching a stable value of around $684\,mAh\,g^{-1}$ after 500 cycles, at a current rate of $100\,mA\,g^{-1}$.

Another pathway is that of using graphene to create composites. Tin and silicon for instance can host large amounts of lithium ions within their structure by creating alloys ($994\,mAh\,g^{-1}$ for Li–Sn and $4200\,mAh\,g^{-1}$ for Li–Si), but they suffer from large volumetric expansions during the alloying process, a phenomenon which then leads to cracking and pulverization during the delithiation step. This limits the possibility of using such materials in rechargeable batteries. Graphene can be used jointly to buffer the volume changes without sacrificing the anode's energy density. The reaction of tin with lithium can be summarized as follows:

$$4.4\,Li^+ + Sn + 4.4e \leftrightarrow Li_{4.4}Sn$$

However, the Sn atom aggregation rapidly induces a loss of capacity that affects the performance. To prevent this, graphene sheets can be used as a matrix that can be functionalized with Sn atoms; this will not only allow for a reversible alloying–dealloying reaction but also prevent the graphene sheets from restacking. Wang et al. were among the first ones to adopt this approach and were able to obtain graphene–Sn composite electrodes that gave a reversible $508\,mAh\,g^{-1}$ capacity after 100 cycles,

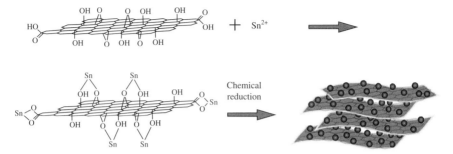

FIGURE 6.10   Sn-decorated graphene sheets. Reprinted with permission from Ref. 162. © RSC.

at a $55\,mA\,g^{-1}$ current density [162]. The improvement is even more remarkable if it is compared to that of pure graphene, which only delivered $255\,mAh\,g^{-1}$, and to that of Sn anodes, which failed working after only 10 cycles. Incremental improvements include depositing hollow carbon structures onto graphene sheets and use these shells to accommodate the alloying products. In this way, it was possible to obtain a reversible capacity of $662\,mAh\,g^{-1}$ after 100 cycles at $100\,mA\,g^{-1}$ [163] (Fig. 6.10).

The same principle can be adopted for SnO_2, which has a theoretical capacity of $782\,mAh\,g^{-1}$ thanks to the conversion reaction:

$$4\,Li^+ + SnO_2 + 4e \leftrightarrow 2\,Li_2O + Sn$$

A SnO_2/C composite could give a stable $625\,mAh\,g^{-1}$ discharge capacity for 100 cycles at $10\,mA\,g^{-1}$ [164]. In this nanocomposite, graphene only contributed with $48\,mAh\,g^{-1}$ to the overall capacity, with the SnO_2 accounting for the largest share of the performance improvement. The synergic effect brought about by the combination of graphene and tin oxide nanoparticles is clear in the work by Yao et al., where SnO_2 alone can only charge reversibly for around 20 cycles before its breakdown, while pure graphene suffers from an early capacity fading that leads to low values of discharge capacity; the combination of the two, however, can buffer the latter effect with a $520\,mAh\,g^{-1}$ reversible capacity while ensuring a long cycle life [165]. A 10-fold boost in current density was achieved by Kim et al., who managed to obtain $634\,mAh\,g^{-1}$ from a similar anode after 50 cycles while retaining comparable values of specific capacity, with a promising outlook on high rate performances [166]. The advantages in fast discharge given by SnO_2/C chemistries are even more visible in the work carried under by Lian and others, who got surprisingly high values of $748\,mAh\,g^{-1}$ at $1\,A\,g^{-1}$ [167]. The high reversible capacity of $1304\,mAh\,g^{-1}$ obtained after 100 cycles led to the formulation of an alternative mixed mechanism, where the alloying reaction of Li with Sn is also taken into account. In this way, the maximum theoretical specific capacity that can be obtained is $1493\,mAh\,g^{-1}$ for SnO_2, with a drop down to $1407\,mAh\,g^{-1}$ when used in combination with graphene. Attempts at increasing SnO_2/C anodes' cyclability are still underway, with a focus on mesoporous structures that control the growth of SnO_2 crystals [168] or the use in combination with N-doped graphene. In the latter case, some researchers managed to have a composite anode that was fully working after 500 cycles, with a reversible charge capacity of 1346 mAh g^{-1}, very close to the theoretical limit, as well as good high rate performances up to 20 A g^{-1} [169].

Silicon is another material that is often investigated as anode. The reaction that describes the alloying process is

$$4.4\,Li^+ + Si + 4e \leftrightarrow Li_{4.4}Si$$

Silicon can expand up to 400% in volume while alloying with lithium, exposing a fresh surface at every cycle that consumes electrolyte to form an SEI; it also sports a low intrinsic conductivity, which is why, like tin, graphene can offer a solution [170]. One of the earliest reports showed that 40 nm Si particles can be mixed with graphene in a 1:1 weight ratio to give electrodes that have a capacity of 1168 mAh g^{-1} after 30 cycles at 100 mA g^{-1} [171]. The low capacity retention (93%) was addressed by other groups, either by encapsulating Si particles in rGO or by using graphene paper, good in order to provide a continuous 3D network that could ensure electrical contact after the Si crumbling. In the former case, a reversible capacity of 786 mAh g^{-1} was retained after 300 cycles at 50 mA g^{-1} [172], while the latter electrodes showed 1500 mAh g^{-1} after 200 cycles [173]. Freeze-drying, a procedure that protects microstructures by sublimating solvents under vacuum, was also experimented with success to encapsulate Si particles in graphene sheets [174]. A lowering of structural defects in graphene via fast heat treatment at 1050°C can also enhance the Si/C composite anodes' performance. After 30 cycles at a current density of 0.3 A g^{-1}, electrodes fabricated by mixing expandable graphite and Si nanoparticles could still deliver 2753 mAh g^{-1} [175]. Another issue that arises during the crumbling and pulverization of Si-based anodes is the continuous formation of an SEI, which involves the reaction of fresh Si with the electrolyte. Preventing this side reaction to take place after the first cycle can guarantee a longer cycle life due to the reduced electrolyte degradation. Wang et al. effectively shielded Si within a carbonaceous matrix by using Si nanowires, which were subsequently sandwiched between two graphene sheets. Contact with the electrolyte is thus avoided, and the chosen morphology ensures a good electrical integrity of the electrode [176].

Other composites that have been investigated include combinations of graphene or GO with metal oxides such as Fe_2O_3 (with a reversible capacity up to 690 mAh g^{-1} after 100 cycles) [177, 178], Fe_3O_4 [179, 180], Co_3O_4 [181], NiO (with up to 1041 mAh g^{-1} after 50 cycles) [182, 183], CuO [184, 185], MnO_2 [186], Mn_3O_4 [187], and TiO_2 [15, 188]. These materials can either have a conversion reaction with lithium or be used to prevent sheet restacking, benefiting at the same time from the presence of graphene in terms of cracking suppression.

6.4.1.2 Cathode Material   The high conductivity as well as structural stability of graphene made it an appealing candidate in the case of lithium-ion battery cathodes too. For instance, materials like lithium iron phosphate (LFP, formula $LiFePO_4$), $Li_2V_2(PO_4)_3$, VO_2, and Li_2MnSiO_4 are well regarded as promising candidates to replace more conventional cathodes, but they all suffer from an intrinsically low conductivity [189]. A conducting agent has to be used to ensure good electronic contact with the particles and proper reversible cycling. Graphene is mostly combined with these cathode materials in the form of GO and then reduced. In one work, LFP, GO, and glucose monohydrate were mixed in a weight ratio of 20:1:2, and stable, reversible performances for up to 1000 cycles at high rates were obtained, in comparison with a reference LFP electrode [190]. The importance of a continuous conducting network is shown in another work, where the performance of LFP deposited on unfolded graphene (UG) was compared to that of a similar electrode using stacked graphene sheets (SG) [191]. In the former case, nearly the theoretical capacity was attained during the first cycle, as opposed to the LFP-SG (166 vs. $77 \, \text{mAh} \, \text{g}^{-1}$). This is because of the higher contact area, a shortening of the Li^+ ion diffusion path, and the ability to direct electrons toward the LFP particles more efficiently. In another paper, VO_2–graphene ribbons were reported that yielded exceptional rate performance, that is, a capacity retention of more than 90% after more than 1000 cycles at a 190C rate. Values of $204 \, \text{mAh} \, \text{g}^{-1}$ were still retained after this fast galvanostatic cycling (time for one discharge: 19 s) [192] (Fig. 6.11). Graphene ribbons enabled the use of this material, which would otherwise suffer from its high charge transfer resistance.

Li_2MnSiO_4 is a promising cathode from which two Li^+ equivalents can be extracted, giving a whopping $332 \, \text{mAh} \, \text{g}^{-1}$ theoretical capacity. Again, the low conductivity hinders this material from practical applications when used on its own, since only one Li^+ equivalent is available during discharge. Zhao et al. synthesized Li_2MnSiO_4 on rGO and obtained $290 \, \text{mAh} \, \text{g}^{-1}$ after 40 cycles at C/20 and $150 \, \text{mAh} \, \text{g}^{-1}$ at 1C after 700 cycles [193], while others used GO to coat $LiNi_{0.5}Mn_{1.5}O_4$, a high-voltage cathode, and galvanostatically cycled it for 1000 times [194]. Another cathode material that has

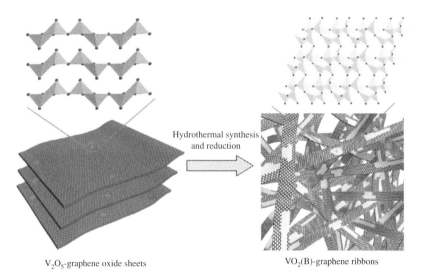

Hydrothermal synthesis and reduction

V_2O_5-graphene oxide sheets

VO_2(B)-graphene ribbons

FIGURE 6.11 Graphene-functionalized VO_2 ribbons for high rate applications. Reprinted with permission from Ref. 192. © ACS.

benefited from its use in combination with graphene is $Li_3V_2(PO_4)_3$ (indicated as LVP), again due to its small intrinsic electronic conductivity ($2.4*10^{-7}\,S\,cm^{-1}$) [195].

6.4.2   Li–S and Li–O_2 Batteries

Another application of graphene at the positive electrode is with next-generation energy storage devices. Lithium–oxygen and lithium–sulfur are perhaps the most representative technologies in this respect. Thanks to a conversion reaction of Li into Li_2O_2, LiOH, or Li_2S, high energy densities can be achieved, enough to ensure full electric vehicles a comparable driving range to that of gasoline cars [196]. In these batteries, a Li metal anode is coupled with a porous carbon on the positive electrode, where the discharge products can deposit, in quantities that depend on the carbon's porosity. This is where graphene can benefit the system, by providing a large surface area as well as a tailorable morphology.

Sulfur has a high specific capacity of $1672\,mAh\,g^{-1}$, but again it suffers from low electronic conductivity, plus it can easily form undesired polysulfides through a shuttle mechanism between the anode and the cathode. Sulfur particles were decorated with polyethylene glycol (PEG) and then wrapped in graphene–carbon black sheets. In the authors' view, both PEG and graphene can trap the polysulfides, while at the same time PEG can buffer volumetric expansions, and graphene ensures a good electrical connectivity [197]. The promising results led to the fabrication of a cell, which showed a $600\,mAh\,g^{-1}$ reversible capacity over 100 cycles. Others achieved smaller capacity values ($505\,mAh\,g^{-1}$) with a sandwich-type architecture over a comparable number of cycles [198], while the promotion of sulfur infiltration into the graphene planes via heat treatment is another viable option to obtain a graphene–sulfur composite for Li–S batteries [199]. The possibility to load such cathodes with as much active material as possible was the idea behind Nazar's work [200] (Fig. 6.12). A sulfur loading of 87% was achieved via a one-pot reaction of Na_2S_x with GO and tested over 50 cycles.

Molecular oxygen is the ultimate cathode material when it comes to lithium-based energy storage. The main advantage is that it does not need to be stored in the cell but rather harvested from the air, thus significantly increasing the battery's theoretical energy density to $3500\,Wh\,kg^{-1}$ by lowering its weight; this value is nearly 10 times as much as that of the theoretical energy density of the Li-ion technology. In this field, graphene has been both used as conducting agent and as an electrocatalyst: the reaction that leads from Li^+ and O_2 to Li_2O_2 (oxygen reduction reaction or ORR) undergoes one intermediate step, where Li_2O is formed. Moreover, Li_2O_2 can precipitate on the cathode and clog the porous structure, therefore preventing further oxygen molecules to reduce. A catalyst is then needed to facilitate the reaction for a reversible cycling.

The 3D structures that can be formed from graphene have already proven the possibility to deliver high capacities of nearly $9000\,mAh\,g^{-1}$ [202]. A graphene air electrode with a hierarchical "broken egg" structure was prepared by Xiao et al., where the high degree of defects, as well as the low C/O ratio (which implies a high level of surface functionalization), led to the fabrication of a porous structure that could deliver an outstanding $15,000\,mAh\,g^{-1}$ capacity upon discharge. While the defects provide an ideal place for Li_2O_2 to deposit, large tunnels within the structure ensure

$$2Li^+ + O_2 + 2e^- \longrightarrow Li_2O_2$$

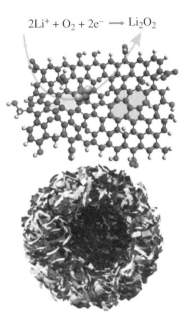

FIGURE 6.12 Broken egg conformation of graphene porous cathodes for enhanced Li_2O_2 deposition. Reprinted with permission from Ref. 201. © ACS. (*See insert for color representation of the figure.*)

a continuous O_2 flow without incurring the risk of pore clogging [201] (Fig. 6.12). Exploiting graphene's defects was also the idea behind two other works, where nitrogen and sulfur atoms were used to dope graphene and, in the case of N doping, increase its capacity to well over $11,000 \, mAh \, g^{-1}$ [203, 204].

Structural defects are not always beneficial toward a stable battery operation. Due to its unique properties, graphene can be used as a cathode with catalytic properties, thus removing the need to decorate carbon electrodes with α-MnO_2, Co_3O_4, Mn_3O_4, or Pt–Au. However, although initial results showed that graphene can compete with conventional carbon + catalyst cathodes, the constant increase of surface defects due to O_2 corrosion led to an overall voltage gap rise between charge and discharge over time. A heat treatment was successfully proposed to address this issue, which can help getting rid of surface groups and increase durability [205]. Studies on the combination of graphene with electrocatalysts like Co_3O_4 [206], Fe_2O_3 [207], and other metal oxides [208, 209] are also under development.

6.5 SUPERCAPACITORS

Electric double-layer capacitors (EDLCs), also found in literature as ultracapacitors or, more commonly, supercapacitors, constitute an interesting family of devices that are well suited for short-/medium-term energy storage. They are somewhat complementary to lithium-ion batteries, in that on average they have a higher specific power but lower-energy density. Thus, they can supply the required amount of energy in an extremely short time, a feature that has already found applications in regenerative braking systems, backup memory devices, and peak power management, with a high lifetime of several hundreds of thousands of cycles.

The basic setup of a supercapacitor is very similar to that of a battery, that is, two electrodes immersed in an electrolytic solution and a separator in between. The energy and power that can be obtained are calculated from the following equations:

$$E = \frac{1}{2}CV^2$$

$$P_{max} = \frac{V^2}{4R}$$

with V being the nominal voltage and R the electrochemical series resistance (ESR). The cell voltage depends on the electrochemical stability window of the electrolyte used, while the ESR is related both to the cell chemistry (e.g., electrolyte resistance) and the cell pack. The cell capacitance "C" is given by

$$\frac{1}{C} = \frac{1}{C_1} + \frac{1}{C_2}$$

where C_1 and C_2 correspond to the capacitance of the first and second electrode. The capacitance directly depends on the surface area and the pore size distribution of each electrode. The enhancement of these two parameters was actually the key enabler for the evolution of conventional double-layer capacitors into modern supercaps, thanks to the use of nanostructured materials, while nonaqueous electrolytes made it possible to extend the cell voltage beyond 1.23 V.

6.5.1 Graphene and Supercapacitors

The use of carbon materials has been extremely popular in supercapacitors, because of their high conductivity, high surface area, resistance to corrosion, temperature stability, ease of pore size control, and low cost [210]. However, most of the surface area increase comes from the porous structure, and a technical limit arises when the pore diameter is smaller than 1 nm (corresponding to $1000\,m^2\,g^{-1}$ specific surface area also indicated as SSA), since the pores become inaccessible to the electrolyte [211]. This is not the case of graphene materials, where a high surface area (and therefore a high capacitance) can be achieved by properly stacking and distributing the active material in the electrode [148]. Graphene's high theoretical SSA ($2630\,m^2\,g^{-1}$) can be exploited to obtain a chemically modified form, like it was done in a pioneering work by Ruoff and his group. They tested chemically obtained graphene in combination with three different electrolytes, KOH, propylene carbonate (PC), and acetonitrile (AN) [13]. The encouraging results of $135\,F\,g^{-1}$ with KOH and $99\,F\,g^{-1}$ with AN were further enhanced by producing GO through microwave exfoliation; the expanded material had a surface area of $463\,m^2\,g^{-1}$ and yielded a capacitance of $191\,F\,g^{-1}$ when used in combination with KOH as an electrolyte [212]. In a following experiment, the values of SSA have been increased up to $3100\,m^2\,g^{-1}$ by adding a KOH chemical activation step while ensuring a low O and H content, fundamental for having a predominant sp^2 hybridization and thus a high specific conductivity. A capacitance of $166\,F\,g^{-1}$ was calculated for this particular device [213]. Graphene-based electrodes with a lower SSA of $2400\,m^2\,g^{-1}$ were also fabricated in the same group. In this work, rGO sheets were thickened into an ink-like paste by evaporating the KOH solution they had been treated with and then thermally activated at 800°C under Ar. These "activated rGO" sheets were then evaluated in $TEABF_4/AN$ up to $400\,mV\,s^{-1}$ scan rates to give a reversible capacitance of $120\,F\,g^{-1}$ after 2000 cycles [214].

Graphene with a nanosheet morphology has also been analyzed, so far achieving only a fifth of its theoretical specific area ($540\,m^2\,g^{-1}$) and capacitance values of $150\,F\,g^{-1}$ [215]. This might be due to the sheet agglomeration and not to the sheets' intrinsic properties. Intersheet restacking was prevented by Yang et al. by drawing from biochemistry knowledge on hydration, which can provide repulsive forces strong enough to keep tissues and cells from collapsing. A similar concept was successfully transferred onto graphene sheets that were produced so as to be solvated by water molecules. The produced electrodes showed a strong resilience to high scan rates ($2–10\,V\,s^{-1}$) while giving still $156\,F\,g^{-1}$ capacitance values, a charging speed that is in the millisecond range. Moreover, 97% of the capacitance was retained after 10,000 cycles, even at high operation currents [216].

Composite materials have also been explored. One example is graphene/polyaniline (PANI), which was first produced via in situ polymerization. PANI is easy and cheap to synthesize, it has excellent conducting properties and has a good environmental record in terms of biocompatibility. A high $1046\,F\,g^{-1}$ capacitance was obtained at a low scan rate ($1\,mV\,s^{-1}$), thanks to the high surface area that graphene can offer to PANI particles [217] (Fig. 6.13). A similar in situ approach was followed by others to create graphene/polypyrrole composites, which showed $223\,F\,g^{-1}$ at a

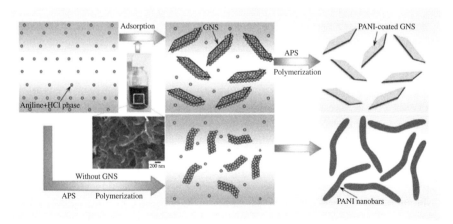

FIGURE 6.13 PANI coating on graphene electrodes for supercapacitor applications. Reprinted with permission from Ref. 217. © Elsevier.

0.5 A g^{-1} current density [218]. In both cases, a synergistic effect between the two materials could be observed, since PANI and polypyrrole are usually associated with a capacitance of 115 and 201 F g^{-1}, respectively. Further improvements have pushed the performances of PANI composites up to 250 F g^{-1} at 100 mV s^{-1}, by grafting it onto rGO sheets [219], and to 300–500 F g^{-1} when in tetrabutylammonium tetrafluoroborate (TBABF$_4$) [220].

MnO$_2$ is an oxide often encountered in combination with graphene. It is cheap, abundant, and more importantly it can enhance the voltage in combination with aqueous electrolytes, due to the overpotential it induces on water's decomposition. Wu et al. produced a composite from these two materials by using α-MnO$_2$ nanowires and assembling asymmetric supercapacitors. This new class of energy storage devices consists of a battery-like faradic electrode acting as the energy source, coupled to a capacitor-like electrode that provides for the power. By making full use of the potential windows of each electrode, they can increase the operation voltage and bridge the gap between EDLCs and pseudocapacitors, finally giving hybrid systems with higher cyclability, higher rate, and more energy density at the same time. A 10-fold increase in energy density was achieved with respect to analogous graphene supercapacitors (30.4 vs. 2.8 Wh kg^{-1}), as well as an acceptable capacitance retention [221]. Another improvement was brought about by covering textile fabrics with graphene sheets and then depositing MnO$_2$ particles on top, to get a positive electrode that would yield 315 F g^{-1}, when used in combination with a CNT-based negative electrode in an aqueous Na$_2$SO$_4$ electrolyte. A good cycling performance and an energy density of 12 Wh kg^{-1} were also recorded [222]. The poor electrical conductivity of MnO$_2$ was addressed by the same group in another work, where a conductive wrapping, consisting of either CNTs or PEDOT:PSS, was developed [223]. 380 F g^{-1} capacitance was achieved at a current density of 0.1 mA cm^{-2}, together with up to 96% capacitance retention after 3000 cycles. Mn$_3$O$_4$ has also been used in combination with graphene through a hydrothermal process that gave finely dispersed nanorods. A nearly unitarian capacitance retention was still observed after 10,000 cycles at a charging rate of 5 A g^{-1}, a neat improvement over pure Mn$_3$O$_4$-based pseudocapacitors [224]. A Ni(OH)$_2$/graphene composite was also investigated. Ni(OH)$_2$ sports a high specific capacitance of 2082 F g^{-1}, as well as a short diffusion path length for both ions and electrons when synthesized in a flowerlike nanostructure. Yan et al. [225] assembled a high rate device with flowerlike composite structures, which spontaneously form to minimize the surface energy. With an average diameter of 200–250 nm, each nanoflower consisted of around 5 nm thick "nanopetals." The group managed to reach a capacitance of 1735 F g^{-1}, a value that dropped to 218.4 F g^{-1} when the positive electrode was coupled to a graphene porous negative electrode.

Finally, the advancements made with graphene in the field of supercaps have also brought some devices close to commercialization. EDLCs with vertically oriented graphene were used in the miniaturization of AC filtering capacitors. After screening the possibility of using 1–7 layers of graphene into a supercap [226], the device that was finally fabricated could store 5.5 FV cm^{-3} with organic electrolytes, in contrast with aluminum electrolytic capacitors that normally have CV/volume ratios of only up to 0.14 FV cm^{-3} [227]. The advantage in size reduction by adopting graphene in

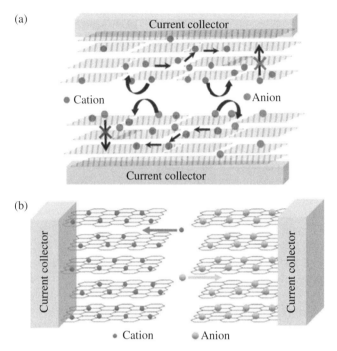

FIGURE 6.14   Surface utilization with (a) parallel and (b) normally aligned graphene sheets. Reprinted with permission from Ref. 228. © ACS.

electronic devices thanks to graphene was considered by Yoo et al. as well, who designed an ultrathin supercap where graphene planes are aligned with the ion movement. This ensures a better surface utilization thanks to a higher electrolyte percolation between the layers [228] (Fig. 6.14).

Others devised rGO paper electrodes with outstanding properties ($199\,F\,g^{-1}$ at $0.1\,A\,g^{-1}$ and $145\,F\,g^{-1}$ at $10\,A\,g^{-1}$) that may also soon find practical applications due to their ease of processing [229]. In terms of practical applications, a special mention has to be made to the work carried by Liu et al.: they achieved an incredibly high energy density of $85.6\,Wh\,kg^{-1}$ at room temperature (which increased to 136 at $80°C$) with a mesoporous graphene electrode. These values, which are comparable to those of modern nickel–metal hydride batteries, are coupled with the advantage of being able to fully recharge the storage device within only a couple of minutes [230].

6.6   GRAPHENE INKS

One of the most promising applications of graphene is expected to be in the field of printed electronics. It is considered as highly promising for replacing tin-doped indium oxide (ITO) and fluorine-doped tin oxide (FTO) materials as a transparent conductor [59, 231]. It may represent a potential breakthrough in this area by the production of conductive inks that provide high performance at low cost compared to expensive metallic inks that dominate the market currently.

Graphene has also been shown to have outstanding dispersibility and stability in a number of solvents [46] and can be manufactured cheaply using chemical reduction of graphite oxide [232–235], liquid-phase exfoliation of graphite [236–239], and electrochemical exfoliation method [56, 57, 59, 240]. Therefore, it shows promise to offer low-cost, high-performance solution for many applications in printed electronics. At the moment, graphene inks are being introduced as a cheaper alternative to silver inks by the start-up firms selling these inks. There are certain applications where silver or copper inks are not practically feasible owing to their prohibitive costs. Graphene inks may enter through this space owing to the need for moderately conductive but extremely low-cost ink. On the other hand, there are certain applications such as flexible displays where a certain amount of sheet resistance is mandatory. Without the required performance, graphene inks will not be able to penetrate that market segment no matter how cheap they are. So both cost and performance are important in order to win market share in the long term. The exact application determines the trade-off between cost and performance. Graphene inks have to win on both cost and performance to play a role in the printed electronics market in the long run.

ITO replacement is one of the most important challenges the electronics and display industry is facing today. ITO is a transparent conductive oxide film that is commonly used in liquid crystal displays, flat panel displays, plasma displays, touch panels, electronic ink applications, organic light-emitting diodes, solar cells, anti-static coatings, and EMI shielding. It has a sheet resistance of less than $100\,\Omega\,sq^{-1}$, optical transparency of nearly 90%, and unlimited scalability. However, indium reserves are running out leading to big price swings and shaky supply chains. In addition, future electronics are expected to be flexible where they can be rolled and bent. ITO is incapable of providing this feature due to its brittle character (cracking and degrading can occur over time when bent many times). The most important features for any ITO replacement aspirant are conductivity, transparency, flexibility, and price. Graphene is expected to be one of the promising materials for ITO replacement. However, it is important to note that there are two approaches being adopted to realize this goal. The first and more promising one is CVD of graphene. The other approach is using graphene inks. These two approaches have very different performance and cost metrics, and it is important to understand these differences. These will be clearly highlighted in the sections to come. In the following is a table comparing different technologies competing for ITO replacement [241, 242] (Table 6.1).

CVD and solution-based methods are both capable of producing graphene on a large scale. However, they differ in the size and quality of the graphene samples. The CVD method has been shown to produce high-quality monolayer graphene sheets of the order of several centimeters. On the other hand, solution-based exfoliation methods have been shown to produce sheets of few microns. The most important factor affecting conductivity of these sheets is the interflake tunnel barriers. In the case of CVD-grown films, very few tunneling barriers are observed, and hence,

Table 6.1 Comparison of Strengths and Drawbacks of Various ITO Replacement Technology Aspirants

	ITO	Graphene CVD Film	CNT	Metal Nanoparticles	Silver Nanowire Mesh	Conductive Polymers
Sheet resistance ($\Omega\,sq^{-1}$)	10–350	30–2000	200–2000	1–150	10–220	100–400
Transmittance (%)	88	>90	82–88	88	90	84–90
Flexibility	Inferior	Good	Good	Superior	Superior	Good
Cost	High	Very high ($10,000\,m^{-2}$)	Very high	Moderate ($10\,m^{-2}$)	High ($30–70\,m^{-2}$)	Moderate
Commercial process	High volume	Lab scale	Lab scale	High volume	High volume	High volume
Environmental effects	Good	Good	Good	Average	Average	Average
Color	Slightly yellow or brown	Colorless	Colorless	Colorless	Colorless	Slightly gray
Key developers	American Elements, Diamond coatings	Samsung, Graphene Laboratories, Stanford, UT Austin	Unidym, Eikos, Canatu, Brewer Sciences, Toray	Cima Nanotech, Applied Nanotech, Fujifilm, Five Star, PolyIC	Cambrios, Carestream Advanced Materials	Agfa, Heraeus, Fibron, Polyera, Plextronics
Drawbacks	Brittle and expensive	Extremely sensitive to defects and impurities	Resistance spiking at junctions of tubes	Needs sintering at high temperature	Challenging to fabricate	Rapid film degradation due to humidity

TABLE 6.2 Comparison of Performance of CVD Graphene with Graphene Inks

	Sheet Resistance ($\Omega\,sq^{-1}$)	Transparency (%)	Reference
Chemical	1,000–70,000	<80	[59, 244]
reduction of GO	31,000–19M	<95	
Liquid-phase	520–3,110	63–90	[59, 60]
exfoliation	5,000–8,000		
Electrochemical	210–43,000 (210 after thermal	96	[57, 59]
exfoliation of	annealing at 450 °C)		
graphite			
CVD graphene	30–2,000	90	[241, 242]

sheets with higher electronic quality are obtained. However, the quantities of tunnel barriers vary greatly in the solution-deposited films, due to variations in flake size and degree of aggregation/exfoliation. Hence, the resulting sheets have higher resistance than CVD-grown sheets [243].

Table 6.2 depicts the state-of-the-art performance values of graphene inks from the three different solution-based synthesis methods. Clearly, graphene inks have to improve their performance significantly in order to aspire to replace ITO at the highest end of spectrum ($10–30\,\Omega\,sq^{-1}$ at over 85%). This level of performance is more likely to be achieved by the CVD method. The advantage of solution-based methods is that they are much more economical than the CVD method. It is important to stress that graphene produced by these two methods have different end applications. Solution-cast graphene films may be suitable for some transparent conductor applications such as electrostatic dissipation, electromagnetic interference shielding, etc., which do not require excellent values of transmittance and sheet resistance. For such applications, graphitic films deposited from solution may play an important role particularly due to their cost effectiveness. Currently, electrochemical exfoliation synthesis method offers the best performance and is scalable.

Graphene inks have been shown to work on a wide variety of screen, gravure, flexographic, and industrial ink-jet printers [245]. These inks have been able to achieve good conductivities at low film thicknesses with vastly improved flexibility and handling characteristics relative to carbon inks, thereby offering solutions that bridge the price–performance gap between silver and traditional carbon inks [245].

6.7 CONCLUSIONS

With respect to other nanocarbon materials like CNTs and fullerenes, graphene and GO are quickly gaining ground for large-scale applications and have already established a position in academia and in the industry, thanks to the disruptive innovation their peculiarities might bring. In the field of energy harvesting and storage, the implementation of graphene and GO into organic solar cells, batteries, and supercaps is still at the developmental stage, but some performance enhancements have already been observed in a few specific areas.

Scientists have proven the material's technical superiority in the replacement of rare and expensive materials like platinum and ITO (for organic solar cells) as well as the synergic effect brought about by the synthesis of graphene composites obtained with metals, semiconductors, or polymers. Moreover, graphene's electronic and quantum peculiarities can be exploited when used as an electrolyte additive. DSSCs and OPVs are two competing technologies that find a common ground when it comes to integrating graphene-based materials within their cell chemistry, and important lessons can be learned from either field.

In lithium-ion batteries, the first and foremost application that can be found for graphene is as a substitute to graphite anodes, thanks to the higher energy density that arises from its higher stoichiometric lithium-ion insertion. Also, it is commonly found in cathode materials as an alternative to commonly used conductive agents. Uses in novel battery chemistries like Li–S and Li–O$_2$ is also being investigated, although these technologies are far from commercialization even if common materials are used.

Finally, graphene's appeal cannot be neglected in the field of EDLCs, where its high SSA makes it an ideal candidate to enhance the devices' power and energy density. Both conventional EDLCs and pseudocapacitors can rely both on graphene per se and in combination with other materials to obtain composites.

The global printed electronics market is expected to grow rapidly in the future. Graphene-based conductive inks have the potential to play a key role in specific parts of the overall market. These inks offer reasonably good functionality at very competitive costs along with the element of flexibility. They are also compatible with all existing printing technologies. The inks are likely to enter the market in stages—gradually moving from low-cost, low-functionality to high-cost, high-functionality applications. This will enable energy solutions scaled up in an effective printable way.

REFERENCES

[1]   A. Geim and A. MacDonald, "Graphene: Exploring carbon flatland," *Physics Today*, vol. **60**, p. 35, 2007.

[2]   Y. Si and E. Samulski, "Synthesis of water soluble graphene," *Nano Letters*, vol. **8**, no. 6, pp. 1679–1682, 2008.

[3]   T. Kuilla, S. Bhadra, D. Yao, N. H. Kim, S. Bose, and J. H. Lee, "Recent advances in graphene based polymer composites," *Progress in Polymer Science*, vol. **35**, no. 11, pp. 1350–1375, July 2010.

[4]   A. K. Geim and K. S. Novoselov, "The rise of graphene," *Nature Materials*, vol. **6**, no. 3, pp. 183–191, 2007.

[5]   T. Seyller, A. Bostwick, and K. Emtsev, "Epitaxial graphene: A new material," *Physica Status Solidi (B)*, vol. **245**, no. 7, pp. 1436–1446, 2008.

[6]   C. Rao and A. Sood, "Graphene: The new two dimensional nanomaterial," *Angewandte Chemie (International ed. in English)*, vol. **48**, no. 42, pp. 7752–7777, 2009.

[7] A. K. Geim, "Graphene: Status and prospects," *Science*, vol. **324**, no. 5934, p. 1530, June 2009.

[8] S. Park and R. S. Ruoff, "Chemical methods for the production of graphenes," *Nature Nanotechnology*, vol. **4**, no. 4, pp. 217–224, 2009.

[9] M. Pumera, "Electrochemistry of graphene: New horizons for sensing and energy storage," *Chemical Record (New York, NY)*, vol. **9**, no. 4, pp. 211–223, January 2009.

[10] Y. Shao, J. Wang, H. Wu, J. Liu, I. A. Aksay, and Y. Lin, "Graphene based electrochemical sensors and biosensors: A review," *Electroanalysis*, vol. **22**, no. 10, pp. 1027–1036, March 2010.

[11] K. Novoselov, A. Geim, S. Morozov, D. Jiang, M. I. Katsnelson, I. V. Grigorieva, S. Dubonos, and A. Firsov, "Two-dimensional gas of massless Dirac fermions in graphene," *Arxiv Preprint Cond-Mat/0509330*, vol. **438**, pp. 197–200, 2005.

[12] J. Hass, W. De Heer, and E. Conrad, "The growth and morphology of epitaxial multilayer graphene," *Journal of Physics: Condensed Matter*, vol. **20**, no. 32, pp. 1–27, 2008.

[13] M. D. Stoller, S. Park, Y. Zhu, J. An, R. S. Ruoff, and S. Murali, "Graphene-based ultracapacitors," *Nano Letters*, vol. **8**, no. 10, pp. 3498–3502, October 2008.

[14] E. J. Yoo, J. Kim, E. Hosono, H. Zhou, T. Kudo, and I. Honma, "Large reversible Li storage of graphene nanosheet families for use in rechargeable lithium ion batteries,," *Nano Letters*, vol. **8**, no. 8, pp. 2277–2282, August 2008.

[15] D. Wang, D. Choi, J. Li, Z. Yang, Z. Nie, and R. Kou, "Self-assembled TiO_2–graphene hybrid nanostructures for enhanced Li-ion insertion," *ACS Nano*, vol. **3**, no. 4, pp. 907–914, 2009.

[16] B. Seger and P. Kamat, "Electrocatalytically active graphene-platinum nanocomposites. Role of 2-D carbon support in PEM fuel cells," *The Journal of Physical Chemistry C*, vol. **113**, no. 19, pp. 7990–7995, 2009.

[17] E. Yoo, T. Okata, T. Akita, and M. Kohyama, "Enhanced electrocatalytic activity of Pt subnanoclusters on graphene nanosheet surface," *Nano Letters*, vol. **9**, no. 6, pp. 2255–2259, 2009.

[18] R. Kou, Y. Shao, and D. Wang, "Enhanced activity and stability of Pt catalysts on functionalized graphene sheets for electrocatalytic oxygen reduction," *Electrochemistry Communications*, vol. **11**, pp. 954–957, 2009.

[19] Y. Si and E. Samulski, "Exfoliated graphene separated by platinum nanoparticles," *Chemistry of Materials*, vol. **20**, no. 21, pp. 6792–6797, 2008.

[20] Y. Li, L. Tang, and J. Li, "Preparation and electrochemical performance for methanol oxidation of pt/graphene nanocomposites," *Electrochemistry Communications*, Vol. **11**, no. 4, pp. 846–849, 2009.

[21] X. Wang, L. Zhi, N. Tsao, Ž. Tomović, J. Li, K. Müllen, and Z. Tomović, "Transparent carbon films as electrodes in organic solar cells," *Angewandte Chemie (International ed. in English)*, vol. **47**, no. 16, pp. 2990–2992, January 2008.

[22] J. Wu, H. A. Becerril, Z. Bao, Z. Liu, Y. Chen, and P. Peumans, "Organic solar cells with solution-processed graphene transparent electrodes," *Applied Physics Letters*, vol. **92**, no. 26, p. 263302, 2008.

[23] Z. Liu, J. Robinson, X. Sun, and H. Dai, "PEGylated nanographene oxide for delivery of water-insoluble cancer drugs," *Journal of the American Chemical Society*, vol. **130**, no. 33, pp. 10876–10877, 2008.

[24] H. Chen and M. Müller, "Mechanically strong, electrically conductive, and biocompatible graphene paper," *Advanced Materials*, vol. **20**, no. 18, pp. 3557–3561, 2008.

[25] C. Shan, H. Yang, J. Song, D. Han, A. Ivaska, and L. Niu, "Direct electrochemistry of glucose oxidase and biosensing for glucose based on graphene," *Analytical Chemistry*, vol. **81**, no. 6, pp. 2378–2382, March 2009.

[26] Z. Wang, X. Zhou, and J. Zhang, "Direct electrochemical reduction of single-layer graphene oxide and subsequent functionalization with glucose oxidase," *The Journal of Physical Chemistry C*, vol. **113**, no. 32, pp. 14071–14075, 2009.

[27] C. Lu, H. Yang, and C. Zhu, "A graphene platform for sensing biomolecules," *Angewandte Chemie*, vol. **48**, pp. 4785–4787, 2009.

[28] Y. Wang, J. Lu, L. Tang, H. Chang, and J. Li, "Graphene oxide amplified electrogenerated chemiluminescence of quantum dots and its selective sensing for glutathione from thiol-containing compounds," *Analytical Chemistry*, vol. **81**, no. 23, pp. 9710–9715, 2009.

[29] A. Balandin, S. Ghosh, and W. Bao, "Superior thermal conductivity of single-layer graphene," *Nano Letters*, vol. **8**, no. 3, pp. 902–907, 2008.

[30] K. Bolotin, K. Sikes, and Z. Jiang, "Ultrahigh electron mobility in suspended graphene," *Solid State Communications*, vol. **146**, no. 9–10, pp. 351–355, 2008.

[31] J. Meyer, "Carbon sheets an atom thick give rise to graphene dreams," *Science*, vol. **324**, no. 5929, pp. 875–877, 2009.

[32] C. Lee, X. Wei, J. W. Kysar, and J. Hone, "Measurement of the elastic properties and intrinsic strength of monolayer graphene," *Science*, vol. **321**, no. 5887, p. 385, July 2008.

[33] R. R. Nair, P. Blake, A. N. Grigorenko, K. S. Novoselov, T. J. Booth, T. Stauber, N. M. R. Peres, and A. K. Geim, "Fine structure constant defines visual transparency of graphene," *Science* (New York, NY), vol. **320**, no. 5881, p. 1308, June 2008.

[34] C. Rao and A. Sood, "Graphene: The new two dimensional nanomaterial," *Angewandte Chemie*, vol. **48**, no. 42, pp. 7752–7777, January 2009.

[35] K. S. Novoselov, A. K. Geim, S. V. Morozov, D. Jiang, Y. Zhang, S. V. Dubonos, I. V. Grigorieva, and A. A. Firsov, "Electric field effect in atomically thin carbon films," *Science*, vol. **306**, no. 5696, pp. 666–669, October 2004.

[36] W. A. de Heer, C. Berger, X. Wu, P. N. First, E. H. Conrad, X. Li, T. Li, M. Sprinkle, J. Hass, M. L. Sadowski, M. Potemski, and G. Martinez, "Epitaxial graphene," *Solid State Communications*, vol. **143**, no. 1–2, pp. 92–100, July 2007.

[37] E. Rollings, G.-H. Gweon, and S. Y. Zhou, "Synthesis and characterization of atomically thin graphite films on a silicon carbide substrate," *Journal of Physics and Chemistry of Solids*, vol. **67**, no. 9–10, pp. 2172–2177, September 2006.

[38] C. Berger, Z. Song, X. Li, X. Wu, N. Brown, C. Naud, D. Mayou, T. Li, J. Hass, A. N. Marchenkov, E. H. Conrad, P. N. First, and W. A. de Heer, "Electronic confinement and coherence in patterned epitaxial graphene," *Science* (New York, NY), vol. **312**, no. 5777, pp. 1191–1196, May 2006.

[39] P. Sutter, J. Flege, and E. Sutter, "Epitaxial graphene on ruthenium," *Nature Materials*, vol. **7**, no. 5, pp. 406–411, May 2008.

[40] K. Kim, Y. Zhao, H. Jang, S. Lee, and J. Kim, "Large-scale pattern growth of graphene films for stretchable transparent electrodes," *Nature*, vol. **457**, no. 7230, pp. 706–710, February 2009.

[41] A. Reina, X. Jia, J. Ho, D. Nezich, H. Son, V. Bulovic, M. S. Dresselhaus, and J. Kong, "Large area, few-layer graphene films on arbitrary substrates by chemical vapor deposition," *Nano Letters*, vol. **9**, no. 1, pp. 30–35, January 2009.

[42] S.-Y. Kwon, C. V Ciobanu, V. Petrova, V. B. Shenoy, J. Bareño, V. Gambin, I. Petrov, and S. Kodambaka, "Growth of semiconducting graphene on palladium," *Nano Letters*, vol. **9**, no. 12, pp. 3985–3990, December 2009.

[43] J. Coraux, A. N'Diaye, C. Busse, and T. Michely, "Structural coherency of graphene on Ir (111)," *Nano Letters*, vol. **8**, no. 2, pp. 565–570, February 2008.

[44] J. Hamilton and J. Blakely, "Carbon segregation to single crystal surfaces of Pt, Pd and Co," *Surface Science*, vol. **91**, pp. 199–217, 1980.

[45] N. Gall, S. Mikhailov, E. Rut'kov, and A. Tontegode, "Nature of the adsorption binding between a graphite monolayer and rhenium surface," *Soviet Physics, Solid State*, vol. **27**, p. 1410–1414, 1985.

[46] D. Wei, H. Li, D. Han, Q. Zhang, L. Niu, H. Yang, C. Bower, P. Andrew, and T. Ryhänen, "Properties of graphene inks stabilized by different functional groups," *Nanotechnology*, vol. **22**, no. 24, p. 245702, April 2011.

[47] X. Li, W. Cai, J. An, S. Kim, J. Nah, and D. Yang, "Large-area synthesis of high-quality and uniform graphene films on copper foils," *Science*, vol. **324**, no. 5932, pp. 1312–1314, June 2009.

[48] A. Dato, V. Radmilovic, and Z. Lee, "Substrate-free gas-phase synthesis of graphene sheets," *Nano Letters*, vol. **8**, no. 7, pp. 2012–2016, July 2008.

[49] A. Obraztsov, "Chemical vapour deposition: Making graphene on a large scale," *Nature Nanotechnology*, vol. **4**, no. 4, pp. 212–213, April 2009.

[50] S. Bae, H. Kim, Y. Lee, X. Xu, J. S. Park, Y. Zheng, J. Balakrishnan, T. Lei, H. R. Kim, Y. I. Song, and others, "Roll-to-roll production of 30-inch graphene films for transparent electrodes," *Nature Nanotechnology*, vol. **5**, pp. 574–578, 2010.

[51] C. Mattevi, H. Kim, and M. Chhowalla, "A review of chemical vapour deposition of graphene on copper," *Journal of Materials Chemistry*, vol. **21**, no. 10, p. 3324, February 2011.

[52] V. Pham, H. Pham, and T. Dang, "Chemical reduction of an aqueous suspension of graphene oxide by nascent hydrogen," *Journal of Materials Chemistry*, vol. **22**, no. 207890, pp. 10530–10536, 2012.

[53] W. Chen, L. Yan, and P. Bangal, "Chemical reduction of graphene oxide to graphene by sulfur-containing compounds," *The Journal of Physical Chemistry C*, vol. **114**, pp. 19885–19890, 2010.

[54] S. Pei and H.-M. Cheng, "The reduction of graphene oxide," *Carbon*, vol. **50**, no. 9, pp. 3210–3228, August 2012.

[55] H. Feng, R. Cheng, X. Zhao, X. Duan, and J. Li, "A low-temperature method to produce highly reduced graphene oxide," *Nature Communications*, vol. **4**, p. 1539, February 2013.

[56] J. Lu, J. Yang, J. Wang, A. Lim, S. Wang, and K. P. Loh, "One-pot synthesis of fluorescent carbon graphene by the exfoliation of graphite in ionic liquids," *ACS Nano*, vol. **3**, no. 8, pp. 2367–2375, 2009.

[57] N. Liu, F. Luo, H. Wu, and Y. Liu, "One step ionic liquid assisted electrochemical synthesis of ionic liquid functionalized graphene sheets directly from graphite," *Advanced Functional Materials*, vol. **18**, no. 10, pp. 1518–1525, May 2008.

[58] D. Wei, L. Grande, V. Chundi, R. White, C. Bower, P. Andrew, and T. Ryhänen, "Graphene from electrochemical exfoliation and its direct applications in enhanced energy storage devices," *Chemical Communications (Cambridge, England)*, vol. **48**, no. 9, pp. 1239–1241, December 2012.

[59] C.-Y. Su, A.-Y. Lu, Y. Xu, F.-R. Chen, A. N. Khlobystov, and L.-J. Li, "High-quality thin graphene films from fast electrochemical exfoliation," *ACS Nano*, vol. **5**, no. 3, pp. 2332–2339, March 2011.

[60] S. Bae, I. Jeon, J. Yang, N. Park, and H. Shin, "Large-area graphene films by simple solution casting of edge-selectively functionalized graphite," *ACS Nano*, vol. **5**, no. 6, pp. 4974–4980, June 2011.

[61] S. J. Jung, F. Bonaccorso, P. J. Paul, D. P. Chu, and A. C. Ferrari, "Inkjet-printed graphene electronics," *ACS Nano*, vol. **6**, pp. 1–12.

[62] M. Grätzel, "Photoelectrochemical cells," *Nature*, vol. **414**, no. November, pp. 338–344, 2001.

[63] H. Spanggaard and F. C. Krebs, "A brief history of the development of organic and polymeric photovoltaics," *Solar Energy Materials and Solar Cells*, vol. **83**, no. 2–3, pp. 125–146, June 2004.

[64] T. W. Hamann, R. A. Jensen, A. B. F. Martinson, H. Van Ryswyk, and J. T. Hupp, "Advancing beyond current generation dye-sensitized solar cells," *Energy & Environmental Science*, vol. **1**, no. 1, p. 66, 2008.

[65] Panasonic, "HIT Power 220A." [Online]. Available: http://www.panasonic.com/business/pesna/includes/pdf/eco-construction-solution/HIT_Power_220A_Datasheet.pdf. [Accessed: May 5, 2013].

[66] S. Chu and A. Majumdar, "Opportunities and challenges for a sustainable energy future," *Nature*, vol. **488**, no. 7411, pp. 294–303, August 2012.

[67] M. Grätzel, R. A. J. Janssen, D. B. Mitzi, and E. H. Sargent, "Materials interface engineering for solution-processed photovoltaics," *Nature*, vol. **488**, no. 7411, pp. 304–312, August 2012.

[68] M. Grätzel, "Photovoltaic and photoelectrochemical conversion of solar energy," *Philosophical Transactions of the Royal Society A*, vol. **365**, no. 1853, p. 993, April 2007.

[69] B. O'Regan and M. Grätzel, "A low-cost, high-efficiency solar cell based on dye-sensitized colloidal TiO_2 films," *Nature*, vol. **353**, no. 6346, pp. 737–740, 1991.

[70] D. Wei, "Dye sensitized solar cells," *International Journal of Molecular Sciences*, vol. **11**, no. 3, pp. 1103–1113, January 2010.

[71] B. Gregg, "Excitonic solar cells," *The Journal of Physical Chemistry B*, vol. **107**, no. 20, pp. 4688–4698, 2003.

[72] F.-T. Kong, S.-Y. Dai, and K.-J. Wang, "Review of recent progress in dye-sensitized solar cells," *Advances in OptoElectronics*, vol. **2007**, pp. 1–13, 2007.

[73] A. Hagfeldt, G. Boschloo, L. Sun, L. Kloo, and H. Pettersson, "Dye-sensitized solar cells," *Chemical Reviews*, vol. **110**, no. 11, pp. 6595–6663, November 2010.

[74] D. Wei, P. Andrew, and T. Ryhänen, "Electrochemical photovoltaic cells-review of recent developments," *Journal of Chemical Technology & Biotechnology*, vol. **85**, no. 12, pp. 1547–1552, December 2010.

[75] A. Kay and M. Grätzel, "Low cost photovoltaic modules based on dye sensitized nanocrystalline titanium dioxide and carbon powder," *Solar Energy Materials and Solar Cells*, vol. **44**, pp. 99–117, 1996.

[76] T. N. Murakami, S. Ito, Q. Wang, M. K. Nazeeruddin, T. Bessho, I. Cesar, P. Liska, R. Humphry-Baker, P. Comte, P. Péchy, and M. Grätzel, "Highly efficient dye-sensitized solar cells based on carbon black counter electrodes," *Journal of the Electrochemical Society*, vol. **153**, no. 12, p. A2255, 2006.

[77] D. Wei, H. E. Unalan, D. Han, Q. Zhang, L. Niu, G. Amaratunga, and T. Ryhanen, "A solid-state dye-sensitized solar cell based on a novel ionic liquid gel and ZnO nanoparticles on a flexible polymer substrate," *Nanotechnology*, vol. **19**, no. 42, p. 424006, October 2008.

[78] H. E. Unalan, D. Wei, K. Suzuki, S. Dalal, P. Hiralal, H. Matsumoto, S. Imaizumi, M. Minagawa, A. Tanioka, A. J. Flewitt, W. I. Milne, and G. A. J. Amaratunga, "Photoelectrochemical cell using dye sensitized zinc oxide nanowires grown on carbon fibers," *Applied Physics Letters*, vol. **93**, no. 13, p. 133116, 2008.

[79] W. J. Lee, E. Ramasamy, D. Y. Lee, and J. S. Song, "Efficient dye-sensitized solar cells with catalytic multiwall carbon nanotube counter electrodes," *ACS Applied Materials & Interfaces*, vol. **1**, no. 6, pp. 1145–1149, June 2009.

[80] B. Lee, D. B. Buchholz, and R. P. H. Chang, "An all carbon counter electrode for dye sensitized solar cells," *Energy & Environmental Science*, vol. **5**, no. 5, p. 6941, 2012.

[81] K. Hung, Y. Li, and H. Wang, "Dye-sensitized solar cells using graphene-based counter electrode," *Nanotechnology (IEEE-NANO)*, p. 1–12, 2012. doi: 10.1109/NANO.2012.6322228.

[82] S. H. Huh, S.-H. Choi, and H.-M. Ju, "Thickness-dependent solar power conversion efficiencies of catalytic graphene oxide films in dye-sensitized solar cells," *Current Applied Physics*, vol. **11**, no. 3, pp. S352–S355, May 2011.

[83] L. Wan, S. Wang, X. Wang, B. Dong, Z. Xu, X. Zhang, B. Yang, S. Peng, J. Wang, and C. Xu, "Room-temperature fabrication of graphene films on variable substrates and its use as counter electrodes for dye-sensitized solar cells," *Solid State Sciences*, vol. **13**, no. 2, pp. 468–475, February 2011.

[84] D. W. Zhang, X. D. Li, H. B. Li, S. Chen, Z. Sun, X. J. Yin, and S. M. Huang, "Graphene-based counter electrode for dye-sensitized solar cells," *Carbon*, vol. **49**, no. 15, pp. 5382–5388, December 2011.

[85] Z. Huang, X. Liu, K. Li, D. Li, Y. Luo, H. Li, W. Song, L. Chen, and Q. Meng, "Application of carbon materials as counter electrodes of dye-sensitized solar cells," *Electrochemistry Communications*, vol. **9**, no. 4, pp. 596–598, April 2007.

[86] Y. Xu, H. Bai, G. Lu, C. Li, and G. Shi, "Flexible graphene films via the filtration of water-soluble noncovalent functionalized graphene sheets," *Journal of the American Chemical Society*, vol. **130**, no. 18, pp. 5856–5857, 2008.

[87] W. Hong, Y. Xu, G. Lu, C. Li, and G. Shi, "Transparent graphene/PEDOT–PSS composite films as counter electrodes of dye-sensitized solar cells," *Electrochemistry Communications*, vol. **10**, no. 10, pp. 1555–1558, October 2008.

[88] G. Yue, J. Wu, Y. Xiao, J. Lin, M. Huang, Z. Lan, and L. Fan, "Functionalized graphene/poly(3,4-ethylenedioxythiophene):polystyrenesulfonate as counter electrode catalyst for dye-sensitized solar cells," *Energy*, pp. 1–7, February 2013.

[89] M. Song, S. Ameen, S. Akhtar, H.-K. Seo, and H. S. Shin, "New counter electrode of hot filament chemical vapor deposited graphene thin film for dye sensitized solar cell," *Chemical Engineering Journal*, pp. 464–471, February 2013.

[90]   H. Wang and Y. H. Hu, "Graphene as a counter electrode material for dye-sensitized solar cells," *Energy & Environmental Science*, vol. **5**, no. 8, p. 8182, 2012.

[91]   M.-Y. Yen, C.-C. Teng, M.-C. Hsiao, P.-I. Liu, W.-P. Chuang, C.-C. M. Ma, C.-K. Hsieh, M.-C. Tsai, and C.-H. Tsai, "Platinum nanoparticles/graphene composite catalyst as a novel composite counter electrode for high performance dye-sensitized solar cells," *Journal of Materials Chemistry*, vol. **21**, no. 34, p. 12880, 2011.

[92]   R. Bajpai, S. Roy, P. Kumar, P. Bajpai, N. Kulshrestha, J. Rafiee, N. Koratkar, and D. S. Misra, "Graphene supported platinum nanoparticle counter-electrode for enhanced performance of dye-sensitized solar cells," *ACS Applied Materials & Interfaces*, vol. **3**, no. 10, pp. 3884–3889, October 2011.

[93]   Y.-G. Kim, Z. A. Akbar, D. Y. Kim, S. M. Jo, and S.-Y. Jang, "Aqueous dispersible graphene/Pt nanohybrids by green chemistry: Application as cathodes for dye-sensitized solar cells," *ACS Applied Materials & Interfaces*, vol. **5**, no. 6, pp. 2053–2061, February 2013.

[94]   Y. Y. Dou, G. R. Li, J. Song, and X. P. Gao, "Nickel phosphide-embedded graphene as counter electrode for dye-sensitized solar cells," *Physical Chemistry Chemical Physics*, vol. **14**, no. 4, pp. 1339–1342, January 2012.

[95]   Z.-S. Wang, Z. Li, F. Gong, and G. Zhou, "NiS 2/reduced graphene oxide nanocomposites for efficient dye-sensitized solar cells," *The Journal of Physical Chemistry C*, vol. **117**, no. 13, pp. 6561–6566, March 2013.

[96]   J. Roy-Mayhew, D. Bozym, C. Punckt, and I. Aksay, "Functionalized graphene as a catalytic counter electrode in dye-sensitized solar cells," *ACS Nano*, vol. **4**, no. 10, pp. 6203–6211, 2010.

[97]   K. S. Lee, Y. Lee, J. Y. Lee, J.-H. Ahn, and J. H. Park, "Flexible and platinum-free dye-sensitized solar cells with conducting-polymer-coated graphene counter electrodes," *ChemSusChem*, vol. **5**, no. 2, pp. 379–382, February 2012.

[98]   H. Choi, H. Kim, S. Hwang, W. Choi, and M. Jeon, "Dye-sensitized solar cells using graphene-based carbon nano composite as counter electrode," *Solar Energy Materials and Solar Cells*, vol. **95**, no. 1, pp. 323–325, January 2011.

[99]   J. Velten, J. Carretero-González, E. Castillo-Martínez, J. Bykova, A. Cook, R. Baughman, and A. Zakhidov, "Photoinduced optical transparency in dye-sensitized solar cells containing graphene nanoribbons," *The Journal of Physical Chemistry C*, vol. **115**, pp. 25125–25131, 2011.

[100]   A. Kaniyoor and S. Ramaprabhu, "Hydrogen exfoliated graphene as counter electrode for dye sensitized solar cells," *Nanoscience, Technology and Societal Implications*, vol. **109**, no. 12, pp. 1–4, December 2011.

[101]   A. Kaniyoor and S. Ramaprabhu, "Thermally exfoliated graphene based counter electrode for low cost dye sensitized solar cells," *Journal of Applied Physics*, vol. **109**, no. 12, p. 124308, 2011.

[102]   H. Choi, S. Hwang, H. Bae, S. Kim, H. Kim, and M. Jeon, "Electrophoretic graphene for transparent counter electrodes in dye-sensitised solar cells," *Electronics Letters*, vol. **47**, no. 4, p. 281, 2011.

[103]   K. Zhou, Y. Zhu, X. Yang, X. Jiang, and C. Li, "Preparation of graphene–TiO_2 composites with enhanced photocatalytic activity," *New Journal of Chemistry*, vol. **35**, no. 2, p. 353, 2011.

[104] H. Zhang, X. Lv, Y. Li, Y. Wang, and J. Li, "P25-graphene composite as a high performance photocatalyst," *ACS Nano*, vol. **4**, no. 1, pp. 380–386, January 2010.

[105] A. Anish Madhavan, S. Kalluri, D. K. Chacko, T. A. Arun, S. Nagarajan, K. R. V. Subramanian, A. Sreekumaran Nair, S. V. Nair, and A. Balakrishnan, "Electrical and optical properties of electrospun TiO_2-graphene composite nanofibers and its application as DSSC photo-anodes," *RSC Advances*, vol. **2**, no. 33, p. 13032, 2012.

[106] X. Fang, M. Li, K. Guo, Y. Zhu, Z. Hu, X. Liu, B. Chen, and X. Zhao, "Improved properties of dye-sensitized solar cells by incorporation of graphene into the photo-electrodes," *Electrochimica Acta*, vol. **65**, pp. 174–178, March 2012.

[107] S. Sun, L. Gao, and Y. Liu, "Enhanced dye-sensitized solar cell using graphene-TiO_2 photoanode prepared by heterogeneous coagulation," *Applied Physics Letters*, vol. **96**, no. 8, p. 083113, 2010.

[108] J. Song, Z. Yin, Z. Yang, P. Amaladass, S. Wu, J. Ye, Y. Zhao, W.-Q. Deng, H. Zhang, and X.-W. Liu, "Enhancement of photogenerated electron transport in dye-sensitized solar cells with introduction of a reduced graphene oxide-TiO_2 junction," *Chemistry (Weinheim an der Bergstrasse, Germany)*, vol. **17**, no. 39, pp. 10832–10837, October 2011.

[109] Y. Zhou, L. Chen, W. Tu, C. Bao, H. Dai, Y. Tao, and J. LIU, "Enhanced photovoltaic performance of dye-sensitized solar cell using graphene-TiO_2 photoanode prepared by a novel in situ simultaneous reduction-hydrolysis technique," *Nanoscale*, vol. **5**, pp. 3481–3485, November 2013.

[110] Z. He, G. Guai, J. Liu, C. Guo, J. S. C. Loo, C. M. Li, and T. T. Y. Tan, "Nanostructure control of graphene-composited TiO_2 by a one-step solvothermal approach for high performance dye-sensitized solar cells," *Nanoscale*, vol. **3**, no. 11, pp. 4613–4616, November 2011.

[111] K.-S. Ahn, S.-W. Seo, J.-H. Park, B.-K. Min, and W.-S. Jung, "The preparation of alumina particles wrapped in few-layer graphene sheets and their application to dye-sensitized solar cells," *Bulletin of the Korean Chemical Society*, vol. **32**, no. 5, pp. 1579–1582, May 2011.

[112] M.-Y. Yen, M.-C. Hsiao, S.-H. Liao, P.-I. Liu, H.-M. Tsai, C.-C. M. Ma, N.-W. Pu, and M.-D. Ger, "Preparation of graphene/multi-walled carbon nanotube hybrid and its use as photoanodes of dye-sensitized solar cells," *Carbon*, vol. **49**, no. 11, pp. 3597–3606, September 2011.

[113] X. Wang, L. Zhi, and K. Müllen, "Transparent, conductive graphene electrodes for dye-sensitized solar cells," *Nano Letters*, vol. **8**, no. 1, pp. 323–327, January 2008.

[114] T. Chen, W. Hu, J. Song, G. H. Guai, and C. M. Li, "Interface functionalization of photoelectrodes with graphene for high performance dye-sensitized solar cells," *Advanced Functional Materials*, vol. **22**, no. 24, pp. 5245–5250, December 2012.

[115] S. R. Kim, M. K. Parvez, and M. Chhowalla, "UV-reduction of graphene oxide and its application as an interfacial layer to reduce the back-transport reactions in dye-sensitized solar cells," *Chemical Physics Letters*, vol. **483**, no. 1–3, pp. 124–127, November 2009.

[116] I. Ahmad, U. Khan, and Y. K. Gun'ko, "Graphene, carbon nanotube and ionic liquid mixtures: Towards new quasi-solid state electrolytes for dye sensitised solar cells," *Journal of Materials Chemistry*, vol. **21**, no. 42, p. 16990, 2011.

[117] Y.-F. Chan, C.-C. Wang, and C.-Y. Chen, "Quasi-solid DSSC based on a gel-state electrolyte of PAN with 2-D graphenes incorporated," *Journal of Materials Chemistry A*, vol. **1**, pp. 5479–5486, 2013.

[118] J. Nelson, "Organic photovoltaic films," *Current Opinion in Solid State and Materials Science*, vol. **6**, no. 1, pp. 87–95, 2002.

[119] B. Kippelen and J.-L. Brédas, "Organic photovoltaics," *Energy & Environmental Science*, vol. **2**, no. 3, p. 251, 2009.

[120] Heliatek, "Neuer Weltrekord für organische Solarzellen: Heliatek behauptet sich mit 12 % Zelleffizienz als Technologieführer." [Online]. Available: http://www.heliatek. com/newscenter/latest_news/neuer-weltrekord-fur-organische-solarzellen-heliatek-behauptet-sich-mit-12-zelleffizienz-als-technologiefuhrer/. [Accessed: May 5, 2013].

[121] V. C. Tung, L.-M. Chen, M. J. Allen, J. K. Wassei, K. Nelson, R. B. Kaner, and Y. Yang, "Low-temperature solution processing of graphene-carbon nanotube hybrid materials for high-performance transparent conductors," *Nano Letters*, vol. **9**, no. 5, pp. 1949–1955, May 2009.

[122] Y. Wang, X. Chen, Y. Zhong, F. Zhu, and K. P. Loh, "Large area, continuous, few-layered graphene as anodes in organic photovoltaic devices," *Applied Physics Letters*, vol. **95**, no. 6, p. 063302, 2009.

[123] Y. Xu, G. Long, L. Huang, Y. Huang, X. Wan, Y. Ma, and Y. Chen, "Polymer photovoltaic devices with transparent graphene electrodes produced by spin-casting," *Carbon*, vol. **48**, no. 11, pp. 3308–3311, September 2010.

[124] G. Kalita, M. Matsushima, H. Uchida, K. Wakita, and M. Umeno, "Graphene constructed carbon thin films as transparent electrodes for solar cell applications," *Journal of Materials Chemistry*, vol. **20**, no. 43, p. 9713, 2010.

[125] L. Valentini, M. Cardinali, S. B. Bon, D. Bagnis, R. Verdejo, M. A. Lopez-Manchado, J. M. Kenny, and S. Bittolo Bon, "Use of butylamine modified graphene sheets in polymer solar cells," *Journal of Materials Chemistry*, vol. **20**, no. 5, pp. 995–1000, 2010.

[126] H. Park, J. A. Rowehl, K. K. Kim, V. Bulovic, and J. Kong, "Doped graphene electrodes for organic solar cells," *Nanotechnology*, vol. **21**, p. 505204, 2010.

[127] L. Gomez De Arco, Y. Zhang, C. W. Schlenker, K. Ryu, M. E. Thompson, and C. Zhou, "Continuous, highly flexible, and transparent graphene films by chemical vapor deposition for organic photovoltaics," *ACS Nano*, vol. **4**, no. 5, pp. 2865–2873, May 2010.

[128] Y. Wang, S. W. Tong, X. F. Xu, B. Ozyilmaz, K. P. Loh, and B. Özyilmaz, "Interface engineering of layer-by-layer stacked graphene anodes for high-performance organic solar cells," *Advanced Materials*, vol. **23**, no. 13, pp. 1514–1518, April 2011.

[129] Z. Yin, S. Sun, T. Salim, S. Wu, X. Huang, Q. He, Y. M. Lam, and H. Zhang, "Organic photovoltaic devices using highly flexible reduced graphene oxide films as transparent electrodes," *ACS Nano*, vol. **4**, no. 9, pp. 5263–5268, September 2010.

[130] Z. Zhao, J. D. Fite, P. Haldar, and J. Ung Lee, "Enhanced ultraviolet response using graphene electrodes in organic solar cells," *Applied Physics Letters*, vol. **101**, no. 6, p. 063305, 2012.

[131] M. Choe, B. H. Lee, G. Jo, J. Park, W. Park, S. Lee, W.-K. Hong, M.-J. Seong, Y. H. Kahng, and K. Lee, "Efficient bulk-heterojunction photovoltaic cells with transparent multi-layer graphene electrodes," *Organic Electronics*, vol. **11**, no. 11, pp. 1864–1869, November 2010.

[132]   Z. Liu, J. Li, Z.-H. Sun, G. Tai, S.-P. Lau, and F. Yan, "The application of highly doped single-layer graphene as the top electrodes of semitransparent organic solar cells," *ACS Nano*, vol. **6**, no. 1, pp. 810–818, January 2012.

[133]   Z. Liu, Q. Liu, Y. Huang, Y. Ma, S. Yin, X. Zhang, W. Sun, and Y. Chen, "Organic photovoltaic devices based on a novel acceptor material: Graphene," *Advanced Materials*, vol. **20**, no. 20, pp. 3924–3930, October 2008.

[134]   Q. Liu, Z. Liu, X. Zhang, N. Zhang, L. Yang, S. Yin, and Y. Chen, "Organic photovoltaic cells based on an acceptor of soluble graphene," *Applied Physics Letters*, vol. **92**, no. 22, p. 223303, 2008.

[135]   Q. Liu, Z. Liu, X. Zhang, L. Yang, N. Zhang, G. Pan, S. Yin, Y. Chen, and J. Wei, "Polymer photovoltaic cells based on solution-processable graphene and P3HT," *Advanced Functional Materials*, vol. **19**, no. 6, pp. 894–904, March 2009.

[136]   Z. Liu, D. He, Y. Wang, H. Wu, and J. Wang, "Solution-processable functionalized graphene in donor/acceptor-type organic photovoltaic cells," *Solar Energy Materials and Solar Cells*, vol. **94**, no. 7, pp. 1196–1200, July 2010.

[137]   Z. Liu, D. He, Y. Wang, H. Wu, J. Wang, and H. Wang, "Improving photovoltaic properties by incorporating both SPFGraphene and functionalized multiwalled carbon nanotubes," *Solar Energy Materials and Solar Cells*, vol. **94**, no. 12, pp. 2148–2153, July 2010.

[138]   Y. Gao, H.-L. Yip, K.-S. Chen, K. M. O'Malley, O. Acton, Y. Sun, G. Ting, H. Chen, and A. K.-Y. Jen, "Surface doping of conjugated polymers by graphene oxide and its application for organic electronic devices," *Advanced Materials*, vol. **23**, no. 16, pp. 1903–1908, April 2011.

[139]   S.-S. Li, K.-H. Tu, C.-C. Lin, C.-W. Chen, and M. Chhowalla, "Solution-processable graphene oxide as an efficient hole transport layer in polymer solar cells," *ACS Nano*, vol. **4**, no. 6, pp. 3169–3174, June 2010.

[140]   Y. Gao, H.-L. Yip, S. K. Hau, K. M. O'Malley, N. C. Cho, H. Chen, and A. K.-Y. Jen, "Anode modification of inverted polymer solar cells using graphene oxide," *Applied Physics Letters*, vol. **97**, no. 20, p. 203306, 2010.

[141]   Q. Zheng, G. Fang, F. Cheng, H. Lei, P. Qin, and C. Zhan, "Low-temperature solution-processed graphene oxide derivative hole transport layer for organic solar cells," *Journal of Physics D*, vol. **46**, no. 13, p. 135101, April 2013.

[142]   M. S. Ryu and J. Jang, "Effect of solution processed graphene oxide/nickel oxide bilayer on cell performance of bulk-heterojunction organic photovoltaic," *Solar Energy Materials and Solar Cells*, vol. **95**, no. 10, pp. 2893–2896, October 2011.

[143]   J.-M. Yun, J.-S. Yeo, J. Kim, H.-G. Jeong, D.-Y. Kim, Y.-J. Noh, S.-S. Kim, B.-C. Ku, and S.-I. Na, "Solution-processable reduced graphene oxide as a novel alternative to PEDOT:PSS hole transport layers for highly efficient and stable polymer solar cells," *Advanced Materials*, vol. **23**, no. 42, pp. 4923–4928, November 2011.

[144]   B. Yin, Q. Liu, L. Yang, X. Wu, Z. Liu, Y. Hua, S. Yin, and Y. Chen, "Buffer layer of PEDOT:PSS/graphene composite for polymer solar cells," *Journal of Nanoscience and Nanotechnology*, vol. **10**, no. 3, pp. 1934–1938, March 2010.

[145]   B. Scrosati and J. Garche, "Lithium batteries: Status, prospects and future," *Journal of Power Sources*, vol. **195**, no. 9, pp. 2419–2430, May 2010.

[146]   P. G. Bruce, "Energy storage beyond the horizon: Rechargeable lithium batteries," *Solid State Ionics*, vol. **179**, no. 21–26, pp. 752–760, September 2008.

[147] D. A. C. Brownson, D. K. Kampouris, and C. E. Banks, "An overview of graphene in energy production and storage applications," *Journal of Power Sources*, vol. **196**, no. 11, pp. 4873–4885, June 2011.

[148] X. Zhang, B. Wang, J. Sunarso, S. Liu, and L. Zhi, "Graphene nanostructures toward clean energy technology applications," *Wiley Interdisciplinary Reviews: Energy and Environment*, vol. **1**, no. 3, pp. 317–336, November 2012.

[149] S. L. Cheekati, "Graphene based anode materials for lithium-ion batteries," Wright State University, 2011.

[150] S. Han, D. Wu, S. Li, F. Zhang, and X. Feng, "Graphene: A two-dimensional platform for lithium storage," *Small (Weinheim an der Bergstrasse, Germany)*, vol. **9**, no. 8, pp. 1–15, March 2013.

[151] J. Hou, Y. Shao, M. W. Ellis, R. B. Moore, and B. Yi, "Graphene-based electrochemical energy conversion and storage: Fuel cells, supercapacitors and lithium ion batteries," *Physical Chemistry Chemical Physics*, vol. **13**, no. 34, pp. 15384–15402, September 2011.

[152] K. Sato, M. Noguchi, A. Demachi, N. Oki, and M. Endo, "A mechanism of lithium storage in disordered carbons," *Science*, vol. **264**, no. 5158, pp. 556–558, April 1994.

[153] D. Wei, L. Grande, V. Chundi, R. White, C. Bower, P. Andrew, and T. Ryhänen, "Graphene from electrochemical exfoliation and its direct applications in enhanced energy storage devices," *Chemical Communications*, vol. **48**, no. 9, pp. 1239–1241, December 2012.

[154] A. Gerouki, "Density of states calculations of small diameter single graphene sheets," *Journal of The Electrochemical Society*, vol. **143**, no. 11, p. L262, 1996.

[155] C. Wang, D. Li, C. O. Too, and G. G. Wallace, "Electrochemical properties of graphene paper electrodes used in lithium batteries," *Chemistry of Materials*, vol. **21**, no. 13, pp. 2604–2606, July 2009.

[156] D. Pan, S. Wang, B. Zhao, M. Wu, H. Zhang, Y. Wang, and Z. Jiao, "Li storage properties of disordered graphene nanosheets," *Chemistry of Materials*, vol. **21**, no. 14, pp. 3136–3142, July 2009.

[157] S. Yang, X. Feng, L. Zhi, Q. Cao, J. Maier, and K. Müllen, "Nanographene-constructed hollow carbon spheres and their favorable electroactivity with respect to lithium storage," *Advanced Materials (Deerfield Beach, Fla.)*, vol. **22**, no. 7, pp. 838–842, February 2010.

[158] G. Wang, X. Shen, B. Wang, J. Yao, and J. Park, "Graphene nanosheets for enhanced lithium storage in lithium ion batteries," *Carbon*, vol. **47**, no. 8, pp. 2049–2053, April 2009.

[159] P. Guo, H. Song, and X. Chen, "Electrochemical performance of graphene nanosheets as anode material for lithium-ion batteries," *Electrochemistry Communications*, vol. **11**, no. 6, pp. 1320–1324, June 2009.

[160] Z.-S. Wu, W. Ren, L. Xu, F. Li, and H.-M. Cheng, "Doped graphene sheets as anode materials with superhigh rate and large capacity for lithium ion batteries," *ACS Nano*, vol. **5**, no. 7, pp. 5463–5471, July 2011.

[161] X. Li, D. Geng, Y. Zhang, X. Meng, R. Li, and X. Sun, "Superior cycle stability of nitrogen-doped graphene nanosheets as anodes for lithium ion batteries," *Electrochemistry Communications*, vol. **13**, no. 8, pp. 822–825, August 2011.

[162] G. Wang, B. Wang, X. Wang, J. Park, S. Dou, H. J. Ahn, and K. Kim, "Sn/graphene nanocomposite with 3D architecture for enhanced reversible lithium storage in lithium ion batteries," *Journal of Materials Chemistry*, vol. **19**, no. 44, p. 8378, 2009.

[163] S. Liang, X. Zhu, P. Lian, W. Yang, and H. Wang, "Superior cycle performance of Sn@C/graphene nanocomposite as an anode material for lithium-ion batteries," *Journal of Solid State Chemistry*, vol. **184**, no. 6, pp. 1400–1404, April 2011.

[164] D. Wang, R. Kou, D. Choi, Z. Yang, Z. Nie, J. Li, L. V Saraf, D. Hu, J. Zhang, G. L. Graff, J. Liu, M. A. Pope, and I. A. Aksay, "Ternary self-assembly of ordered metal oxide-graphene nanocomposites for electrochemical energy storage," *ACS Nano*, vol. **4**, no. 3, pp. 1587–1595, March 2010.

[165] J. Yao, X. Shen, B. Wang, H. Liu, and G. Wang, "In situ chemical synthesis of SnO_2– graphene nanocomposite as anode materials for lithium-ion batteries," *Electrochemistry Communications*, vol. **11**, no. 10, pp. 1849–1852, October 2009.

[166] H. Kim, S.-W. Kim, Y.-U. Park, H. Gwon, D.-H. Seo, Y. Kim, and K. Kang, "SnO_2/ graphene composite with high lithium storage capability for lithium rechargeable batteries," *Nano Research*, vol. **3**, no. 11, pp. 813–821, October 2010.

[167] P. Lian, X. Zhu, S. Liang, Z. Li, W. Yang, and H. Wang, "High reversible capacity of SnO_2/graphene nanocomposite as an anode material for lithium-ion batteries," *Electrochimica Acta*, vol. **56**, no. 12, pp. 4532–4539, February 2011.

[168] S. Yang, W. Yue, J. Zhu, Y. Ren, and X. Yang, "Graphene-based mesoporous SnO_2 with enhanced electrochemical performance for lithium-ion batteries," *Advanced Functional Materials*, vol. **23**, no. 28, pp. 3570–3576, February 2013.

[169] X. Zhou, L. Wan, and Y. Guo, "Binding SnO_2 nanocrystals in nitrogen-doped graphene sheets as anode materials for lithium-ion batteries," *Advanced Materials*, vol. **25**, no. 15, pp. 2152–2157, February 2013.

[170] H. K. Liu, Z. P. Guo, J. Z. Wang, and K. Konstantinov, "Si-based anode materials for lithium rechargeable batteries," *Journal of Materials Chemistry*, vol. **20**, no. 45, p. 10055, 2010.

[171] S.-L. Chou, J.-Z. Wang, M. Choucair, H.-K. Liu, J. A. Stride, and S.-X. Dou, "Enhanced reversible lithium storage in a nanosize silicon/graphene composite," *Electrochemistry Communications*, vol. **12**, no. 2, pp. 303–306, February 2010.

[172] H.-C. Tao, L.-Z. Fan, Y. Mei, and X. Qu, "Self-supporting Si/Reduced Graphene Oxide nanocomposite films as anode for lithium ion batteries," *Electrochemistry Communications*, vol. **13**, no. 12, pp. 1332–1335, December 2011.

[173] J. K. Lee, K. B. Smith, C. M. Hayner, and H. H. Kung, "Silicon nanoparticles-graphene paper composites for Li ion battery anodes," *Chemical Communications (Cambridge, England)*, vol. **46**, no. 12, pp. 2025–2027, March 2010.

[174] X. Zhou, Y.-X. Yin, L.-J. Wan, and Y.-G. Guo, "Facile synthesis of silicon nanoparticles inserted into graphene sheets as improved anode materials for lithium-ion batteries," *Chemical Communications (Cambridge, England)*, vol. **48**, no. 16, pp. 2198–2200, February 2012.

[175] H. Xiang, K. Zhang, G. Ji, J. Y. Lee, C. Zou, X. Chen, and J. Wu, "Graphene/nanosized silicon composites for lithium battery anodes with improved cycling stability," *Carbon*, vol. **49**, no. 5, pp. 1787–1796, April 2011.

[176] B. Wang, X. Li, X. Zhang, B. Luo, M. Jin, M. Liang, S. A. Dayeh, S. T. Picraux, and L. Zhi, "Adaptable silicon-carbon nanocables sandwiched between reduced graphene oxide sheets as lithium ion battery anodes," *ACS Nano*, vol. **7**, no. 2, pp. 1437–1445, February 2013.

[177] G. Wang, T. Liu, Y. Luo, Y. Zhao, Z. Ren, J. Bai, and H. Wang, "Preparation of Fe_2O_3/ graphene composite and its electrochemical performance as an anode material for lithium ion batteries," *Journal of Alloys and Compounds*, vol. **509**, pp. 0–4, April 2011.

[178] I. T. Kim, A. Magasinski, K. Jacob, G. Yushin, and R. Tannenbaum, "Synthesis and electrochemical performance of reduced graphene oxide/maghemite composite anode for lithium ion batteries," *Carbon*, vol. **52**, pp. 56–64, February 2013.

[179] G. Zhou, D.-W. Wang, F. Li, L. Zhang, N. Li, Z.-S. Wu, L. Wen, G. Q. (Max) Lu, and H.-M. Cheng, "Graphene-wrapped Fe_3O_4 anode material with improved reversible capacity and cyclic stability for lithium ion batteries," *Chemistry of Materials*, vol. **22**, no. 18, pp. 5306–5313, September 2010.

[180] G. Wang, T. Liu, X. Xie, Z. Ren, J. Bai, and H. Wang, "Structure and electrochemical performance of Fe_3O_4/graphene nanocomposite as anode material for lithium-ion batteries," *Materials Chemistry and Physics*, vol. **128**, no. 3, pp. 336–340, May 2011.

[181] Z. Wu, W. Ren, L. Wen, L. Gao, J. Zhao, Z. Chen, G. Zhou, F. Li, and H. Cheng, "Graphene anchored with Co_3O_4 nanoparticles as anode of lithium ion batteries with enhanced reversible capacity and cyclic performance," *ACS Nano*, vol. **4**, no. 6, pp. 3187–3194, 2010.

[182] D. Qiu, Z. Xu, M. Zheng, B. Zhao, L. Pan, L. Pu, and Y. Shi, "Graphene anchored with mesoporous NiO nanoplates as anode material for lithium-ion batteries," *Journal of Solid State Electrochemistry*, vol. **16**, no. 5, pp. 1889–1892, June 2011.

[183] X.-J. Zhu, J. Hu, H.-L. Dai, L. Ding, and L. Jiang, "Reduced graphene oxide and nanosheet-based nickel oxide microsphere composite as an anode material for lithium ion battery," *Electrochimica Acta*, vol. **64**, pp. 23–28, March 2012.

[184] Y. J. Mai, X. L. Wang, J. Y. Xiang, Y. Q. Qiao, D. Zhang, C. D. Gu, and J. P. Tu, "CuO/graphene composite as anode materials for lithium-ion batteries," *Electrochimica Acta*, vol. **56**, no. 5, pp. 2306–2311, February 2011.

[185] B. Wang, X. L. Wu, C. Y. Shu, Y. G. Guo, and C. R. Wang, "Synthesis of CuO/graphene nanocomposite as a high-performance anode material for lithium-ion batteries," *Journal of Materials Chemistry*, vol. **20**, pp. 10661–10664, 2010.

[186] A. Yu, H. Park, and A. Davies, "Free-standing layer-by-layer hybrid thin film of graphene-MnO_2 nanotube as anode for lithium ion batteries," *The Journal of Physical Chemistry Letters*, vol. **2**, pp. 1855–1860, 2011.

[187] H. Wang, L.-F. Cui, Y. Yang, H. Sanchez Casalongue, J. T. Robinson, Y. Liang, Y. Cui, and H. Dai, "Mn_3O_4-graphene hybrid as a high-capacity anode material for lithium ion batteries," *Journal of the American Chemical Society*, vol. **132**, no. 40, pp. 13978–13980, October 2010.

[188] D. Choi, D. Wang, V. V. Viswanathan, I.-T. Bae, W. Wang, Z. Nie, J.-G. Zhang, G. L. Graff, J. Liu, Z. Yang, and T. Duong, "Li-ion batteries from $LiFePO_4$ cathode and anatase/graphene composite anode for stationary energy storage," *Electrochemistry Communications*, vol. **12**, no. 3, pp. 378–381, March 2010.

[189] G. Kucinskis, G. Bajars, and J. Kleperis, "Graphene in lithium ion battery cathode materials: A review," *Journal of Power Sources*, vol. **240**, pp. 66–79, October 2013.

[190] X. Zhou, F. Wang, Y. Zhu, and Z. Liu, "Graphene modified $LiFePO_4$ cathode materials for high power lithium ion batteries," *Journal of Materials Chemistry*, vol. **21**, no. 10, p. 3353, 2011.

[191] J. Yang, J. Wang, Y. Tang, D. Wang, X. Li, and Y. Hu, "$LiFePO_4$/graphene as a superior cathode material for rechargeable lithium batteries: Impact of stacked graphene and unfolded graphene," *Energy & Environmental Science*, vol. **6**, no. 207890, pp. 1521–1528, 2013.

[192] S. Yang, Y. Gong, Z. Liu, L. Zhan, D. P. Hashim, L. Ma, R. Vajtai, and P. M. Ajayan, "Bottom-up approach toward single-crystalline VO_2-graphene ribbons as cathodes for ultrafast lithium storage," *Nano Letters*, vol. **13**, no. 4, pp. 1596–1601, March 2013.

[193] Y. Zhao, C. Wu, J. Li, and L. Guan, "Long cycling life of Li_2MnSiO_4 lithium battery cathodes under the double protection from carbon coating and graphene network," *Journal of Materials Chemistry A*, vol. **1**, no. 12, p. 3856, 2013.

[194] X. Fang, M. Ge, J. Rong, and C. Zhou, "Graphene-oxide-coated $LiNi_{0.5}Mn_{1.5}O_4$ as high voltage cathode for lithium ion batteries with high energy density and long cycle life," *Journal of Materials Chemistry A*, vol. **1**, no. 12, p. 4083, 2013.

[195] H. Liu, P. Gao, J. Fang, and G. Yang, "$Li_3V_2(PO_4)_3$/graphene nanocomposites as cathode material for lithium ion batteries," *Chemical Communications*, vol. **47**, pp. 1–3, July 2011.

[196] P. G. Bruce, S. A. Freunberger, L. J. Hardwick, and J.-M. Tarascon, "$Li–O_2$ and Li–S batteries with high energy storage," *Nature Materials*, vol. **11**, no. 1, pp. 19–29, December 2011.

[197] H. Wang, Y. Yang, Y. Liang, J. T. Robinson, Y. Li, A. Jackson, Y. Cui, and H. Dai, "Graphene-wrapped sulfur particles as a rechargeable lithium-sulfur battery cathode material with high capacity and cycling stability," *Nano Letters*, vol. **11**, no. 7, pp. 2644–2647, July 2011.

[198] Y. Cao, X. Li, I. A. Aksay, J. Lemmon, Z. Nie, Z. Yang, and J. Liu, "Sandwich-type functionalized graphene sheet-sulfur nanocomposite for rechargeable lithium batteries," *Physical Chemistry Chemical Physics*, vol. **13**, no. 17, pp. 7660–7665, March 2011.

[199] J.-Z. Wang, L. Lu, M. Choucair, J. A. Stride, X. Xu, and H.-K. Liu, "Sulfur-graphene composite for rechargeable lithium batteries," *Journal of Power Sources*, vol. **196**, no. 16, pp. 7030–7034, August 2011.

[200] S. Evers and L. F. Nazar, "Graphene-enveloped sulfur in a one pot reaction: A cathode with good coulombic efficiency and high practical sulfur content," *Chemical Communications (Cambridge, England)*, vol. **48**, no. 9, pp. 1233–1235, January 2012.

[201] J. Xiao, D. Mei, X. Li, W. Xu, D. Wang, G. L. Graff, W. D. Bennett, Z. Nie, L. V Saraf, I. A. Aksay, J. Liu, and J.-G. Zhang, "Hierarchically porous graphene as a lithium-air battery electrode," *Nano Letters*, vol. **11**, no. 11, pp. 5071–5078, November 2011.

[202] Y. Li, J. Wang, X. Li, D. Geng, R. Li, and X. Sun, "Superior energy capacity of graphene nanosheets for a nonaqueous lithium-oxygen battery," *Chemical Communications (Cambridge, England)*, vol. **47**, no. 33, pp. 9438–9440, September 2011.

[203] Y. Li, J. Wang, X. Li, D. Geng, M. N. Banis, R. Li, and X. Sun, "Nitrogen-doped graphene nanosheets as cathode materials with excellent electrocatalytic activity for high capacity lithium-oxygen batteries," *Electrochemistry Communications*, vol. **18**, pp. 12–15, January 2012.

[204] Y. Li, J. Wang, X. Li, D. Geng, M. N. Banis, Y. Tang, D. Wang, R. Li, T.-K. Sham, and X. Sun, "Discharge product morphology and increased charge performance of lithium–oxygen batteries with graphene nanosheet electrodes: The effect of sulphur doping," *Journal of Materials Chemistry*, pp. 20170–20174, 2012.

[205] E. Yoo and H. Zhou, "Li−air rechargeable battery based on metal-free graphene nanosheet catalysts," *ACS Nano*, vol. **5**, no. 4, pp. 3020–3026, 2011.

[206]   R. Black, J.-H. Lee, B. Adams, C. A. Mims, and L. F. Nazar, "The role of catalysts and peroxide oxidation in lithium-oxygen batteries," *Angewandte Chemie (International ed. in English)*, vol. **52**, pp. 392–396, November 2012.

[207]   Q. Yan, W. Zhang, Y. Zeng, C. Xu, and H. Tan, "Fe_2O_3 nanoclusters decorated graphene as O_2 electrode for high energy Li-O_2 batteries," *RSC Advances*, vol. **2**, no. 207890, pp. 8508–8514, 2012.

[208]   H. Wang, Y. Yang, Y. Liang, G. Zheng, Y. Li, Y. Cui, and H. Dai, "Rechargeable Li–O_2 batteries with a covalently coupled $MnCo_2O_4$–graphene hybrid as an oxygen cathode catalyst," *Energy & Environmental Science*, vol. **5**, no. 7, p. 7931, 2012.

[209]   R. Black, J. Lee, and B. Adams, "Graphene-metal oxide catalysts for Li-O_2 batteries," *Meeting Abstracts*, vol. **158**, no. 2011, p. 12171, 2012.

[210]   A. G. Pandolfo and A. F. Hollenkamp, "Carbon properties and their role in supercapacitors," *Journal of Power Sources*, vol. **157**, no. 1, pp. 11–27, June 2006.

[211]   O. Barbieri, M. Hahn, A. Herzog, and R. Kötz, "Capacitance limits of high surface area activated carbons for double layer capacitors," *Carbon*, vol. **43**, no. 6, pp. 1303–1310, May 2005.

[212]   Y. Zhu, S. Murali, M. D. Stoller, A. Velamakanni, R. D. Piner, and R. S. Ruoff, "Microwave assisted exfoliation and reduction of graphite oxide for ultracapacitors," *Carbon*, vol. **48**, no. 7, pp. 2118–2122, 2010.

[213]   Y. Zhu, S. Murali, M. D. Stoller, K. J. Ganesh, W. Cai, P. J. Ferreira, A. Pirkle, R. M. Wallace, K. A. Cychosz, M. Thommes, D. Su, E. A. Stach, and R. S. Ruoff, "Carbon-based supercapacitors produced by activation of graphene," *Science*, vol. **332**, no. 6037, pp. 1537–1541, June 2011.

[214]   L. L. Zhang, X. Zhao, M. D. Stoller, Y. Zhu, H. Ji, S. Murali, Y. Wu, S. Perales, B. Clevenger, and R. S. Ruoff, "Highly conductive and porous activated reduced graphene oxide films for high-power supercapacitors," *Nano Letters*, vol. **12**, no. 4, pp. 1806–1812, April 2012.

[215]   X. Du, P. Guo, H. Song, and X. Chen, "Graphene nanosheets as electrode material for electric double-layer capacitors," *Electrochimica Acta*, vol. **55**, no. 16, pp. 4812–4819, June 2010.

[216]   X. Yang, J. Zhu, L. Qiu, and D. Li, "Bioinspired effective prevention of restacking in multilayered graphene films: Towards the next generation of high-performance supercapacitors," *Advanced Materials (Deerfield Beach, Fla.)*, vol. **23**, no. 25, pp. 2833–2838, July 2011.

[217]   J. Yan, T. Wei, B. Shao, Z. Fan, W. Qian, M. Zhang, and F. Wei, "Preparation of a graphene nanosheet/polyaniline composite with high specific capacitance," *Carbon*, vol. **48**, no. 2, pp. 487–493, February 2010.

[218]   Y. Han, B. Ding, and X. Zhang, "Preparation of graphene/polypyrrole composites for electrochemical capacitors," *Journal of New Materials for Electrochemical Systems*, vol. **13**, pp. 315–320, 2010.

[219]   N. Kumar, H. Choi, Y. Shin, and D. Chang, "Polyaniline-grafted reduced graphene oxide for efficient electrochemical supercapacitors," *ACS Nano*, vol. **6**, no. 2, pp. 1715–1723, 2012.

[220]   H. Gómez, M. K. Ram, F. Alvi, P. Villalba, E. (Lee) Stefanakos, and A. Kumar, "Graphene-conducting polymer nanocomposite as novel electrode for supercapacitors," *Journal of Power Sources*, vol. **196**, no. 8, pp. 4102–4108, April 2011.

[221] Z.-S. Wu, W. Ren, D.-W. Wang, F. Li, B. Liu, and H.-M. Cheng, "High-energy MnO_2 nanowire/graphene and graphene asymmetric electrochemical capacitors," *ACS Nano*, vol. **4**, no. 10, pp. 5835–5842, October 2010.

[222] G. Yu, L. Hu, M. Vosgueritchian, H. Wang, X. Xie, J. R. McDonough, X. Cui, Y. Cui, and Z. Bao, "Solution-processed graphene/MnO_2 nanostructured textiles for high-performance electrochemical capacitors," *Nano Letters*, vol. **11**, no. 7, pp. 2905–2911, June 2011.

[223] G. Yu, L. Hu, N. Liu, H. Wang, M. Vosgueritchian, Y. Yang, Y. Cui, and Z. Bao, "Enhancing the supercapacitor performance of graphene/MnO_2 nanostructured electrodes by conductive wrapping," *Nano Letters*, vol. **11**, no. 10, pp. 4438–4442, October 2011.

[224] J. Lee, A. Hall, J. Kim, and T. Mallouk, "A facile and template-free hydrothermal synthesis of Mn_3O_4 nanorods on graphene sheets for supercapacitor electrodes with long cycle stability," *Chemistry of Materials*, vol. **24**, no. 6, pp. 1158–1164, 2012.

[225] J. Yan, Z. Fan, W. Sun, G. Ning, T. Wei, Q. Zhang, R. Zhang, L. Zhi, and F. Wei, "Advanced asymmetric supercapacitors based on $Ni(OH)_2$/graphene and porous graphene electrodes with high energy density," *Advanced Functional Materials*, vol. **22**, no. 12, pp. 2632–2641, June 2012.

[226] X. Zhao, H. Tian, M. Zhu, K. Tian, J. J. Wang, F. Kang, and R. A. Outlaw, "Carbon nanosheets as the electrode material in supercapacitors," *Journal of Power Sources*, vol. **194**, no. 2, pp. 1208–1212, December 2009.

[227] J. R. Miller, R. A. Outlaw, and B. C. Holloway, "Graphene double-layer capacitor with ac line-filtering performance," *Science*, vol. **329**, no. 5999, pp. 1637–1639, September 2010.

[228] J. J. Yoo, K. Balakrishnan, J. Huang, V. Meunier, B. G. Sumpter, A. Srivastava, M. Conway, A. L. M. Reddy, J. Yu, R. Vajtai, and P. M. Ajayan, "Ultrathin planar graphene supercapacitors," *Nano Letters*, vol. **11**, no. 4, pp. 1423–1427, April 2011.

[229] M. M. Hantel, T. Kaspar, R. Nesper, A. Wokaun, and R. Kotz, "Partially reduced graphene oxide paper: A thin film electrode for electrochemical capacitors," *Journal of the Electrochemical Society*, vol. **160**, no. 4, pp. A747–A750, March 2013.

[230] C. Liu, Z. Yu, D. Neff, A. Zhamu, and B. Z. Jang, "Graphene-based supercapacitor with an ultrahigh energy density," *Nano Letters*, vol. **10**, no. 12, pp. 4863–4868, November 2010.

[231] A. Kumar and C. Zhou, "The race to replace tin-doped indium oxide: Which material will win?," *ACS Nano*, vol. **4**, no. 1, pp. 11–14, 2010.

[232] S. Stankovich, D. Dikin, R. Piner, K. Kohlhaas, A. Kleinhammes, Y. Jia, Y. Wu, S. Nguyen, and R. Ruoff, "Synthesis of graphene-based nanosheets via chemical reduction of exfoliated graphite oxide," *Carbon*, vol. **45**, no. 7, pp. 1558–1565, June 2007.

[233] S. Gilje, S. Han, M. Wang, K. L. Wang, and R. B. Kaner, "A chemical route to graphene for device applications," *Nano Letters*, vol. **7**, no. 11, pp. 3394–3398, November 2007.

[234] D. Li, M. B. Müller, S. Gilje, R. B. Kaner, and G. G. Wallace, "Processable aqueous dispersions of graphene nanosheets," *Nature Nanotechnology*, vol. **3**, no. 2, pp. 101–105, March 2008.

[235] X. Fan, W. Peng, Y. Li, X. Li, S. Wang, G. Zhang, and F. Zhang, "Deoxygenation of exfoliated graphite oxide under alkaline conditions: A green route to graphene preparation," *Advanced Materials*, vol. **20**, no. 23, pp. 4490–4493, December 2008.

[236] Y. Hernandez, V. Nicolosi, M. Lotya, F. M. Blighe, Z. Sun, S. De, I. T. McGovern, B. Holland, M. Byrne, Y. K. Gun'Ko, J. J. Boland, P. Niraj, G. Duesberg, S. Krishnamurthy, R. Goodhue, J. Hutchison, V. Scardaci, A. C. Ferrari, and J. N. Coleman, "High-yield production of graphene by liquid-phase exfoliation of graphite," *Nature Nanotechnology*, vol. **3**, no. 9, pp. 563–568, September 2008.

[237] M. Lotya, Y. Hernandez, and P. King, "Liquid phase production of graphene by exfoliation of graphite in surfactant/water solutions," *Journal of the American Chemical Society*, vol. **131**, no. 10, pp. 3611–3620, March 2009.

[238] A. Bourlinos, V. Georgakilas, and R. Zboril, "Liquid-phase exfoliation of graphite towards solubilized graphenes," *Small*, vol. **5**, no. 16, pp. 1841–1845, August 2009.

[239] X. Li, G. Zhang, X. Bai, X. Sun, and X. Wang, "Highly conducting graphene sheets and Langmuir–Blodgett films," *Nature Nanotechnology*, vol. **3**, no. 9, pp. 538–542, September 2008.

[240] J. Wang, K. K. Manga, Q. Bao, and K. P. Loh, "High-yield synthesis of few-layer graphene flakes through electrochemical expansion of graphite in propylene carbonate electrolyte," *Journal of the American Chemical Society*, vol. **133**, no. 23, pp. 8888–8891, June 2011.

[241] R. Sordan, F. Traversi, and V. Russo, "Graphene: Prospects for future electronics talk," 2010. [Online]. Available: http://www.graphene-nanotubes.org/uploads/uploads/pirjo/ Cargese Graphene Pirjo Pasanen_public_reduit2.pdf. [Accessed: February 27, 2015].

[242] Lux Research Inc., "Sorting Hype From Reality in Printed, Organic, and Flexible Display Technologies," 2010. [Online]. Available: http://info.luxresearchinc.com/ Portals/86611/docs/research%20downloads/lux_research_sorting_hype_from_reality_ printed_organic_and_display_technologies%5B1%5D.pdf. [Accessed: May 8, 2015].

[243] S. De and J. Coleman, "Are there fundamental limitations on the sheet resistance and transmittance of thin graphene films?," *ACS Nano*, vol. **4**, no. 5, pp. 2713–2720, May 2010.

[244] M. Chhowalla, "Opto-electronic properties of graphene oxide and partially oxidized graphene," Materials Science. [Online]. Available: http://nanotubes.rutgers.edu/PDFs/ Prof.%20Chhowalla%27s%20Presentation%20on%20Opto-Electronic%20Properties%20 of%20Graphene%20Oxide.pdf. [Accessed: May 8, 2015].

[245] S. Monie, "Developments in conductive inks," 2010. [Online]. Available: http:// industrial-printing.net/content/developments-conductive-inks#.VTwMbPnF9zU. [Accessed: May 8, 2015].

7

QUANTUM DOT AND HETEROJUNCTION SOLAR CELLS CONTAINING CARBON NANOMATERIALS

Xinming Li[1,2], Tianshuo Zhao[3] and Hongwei Zhu[1,4]

[1] School of Materials Science and Engineering, Tsinghua University, Beijing, China
[2] National Center for Nanoscience and Technology, Beijing, China
[3] Department of Material Science and Engineering, University of Pennsylvania, Philadelphia, PA, USA
[4] Center for Nano and Micro Mechanics, Tsinghua University, Beijing, China

7.1 INTRODUCTION

The photovoltaic (PV) cell based on new materials and structure is rising up as a promising alternative for the commercial solar panels in order to meet the increasing great demand of clean energy. The p–n junction is the common and key part for many optoelectronic devices. Single-crystal silicon-based solar cells, based on the PV effect of p–n junctions, have been characterized by high efficiency levels. However, cost and device efficiency are always the key factors determining the practical viability of the technology. As a result, the new generation solar cell, which is helpful to overcome the Shockley–Queisser limit [1] and keep a low cost at the same time, shows a great potential than its predecessors.

Semiconductor nanoparticles, or quantum dots (QDs), emerge as building blocks for new generation solar cells. These novel materials own outstanding features in the PV applications: (i) Synthesized from the solution phase, colloidal QDs can be

Carbon Nanomaterials for Advanced Energy Systems: Advances in Materials Synthesis and Device Applications, First Edition. Edited by Wen Lu, Jong-Beom Baek and Liming Dai.
© 2015 John Wiley & Sons, Inc. Published 2015 by John Wiley & Sons, Inc.

precisely controlled in terms of its size and shape as well as its optical properties. (ii) As direct band gap material, QDs offer strong absorption at certain wavelengths according to its tunable band gap. (iii) The size-dependent quantization of QDs provides opportunities for band offset modulation, hot-electron utilization [2], and multiple-carrier generation [3] in the devices.

Heterojunction solar cells are the typical PV devices that convert solar energy to electricity. Recently, heterojunction structures based on carbon nanomaterials also have attracted a great deal of interest for both scientific fundamentals and potential applications in various new optoelectronic devices, for example, PV solar cells. In the heterojunction solar cells, the interface between carbon nanomaterials and semiconductor plays the key role in energy conversion [4]. The operation mechanism of the carbon nanomaterial/semiconductor solar cells could be considered as Schottky (metal/semiconductor) junction and p–n (semiconductor/semiconductor) junction.

Carbon nanomaterials exist in a variety of stable forms ranging from amorphous carbon (a-C), fullerene (C_{60}), and carbon nanotubes (CNTs) to graphene. These various forms of carbon nanomaterials have attracted great attention because of their unique structures, properties, and potential applications. In the case of PVs, carbon family owns advantages in flexibility, surface area, carrier mobility, chemical stability, and optoelectronic properties, which fulfill the requirement of QDs and heterojunction-based solar cells very well.

7.2 QD SOLAR CELLS CONTAINING CARBON NANOMATERIALS

Using QDs with tuned band gaps, the single-material-system solar cell is able to contain multiple junctions, having a better use of the sun spectrum even challenging the single-junction limitation. Figure 7.1 illustrates the structure and band diagram of two types of QD-based devices [5].

However, due to the organic ligands and defect states on the surface of QDs, the mobility of photogenerated electrons is very limited, while the recombination of electrons and holes occurs easily on the QD surface. Therefore, the major problems that need to be solved are the separation of photogenerated hole/electron pairs and the transport of carriers from QDs to the electrode. Therefore, C_{60}, CNTs, and graphene have been applied as transparent conductive electrodes (TCE), electron acceptors, and light absorbers to improve the device performance. Herein, CNTs and graphene as TCE, carbon nanomaterial/QD composites, and graphene QDs (GQDs) will be discussed in the context of QD-based PVs.

7.2.1 CNTs and Graphene as TCE in QD Solar Cells

As one of the most important components in various optoelectronic devices, TCE requires the excellence of transparency and conductivity. Doped metal oxide films, such as tin-doped indium oxide (ITO), have been the conventional choice for TCE. However, the unbending nature of ITO and scarcity of indium resource make it more and more difficult to reach the additional requirements for the new generation solar

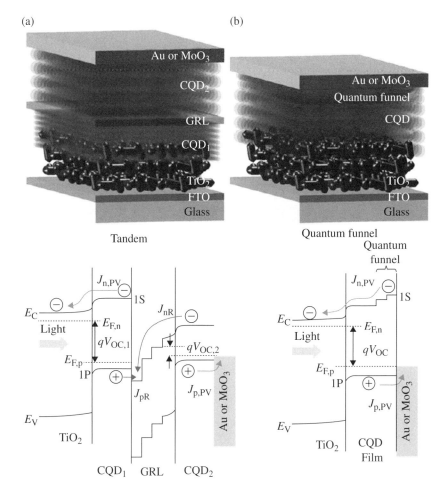

FIGURE 7.1 Device structure and band diagram of the tandem cell and the quantum funnel solar cells. (a) Device structure and band diagram of the tandem cell. The two junctions contain two layers of QDs of different sizes and a graded recombination layer (GRL) lies in between. (b) Device structure and band diagram of the quantum funnel solar cell. The quantum funnel is formed by the same QDs but in different sizes. Reprinted with permission from Ref. 5. © 2011 American Chemical Society.

cells: flexible, low-cost, and large-scale manufacture. As a result, one question arises: what material can replace ITO? In the search of alternatives, CNT films and graphene are considered as promising candidates besides other approaches, such as metal gratings and nanowire networks [6].

7.2.1.1 CNTs as TCE Material in QD Solar Cells

CNTs including single-walled CNTs (SWCNTs) and multiwalled CNTs (MWCNTs) are known to have outstanding conductivity individually, but the resistance between nanotube junctions signifies the sheet resistance of CNT films. In addition, due to strong van der Waals interactions between single nanotubes, CNTs are hard to disperse in many solvents. However, in order to form a percolation network, the density of CNTs should be above a threshold level [7]. Also, the thickness of the CNT film determines the transparency of TCE at the same time. Therefore, effective control of the CNT density remains a big challenge. Many methods have been reported to enhance the solubility of CNTs in polymer metrics [8], including adding covalent functional groups to the side walls of CNTs [9] and noncovalent functionalizing CNTs with surfactants [10]. Recently, Li et al. have reported a new technique that can achieve multifunctional freestanding CNT/polymer composite thin film with a roughness of 3.8 nm [11]. Different CNT and polymer composite films have been studied in organic solar cell system, attempting to replace the role of ITO [12–16]. The application of CNT films in QD-based solar cells has rarely been reported, and one of the main reasons is that its optoelectronic properties remain to be optimized before competing with the ITO counterpart.

7.2.1.2 Graphene as TCE Material in QD Solar Cells

Graphene is another promising choice to replace ITO as the TCE. As a result of its unique electronic structure, graphene is able to have extremely high carrier mobility [17]. Since the rise of graphene, numerous researches and studies have found different methods to synthesize large-area graphene films, namely, solution-based exfoliation and chemical vapor deposition (CVD).

The first attempt of graphene as TCE can date back to the year 2008, researchers have demonstrated a solution-based method that can deposit reduced graphene oxide (rGO) into thin film [18], and several other groups followed up based on the similar approach [19–21]. Although the performance of rGO/polymer composite film can be comparable to ITO in some devices [20], this method still suffers from innate drawbacks. Neither thermal nor chemical treatment can reduce the GO completely and there are always defects remaining in rGO, which prevent it from succeeding the excellent conductivity of pristine graphene. On the other hand, graphene grown from CVD method has higher quality over large area, which is desirable for TCE fabrication. Furthermore, with effective doping, the graphene film can achieve a sheet resistance as low as 30 Ω/sq at 90% transparency [22] comparable to that of ITO. However, the high cost of CVD synthesis is a limiting factor to large-scale manufacture. Therefore, the controversy between quality and cost has to be well balanced before the further application of graphene as TCE material.

There have been some efforts on polymer solar cells with graphene electrode instead of ITO. It is worth noting that graphene has superior properties than ITO as TCE in QD-based solar cells. In order to better utilize the sun spectrum, the device is designed to harvest light with a wavelength of over 1000 nm, or infrared (IR) light, which can take up one third of the Sun's power. As introduced previously, some inorganic nanocrystals with small band gaps, such as PbS and PbSe, are suitable for IR solar cells. However, ITO electrode cannot give satisfying transparency beyond 1000 nm, while graphene can persist highly transparent from visible to IR region [23]. Based on these reasons, Lin et al. have explored the IR photoresponse of PV device consisting of PbS QDs deposited on the graphene electrode [24]. For the

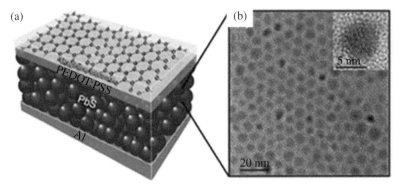

FIGURE 7.2 The device structure of the IR solar cell. (a) Al/PbS/PEDOT:PSS/graphene electrode. (b) TEM and HRTEM (inset) images of PbS QDs. Reproduced with permission from Ref. 24. © 2011 American Institute of Physics.

purpose of optimizing conductivity and transparency, multiple layers of CVD-grown graphene were stacked on glass electrode and a p-type doping was done by soaking into nitric acid. The three-layer graphene structure was selected with a sheet resistance of 165 Ω/sq and 85% transmittance at 1500 nm. Then, a 100 nm thick layer of PbS QDs was deposited on graphene with the ligand exchange step by 1,2-ethanedithiol. To further improve the interface condition, the PEDOT:PSS layer was put in between the graphene and QD layer (Fig. 7.2). Compared to the control cell with ITO electrode replacing graphene, it is encouraging to see that the graphene-based solar cell has higher external quantum efficiency (EQE) and photovoltaic conversion efficiency (PCE) at longer wavelength in the IR region, although the overall PCE of the ITO case is better than the graphene one under AM 1.5 (Fig. 7.3). The lagging performance of the latter at short wavelength results from a larger sheet resistance of the graphene electrode. Nevertheless, the graphene TCE still exhibits huge potentials in cooperating with QDs for the IR optoelectronic applications.

7.2.2 Carbon Nanomaterials and QD Composites in Solar Cells

As mentioned earlier in Section 7.2, owing to surface defect states, QD solar cells fail to provide as high efficiency as expected. To solve this problem, a kind of material acting as electron acceptor needs to be found with high electron mobility, suitable energy levels, and compatibility with QDs, so that the photogenerated carriers can be transported to the electrode through the acceptor. Carbon nanomaterials meet the demand perfectly, since they have unique electronic properties and can be functionalized by different chemical groups to interact with QDs. In the next paragraphs, the composite of C_{60}, CNTs, and graphene with QDs will be discussed, respectively, from the aspects of synthesis, conjugation, and device performance.

7.2.2.1 C_{60} and QD Composites

C_{60} has been widely used in organic PVs as electron acceptor [25] because of its low work function and high electron mobility. First, the charge transport effect between QDs and C_{60} has been studied. In 2006,

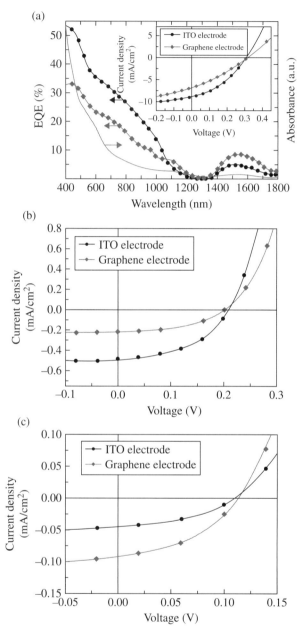

FIGURE 7.3 EQE and current density–voltage (*J–V*) characteristics of graphene electrode and ITO electrode solar cells. (a) EQEs of PbS solar cell based on graphene electrode and ITO electrode. The blue line represents the absorption of PbS QDs. The inset shows the *J–V* curves of two samples under AM 1.5 (100 mW/cm²). (b) *J–V* curves under 500 nm illumination (2.519 mW/cm²). (c) *J–V* curves under 1500 nm illumination (0.368 mW/cm²). Reproduced with permission from Ref. 24. © 2011 American Institute of Physics. (*See insert for color representation of the figure.*)

Biebersdorf et al. initialized the research of the photosensitization of C_{60} by nanocrystals [26]. The mixture of trioctylphosphine oxide (TOPO)-/trioctylphosphine (TOP)-capped CdSe QDs and C_{60} in toluene was dropped cast onto metal contact; then as the toluene evaporated, the needlelike C_{60} crystals formed. Under illumination, a three-fold enhancement of photocurrent was observed from the mixed solid compared with single C_{60} crystals. However, this preparation method is too rough to build good interaction between the two components. Consequently, Brown et al. proposed a new method to obtain CdSe–nC_{60} composite clusters by introducing electrophoretic deposition [27]. The formed cluster looks like a CdSe core with C_{60} shell surrounding and numerous clusters ensemble a thin film. With a more efficient structure, the incident photon to photocurrent generation efficiency (IPCE) of the cluster increased to 4%. The photocurrent generated by the composite is two orders of magnitude greater than pristine CdSe film (Fig. 7.4).

Besides the fundamental study, applications of C_{60}/QD composite in PVs have also emerged. Small band gap nanocrystals–PbS QDs cooperating with C_{60} become a mainstream [28, 29], since the introduction of PbS can help organic solar cells capture IR light. In the report by Dissanayake et al., a PbS nanocrystal–C_{60} heterojunction solar cell was fabricated with the structure of ITO/PbS/C_{60}/Al [28]. A heating process has been applied after the deposition of PbS layer, which can partially remove the oleic acid (OA) ligands on the PbS surface to shorten the distance between donors and acceptors, improving the short-circuit current density (J_{sc}) and fill factor (FF). However, the efficiency of the cell is below 0.1%, which indicates the poor carrier transport between and within layers. A better performance calls for a more effective ligand exchange strategy. So soon later, the same team published a better PCE of 0.44% [29]. The OA-capped PbS was exchanged by shorter butylamine ligands, and a PEDOT:PSS layer was added between ITO and PbS QDs. Since the PEDOT:PSS layer has an attractive interaction with butylamine-capped PbS, a smooth film of PbS can be fabricated. Also, the washing step in methanol can help clean the excess ligands as well, further improving J_{sc}, open circuit voltage (V_{oc}), and

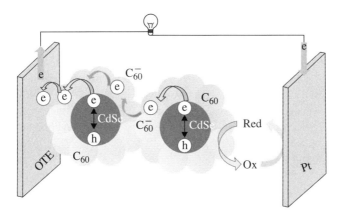

FIGURE 7.4 Photocurrent generation by CdSe–nC_{60} clusters. Reproduced with permission from Ref. 27. © 2008 American Chemical Society.

FF at the same time. Finally, a record of 2.2% has been reached by Tsang et al. by using 1,3-benzenedithiol (BDT) as the new ligand to replace OA on the surface of PbS [30]. The technique used in this chapter ensures a well-passivated PbS surface by shorter ligand, facilitating charge transfer. It is also noticeable that the different thickness of C_{60} was investigated as a parameter for the device and an optimal value has been found, indicating the importance of C_{60} as the electron acceptor.

7.2.2.2 CNTs and QD Composites
One-dimensional nanostructures can not only accept and transport photogenerated electrons from QDs but also provide scaffold for anchoring the light-absorbing QDs. CNTs have been widely studied in combination with various QDs. There are abundant literatures reporting CNT/QD composite synthesis, properties, and its application in solar cells. In general, the preparation methods of CNT/QD composite can be categorized as selective coupling and direct growing or deposit.

The basic idea of the first approach is to attach QDs to functionalized CNTs through the "linker," which is formed by the functional group on CNT walls and the corresponding ligand on QD surfaces. Different interactions can be used to hold the linkers, such as covalent chemical bonds [31] and noncovalent electrostatic interactions [32]. The drawback of covalent bonding is the property of pristine CNT that can be destroyed during the side-wall functionalization, while using noncovalent surfactants can protect the CNT yet sometimes increase the distance between CNTs and QDs, lowering the charge transfer. The second approach follows a complete distinct concept, based on which the QDs are grown directly from the CNTs without any linker in between. There are also covalent [33] and noncovalent [34] interactions in this kind. Among these synthetic mechanisms, the noncovalent direct growth of QDs onto the CNTs seems the most promising, because it has the least modifications to the original CNTs, and the simple one-step wet chemical method has proven to be effective for a variety of QDs.

After studying the fundamental carrier transport between QDs and CNTs through all aforementioned composites, researchers spare their efforts developing PV devices based on this structure. According to the band gap of QDs, the nanostructured composite can be divided into wide-band gap CNTs/QDs and narrow-band gap CNTs/QDs, enhancing the absorption of ultraviolet (UV) region and near-infrared (NIR) region, respectively. In the first group, CNTs have been combined with CdSe [35], CdS [36], TiO_2 [37], CdTe [34], and other nanocrystals. Landi et al. were able to attach CdSe QDs to SWCNTs covalently via the chemical bonds between amino-ethanethiol (AET) ligand of QD and carboxylic acid function group of CNTs [35]. In order to achieve a proper energy band alignment (Fig. 7.5), a poly(3-octylthiophene) (P3OT) layer was incorporated, which is a p-type polymer. Although only limited PCE can be gained due to undesirable recombination, a V_{oc} of 0.75 V still indicates a decent configuration of the device. In the case of CdS CNTs/QDs, Li et al. provided another way of fabricating solar cells (Fig. 7.6) [36]. Based on the heterojunction model of SWCNT/Si solar cell, CdS QDs were successfully grafted onto CNTs to further improve light harvesting. With appropriate CdS density on the surface of CNTs, the device gives out a V_{oc} of 0.47 V, J_{sc} of 6.82 mA/cm^2, FF of 43.7%, and PCE of 1.4% under AM 1.5 (100 mW/cm^2). One more interesting point is that owing to

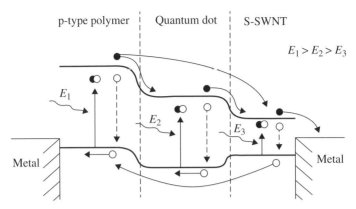

FIGURE 7.5   Illustration of the equilibrated band diagram for SWCNT/QD polymer solar cell. Reproduced with permission from Ref. 35. © 2005 Elsevier.

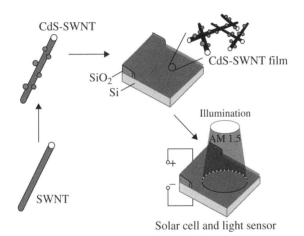

FIGURE 7.6   Fabrication process of CdS/SWCNT/Si solar cells. Reproduced with permission from Ref. 36. © 2010 American Chemical Society.

existence of CdS QDs, the device shows degradation under continuous illumination; however, the mechanism remains unclear.

In addition to wide-band gap semiconductor nanocrystals, PbS QDs have also been attached to MWCNTs in the application of NIR PV cells [38]. To form the donor–acceptor structure, P3HT was employed as the hole-conducting layer. The interaction between QDs and CNTs is strong, proven by the complete quenching of PbS, even though no ligand exchange steps applied. Finally, the optimized P3HT:PbS/MWCNT PV device can achieve a PCE of 3.03%.

7.2.2.3   *Graphene and QD Composites*   Compared to C_{60} and CNTs, graphene has even larger surface area, outstanding electron mobility, and a more proper work function. Therefore, the graphene/QD composite has received great expectations.

Similar to CNTs, the composite material can be synthesized through direct attachment and *in situ* growth or deposition. The QDs can be attached to the graphene or GO via physical attractions [39], electrostatic force [40], and π–π stacking interaction [41]. On the other hand, the QDs can crystallize and grow onto the GO or rGO [42–45]. Usually, the precursor of nanocrystals is added into the GO solution, and then with certain assistance outside the system, such as heating [42, 43] and microwave irradiation [44, 45], the QDs will be formed as the GO is being reduced simultaneously. Furthermore, electrochemical deposition of QDs has also been shown feasible owing to the chemical stability of graphene substrate [46, 47].

The attractive features of individual graphene and QD component make researchers anticipate more on the hybrid material in the solar cell application. In the case of dye-sensitized solar cell and photoelectrochemical cells, the graphene/QD electrode improves the performance effectively, even greater than the CNT/QD combination [48, 49], attributed to the special 2D morphology of graphene providing a better coverage of QDs on the surface and the strong coupling between the two components that facilitate fast transport of photogenerated electrons. Guo et al. demonstrated a layered graphene/CdS QD structure for the application in photoelectrochemical cells [49]. The composite was fabricated layer by layer through a simple bottom-up process (Fig. 7.7), where the graphene layers were deposited from chemically reduced graphene oxide. An IPCE value as high as 16% was observed from layered graphene/CdS compared to that of CNTs/QDs. Later on, the same team dip-coated an additional TiO$_x$ interlayer between the graphene and QD layer to further limit the recombination at the interface [50].

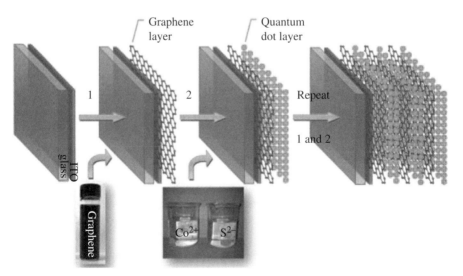

FIGURE 7.7 Fabrication process of layered graphene/CdS QDs on ITO/glass substrate. Reproduced with permission from Ref. 49. © 2010 John Wiley & Sons, Inc.

7.2.3 Graphene QDs Solar Cells

In addition to combine with QDs, graphene can be also tailored into the form of QDs and act as light-absorbing material in PV devices. Since the solution approach has the precise control of graphene nanostructure and ensures large yields at a time, the colloidal GQDs will be discussed in the following aspects: physical properties, synthesis, and PV devices.

7.2.3.1 Physical Properties of GQDs   Due to the unique feature of graphene in zero band gap and zero effective mass carriers at band edges, there are more remarkable properties that can be expected from GQDs, such as strong carrier–carrier interactions and weak spin–orbit coupling [51]. In the zero-dimensional semiconductor system like semiconductor QDs, the electronic energy levels are determined by E_q, E_c, and E_x, where E_q is the quantum-size-dependent energy, E_c is the hole/electron Coulomb interaction energy, and E_x is from exchange interaction [52]. Since the greater carrier–carrier columbic interaction in GQDs, the columbic energy factor E_c is comparable to E_q. However, in other semiconductor QD systems, E_c is only a minor correction to E_q. As a result, the excitonic effects are distinct for GQDs. Another benefit brought by the strong carrier–carrier interaction is the possibility of generating multiple excitons with only one photon absorbed by graphene, which is promising to enhance the efficiency of solar cells [51]. Moreover, the weak spin–orbit coupling leads to the reduction of singlet–triplet splitting and enhancement of intersystem crossing, so that it can compete with internal conversion among different states. Consequently, both fluorescence and phosphorescence can be emitted by GQDs at room temperature, whose relative intensity depends on the excitation wavelength [53].

7.2.3.2 Synthesis of GQDs   The synthetic methods of GQDs can be generalized into top-down and bottom-up approaches. For the top-down methods, the GQDs are formed by cutting large carbon material hydrothermally [54], electrochemically [55], etc. On the contrary, the QDs can also be formed through chemical reactions of small molecules, including solution chemical method [56] and microwave-assisted hydro-thermal approach [57]. Take the solution chemical synthetic method as an example. Yan et al. had developed the capability of synthesizing large and stable colloidal GQDs [58]. Based on oxidative condensation reactions, this approach was able to attach phenyl groups to the edges of the graphene, which gives rise to the twist of the phenyl groups and finally formation of the three-dimensional closing QD struc-ture (Fig. 7.8). This also reduced the interaction between GQDs, thus improving the solubility and stability of GQDs.

7.2.3.3 PV Devices of GQDs   As the synthetic techniques are developing, the GQDs can be controlled to such uniformity and stability that they can be incorpo-rated into solar cells as electron acceptors or light absorbers. The high mobility inherited from graphene makes GQDs promising candidates for electron acceptors. In addition, the tunability of the band gap of GQDs by size control further facilitates

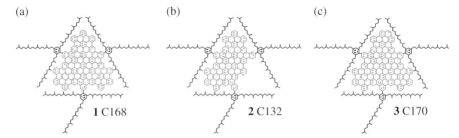

FIGURE 7.8   The structure of colloidal GQDs containing (a) 1:168 carbon atoms, (b) 2:132 carbon atoms, and (c) 3:170 carbon atoms. Reproduced with permission from Ref. 58. © 2010 American Chemical Society.

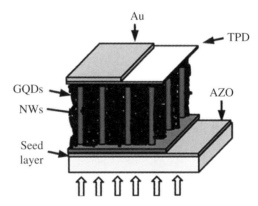

FIGURE 7.9   The structure of ZnO/GQD solid-state solar cell. Reproduced with permission from Ref. 59. © 2012 American Chemical Society.

manipulation of band alignment. Li et al. added the electrochemically synthesized GQDs into P3HT organic solar cells to act as electron acceptor and achieved a PCE of 1.28% [55]. The introduction of GQDs provides more interfaces for dissociation of photogenerated excitons and more effective carrier transport.

Meanwhile, the GQDs persist a strong quantum confinement but without toxicity, which is a vital advantage over the semiconductor QDs, such as CdSe and PbS. Therefore, the potential of GQDs to be light-harvesting material instead of semiconductor QDs has also been investigated. Yan et al. replaced the traditional ruthenium complexes by colloidal GQDs in the dye-sensitized solar cells [56], which can preserve a similar V_{oc} and FF as before. Another example, demonstrated by Dutta et al., is a ZnO/GQD solid-state solar cell [59]. The charge separation and transport took place at ZnO nanowire and GQD interfaces (Fig. 7.9). Although this device has obtained a V_{oc} of 0.8 V, the efficiency is far below expectation. Because the GQDs are mostly in the space between the ZnO nanowires, they may not get good contact with the hole-transporting layer, so the hole collection efficiency at the electrode is not satisfying.

Only part of the unique characteristics of GQDs has been utilized in device applications so far. There should be more novel phenomena and features, which call for a more profound understanding of GQDs, higher-quality materials, and proper structures for the device.

7.3 CARBON NANOMATERIAL/SEMICONDUCTOR HETEROJUNCTION SOLAR CELLS

It is imperative to find a new kind of cheap alternative materials to be introduced to replace silicon in solar PV devices. Carbon nanomaterials are adaptable to offer alternative materials to silicon in solar cells, so it is necessary to further understand the mechanism of the generation, transport, and dissociation of charge carriers occurring at the interface of the heterojunction. The following context will discuss the incorporation of carbon nanomaterials into the heterojunction solar cells.

7.3.1 Principle of Carbon/Semiconductor Heterojunction Solar Cells

The energy band diagrams of the carbon nanomaterials/semiconductor are shown both Schottky (metal/semiconductor) junction and p–n (semiconductor/semiconductor) junction (Fig. 7.10). When intrinsic metallic or p-type carbon nanomaterials and n-type semiconductor with a prominent work function difference are in contact with each other, electrons on the semiconductor side are depleted, which bends the energy band to form a space charge region (SCR). The SCR has a depletion layer of 0.4–0.5 μm estimated from equation [60]

$$W = \left(\frac{2\varepsilon_s V_{bi}}{qN_D} \right)^{1/2}$$

which was established in the n-Si near the heterojunction interface, where N_D is the doping density (3×10^{15}/cm^3). ε is the dielectric permittivity ($\varepsilon_s \approx 1.05 \times 10^{-12}$ F/cm). Photons are absorbed mainly in the n-type semiconductor region with resulting excitons diffusing to the SCR where they are split into free electrons and holes under the action of the built-in potential. Due to the difference of work function of carbon nanomaterials (4.7–5.1 eV) and semiconductor (e.g., Si=4.3 eV), the built-in electric field V_{bi} equals the work function difference of these two materials. If the work function of carbon nanomaterials becomes larger, a stronger electric field will be formed on the semiconductor side of the junction, hence improving the junction's capacity to collect photogenerated carriers. The barrier height (Φ_b) is in agreement with the difference between Φ_G and χ, where χ is the electron affinity of semiconductor (e.g., 4.05 eV for Si). These indicate that the carbon nanomaterials not only serve as a transparent electrode for light illumination but also an active layer for hole/electron separation and hole transport.

With the results mentioned earlier, three factors are found to impact the efficiency of photogenerated excess carriers in this carbon nanomaterial/semiconductor heterojunction: (i) the optical reflectance of the heterojunction solar cell, (ii) the resistances dissipated the electric energy, and (iii) the amplitude of the built-in potential near the

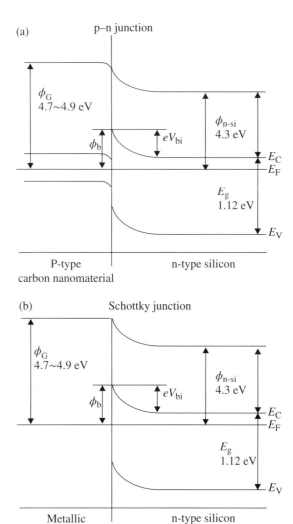

FIGURE 7.10 Energy band diagrams of carbon/semiconductor heterojunction. (a) p–n (semiconductor/semiconductor) junction. (b) Schottky (metal/semiconductor) junction.

heterojunction. Therefore, in order to optimize the efficiency of this heterojunction solar cell, modulation of the work function and reduction of optical reflection should be developed.

7.3.2 a-C/Semiconductor Heterojunction Solar Cells

The earliest successful carbon material to partially replace silicon in the heterojunction solar cell is a-C film [61–67]. Several groups have reported on a-C films in PV devices based on a-C/silicon heterojunction fabricated by CVD [61–64], ion

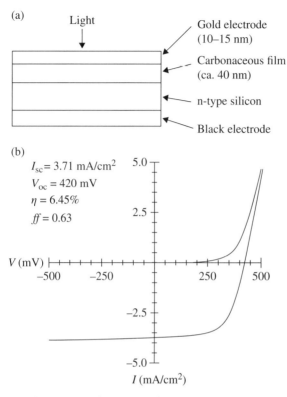

FIGURE 7.11   (a) The structure of the a-C/n-Si heterojunction solar cell. Reproduced with permission from Ref. 61. © 1996 American Institute of Physics. (b) *J–V* characteristics of a-C/n-Si heterojunction solar cells in the dark and under simulated 15 mW/cm² light with wavelengths between 400 and 800 nm. Reproduced with permission from Ref. 62. © 1996 American Institute of Physics.

implantation [65], pulsed laser deposition (PLD) [66], and vacuum deposition [67] techniques. Yu et al. first reported the a-C/silicon heterojunction solar cell in 1996 [61]. A layer of a-C thin film was deposited on an n-type single-crystal silicon (n-Si) substrate by CVD (Fig. 7.11a). The PV cell displayed a near perfect rectifying current–voltage characteristic. Under illumination of 15 mW/cm² light with wavelengths between 400 and 800 nm, a PCE of 3.80% and a FF value of 0.65 were achieved. The conversion efficiency of this a-C/n-Si PV cell increased to 6.45% due to the improvement of light transmission using thinner gold surface electrode [62] (Fig. 7.11b). Similarly, Ma et al. prepared boron-doped a-C/n-Si heterojunction solar cell using an arc discharge plasma CVD technique [63]. The cell shows a high efficiency of 7.9% at AM 1.5 (100 mW/cm²) with a V_{oc} of 0.58 V and a J_{sc} of 32.5 mA/cm². The successful formation of p-type a-C/n-Si junctions represented a first step toward further realizing practical carbon nanomaterial/semiconductor heterojunction solar cells. However, a-C is mainly a unipolar semiconductor and it is difficult to process dopants by diffusion and to remove defects by annealing because of its extreme bond energy and large diffusion energies. These problems have delayed applications of this material, and new carbon nanomaterials have been intensively investigated.

7.3.3 CNT/Semiconductor Heterojunction Solar Cells

Recently, CNT/semiconductor heterojunction has also been widely investigated. The MWCNT/Si heterojunction shows a rectifying J–V behavior for highly doped p-type silicon (p-Si) [68]. Due to the inhomogeneous mixtures of metallic and semiconducting CNTs, it presumes that a PV device constructed with semiconducting nanotubes would behave as a p–n junction solar cell, while one with metallic nanotubes would behave as a Schottky junction solar cell. Wei et al. reported a CNT/Si solar cell with a 1.3% efficient device in 2007 [69]. The semitransparent double-walled CNT (DWCNT) films were deposited on n-Si shown in Figure 7.12a. The CNTs also acted as a charge-collecting and transport layer. By optimizing the cell fabrication, the efficiency of solar cell was further improved to 7.4% with a V_{oc} of 0.54 V, a J_{sc} of 26 mA/cm^2, and a FF of 53% (Fig. 7.12b) [70].

Jia et al. reported that infiltration of CNTs by HNO$_3$ boosted the conversion efficiency from 6.8 to 13.8% (Fig. 7.13a and b) [71]. This enhancement benefits from the two possible reasons: (i) increment of the conductivity of CNTs and (ii) forming photoelectrochemical units in addition to CNT/Si heterojunctions at the interface that enhance the charge separation and transport. To reduce of the optical reflection, Shi et al. adopted the TiO$_2$ antireflection layer on the front of the heterojunction solar cell and combined with the acid doping (Fig. 7.13c) [72]. A considerable high conversion efficiency of 15% was achieved (Fig. 7.13d).

The arrangement pattern of CNT networks has a certain influence at the interface of the heterojunction. Di et al. reported that the aligned CNT films can act as a promising window layer for CNT/Si Schottky solar cells (Fig. 7.14a, b, and c) [73]. The solar cell exhibits a J_{sc} of 33.4 mA/cm^2, a V_{oc} of 0.54 V, a high FF of 58.3%, and a PCE of 10.5% without any treatment (Fig. 7.14d), which is higher than those of solar cells fabricated using pristine and random CNT networks. The improved PCE is attributed to the dense CNT coating on the Si substrate, forming high density CNT/Si junctions. Furthermore, the aligned CNTs are as long as 200 μm and are expected to provide more direct charge transfer paths.

Besides bulk silicon, CNTs can also form a rectifying heterojunction with other bulk or nanostructure semiconductor. Zhang et al. designed the Schottky junction solar cells by coating CNT films on individual CdSe nanobelts with a V_{oc} of 0.5–0.6 V and a PCE of 0.45–0.72% under 100 mW/cm^2 light condition (Fig. 7.15a and b) [74]. In this device structure, the CdSe nanobelt serves as a flat substrate to sustain a network of CNTs. This nanotube/nanobelt solar cell can work either in both sides of illumination. Furthermore, Shi et al. developed this nanostructure heterojunction solar cell and demonstrated individual CNTs and CdSe nanobelts arranged in simple cross-junction configurations [75]. Multiple CNTs in parallel with a single nanobelt could be constructed in an array of cross-junction solar

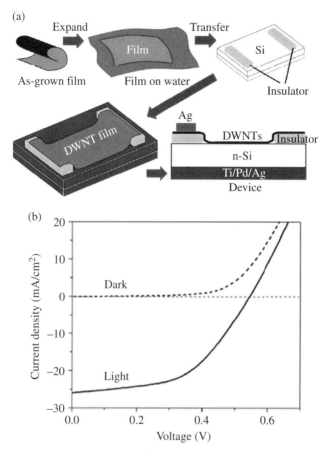

FIGURE 7.12   (a) Illustration of the fabrication process of DWCNT/n-Si solar cell. As-grown DWCNT films were conformally transferred to a patterned Si substrate. Reproduced with permission from Ref. 69. © 2007 American Chemical Society. (b) Dark and light (100 mW/cm² illumination) J–V curves of a heterojunction solar cell, showing a PCE of 7.4%. Reproduced with permission from Ref. 70. © 2008 John Wiley & Sons, Inc.

cells, indicating the possibility of parallel device connection and large-scale production (Fig. 7.15c and d).

7.3.4   Graphene/Semiconductor Heterojunction Solar Cells

Though CNT/semiconductor heterojunction solar cells have been promoted a lot, CNT networks suffer from bundling between tubes/bundles and thus reduce the film connectivity and conductivity [76]. In contrast, graphene films can be prepared with controlled thickness and good surface continuity. Li et al. first reported the graphene/Si heterojunction solar cells in 2010 [77]. CVD-grown graphene films have been conformally deposited on n-Si with 100% coverage to make the Schottky junction solar

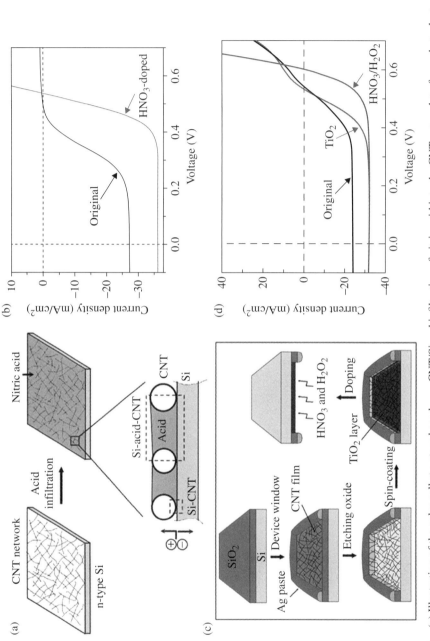

FIGURE 7.13 (a) Illustration of the solar cell structure based on a CNT/Si and infiltration of nitric acid into the CNT network to form photoelectrochemical units. Reproduced with permission from Ref. 71. © 2011 American Chemical Society. (b) *J–V* curves of the solar cell before (black curve) and after (red) infiltration of dilute HNO₃. Reproduced with permission from Ref. 71. © 2011 American Chemical Society. (c) Illustration of the fabrication process of a TiO₂/CNT/Si solar. (d) *J–V* curves of a CNT/Si cell recorded in original state (without coating), with a TiO₂ antireflection layer, and after HNO₃/H₂O₂

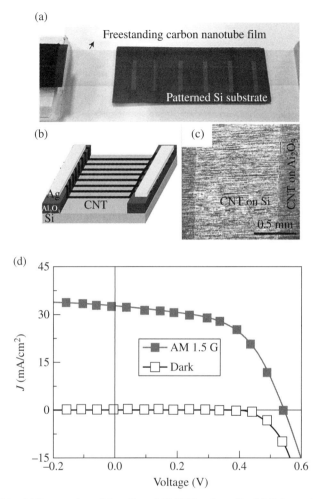

FIGURE 7.14   (a) Preparation of the aligned CNT/Si solar cells. (b) Schematic showing the structure of the PV device. (c) Optical image showing aligned CNT/Si solar cell. (d) *J–V* curves of the aligned CNT/Si solar cell under AM 1.5 illumination and in the dark. Reproduced with permission from Ref. 73. © 2013 John Wiley & Sons, Inc.

cell, which shows a PCE of 1.5% with a FF of 56% (Fig. 7.16). This low cost and simple fabrication of Schottky junctions could be enhanced with surface passivation, doping, and junction formation.

Several strategies have been proposed to improve this graphene/Si Schottky solar cell. The silicon nanowire arrays (SNA) (Fig. 7.17a and b) [78] and silicon pillar arrays (SPA) (Fig. 7.17d and e) [79] were adopted instead of planar Si. The geometry of SNA and SPA leads to a drastically suppressed reflectance (Fig. 7.17c) and enhanced light harvesting. A typical Schottky solar cell exhibited a PCE of 1.25–2.90% (Fig. 7.17f).

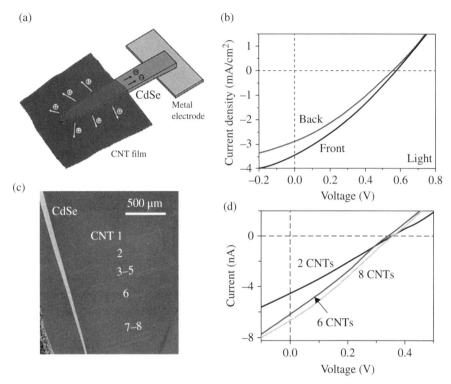

FIGURE 7.15 (a) Schematic illustration of a CdSe nanobelt covered by a CNT film. Arrows show the flow direction of charge carriers when the device is illuminated. (b) *J–V* curves recorded when the solar cell is illuminated from the front (CNT film) or back (CdSe nanobelt) sides, with PCE of 0.59 and 0.49%, respectively. Reproduced with permission from Ref. 74. © 2010 American Chemical Society. (c) SEM image of a CdSe nanobelt transferred to an array of 8 CNTs in parallel. (d) *J–V* curves (75 mW/cm²) of the three solar cells containing 2, 6, and 8 CNTs. Reproduced with permission from Ref. 75. © 2012 The Royal Society of Chemistry.

Moreover, the tunable work function of graphene affords the advantage to the efficient carrier injection at the interface of graphene/silicon solar cells [4]. The effective way to control the work function of graphene is by p-type chemical doping, leading to the increment of not only the conductivity but also the built-in field at the interface (Fig. 7.18a). For example, a high PCE was achieved in the p-doped graphene-/Si-based solar cell devices by spin-casting a layer of bis(trifluoromethanesulfonyl)amide (TFSA) on the pristine graphene/n-Si layer (Fig. 7.18b) [80]. Upon doping with TFSA, the V_{oc}, J_{sc}, and FF all increased from 0.43 to 0.54V, 14.2 to 25.3 mA/cm², and 0.32 to 0.63. As a result, the PCE was boosted from 1.9 to 8.6% (Fig. 7.18c). Li et al. optimized the solar cell fabrication, and the PCE of solar cell was further improved to 5.0–7.4% [82]. P-type

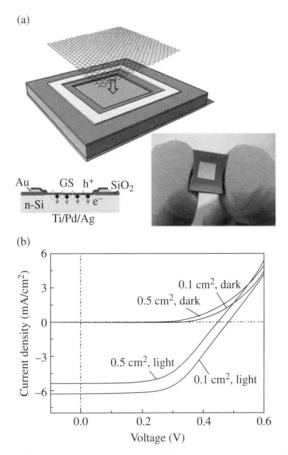

FIGURE 7.16 (a) Schematic illustration of the PV device configuration. Bottom-left inset: photogenerated holes (h⁺) and electrons (e⁻) are driven into the graphene and Si, respectively. Bottom-right inset: photograph of a graphene/Si Schottky solar cell. (b) *J–V* curves of the 0.1 and 0.5 cm² solar cells illuminated with simulated AM 1.5. Reproduced with permission from Ref. 77. © 2010 John Wiley & Sons, Inc.

chemical doping by chlorine and nitrate anions enhanced the Schottky junction and improved the V_{oc} of the graphene/Si device from 0.45 to 0.55 V and its FF from 0.45 to 0.73, which nearly doubled its PCE from 4.98 to 9.63% (Fig. 7.18d and e) [81]. P-type chemical doping was found to be effective for reducing the sheet resistance and enlarging the work function of graphene for effective charge separation and transport. In general, the PCE of the doped graphene/Si solar cells ranged from 8.0 to 9.6% [82].

Employing a similar structure to TiO_2/CNT/Si heterojunction solar cells described earlier, a TiO_2 film directly transferred onto the front of graphene/Si solar cell (Fig. 7.19a and b) [83]. After coating TiO_2, the TiO_2/graphene/Si heterojunction showed a much reduced light reflectance in the visible region (500–800 nm) from

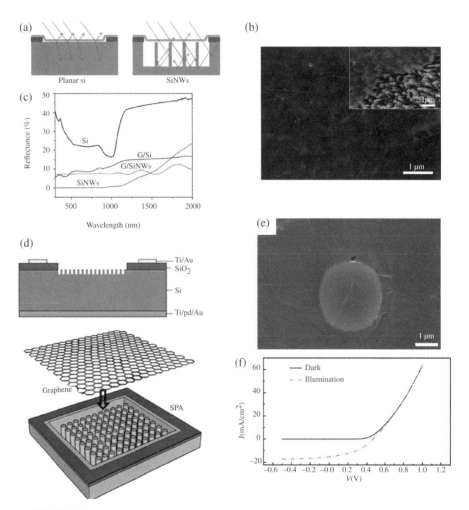

FIGURE 7.17 (a) Schematics of graphene/planar Si and graphene/SiNW junctions. (b) Top-view SEM images of graphene/SiNW junction. (c) Reflection spectra of planar Si, SiNWs, graphene/Si, and graphene/SiNWs. Reproduced with permission from Ref. 78. © 2011 American Chemical Society. (d) Schematic structure of the cross-sectional SPA substrate. (e) A top-view SEM microstructure of graphene film covering on one silicon pillar. (f) *J–V* curves of graphene/SPA solar cell in dark and under illumination. Reproduced with permission from Ref. 79. © 2011 American Institute of Physics.

nearly 40 to 10% (Fig. 7.19c). By transferring the antireflection layer, a PCE of solar cell can be enhanced to 14.5% (Fig. 7.19d).

Semiconducting monolayer transition metal dichalcogenides (TMDs), particularly MoS_2, $MoSe_2$, WS_2, and WSe_2, are promising materials for next-generation ultrathin optoelectronic devices [84–87]. Current rapid progress in growth and

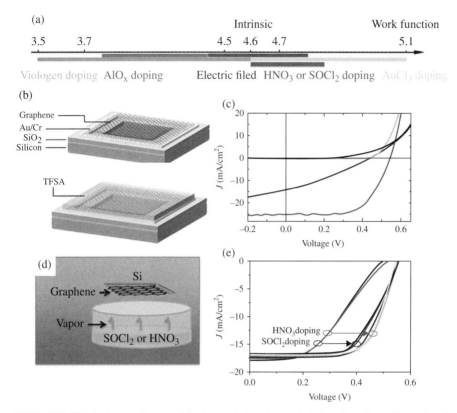

FIGURE 7.18   (a) Approaches to achieving work function modulation of graphene. Reproduced with permission from Ref. 4. © 2013 The Royal Society of Chemistry. (b) Schematic structure of TFSA-doped graphene/Si Schottky solar cell. (c) *J–V* curves of graphene/Si (blue) and doped graphene/Si (red) Schottky solar cells in dark and after illumination. Reproduced with permission from Ref. 80. © 2012 American Chemical Society. (d) Schematic illustration of the vapor-doping process. (e) Light *J–V* curves of the pristine and doped cells with increasing doping time. Reproduced with permission from Ref. 81. © 2013 The Royal Society of Chemistry. (*See insert for color representation of the figure.*)

synthesis of TMDs has opened up a new possibility for the implementation of TMDs into PV devices [88–91]. Britnell et al. reported this simple stacking of TMD layer (WS_2, MoS_2, and GaSe) sandwiched between graphene sheet heterojunction ultrathin PV devices (Fig. 7.20a) [87]. This heterostructure device exhibited the PV effects with external quantum efficiencies as high as 30% (Fig. 7.20b). Because of the flat and dangling-bond-free surfaces of 2D crystals, this device can be fabricated on flexible substrates. Bernardi et al. demonstrated that the Schottky junction solar cell between graphene and MoS_2 can attain a PCE of up to 1% (Fig. 7.20c and d) [92]. This work shows the potential for graphene-based heterostructure solar cell at the nanoscale.

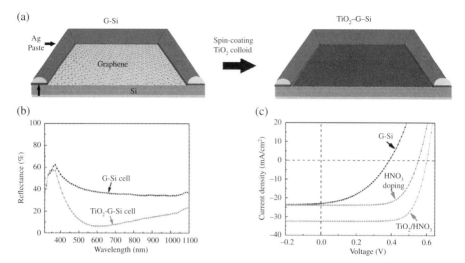

FIGURE 7.19 (a) Schematic illustration of the spin-coating process in which a colloidal TiO$_2$ was applied to a graphene/Si cell as antireflection coating. (b) Light reflection spectra of a graphene/Si solar cell before (black) and after (red) coating the TiO$_2$ colloid. (c) Light *J–V* curves of graphene/Si solar cell, after HNO$_3$ doping and after TiO$_2$ coating, respectively. Reproduced with permission from Ref. 83. © 2013 American Chemical Society.

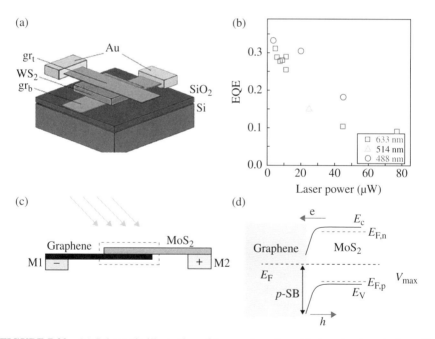

FIGURE 7.20 (a) Schematic illustration of the graphene-based heterostructure device with the principal layers shown. (b) The external quantum efficiency of the devices is the ratio of the number of measured charge pairs to absorbed incident photons. Reproduced with permission from Ref. 87. © 2013 The American Association for the Advancement of Science. (c) The MoS$_2$/graphene solar cell. (d) Band alignment at a MoS$_2$/graphene interface. Reproduced with permission from Ref. 92. © 2013 American Chemical Society.

7.4 SUMMARY

Both QDs and heterostructures containing carbon nanomaterial solar cell possess the interesting and useful optical and electronic properties. These kinds of new generation solar cell are promising in the low-cost and high-efficiency nanoscale power source, which are desired in the integrated nanosystems. However, challenges remain in this filed, like developing the structure of PV devices and improving the properties of carbon nanomaterials. Nonetheless, the research of carbon nanomaterial-based solar cells still remains an intriguing issue that needs to be explored further.

REFERENCES

[1] W. Shockley, H. J. Queisser, Detailed Balance Limit of Efficiency of p-n Junction Solar Cells. *J. Appl. Phys.* 1961, **32**, 510.

[2] R. T. Ross, A. J. Nozik, Efficiency of Hot-Carrier Solar Energy Converters. *J. Appl. Phys.* 1982, **53**, 3813.

[3] R. D. Schaller, V. I. Klimov, High Efficiency Carrier Multiplication in PbSe Nanocrystals: Implications for Solar Energy Conversion. *Phys. Rev. Lett.* 2004, **92**, 186601.

[4] Y. X. Lin, X. M. Li, D. Xie, T. T. Feng, Y. Chen, R. Song, H. Tian, T. L. Ren, K. L. Wang, H. W. Zhu, Graphene/Semiconductor Heterojunction Solar Cells with Modulated Antireflection and Graphene Work Function. *Energy Environ. Sci.* 2013, **6**, 108.

[5] I. J. Kramer, E. H. Sargent, Colloidal Quantum Dot Photovoltaics: A Path Forward. *ACS Nano* 2011, **5**, 8506.

[6] A. Kumar, C. W. Zhou, The Race to Replace Tin-Doped Indium Oxide: Which Material Will Win? *ACS Nano* 2010, **4**, 11.

[7] L. Hu, D. S. Hecht, G. Gruner, Percolation in Transparent and Conducting Carbon Nanotube Networks. *Nano Lett.* 2004, **4**, 2513.

[8] P. M. Ajayan, J. M. Tour, Materials Science: Nanotube Composites. *Nature* 2007, **447**, 1066.

[9] J. N. Coleman, M. Cadek, R. Blake, V. Nicolosi, K. P. Ryan, C. Belton, A. Fonseca, J. B. Nagy, Y. K. Gun'Ko, W. J. Blau, High Performance Nanotube-Reinforced Plastics: Understanding the Mechanism of Strength Increase. *Adv. Funct. Mater.* 2004, **14**, 791.

[10] H. Z. Geng, D. S. Lee, K. K. Kim, G. H. Han, H. K. Park, Y. H. Lee, Absorption Spectroscopy of Surfactant-Dispersed Carbon Nanotube Film: Modulation of Electronic Structures. *Chem. Phys. Lett.* 2008, **455**, 275.

[11] X. K. Li, F. Gittleson, M. Carmo, R. C. Sekol, A. D. Taylor, Scalable Fabrication of Multifunctional Freestanding Carbon Nanotube/Polymer Composite Thin Films for Energy Conversion. *ACS Nano* 2012, **6**, 1347.

[12] E. Kymakis, E. Stratakis, E. Koudoumas, Integration of Carbon Nanotubes as Hole Transport Electrode in Polymer/Fullerene Bulk Heterojunction Solar Cells. *Thin Solid Films* 2007, **515**, 8598.

[13] J. van de Lagemaat, T. M. Barnes, G. Rumbles, S. E. Shaheen, T. J. Coutts, C. Weeks, I. Levitsky, J. Peltola, P. Glatkowski, Organic Solar Cells with Carbon Nanotubes Replacing In_2O_3:Sn as the Transparent Electrode. *Appl. Phys. Lett.* 2006, **88**, 233503.

[14] A. D. Pasquier, H. E. Unalan, A. Kanwal, S. Miller, M. Chhowalla, Conducting and Transparent Single-wall Carbon Nanotube Electrodes for Polymer-Fullerene Solar Cells. *Appl. Phys. Lett.* 2005, **87**, 203511.

[15] M. W. Rowell, M. A. Topinka, M. D. McGehee, H. J. Prall, G. Dennler, N. S. Sariciftci, L. B. Hu, G. Gruner, Organic Solar Cells with Carbon Nanotube Network Electrodes. *Appl. Phys. Lett.* 2006, **88**, 233506.

[16] K. Sears, G. Fanchini, S. E. Watkins, C. P. Huynh, S. C. Hawkins, Aligned Carbon Nanotube Webs as a Replacement for Indium Tin Oxide in Organic Solar Cells. *Thin Solid Films* 2013, **531**, 525.

[17] A. K. Geim, K. S. Novoselov, The Rise of Graphene. *Nat. Mater.* 2007, **6**, 183.

[18] G. Eda, G. Fanchini, M. Chhowalla, Large-Area Ultrathin Films of Reduced Graphene Oxide as a Transparent and Flexible Electronic Material. *Nat. Nanotechnol.* 2008, **3**, 270.

[19] H. A. Becerril, J. Mao, Z. Liu, R. M. Stoltenberg, Z. Bao, Y. Chen, Evaluation of Solution-Processed Reduced Graphene Oxide Films as Transparent Conductors. *ACS Nano* 2008, **2**, 463.

[20] H. X. Chang, G. F. Wang, A. Yang, X. M. Tao, X. Q. Liu, Y. D. Shen, Z. J. Zheng, A Transparent, Flexible, Low-Temperature, and Solution-Processible Graphene Composite Electrode. *Adv. Funct. Mater.* 2010, **20**, 2893.

[21] J. P. Zhao, S. F. Pei, W. C. Ren, L. B. Gao, H. M. Cheng, Efficient Preparation of Large-Area Graphene Oxide Sheets for Transparent Conductive Films. *ACS Nano* 2010, **4**, 5245.

[22] S. Bae, H. Kim, Y. Lee, X. F. Xu, J. S. Park, Y. Zheng, J. Balakrishnan, T. Lei, H. R. Kim, Y. I. Song, Y. J. Kim, K. S. Kim, B. Ozyilmaz, J. H. Ahn, B. H. Hong, S. Iijima, Roll-to-Roll Production of 30-Inch Graphene Films for Transparent Electrodes. *Nat. Nanotechnol.* 2010, **5**, 574.

[23] L. B. Hu, D. S. Hecht, G. Gruner, Infrared Transparent Carbon Nanotube Thin Films. *Appl. Phys. Lett.* 2009, **94**, 811038.

[24] C. C. Lin, D. Y. Wang, K. H. Tu, Y. T. Jiang, M. H. Hsieh, C. C. Chen, C. W. Chen, Enhanced Infrared Light Harvesting of Inorganic Nanocrystal Photovoltaic and Photodetector on Graphene Electrode. *Appl. Phys. Lett.* 2011, **98**, 263509.

[25] G. Yu, A. J. Heeger, Charge Separation and Photovoltaic Conversion in Polymer Composites with Internal Donor/Acceptor Heterojunctions. *J. Appl. Phys.* 1995, **78**, 4510.

[26] A. Biebersdorf, R. Dietmuller, A. S. Susha, A. L. Rogach, S. K. Poznyak, D. V. Talapin, H. Weller, T. A. Klar, J. Feldmann, Semiconductor Nanocrystals Photosensitize C_{60} Crystals. *Nano Lett.* 2006, **6**, 1559.

[27] P. Brown, P. V. Kamat, Quantum Dot Solar Cells. Electrophoretic Deposition of CdSe$-C_{60}$ Composite Films and Capture of Photogenerated Electrons with nC_{60} Cluster Shell. *J. Am. Chem. Soc.* 2008, **130**, 8890.

[28] D. Dissanayake, R. A. Hatton, T. Lutz, C. E. Giusca, R. J. Curry, S. Silva, A PbS Nanocrystal-C_{60} Photovoltaic Device for Infrared Light Harvesting. *Appl. Phys. Lett.* 2007, **91**, 133506.

[29] D. Dissanayake, R. A. Hatton, T. Lutz, R. J. Curry, S. Silva, The Fabrication and Analysis of a PbS Nanocrystal:C_{60} Bilayer Hybrid Photovoltaic System. *Nanotechnology* 2009, **20**, 245202.

[30] S. W. Tsang, H. Fu, R. Wang, J. Lu, K. Yu, Y. Tao, Highly Efficient Cross-linked PbS Nanocrystal/C_{60} Hybrid Heterojunction Photovoltaic Cells. *Appl. Phys. Lett.* 2009, **95**, 183505.

[31] S. Banerjee, S. S. Wong, Synthesis and Characterization of Carbon Nanotube–Nanocrystal Heterostructures. *Nano Lett.* 2002, **2**, 195.

[32] D. M. Guldi, G. Rahman, V. Sgobba, N. A. Kotov, D. Bonifazi, M. Prato, CNT–CdTe Versatile Donor–Acceptor Nanohybrids. *J. Am. Chem. Soc.* 2006, **128**, 2315.

[33] I. Robel, B. A. Bunker, P. V. Kamat, Single-Walled Carbon Nanotube–CdS Nanocomposites as Light-Harvesting Assemblies: Photoinduced Charge-Transfer Interactions. *Adv. Mater.* 2005, **17**, 2458.

[34] S. Banerjee, S. S. Wong, In Situ Quantum Dot Growth on Multiwalled Carbon Nanotubes. *J. Am. Chem. Soc.* 2003, **125**, 10342.

[35] B. J. Landi, S. L. Castro, H. J. Ruf, C. M. Evans, S. G. Bailey, R. P. Raffaelle, CdSe Quantum Dot-Single Wall Carbon Nanotube Complexes for Polymeric Solar Cells. *Sol. Energy Mater. Sol. Cells* 2005, **87**, 733.

[36] X. L. Li, Y. Jia, J. Q. Wei, H. W. Zhu, K. L. Wang, D. H. Wu, A. Y. Cao, Solar Cells and Light Sensors Based on Nanoparticle-Grafted Carbon Nanotube Films. *ACS Nano* 2010, **4**, 2142.

[37] A. Kongkanand, R. M. Dominguez, P. V. Kamat, Single Wall Carbon Nanotube Scaffolds for Photoelectrochemical Solar Cells. Capture and Transport of Photogenerated Electrons. *Nano Lett.* 2007, **7**, 676.

[38] D. F. Wang, J. K. Baral, H. G. Zhao, B. A. Gonfa, V. V. Truong, M. A. El Khakani, R. Izquierdo, D. L. Ma, Controlled Fabrication of PbS Quantum-Dot/Carbon-Nanotube Nanoarchitecture and its Significant Contribution to Near-Infrared Photon-to-Current Conversion. *Adv. Funct. Mater.* 2011, **21**, 4010.

[39] N. L. Yang, J. Zhai, D. Wang, Y. S. Chen, L. Jiang, Two-Dimensional Graphene Bridges Enhanced Photoinduced Charge Transport in Dye-Sensitized Solar Cells. *ACS Nano* 2010, **4**, 887.

[40] S. B. Yang, X. L. Feng, S. Ivanovici, K. Mullen, Fabrication of Graphene-Encapsulated Oxide Nanoparticles: Towards High-Performance Anode Materials for Lithium Storage. *Angew. Chem. Int. Edit.* 2010, **49**, 8408.

[41] X. M. Geng, L. Niu, Z. Y. Xing, R. S. Song, G. T. Liu, M. T. Sun, G. S. Cheng, H. J. Zhong, Z. H. Liu, Z. J. Zhang, L. F. Sun, H. X. Xu, L. Lu, L. W. Liu, Aqueous-Processable Noncovalent Chemically Converted Graphene–Quantum Dot Composites for Flexible and Transparent Optoelectronic Films. *Adv. Mater.* 2010, **22**, 638.

[42] D. H. Wang, D. W. Choi, J. Li, Z. G. Yang, Z. M. Nie, R. Kou, D. H. Hu, C. M. Wang, L. V. Saraf, J. G. Zhang, I. A. Aksay, J. Liu, Self-Assembled TiO_2–Graphene Hybrid Nanostructures for Enhanced Li-Ion Insertion. *ACS Nano* 2009, **3**, 907.

[43] A. N. Cao, Z. Liu, S. S. Chu, M. H. Wu, Z. M. Ye, Z. W. Cai, Y. L. Chang, S. F. Wang, Q. H. Gong, Y. F. Liu, A Facile One-step Method to Produce Graphene–CdS Quantum Dot Nanocomposites as Promising Optoelectronic Materials. *Adv. Mater.* 2010, **22**, 103.

[44] J. Yan, Z. J. Fan, T. Wei, W. Z. Qian, M. L. Zhang, F. Wei, Fast and Reversible Surface Redox Reaction of Graphene–MnO_2 Composites as Supercapacitor Electrodes. *Carbon* 2010, **48**, 3825.

[45] J. Yan, T. Wei, W. M. Qiao, B. Shao, Q. K. Zhao, L. J. Zhang, Z. J. Fan, Rapid Microwave-Assisted Synthesis of Graphene Nanosheet/Co_3O_4 Composite for Supercapacitors. *Electrochim. Acta* 2010, **55**, 6973.

[46] Y. T. Kim, J. H. Han, B. H. Hong, Y. U. Kwon, Electrochemical Synthesis of CdSe Quantum-Dot Arrays on a Graphene Basal Plane Using Mesoporous Silica Thin-Film Templates. *Adv. Mater.* 2010, **22**, 515.

[47]   Y. B. Tang, C. S. Lee, J. Xu, Z. T. Liu, Z. H. Chen, Z. B. He, Y. L. Cao, G. D. Yuan, H. S. Song, L. M. Chen, L. B. Luo, H. M. Cheng, W. J. Zhang, I. Bello, S. T. Lee, Ncorporation of Graphenes in Nanostructured TiO_2 Films via Molecular Grafting for Dye-Sensitized Solar Cell Application. *ACS Nano* 2010, **4**, 3482.

[48]   H. X. Chang, X. J. Lv, H. Zhang, J. H. Li, Quantum Dots Sensitized Graphene: In Situ Growth and Application in Photoelectrochemical Cells. *Electrochem. Commun.* 2010, **12**, 483.

[49]   C. X. Guo, H. B. Yang, Z. M. Sheng, Z. S. Lu, Q. L. Song, C. M. Li, Layered Graphene/ Quantum Dots for Photovoltaic Devices. *Angew. Chem. Int. Edit.* 2010, **49**, 3014.

[50]   H. B. Yang, C. X. Guo, G. H. Guai, Q. L. Song, S. P. Jiang, C. M. Li, Reduction of Charge Recombination by an Amorphous Titanium Oxide Interlayer in Layered Graphene/ Quantum Dots Photochemical Cells. *ACS Appl. Mater. Interfaces* 2011, **3**, 1940.

[51]   L. S. Li, X. Yan, Colloidal Graphene Quantum Dots. *J. Phys. Chem. Lett.* 2010, **1**, 2572.

[52]   A. L. Efros, M. Rosen, The Electronic Structure of Semiconductor Nanocrystals. *Ann. Rev. Mater. Sci.* 2000, **30**, 475.

[53]   M. L. Mueller, X. Yan, J. A. McGuire, L. S. Li, Triplet States and Electronic Relaxation in Photoexcited Graphene Quantum Dots. *Nano Lett.* 2010, **10**, 2679.

[54]   D. Y. Pan, L. Guo, J. C. Zhang, C. Xi, Q. Xue, H. Huang, J. H. Li, Z. W. Zhang, W. J. Yu, Z. W. Chen, Z. Li, M. H. Wu, Cutting sp^2 Clusters in Graphene Sheets into Colloidal Graphene Quantum Dots with Strong Green Fluorescence. *J. Mater. Chem.* 2012, **22**, 3314.

[55]   Y. Li, Y. Hu, Y. Zhao, G. Q. Shi, L. E. Deng, Y. B. Hou, L. T. Qu, An Electrochemical Avenue to Green-Luminescent Graphene Quantum Dots as Potential Electron-Acceptors for Photovoltaics. *Adv. Mater.* 2011, **23**, 776.

[56]   X. Yan, X. Cui, B. S. Li, L. S. Li, Large, Solution-Processable Graphene Quantum Dots as Light Absorbers for Photovoltaics. *Nano Lett.* 2010, **10**, 1869.

[57]   L. B. Tang, R. B. Ji, X. M. Li, K. S. Teng, S. P. Lau, Size-Dependent Structural and Optical Characteristics of Glucose-Derived Graphene Quantum Dots. *Part. Part. Syst. Char.* 2013, **30**, 523.

[58]   X. Yan, X. Cui, L. S. Li, Synthesis of Large, Stable Colloidal Graphene Quantum Dots with Tunable Size. *J. Am. Chem. Soc.* 2010, **132**, 5944.

[59]   M. Dutta, S. Sarkar, T. Ghosh, D. Basak, ZnO/Graphene Quantum Dot Solid-State Solar Cell. *J. Phys. Chem. C* 2012, **116**, 20127.

[60]   S. M. Sze, K. K. Ng, *Physics of Semiconductor Devices*, John Wiley & Sons, Inc., Hoboken 2007.

[61]   H. A. Yu, Y. Kaneko, S. Yoshimura, S. Otani, Photovoltaic Cell of Carbonaceous Film/ n-Type Silicon. *Appl. Phys. Lett.* 1996, **68**, 547.

[62]   H. A. Yu, T. Kaneko, S. Yoshimura, Y. Suhng, S. Otani, Y. Sasaki, The Spectro-Photovoltaic Characteristics of a Carbonaceous Film/n-Type Silicon (C/n-Si) Photovoltaic Cell. *Appl. Phys. Lett.* 1996, **69**, 4078.

[63]   Z. Q. Ma, B. X. Liu, Boron-Doped Diamond-Like Amorphous Carbon as Photovoltaic Films in Solar Cell. *Sol. Energy Mater. Sol. Cells* 2001, **69**, 339.

[64]   H. A. Yu, T. Kaneko, S. Yoshimura, Y. Suhng, Y. Sasaki, S. Otani, The Junction Characteristics of Carbonaceous Film/n-Type Silicon (C/n-Si) Layer Photovoltaic Cell. *Appl. Phys. Lett.* 1996, **69**, 3042.

[65]   K. M. Krishna, Y. Nukaya, T. Soga, T. Jimbo, M. Umeno, Solar Cells Based on Carbon Thin Films. *Sol. Energy Mater. Sol. Cells* 2001, **65**, 163.

[66] X. M. Tian, M. Rusop, Y. Hayashi, T. Soga, T. Jimbo, M. Umeno, Boron-Incorporated Amorphous Carbon Films Deposited by Pulsed Laser Deposition. *Jpn. J. Appl. Phys.* 2002, **41**, L970.

[67] K. Mukhopadhyay, K. M. Krishna, M. Sharon, A Simple Method and New Source for Getting Diamond-like Carbon Film and Polycrystalline Diamond Film. *Mater. Chem. Phys.* 1997, **49**, 252.

[68] J. T. Hu, M. Ouyang, P. D. Yang, C. M. Lieber, Controlled Growth and Electrical Properties of Heterojunctions of Carbon Nanotubes and Silicon Nanowires. *Nature* 1999, **399**, 48.

[69] J. Q. Wei, Y. Jia, Q. K. Shu, Z. Y. Gu, K. L. Wang, D. M. Zhuang, G. Zhang, Z. C. Wang, J. B. Luo, A. Y. Cao, D. H. Wu, Double-Walled Carbon Nanotube Solar Cells. *Nano Lett.* 2007, **7**, 2317.

[70] Y. Jia, J. Q. Wei, K. L. Wang, A. Y. Cao, Q. K. Shu, X. C. Gui, Y. Q. Zhu, D. M. Zhuang, G. Zhang, B. B. Ma, L. D. Wang, W. J. Liu, Z. C. Wang, J. B. Luo, D. Wu, Nanotube–Silicon Heterojunction Solar Cells. *Adv. Mater.* 2008, **20**, 4594.

[71] Y. Jia, A. Y. Cao, X. Bai, Z. Li, L. H. Zhang, N. Guo, J. Q. Wei, K. L. Wang, H. W. Zhu, D. H. Wu, P. M. Ajayan, Achieving High Efficiency Silicon-Carbon Nanotube Heterojunction Solar Cells by Acid Doping. *Nano Lett.* 2011, **11**, 1901.

[72] E. Z. Shi, L. H. Zhang, Z. Li, P. X. Li, Y. Y. Shang, Y. Jia, J. Q. Wei, K. L. Wang, H. W. Zhu, D. H. Wu, S. Zhang, A. Y. Cao, TiO_2-Coated Carbon Nanotube-Silicon Solar Cells with Efficiency of 15%. *Sci. Rep.* 2012, **2**, 884.

[73] J. T. Di, Z. Z. Yong, X. H. Zheng, B. Q. Sun, Q. W. Li, Aligned Carbon Nanotubes for High-Efficiency Schottky Solar Cells. *Small* 2013, **9**, 1367.

[74] L. H. Zhang, Y. Jia, S. S. Wang, Z. Li, C. Y. Ji, J. Q. Wei, H. W. Zhu, K. L. Wang, D. H. Wu, E. Z. Shi, Y. Fang, A. Y. Cao, Carbon Nanotube and CdSe Nanobelt Schottky Junction Solar Cells. *Nano Lett.* 2010, **10**, 3583.

[75] E. Z. Shi, J. Q. Nie, X. J. Qin, Z. J. Li, L. H. Zhang, Z. Li, P. X. Li, Y. Jia, C. Y. Ji, J. Q. Wei, K. L. Wang, H. W. Zhu, D. H. Wu, Y. Li, Y. Fang, W. Z. Qian, F. Wei, A. Y. Cao, Nanobelt–Carbon Nanotube Cross-Junction Solar Cells. *Energy Environ. Sci.* 2012, **5**, 6119.

[76] L. Pereira, C. G. Rocha, A. Latge, J. N. Coleman, M. S. Ferreira, Upper Bound for the Conductivity of Nanotube Networks. *Appl. Phys. Lett.* 2009, **95**, 123106.

[77] X. M. Li, H. W. Zhu, K. L. Wang, A. Y. Cao, J. Q. Wei, C. Y. Li, Y. Jia, Z. Li, X. Li, D. H. Wu, Graphene-On-Silicon Schottky Junction Solar Cells. *Adv. Mater.* 2010, **22**, 2743.

[78] G. F. Fan, H. W. Zhu, K. L. Wang, J. Q. Wei, X. M. Li, Q. K. Shu, N. Guo, D. H. Wu, Graphene/Silicon Nanowire Schottky Junction for Enhanced Light Harvesting. *ACS Appl. Mater. Interfaces* 2011, **3**, 721.

[79] T. T. Feng, D. Xie, Y. X. Lin, Y. Y. Zang, T. L. Ren, R. Song, H. M. Zhao, H. Tian, X. Li, H. W. Zhu, L. T. Liu, Graphene Based Schottky Junction Solar Cells on Patterned Silicon-Pillar-Array Substrate. *Appl. Phys. Lett.* 2011, **99**, 233505.

[80] X. C. Miao, S. Tongay, M. K. Petterson, K. Berke, A. G. Rinzler, B. R. Appleton, A. F. Hebard, High Efficiency Graphene Solar Cells by Chemical Doping. *Nano Lett.* 2012, **12**, 2745.

[81] X. M. Li, D. Xie, H. Park, Z. Miao, T. Y. Zeng, K. L. Wang, J. Q. Wei, D. H. Wu, J. Kong, H. W. Zhu, Ion Doping of Graphene for High-Efficiency Heterojunction Solar Cells. *Nanoscale* 2013, **5**, 1945.

[82] X. M. Li, D. Xie, H. S. Park, T. Y. Zeng, K. L. Wang, J. Q. Wei, M. L. Zhong, D. H. Wu, J. Kong, H. W. Zhu, Anomalous Behaviors of Graphene Transparent Conductors in Graphene–Silicon Heterojunction Solar Cells. *Adv. Energy Mater.* 2013, **3**, 1029.

[83] E. Z. Shi, H. B. Li, L. Yang, L. H. Zhang, Z. Li, P. X. Li, Y. Y. Shang, S. T. Wu, X. M. Li, J. Q. Wei, K. L. Wang, H. W. Zhu, D. H. Wu, Y. Fang, A. Y. Cao, Colloidal Antireflection Coating Improves Graphene–Silicon Solar Cells. *Nano Lett.* 2013, **13**, 1776.

[84] Q. H. Wang, K. Kalantar-Zadeh, A. Kis, J. N. Coleman, M. S. Strano, Electronics and Optoelectronics of Two-Dimensional Transition Metal Dichalcogenides. *Nat. Nanotechnol.* 2012, **7**, 699.

[85] M. S. Xu, T. Liang, M. M. Shi, H. Z. Chen, Graphene-Like Two-Dimensional Materials. *Chem. Rev.* 2013, **113**, 3766.

[86] V. Nicolosi, M. Chhowalla, M. G. Kanatzidis, M. S. Strano, J. N. Coleman, Liquid Exfoliation of Layered Materials. *Science* 2013, **340**, 1420.

[87] L. Britnell, R. M. Ribeiro, A. Eckmann, R. Jalil, B. D. Belle, A. Mishchenko, Y. J. Kim, R. V. Gorbachev, T. Georgiou, S. V. Morozov, A. N. Grigorenko, A. K. Geim, C. Casiraghi, A. Neto, K. S. Novoselov, Strong Light-Matter Interactions in Heterostructures of Atomically Thin Films. *Science* 2013, **340**, 1311.

[88] A. L. Elias, N. Perea-Lopez, A. Castro-Beltran, A. Berkdemir, R. T. Lv, S. M. Feng, A. D. Long, T. Hayashi, Y. A. Kim, M. Endo, H. R. Gutierrez, N. R. Pradhan, L. Balicas, T. Houk, F. Lopez-Urias, H. Terrones, M. Terrones, Controlled Synthesis and Transfer of Large-Area WS2 Sheets: From Single Layer to Few Layers. *ACS Nano* 2013, **7**, 5235.

[89] D. S. Kong, H. T. Wang, J. J. Cha, M. Pasta, K. J. Koski, J. Yao, Y. Cui, Synthesis of MoS_2 and $MoSe_2$ Films with Vertically Aligned Layers. *Nano Lett.* 2013, **13**, 1341.

[90] Y. H. Lee, L. L. Yu, H. Wang, W. J. Fang, X. Ling, Y. M. Shi, C. T. Lin, J. K. Huang, M. T. Chang, C. S. Chang, M. Dresselhaus, T. Palacios, L. J. Li, J. Kong, Synthesis and Transfer of Single-Layer Transition Metal Disulfides on Diverse Surfaces. *Nano Lett.* 2013, **13**, 1852.

[91] Y. H. Lee, X. Q. Zhang, W. J. Zhang, M. T. Chang, C. T. Lin, K. D. Chang, Y. C. Yu, J. Wang, C. S. Chang, L. J. Li, T. W. Lin, Synthesis of Large-Area MoS_2 Atomic Layers with Chemical Vapor Deposition. *Adv. Mater.* 2012, **24**, 2320.

[92] M. Bernardi, M. Palummo, J. C. Grossman, Extraordinary Sunlight Absorption and One Nanometer Thick Photovoltaics Using Two-Dimensional Monolayer Materials. *Nano Lett.* 2013, **13**, 3664.

8

FUEL CELL CATALYSTS BASED ON CARBON NANOMATERIALS

JUNJI NAKAMURA AND TAKAHIRO KONDO

Faculty of Pure and Applied Sciences, University of Tsukuba, Tsukuba, Japan

8.1 INTRODUCTION

Recently, nanocarbons have been paid attention as the electrocatalyst materials for fuel cells [1]. Carbon nanotubes (CNTs) and graphene are considered as new catalyst materials in polymer electrolyte membrane fuel cells (PEMFC). In the present commercial fuel cell catalysts, Pt particles with sizes of 3–10 nm are supported on conductive carbons such as carbon black (CB). The Pt costs 40–50 US\$/g and a fuel cell car requires 80–100 g Pt. The price of Pt in a fuel cell thus costs 3200–5000 US\$ for the Pt catalyst. To reduce the cost of Pt catalysts, (i) the improvement of catalytic activity or (ii) the development of non-Pt catalysts is required. One of the promising ways to reduce the Pt usage is the application of nanocarbons to electrocatalysts. We have firstly found that Pt electrocatalysts supported by CNTs are active in a H_2–O_2 PEMFC [2]. Later, we have used graphene as an anode catalyst support [3], in which we found that Pt subnanoclusters are formed, showing unique catalytic properties in terms of CO tolerance and electro-oxidation. It is quite attractive for not only catalytic researchers but also material scientists to control the electronic structure or catalytic activity of Pt by the support carbon materials. That is because cheap material of carbon can promote the activity of the expensive material of Pt. Generally speaking, the surface of carbon is chemically inactive due to the π-conjugated system, which is why we were initially skeptical about the significant support effect due to the interface interaction between carbon surfaces and Pt particles. However, we have become

Carbon Nanomaterials for Advanced Energy Systems: Advances in Materials Synthesis and Device Applications, First Edition. Edited by Wen Lu, Jong-Beom Baek and Liming Dai.
© 2015 John Wiley & Sons, Inc. Published 2015 by John Wiley & Sons, Inc.

confident about the support effect by observing several experimental data in our group. Many literatures have also reported the improved support effect of nanocarbons on the electrocatalytic activities. On the other hand, non-Pt catalysts have been extensively studied particularly for oxygen reduction reaction (ORR) in PEMFC. Carbon catalysts doped with nitrogen or metals such as Fe and/or Co have been reported as the active ORR catalyst. The carbon catalysts are called as carbon alloy catalysts. However, the active sites of carbon alloy catalysts have not been clarified yet.

As described previously, catalytic functions of nanocarbons have been paid attention recently. It is also interesting from the viewpoints of surface chemistry. That is, the relationships among the origin of the reactivity of carbons, the electronic structure, and the properties of the π-conjugated system with dopants, defects, and so on are an interesting topic in surface chemistry. The surface chemistry of the graphitic materials has been gradually understood using various surface science techniques such as scanning tunneling microscopy (STM) and photoelectron spectroscopies (PES).

In this chapter, the recent catalyst studies of nanocarbons are introduced with our studies.

8.2   NANOCARBON-SUPPORTED CATALYSTS

8.2.1   CNT-Supported Catalysts

Figure 8.1 shows the cell voltage and power density results of Pt/CNT and Pt/CB catalysts as a function of current density in H_2–O_2 PEMFC. Here, the difference is the loading of Pt. The loading of Pt is smaller for Pt/CNT (12 wt%) than that for Pt/CB (29 wt%). However, it is found that voltage drop below $500\,mA\,cm^{-2}$ is smaller for 12 wt% Pt/CNT compared to that for 29 wt% Pt/CB. That is, the Pt/CNT is more active than Pt/CB. The power density shown in Figure 8.1b is comparable or somewhat larger for Pt/CNT, although the loading of Pt is smaller for Pt/CNT. Many papers report that Pt electrocatalysts using CNT supports show higher catalytic activities. In particular, Villers have reported significant higher catalytic activities [4]. On the other hand, some literatures have reported that no significant support effect of CNT has been observed in catalytic activities in PEMFC [5]. This may be due to the size of Pt particles on the carbon surface. When the Pt size is smaller, the number of Pt atoms at the interface between a Pt particle and carbon surface increases relatively compared to the total number of Pt atoms in a Pt particle. That is why the support effect of carbon is significant for smaller Pt particles. The support effect seems to appear when the particle size of Pt is below 2 nm. The sizes of Pt particles in commercial electrocatalysts range from 2 to 5 nm. As for the cathode catalysts of PEMFC, Pt particles about 5 nm show high catalytic activities. It is considered that deactivation by the bulk oxidation of Pt particles is significant for smaller Pt particles. On the other hand, Pt/CNT catalysts with small Pt particles below 5 nm show higher catalytic activity, which is also a support effect of nanocarbons.

Surface flatness of carbon is also very important for the support effect of nanocarbons. Figure 8.2 shows the TEM image of PtRu catalysts supported on CNT. The small

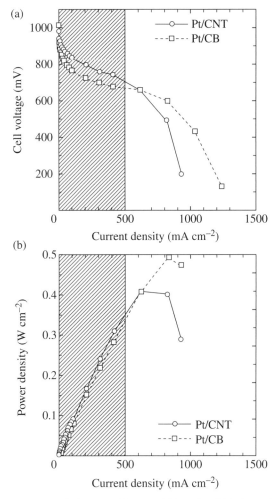

FIGURE 8.1 Performance of 12 wt% Pt/CNT and 29 wt% Pt/CB electrodes. (a) *I–V* curves. (b) Power density–current density curves [2].

PtRu particles are well dispersed on the flat CNT surface. The size of PtRu particles is about 1.5 nm and the diameter of multiwalled CNT (MWCNT) ranges from 10 to 20 nm. The support effect of CNT was thus observed as an improved CO-tolerant and methanol oxidation anode catalyst [6, 7]. In the support effect, the π-conjugated orbitals of the flat carbon surface interact with the *d* orbitals of metal catalysts. On the other hand, the support effect is weak if the surface of CNT is composed of graphene edges like the surfaces of fish bone-type CNTs.

Generally, the surface flatness is further determined by the diameter of MWCNTs. The surfaces of MWCNTs with diameters of 10–100 nm are flat, and the surfaces are thus composed of sp^2 carbons with lower densities of defects. On the other hand, thicker MWCNTs with diameters over 100 nm tend to have many defects on the tube

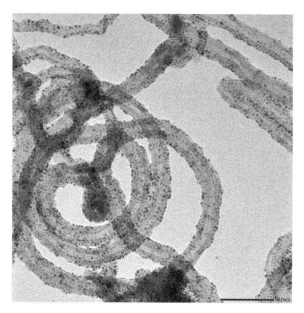

FIGURE 8.2 TEM images of PtRu/CNT catalysts. The loadings of Pt and Ru are 20 and 10 wt%, respectively.

surface. The difference in the density of defects originated from the CNT growth mechanism. One of the widely used methods to produce MWCNTs is chemical vapor deposition (CVD) using Mo, Co, Ni, or Fe catalysts with carbon sources of CO or hydrocarbons such as methane or acetylene. In the mechanism of CNT growth, segregated carbons at the surface of catalyst particles form the graphene sheets of CNT so that the diameter of CNT is determined by the size of catalyst particles. If the diameter of catalyst particles is small, the graphene sheet is rolled up easily and the surface of CNT is composed of well-developed graphene plane. On the other hand, if the diameter of catalyst particles is larger than 100 nm, the surface of MWCNTs tends to be composed of edges of graphene sheets.

The support effects of CNT in Pt/CNT or PtRu/CNT catalysts have been thus reported for MWCNTs with diameter of 10–100 nm. Single-walled CNT (SWCNT) with diameter of 0.8–2 nm is too thin to support Pt particles because the size of SWCNT is comparable with that of Pt particles. Cheaper MWCNTs are rather better carbon supports. The price of MWCNT is currently about 10 US cents/g, which is much lower compared to that of Pt. So the cost of CNT is not a problem.

However, it is not always easy to prepare small Pt particles on the flat carbon surfaces because the π-conjugated surface is basically inactive. On the other hand, it is easy to deposit small nanocatalyst particles on the thicker MWCNT, although no significant support effect can be expected. Elaborated preparation methods are thus required on the flat carbon surface of thinner MWCNT. The small PtRu particles shown in Figure 8.2 were to be successfully prepared on thin MWCNTs with diameters of 10–20 nm. One of the effective preparation methods of Pt nanoparticles on the flat surface is the reduction of H_2PtCl_6 by $NaBH_4$ or ethanol in a liquid phase.

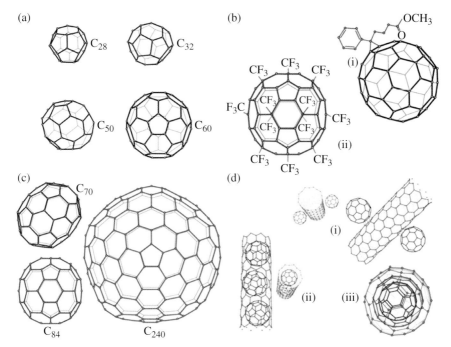

FIGURE 1.1 (a) Fullerenes. (b) Fullerene derivatives: (i) C_{60} derivative [6,6]-phenyl-C_{61}-butyric acid methyl ester (PCBM) and (ii) trifluoromethyl derivative of C_{84} ([C_{84}](CF$_3$)$_{12}$). (c) Higher-order fullerenes. (d) (i) Nanobud (fullerenes covalently bound to the outer sidewalls of single-walled carbon nanotube), (ii) peapod (fullerenes encapsulated inside a single-walled carbon nanotube), and (iii) nano-onion (multishelled fullerenes).

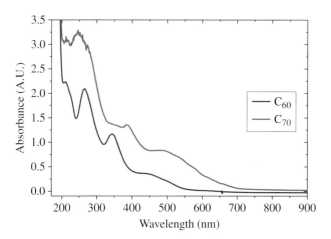

FIGURE 1.5 UV–visible spectra of Pluronic modified C_{60} and C_{70} aqueous suspensions.

Carbon Nanomaterials for Advanced Energy Systems: Advances in Materials Synthesis and Device Applications, First Edition. Edited by Wen Lu, Jong-Beom Baek and Liming Dai.
© 2015 John Wiley & Sons, Inc. Published 2015 by John Wiley & Sons, Inc.

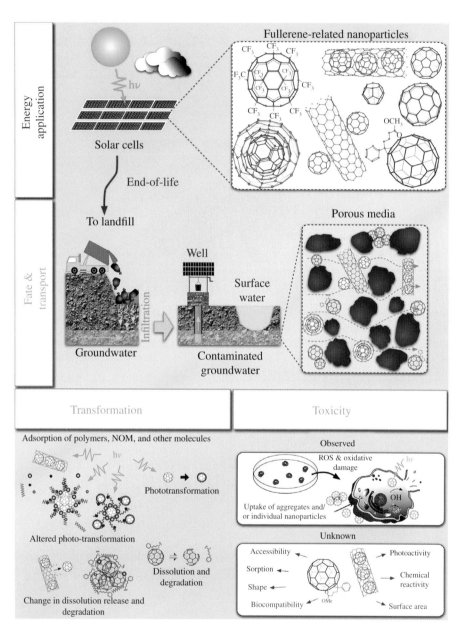

FIGURE 1.9 Likely environmental fate, transport, transformation, and toxicity of fullerenes and related nanomaterials.

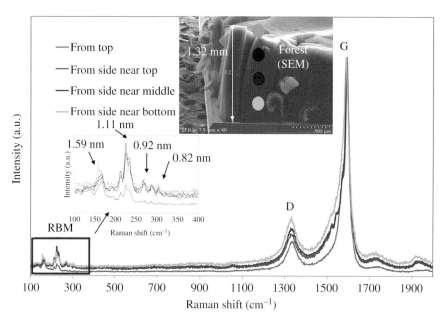

FIGURE 2.7 Raman spectra along the height of a forest grown on a top Al_2O_3/Fe/bottom Al_2O_3 sandwich catalytic substrate with a Ti layer on the top, with nanotube diameters estimated via d $(n, m) = 248/x$ (cm⁻¹) [69]. The upper inset is an SEM image of the forest. The colored circles on the forest show the measurement points of Raman spectroscopy. The lower inset is a magnified image of RBM (for Raman spectra shown in the red square). The laser excitation wavelength was 532 nm. Reprinted from Ref. 68 with permission from Elsevier.

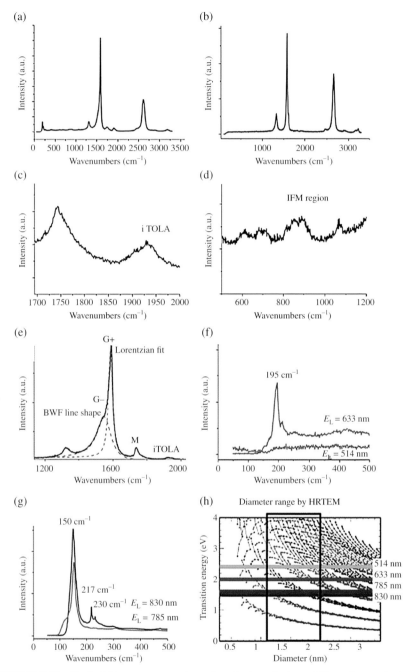

FIGURE 2.11 Typical Raman spectra (wavelength, λ excitation = 633 nm in panels (a–e)) from fibers obtained with (a) carbon disulfide and (b) thiophene as the sulfur precursors. (c) M, iTOLA, and (d) IFM regions in the Raman spectra of the carbon disulfide fibers. (e) The internal structure of the G band, with the Lorentzian G+ and the G− exhibiting the Fano line shape with fit parameters I_0, ω, T, and $q = 2256$, 1556, 49.5, and -0.20, respectively. (f) RBM regions with a peak at $195\,cm^{-1}$ with $\lambda = 633\,nm$ and the absence of the RBM peak with $\lambda = 514\,nm$. (g) RBM regions for $\lambda = 785\,nm$ and $830\,nm$. (h) Kataura plot with mapping of the four wavelengths and the measured diameter range from high-resolution TEM. Black points represent families of metallic nanotubes while red and blue represent semiconducting nanotubes. Reprinted from Ref. 108 with permission from Wiley.

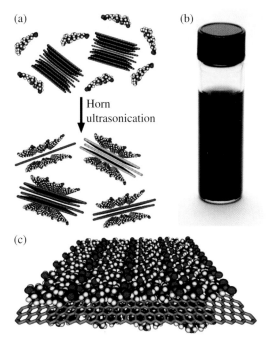

FIGURE 3.10 Flow diagram of the graphene exfoliation process via horn sonication followed by ultracentrifugation. (b) Photograph showing $90\,\mu g\,ml^{-1}$ graphene dispersion in SC 6 weeks after it was prepared. (c) Schematic illustrating an ordered SC monolayer on graphene. Reprinted with permission from Ref. 42.

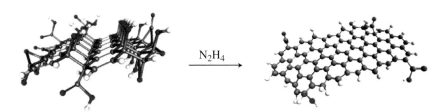

FIGURE 3.15 Schematic showing the 3D GO (carbon in gray, oxygen in red, and hydrogen in white) restoring its planar structure when reduced and dispersed with N_2H_4. Reprinted with permission from Ref. 56.

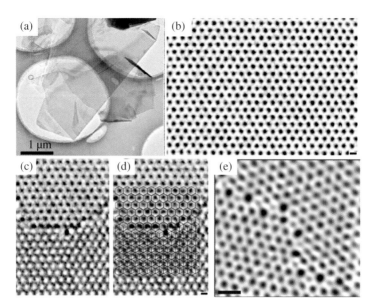

FIGURE 3.27 (a) TEM image showing a graphene flake. Reprinted with permission from Ref. 136. (b) The hexagonal honeycomb lattice of single-layer suspended graphene under HRTEM (scale bar 2 Å). (c) The clear step between single- and bilayer graphene. (d) The same HRTEM image of (c) marked with the two overlay layers (red line, bottom layer; blue line, top layer) of graphene (scale bar 2 Å). Reprinted with permission from Ref. 137. (e) HRTEM image of grain boundary of CVD graphene (scale bar 5 Å). Reprinted with permission from Ref. 105.

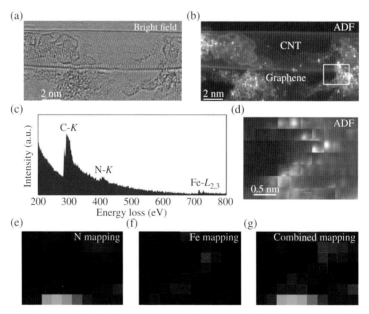

FIGURE 4.5 Longitudinal slices extracted at the same depth and orientation from shape-sensitive reconstructions of EELS energy-filtered images of an N-MWCNT, measured at the C–K edge for the carbon map and the N–K edge for the nitrogen map. The panels show the mean density (left), C and N 3D elemental maps (middle), and C-to-N 3D relative map (right). The relative map was obtained by superimposing the two elemental 3D maps with different colors, nitrogen in green and carbon in red. The presence of two types of arches (i.e., transverses and rounded ones), which is typical for the highly doped MWCNTs, can be observed. Reproduced with permission from Ref. 95. © 2012, American Chemical Society.

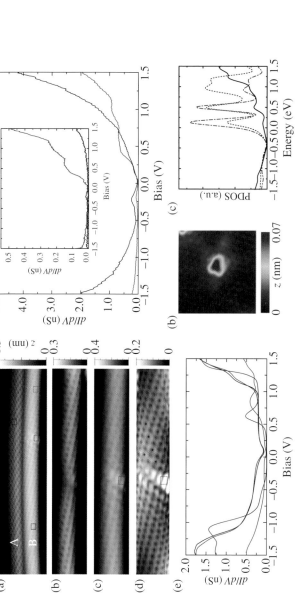

FIGURE 4.6 STM/STS signatures of nitrogen doping. (Left) (a) SWCNTs labeled *A* and *B*, with *B* showing a defect in the scan area. The colored boxes in image (a) represent the points where the spectra in (e) were recorded. (b) ($7 \times 1.2 \, nm^2$) constant current image of the defect observed in image (a) at a bias voltage $V_s = -1.00 \, V$, (c) ($7 \times 1.2 \, nm^2$) constant current image of the defect observed in image (a) at $V_s = +1.00 \, V$. (d) STM image shown in panel (c) corrected by a line-by-line flattening; the red dots indicate the periodicity of a superstructure "donut" pattern. (e) Tunneling spectra measured on the two tubes, with the colors of the lines corresponding to the positions of the colored boxes in (a). Note that the peak around $0.9 \, eV$ in the red curve corresponds to the donor state measured at the defect, while the peak at $0.6 \, eV$ in the blue curve corresponds to the first Van Hove singularity of the pristine tube A assigned with a (12,4) chirality. Reproduced with permission from Ref. 35. © 2013, American Chemical Society. (Right) (a) STM image of a single substitutional N dopant in graphene on copper foil. Inset: Line profile across the dopant shows atomic corrugation and the apparent height of the site. (b) STM image of N-graphene showing 14 substitutional dopants and extended electronic perturbations due to intervalley scattering. Inset: Fast Fourier transform of the topography shows both atomic peaks (outer hexagon) and intervalley scattering peaks (inner hexagon). (c) dI/dV curves taken on a N atom (bottom) and on the bright topographic features near the N atom, offset vertically for clarity. The top curve was taken approximately 2 nm away from the dopant. Inset: positions where the spectra were taken. Adapted from Ref. 104, reproduced with permission. © 2012 by The American Physical Society.

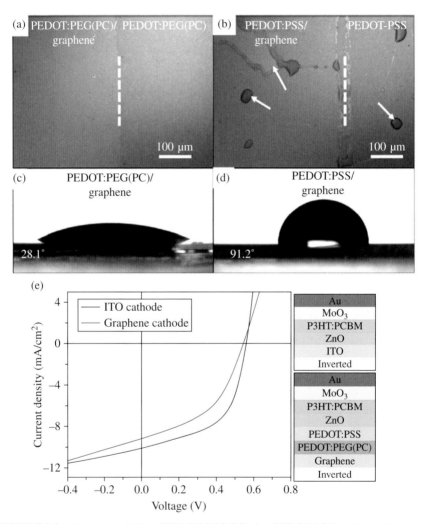

FIGURE 5.4 Better wettability of PEDOT:PEG(PC) than PEDOT:PSS on the graphene sur-face. Optical microscopy images of PEDOT:PEG(PC) (a) and PEDOT:PSS (b) on graphene and bare quartz substrates. The white dotted lines indicate the edge of the graphene, and the arrows denote dewetted PEDOT:PSS. Contact angle images of graphene/PEDOT:PEG(PC) (c) and graphene/PEDOT:PSS (d). (e) Current density–voltage curves for the graphene-based PSC device and the ITO-based PSC device. Adapted from Ref. 32 with permission. © 2013, Nature Publishing Group.

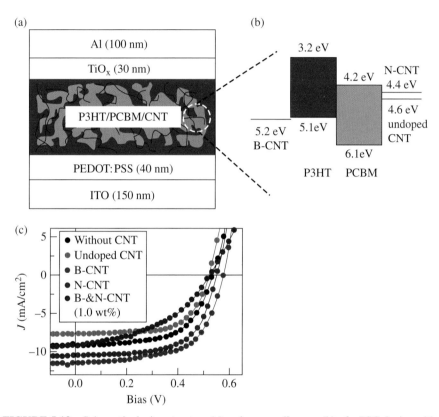

FIGURE 5.12 Schematic device structure (a) and energy diagram (b) of a PSC device with CNTs blended in the active layer. (c) Current density–voltage characteristics of the PSC devices without or with different CNTs. Adapted from Ref. 51 with permission. © 2011, John Wiley & Sons.

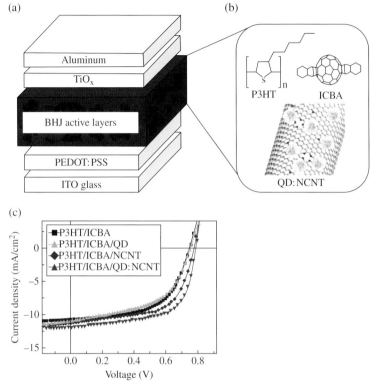

FIGURE 5.13 (a) Schematic of a PSC device architecture with QD:N-CNTs blended in the active layer. (b) Chemical structure of P3HT donor and ICBA acceptor as well as schematic of QD:N-CNTs. (c) Current density–voltage characteristics of the devices with QD, N-CNT, and QD:N-CNT. Adapted from Ref. 53 with permission. © 2013, John Wiley & Sons.

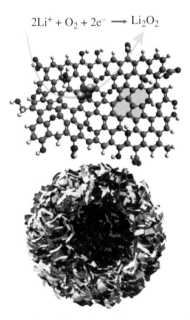

FIGURE 6.12 Broken egg conformation of graphene porous cathodes for enhanced Li_2O_2 deposition. Reprinted with permission from Ref. 201. © ACS.

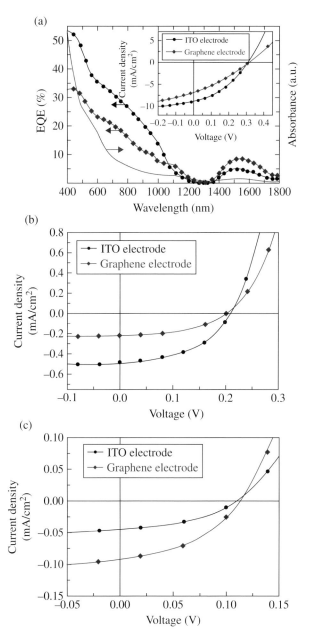

FIGURE 7.3 EQE and current density–voltage (*J–V*) characteristics of graphene electrode and ITO electrode solar cells. (a) EQEs of PbS solar cell based on graphene electrode and ITO electrode. The blue line represents the absorption of PbS QDs. The inset shows the *J–V* curves of two samples under AM 1.5 (100 mW/cm²). (b) *J–V* curves under 500 nm illumination (2.519 mW/cm²). (c) *J–V* curves under 1500 nm illumination (0.368 mW/cm²). Reproduced with permission from Ref. 24. © 2011 American Institute of Physics.

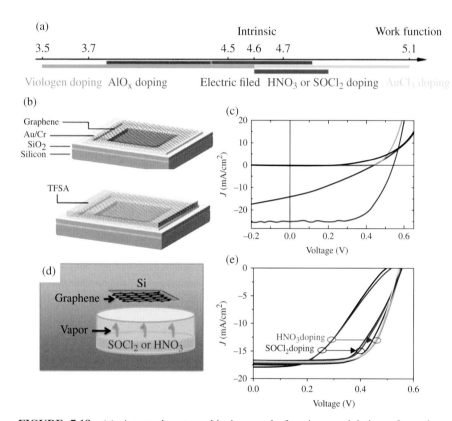

FIGURE 7.18 (a) Approaches to achieving work function modulation of graphene. Reproduced with permission from Ref. 4. © 2013 The Royal Society of Chemistry. (b) Schematic structure of TFSA-doped graphene/Si Schottky solar cell. (c) *J–V* curves of graphene/Si (blue) and doped graphene/Si (red) Schottky solar cells in dark and after illumination. Reproduced with permission from Ref. 80. © 2012 American Chemical Society. (d) Schematic illustration of the vapor-doping process. (e) Light *J–V* curves of the pristine and doped cells with increasing doping time. Reproduced with permission from Ref. 81. © 2013 The Royal Society of Chemistry.

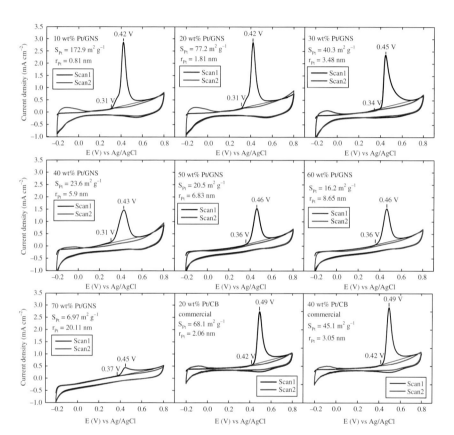

FIGURE 8.7 CO stripping voltammograms of 10–70 wt% Pt/GNS and 20 and 40 wt% Pt/CB commercial catalysts measured in 0.1 M HClO$_4$ at 60°C with the scan rate of 10 mV s^{-1} [11].

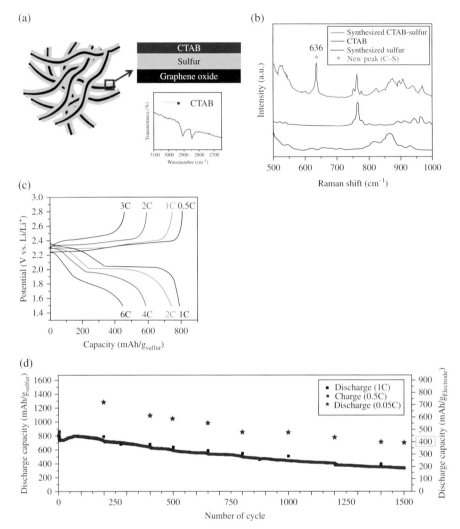

FIGURE 11.7 (a) Schematic of the S–GO nanocomposite structure. The presence of CTAB on the S–GO surface was confirmed by Fourier transform infrared spectroscopy (FTIR) and shown to be critical for achieving improved cycling performance by minimizing the loss of sulfur. (b) Enlarged view of Raman spectra on CTAB, synthesized sulfur, and CTAB-modified sulfur from 500 to 1000 cm^{-1}. It clearly shows the formation of a new peak, which can be assigned as a C–S bond (600–700 cm^{-1}), confirming that there is strong interaction between CTAB and sulfur. (c) Voltage profiles of CTAB-modified S–GO composite cathodes at different rates. (d) Long-term cycling test results of the Li/S cell with CTAB-modified S–GO composite cathodes. This result represents the longest cycle life (exceeding 1500 cycles) with an extremely low decay rate (0.039% per cycle) demonstrated so far for a Li/S cell. The S–GO composite contained 80% S, and elastomeric SBR/CMC binder was used. 1 M LiTFSI in PYR$_{14}$TFSI/DOL/DME mixture (2:1:1 by volume) with 0.1 M LiNO$_3$ was used as the electrolyte (total 60 μl). Reproduced with permission from Ref. 35. © 2013 American Chemical Society.

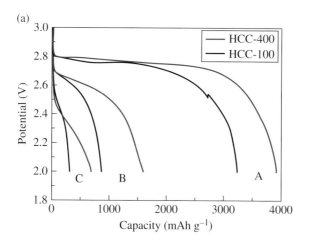

(a)

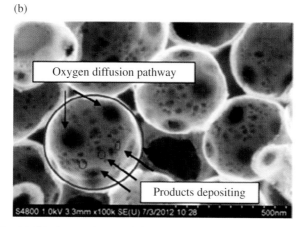

(b)

FIGURE 12.7 (a) Discharge characteristics of the HCC-400 and HCC-100 electrodes at various current densities of (A) $0.05\,mA\,cm^{-2}$, (B) $0.2\,mA\,cm^{-2}$, and (C) $0.5\,mA\,cm^{-2}$. (b) FESEM micrograph of HCC-100 carbon. Reproduced with permission from Ref. 25. © 2013 RSC.

(a)

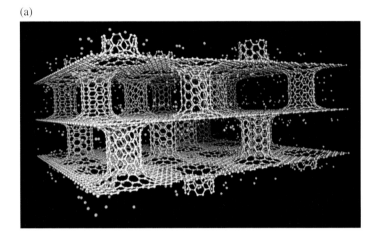

(b)

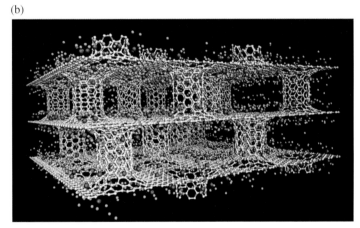

FIGURE 13.14 (a) Snapshot from the GCMC simulations of pure pillared structure at 77 K and 3 bar; (b) snapshot from the GCMC simulations of lithium-doped pillared structure at 77 K and 3 bar. Hydrogen molecules are represented in green, while lithium atoms are in purple. Reproduced with permission from Ref. 147. © 2008 American Chemical Society.

8.2.2   Graphene-Supported Catalysts

Graphene-supported electrocatalysts have been studied extensively since 2008 because graphene is highly conductive and have a large surface area ($2630\,m^2\,g^{-1}$) [3, 8–11]. One can maximize the loading of Pt nanoparticles on graphene sheets. Micrometer-sized graphene sheets are prepared from graphite powders by the well-known Hummers method so that the graphene is sometimes called graphene nanosheet (GNS). Figure 8.3 shows the TEM image of Pt/GNS catalysts prepared by an impregnation method [3]. Subnanometer-sized Pt clusters (~0.3 nm) are seen in addition to larger particles with diameters of about 2 nm. No subnanometer Pt clusters have been observed for CB supports. The number of Pt atoms contained in the Pt subnanoclusters is about 3–20. The surface area is surprisingly high, $170\,m^2\,g^{-1}$, which is two or three times larger than that for CB ($40–80\,m^2\,g^{-1}$). In general, the size of small Pt particles ranges from 2 to 5 nm for Pt/CB. When the size of cluster becomes small,

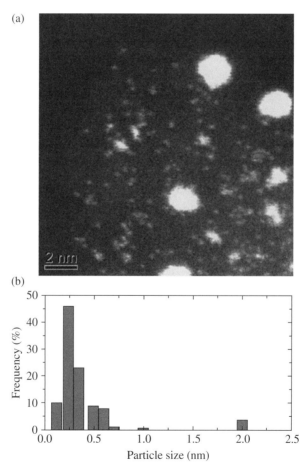

FIGURE 8.3   (a) HAADF-STEM image of 20 wt% Pt/GNS catalyst. (b) The histogram of Pt subnanoclusters of Pt/GNS in a [3].

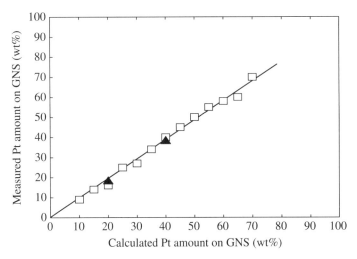

FIGURE 8.4   Pt amount on GNS (wt%) measured by TG/DTA versus calculated Pt amount on GNS (□). The results for 20 and 40 wt% Pt/CB commercial catalysts are also shown (▲) [11].

it is known that quantum effect appears in terms of electronic structures like molecules. More importantly, the support effect of graphene sometimes promotes catalytic activity of Pt. It is possible to control the size of Pt on GNS by controlling the loading of Pt. Figure 8.4 shows the loading of Pt measured by thermogravimetric/differential thermal analysis (TG/DTA) as a function of calculated amount of Pt in the preparation [11]. It is found that the loading of Pt is in good agreement with the calculated amounts of Pt, meaning that all of the Pt atoms in the Pt precursors [$H_2PtCl_6 \cdot 6H_2O$] are supported by GNS after reduction.

Since one can control the loading, the sizes of Pt particles were measured by TEM, X-ray diffraction (XRD), and electrochemical surface area (ECSA) as a function of Pt loading. Figure 8.5 shows TEM images of Pt/GNS, in which the average size increase has diameters of 0.87, 1.03, 1.20, and 1.85 nm with increasing loading of 10, 15, 20, and 30 wt%, respectively. In XRD patterns, no sharp peaks were observed for the Pt/GNS catalysts, supporting that the major parts of the Pt clusters are very small around 1 nm. Figure 8.6 shows the diameter of Pt as a function of Pt loading. It is clearly shown that the particle sizes of Pt increase with increasing Pt loading. At 60 and 70 wt%, the Pt sizes are 3 and 4 nm, respectively. On CB supports shown in Figure 8.5, the sizes are 2 and 3 nm for 20 and 40 wt%, which are larger than those for the GNS support. It is thus possible to see the relationship between the particle size of Pt and the catalytic activity.

Figure 8.7 shows the results of CO stripping voltammetry measured for Pt/GNS and Pt/CB with different Pt loading. Peaks due to electro-oxidation of adsorbed CO are seen around 0.3–0.4 V versus Ag/AgCl, where lower-onset voltages mean higher electrocatalytic activities. As for 10–20 wt% Pt/GNS, the threshold voltages of the oxidation peaks are very low, 0.31 V, and the oxidation peak is located at 0.41 V, which is in contrast with the results of Pt/CB catalysts with the threshold of 0.42 V

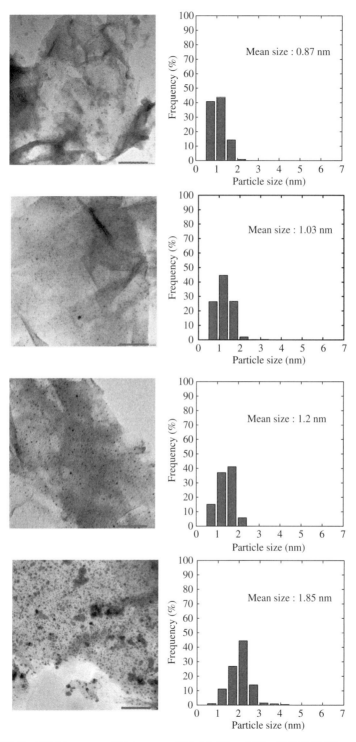

FIGURE 8.5 TEM images and histogram of 10, 15, 20 and 30 wt%Pt [11]. The mean size of Pt cluster is 0.87, 1.03, 1.20 and 1.85 nm, respectively.

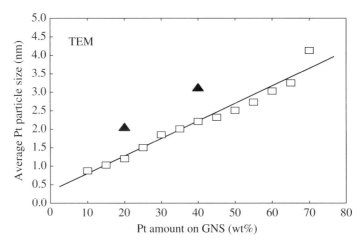

FIGURE 8.6 Average Pt particle size (nm) estimated by TEM versus Pt amount on GNS (□). The results for 20 and 40 wt% Pt/CB commercial catalysts (▲) are also shown [11].

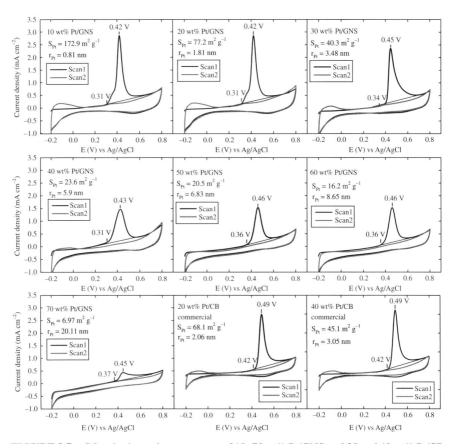

FIGURE 8.7 CO stripping voltammograms of 10–70 wt% Pt/GNS and 20 and 40 wt% Pt/CB commercial catalysts measured in 0.1 M $HClO_4$ at 60°C with the scan rate of 10 mV s^{-1} [11]. (*See insert for color representation of the figure.*)

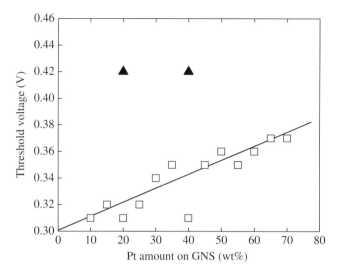

FIGURE 8.8 Threshold voltages of CO electro-oxidation peak in CO stripping voltammo-gram as a function of Pt amount on GNS (wt%) (□). The results for the 20 and 40 wt% Pt/CB commercial catalysts are also shown (▲) [11].

and the peak of 0.49 V versus Ag/AgCl. That is, the lower shift in potential is about 1 V. It is clear that the catalytic properties of Pt supported by GNS are very different from those of Pt/CB catalysts. With increasing Pt loading on GNS, however, the oxidation peaks are shifted to higher voltages. The tendency is clearly shown by Figure 8.8, in which the threshold of oxidation peaks is shifted to higher voltages with increasing Pt loading. Considering that Pt subnanoclusters are formed at low loadings of 10–20 wt%, the shift of voltages in the electro-oxidation of adsorbed CO is due to a superior catalytic activity of Pt subnanoclusters on GNS. That is, the shift to higher voltages for Pt/GNS catalysts is explained by the particle size effect of Pt. The prominent improvement of the electrocatalytic activity can be regarded as the significant support effect of GNS. It is also noticed that the width of the oxidation peaks for Pt/GNS gradually increases with increasing Pt loading from 10 to 70 wt%. This is explained by the coexistence of Pt subnanoclusters and larger Pt particles, where two characters of high and low activities are probably mixed in the CO oxidation. On the other hand, no significant effect of the particle size was observed for Pt/CB catalysts. As shown in Figure 8.7, the thresholds and the peak maxima of the CO oxidation peaks are comparable between 20 and 40 wt% Pt/CB catalysts, although the particle sizes are different—2.1 and 3.3 nm, respectively. Furthermore, the width of the peaks for Pt/CB is relatively sharp compared to those for Pt/GNS, indicating the homogeneous catalytic activity of Pt in Pt/CB. Those results indicate that the catalytic properties of Pt/CB are identical independent of the particle size, further suggesting that the support effect of CB is weak.

Concerning the electronic structure of Pt for Pt/GNS and Pt/CB catalysts, it is found that the Pt 4f peak of Pt/GNS is shifted to higher binding energies with decreasing Pt loading as shown in Figure 8.9. The peak maximum of Pt $4f_{7/2}$ for

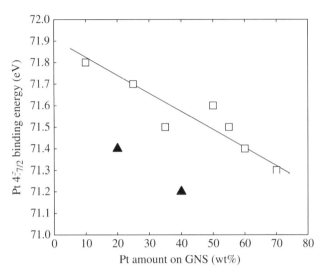

FIGURE 8.9   Pt $4f_{7/2}$ binding energy (eV) versus Pt amount on GNS (wt%) of 10–70 wt% Pt/GNS (□). The results for the 20 and 40 wt% Pt/CB commercial catalysts are also shown (▲) [11].

10 wt% Pt/GNS is located at 71.8 eV, which is much higher than that of bulk Pt, 71.2 eV. The energy shift should be related to the particle size of Pt on GNS. That is, modification in the electronic state of Pt should become significant due to the inter-face interaction between Pt and graphene when Pt subnanoclusters are the majority on GNS. The interface interaction is considered to be π–d hybridization of graphene and Pt. On the other hand, binding energies of Pt $4f_{7/2}$ for 20 and 40 wt% Pt/CB at 71.4 and 71.2 eV are close to the bulk value of 71.2 eV, although the particle sizes of Pt in Pt/CB are small, 2.1 and 3.3 nm. The absence of significant BE shift of Pt $4f_{7/2}$ in Pt/CB suggests that there is no significant interface interaction between Pt and CB.

8.3   INTERFACE INTERACTION BETWEEN Pt CLUSTERS AND GRAPHITIC SURFACE

The support effect of nanocarbons has been studied through surface science methods using a model catalyst, in which Pt particles are deposited on a highly oriented pyro-lytic graphite (HOPG) as a model of CNT surface [12–16]. The curvature of MWCNT is not large so that the surface can be regarded as the basal plane of graphite. Platinum atoms were deposited only on the central part of the HOPG surface through a slit in the ultrahigh vacuum (UHV) at room temperature by resistively heating a Pt wire. On the Pt/HOPG model catalysts, H_2–D_2 exchange reaction is carried out at 24 torr in a reaction cell connected to a UHV STM [12]. Figure 8.10 shows the rate constants for sequential H_2–D_2 exchange reactions as a function of surface temperature (a), STM images before reaction and after the sequential reactions (b), and the particle size dis-tribution estimated by the STM images before and after the sequential reactions (c).

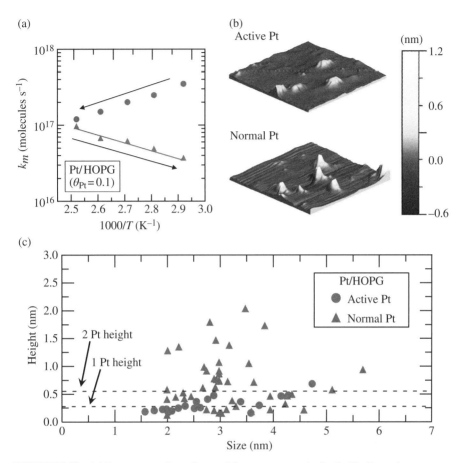

FIGURE 8.10   (a) Temperature dependence of the rate constant k_m for the H_2–D_2 exchange reaction conducted on the Pt/HOPG surface with $\theta_{Pt} = 0.1$. The solid circles and triangles represent the results of the heating mode and those of the cooling mode, respectively (see text). The solid line corresponds to the slope of the cooling mode used for the estimation of the activation energy. (b) STM images (40×40 nm) of Pt/HOPG at $\theta_{Pt} = 0.1$ before and after the series of the H_2–D_2 exchange reaction (active Pt and normal Pt, respectively). Tunneling current I_t and sample bias voltage V_s are 0.55 nA and −0.50 V for active Pt and 0.2 nA and −0.2 V for normal Pt, respectively. (c) Cluster height distribution of Pt clusters on HOPG observed by STM measurements as a function of the corresponding cluster diameter (diameter of cluster in top-view STM image) [12].

In Figure 8.10a, arrows indicate the order of the measurements. That is, the measurement of the rate constant started at 342 K followed by the measurements at higher temperatures up to 396 K using the same sample. In this heating mode, the rate constant monotonically decreases with increasing surface temperature. After the measurement at 396 K, the same sequential measurements were conducted while decreasing the reaction temperature by 10 K, as shown by the triangles. This kind of hysteresis in the rate constant has always been observed for Pt coverage of 0.03–0.11.

Here, we note that the deactivation did not occur for the sample experienced by the 396 K annealing in UHV after the Pt deposition, that is, the hysteresis is reproducibly observed only when we repeat the H_2–D_2 exchange reactions. The hysteresis was found to be originated from the morphology of Pt particles on the HOPG surface. Before the sequential H_2–D_2 exchange reactions, flat Pt monolayer clusters with a height of 0.277 nm were observed as shown in Figure 8.10b. On the other hand, the morphology was changed and taller Pt clusters with two (0.554 nm) or more atomic heights are mainly observed after the reactions as shown in Figure 8.10c. The important point is that the catalytic activity of the Pt monolayer cluster is more active than those of taller Pt clusters. The deactivation is not due to the reduction in the Pt surface area. The higher activity of the Pt monolayer cluster can be regarded as the support effect.

Figure 8.11a shows the atomically resolved STM image of the Pt monolayer, in which Pt atoms tend to be observed on β-site carbons (the carbon atoms located

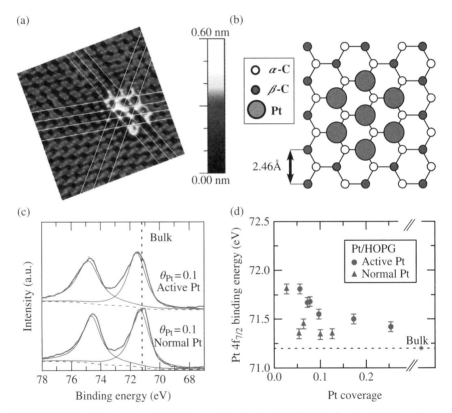

FIGURE 8.11 (a) Atomic-scale STM image (2 × 2 nm) of Pt/HOPG before the H_2–D_2 exchange reaction. I_t and V_s are 0.58 nA and −0.50 V, respectively. (b) Schematic model of the Pt monolayer cluster on HOPG. Pt atoms are located on the β-site carbon. (c) XPS spectra of Pt/HOPG at $\theta_{Pt} = 0.1$ before and after the series of the H_2–D_2 exchange reaction (active Pt and normal Pt, respectively). (d) Pt4f$_{7/2}$ peak position energy of the Pt/HOPG before and after the series of the H_2–D_2 exchange reaction (active Pt and normal Pt, respectively) at various Pt coverages [12].

above the centers of the hexagons of the layer beneath) of the HOPG surface as schematically shown in Figure 8.11b. The formation of flat Pt clusters is clearly found, and the reduction of the Pt–Pt distance is observed as a result of the interface interaction between Pt and graphite. As for the electronic structure of Pt, the XPS spectra in Figure 8.11c for the active and normal Pt clusters on HOPG with $\theta_{Pt} = 0.1$ show that the peak of Pt4f$_{7/2}$ is located at higher binding energies as approximately 71.6 eV for the active Pt monolayer clusters. Since the peak position of the single-crystal Pt is known to appear at approximately 71.1 eV, the observed higher binding energy for active Pt (Fig. 8.11d) indicates a different electronic structure from that of single-crystal Pt.

The higher energy shift of Pt4f is consistent with that observed for Pt/GNS catalysts [11], suggesting the chemical interface interaction such as π–d interaction. In this case, Pt $5d$ filled orbitals and π* empty graphite states may hybridize in the Fermi energy region as reported between Pd $4d$ and graphite π* for the Pd clusters on HOPG. The hybridization results in a shift of the d band center of Pt away from the Fermi level. This is the situation in which the adsorption energy of the molecule on the metal decreases according to the d-band center criterion. The rate-limiting step of the H$_2$–D$_2$ exchange reaction under our experimental conditions is considered to be the associative desorption of HD from the Pt/HOPG sample, because the dissociative chemisorption of hydrogen on Pt is very fast with little barrier. The smaller adsorption energy (i.e., desorption energy) of hydrogen thus leads to a lower barrier of the H$_2$–D$_2$ exchange reaction or higher catalytic activity.

The support effect can be further controlled by nitrogen doping into graphitic materials. This can be evidenced by the results using a nitrogen-doped HOPG (N-HOPG) model catalyst. The nitrogen is doped into HOPG by N$_2^+$ bombardments. In the same experiments of the H$_2$–D$_2$ exchange reaction shown in Figure 8.10, no hysteresis is observed for Pt/N-HOPG as shown in Figure 8.12a. That is, the catalytic activity is kept in the heating mode and cooling mode. After the sequential kinetic measurements, the morphology of Pt clusters was examined by STM observations. As shown in Figure 8.12b, the diameter of Pt clusters ranges from 1 to 5 nm, but they are found to be monolayer clusters with one atomic height. That is, the nitrogen doping stabilizes Pt monolayer clusters in contrast with that on Pt/HOPG.

Temperature programmed desorption (TPD) experiments of CO were carried out on Pt/HOPG model catalysts to examine the adsorption properties of Pt monolayer clusters. Figure 8.13a shows the CO-TPD results at Pt coverage (θ_{Pt}) of 0, 0.12, 0.18, and 0.24. Two main desorption peaks of CO appear at ca. 270 and 440 K, and no adsorption of CO was observed for $\theta_{Pt} = 0$. The desorption peak at 440 K is similar to those observed for Pt single-crystal surfaces: 420–550 K for Pt(111), Pt(110), Pt(100), and Pt(321). However, the CO desorption peak at 270 K has not been reported for any type of Pt single-crystal surface. The significant reduction of adsorption energy of CO cannot be explained by structural effects of Pt clusters. It is considered that the modified adsorption properties are due to the formation of Pt monolayer clusters strongly interacted with graphite. This is a clear indication of the support effect of carbon on the electronic structure of Pt. Contrary to the results of Pt/HOPG, a single CO desorption peak was observed at about 400 K for Pt clusters on

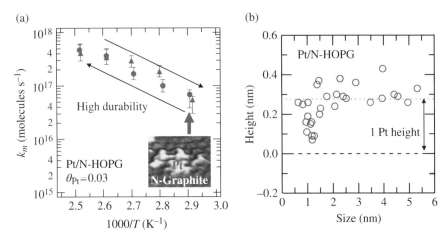

FIGURE 8.12 (a) Rate constant, k_m, for H_2–D_2 exchange reaction on Pt/N-HOPG as a function of inverse temperature. The solid circles and triangles represent the results of the heating and cooling modes, respectively. (b) Height and size distributions of the protrusions, dips, or both in the STM images for Pt/N-HOPG after the reaction cycles [16].

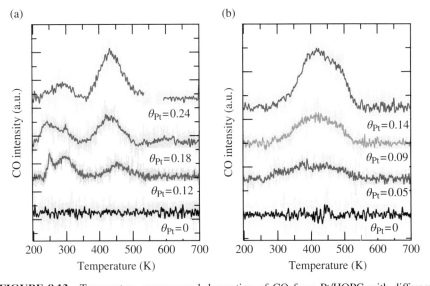

FIGURE 8.13 Temperature programmed desorption of CO from Pt/HOPG with different Pt coverage (θ_{Pt}) [14]. The heating rate is $0.5\,K\,s^{-1}$. (a) The results on the Pt/HOPG, where Pt atoms are dominantly deposited on the terrace of graphite. (b) The results on the Pt/HOPG, where Pt atoms are dominantly deposited on the step and defects of graphite (see text in detail).

the defects or step of HOPG. The Pt clusters tend to adsorb at edges of graphene sheets, which are imaged as tall clusters rather than monolayer clusters by STM. The tall Pt clusters can be compared with the Pt particles supported by CB. This suggests that the interface should be flat between Pt clusters and HOPG, taking suitable atomic positions of Pt and C for the effective π–d interaction.

8.4 CARBON CATALYST

Nitrogen-containing carbon-based catalyst is one of the most promising candidates of nonprecious metal catalysts for ORR. It can be prepared by heating mixtures of transition metal (Fe or Co) precursors, nitrogen-containing molecules, and carbon sources at 800–1000°C. Heating carbon samples in ammonia has been often conducted as a preparation method. The catalytic activity significantly depends on the type of nitrogen and transition metal precursors, heat treatment temperature, carbon support morphology, and synthesis conditions [17–20]. To improve the catalytic activity and stability, it is indispensable to clarify the catalytic active site of nitrogen-containing carbon-based catalysts. Here, the current status of the catalyst development and the discussion on the active sites are reviewed concerning nitrogen-based carbon catalysts.

8.4.1 Catalytic Activity for ORR

Recently, Dodelet's group has reported superior catalytic activity for a membrane electrode assembly (MEA) comprising the cathode catalysts derived from iron(II) acetate (FeAc), 1-10-phenanthroline (Phen), and Zn(II) zeolitic imidazolate framework (ZIF-8) [21]. At 0.6 V in their polarization curve, the peak power density for the Fe/Phen/ZIF-8-derived catalyst is $0.91\,W\,cm^{-2}$, which is the highest density among their nonprecious metal catalysts, for example, a factor of 2.3 higher than their previous best catalyst performance of nonprecious metal catalysts [22]. Moreover, the power performance is very close to that of a state-of-the-art Pt-based cathode with a loading of $0.3\,mgPt\,cm^{-2}$. Although improvement in durability is required for this catalyst, the high catalytic activity indicates that the Fe/Phen/ZIF-8-derived catalyst is one of the best candidates of the Pt-substitute catalysts for ORR in PEFC.

The high performance of nitrogen-containing carbon-based catalysts for the fuel cell cathode in the acidic condition has also been reported by many other groups [17–20]. Zelenay's group has recently reported a preparation method of active materials using polyaniline (PANI), where synthesized catalysts show ORR at potentials very close to, lower within approximately 60 (mV vs. RHE) to, that of the state-of-the-art carbon-supported platinum catalysts with an excellent four-electron selectivity (hydrogen peroxide yield <1.0%) [23]. In their report, the sample named PANI-FeCo-C(2), which consists of C, N, Fe, and Co, shows the best performance among the PANI-derived catalysts shown in fuel cell testing. This catalyst also shows a long-term performance durability, the highest rotating disk electrode (RDE) activity, and the highest maximum power density: $0.55\,W\,cm^{-2}$, reached at 0.38 V in their H_2–O_2 fuel cell polarization plots [23]. The high catalytic activity thus again indicates that PANI-derived catalysts is one of the best candidates of the Pt-substitute catalysts for ORR in PEFC as well as Fe/Phen/ZIF-8-derived catalyst. Here, it is noteworthy that, as well as the high catalytic activity, the PANI-derived catalysts also show remarkable performance stability [23]. A 700 h fuel cell performance test at a constant cell voltage of 0.4 V reveals a very promising performance stability of the sample named PANI-FeCo-C(1) catalyst in the fuel cell cathode [23]. The cell current density in a lifetime test remains nearly constant at approximately $0.340 A cm^{-2}$. The current density declines by only 3%, from the average value of $0.347 A cm^{-2}$ in the first 24 h to $0.337 A cm^{-2}$ in the last 24 h of the test (average current density loss, $18 mA h^{-1}$). The high stability of the PANI-derived catalysts does not apply solely to constant-potential operation of an RDE (nor to constant-voltage operation of a fuel cell); it extends over potential cycling conditions, and hence, it is especially relevant to practical fuel cell systems. High cycling stability of a PANI-Fe-C catalyst at various RDE potentials and fuel cell voltages was reported to be carried out within a potential (RDE) and voltage (fuel cell) range of 0.6 to 1.0 V in nitrogen gas at a scan rate of $50 mV s^{-1}$ (a protocol recommended by the US automotive industry). The catalyst performance loss calculated from linear regression was 10–39% after 10,000 RDE cycles and 3–9% after 30,000 fuel cell cycles, further attesting to the high durability of PANI-derived catalysts, especially in the fuel cell cathode. Based on the TEM measurements, they noted that the graphitization of the onion-like nanoshells and nanofibers results in the formation of graphene sheets throughout the PANI-derived catalysts by heat treatment at 900°C under a N_2 atmosphere. The graphitized carbon shell formation is then pointed out as the possible contributor to the enhancement of the electronic conductivity and corrosion resistance of the carbon-based catalyst [23].

As described earlier, the catalytic activities of the nitrogen-containing carbons synthesized from Fe and/or Co precursors are high for ORR in acidic conditions. It has been thus interpreted that the active site should be related to the metal-coordinated species such as M–N–C (M: Fe or Co) structure [17–20]. On the other hand, it has been reported that the catalytic activity for ORR is high for nitrogen-containing carbons without metal species [24–27] or catalysts thoroughly washed by acid to remove metals such as Fe and Co [28–32]. Figure 8.14 shows the ORR voltammogram of polyimide-derived catalyst. The catalysts labeled as H22 and H24 were both synthesized from polyimide and $FeCl_2$, where polyimide–Fe composites were firstly pyrolyzed followed by HCl washing and then heat treatment in NH_3 with additional HCl washing to remove Fe. Finally, each sample was treated by ball mill. The sample labeled H24 is obtained as the improved one of H22 by optimizing heat treatments and the usage of nitrogen precursors [32]. As shown in Figure 8.14, the sample labeled H24 shows high catalytic activity for ORR. The sample also shows high catalytic activity in a fuel cell testing as 0.55 V at $1.0 A cm^{-2}$ (80°C anode, Pt/C, H_2 0.2 MPa; cathode, H24 sample, O_2 0.2 MPa) [32]. It is noteworthy that the high catalytic activity is achieved even when Fe atoms were removed from the sample. That is, N–C species may be one of the catalytic active sites for ORR. In the following of this review, therefore, we focus on the possible active site related to C–N species for ORR rather than M–N–C (M: Fe or Co) structure.

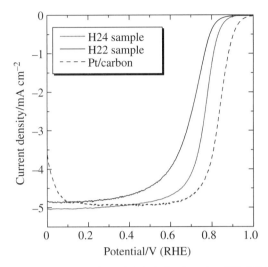

FIGURE 8.14   ORR voltammogram of polyimide-derived catalyst [32].ORR voltammogram of polyimide-derived catalyst (H22 and H24) and Pt/Carbon; temperature, RT; anode, $0.2\,mg\,cm^{-2}$ on glassy carbon; electrolyte, O_2 saturated H_2SO_4 (0.5 M); rotation, 1500 rpm.

8.4.2   Effect of N-Dope on O_2 Adsorption

In the ORR $((1/2)O_2 + H_2 \rightarrow H_2O)$ on the nitrogen-containing carbon-based catalyst, oxygen molecules have to adsorb on the catalyst surface as an initial step of ORR. Therefore, oxygen adsorption is one of the important factors to understand the catalytic activity. Interestingly, nitrogen doping in the carbon materials has been reported to create the oxygen adsorption sites [33–35], and thus, such an adsorption site is considered to be closely related to the active site for ORR. Here, we introduce experimental results reported by Stohr, Boehm, and Schlogl about oxygen adsorption on nitrogen-doped carbon materials [33].

It has been found that reversible O_2 adsorption can take place on the activated carbons after the treatment with NH_3 at elevated temperatures (600–900°C) [33]. Figure 8.15a shows the O1s photoelectron spectra of adsorbed oxygen on the sample named A-NH_3-900°C (activated carbon heated in NH_3 at 900°C) after stepwise heating to the indicated temperatures. The spectrum measured at −190°C after dosing with 1000 L O_2 shows a main peak at 533 eV and a smaller peak near 531 eV. The second peak at 531 eV has also been found on electrochemically oxidized carbon fibers and has been attributed to chemisorbed oxygen in the form of carbonyl groups [36]. The spectrum taken after warming to −60°C shows that there is also a smaller component at higher binding energy (ca. 533 eV) in the chemisorbed oxygen, which is assigned to oxygen linked to carbon by single bonds (e.g., ether groups) [36]. The most interesting feature is the big peak at 533 eV that disappeared after warming to −100°C. Its intensity was already diminished at −110°C. This peak could not be observed with the reference carbon A-N_2-900°C (activated carbon heated in N_2 at 900°C shown in Fig. 8.15b). The peak of reversibly chemisorbed oxygen (533 eV) is

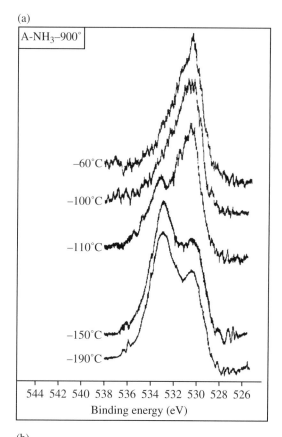

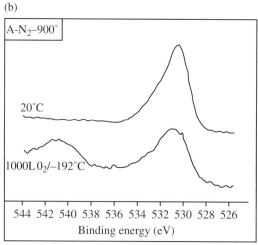

FIGURE 8.15 O1s photoelectron spectra of oxygen adsorbed on (a) A-NH$_3$-900°C and (b) A-N$_2$-900°C [33]. A-NH$_3$-900°C and A-N$_2$-900°C are the samples prepared from the activated carbon of "Anthraur" that were heated in NH$_3$ and N$_2$ atmosphere at 900°C followed by cooling in N$_2$ atmosphere, respectively. Each sample was exposed at 1000 L O$_2$ at ca. 78 K. The temperature was then raised stepwise to the indicated values.

assigned as an ionized O_2 species, very likely superoxide, O_2^-. Here, we note that the binding energy of 533 eV is also observed with adsorbed water, but the authors comments that this can be ruled out on the basis of UPS spectra taken of the valence band region [33]. Therefore, O_2 molecules are concluded to adsorb as ionized species like O_2^- on the surface of the activated carbon at −190°C after the NH_3 treatment at elevated temperature of 900°C.

8.4.3 Effect of N-Dope on the Local Electronic Structure for Pyridinic-N and Graphitic-N

As described previously, nitrogen doping into carbon-based materials create the adsorption site of oxygen molecules, which should be closely related to the active site of ORR because the oxygen molecule needs to adsorb on catalysts as the initial step of the reaction. The question thus arises as to which type of nitrogen species make an oxygen adsorption site in carbon materials and why oxygen can be adsorbed if nitrogen is doped. Here, we describe the effect of the nitrogen doping on the local electronic structure for pyridinic-N and graphitic-N species.

It is known based on the results of X-ray photoelectron spectroscopy (XPS) that the graphitic-N and pyridinic-N species are the dominant species in the nitrogen-containing carbon-based catalysts [19, 20]. As shown in Figure 8.16, the graphitic-N is the nitrogen atom bonded with three carbon atoms, while the pyridinic-N is the nitrogen atom bonded with two carbon atoms. To clarify the effect of each nitrogen species on the local electronic structure of the carbon-based material, we have prepared a model system of nitrogen-doped graphite and examined the sample by XPS and scanning tunneling microscopy/spectroscopy (STM/STS) [37]. Figure 8.17a shows N1s XPS spectra of the nitrogen-doped graphite sample. The results are shown for the surface bombarded with N_2^+ at 200 eV before and after annealing at 900 K. Four N1s peak components at 398.5, 399.9, 401.1, and 403.2 eV have been assigned to pyridinic-N, pyrrolic-N (N part of a pentagon ring connected to two C and one H), graphitic-N, and oxide-N species, respectively [38–41]. Here, we note that the energy

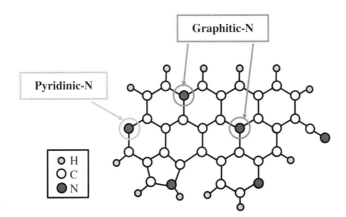

FIGURE 8.16 Schematic of the graphitic-N and pyridinic-N in the carbon-based material.

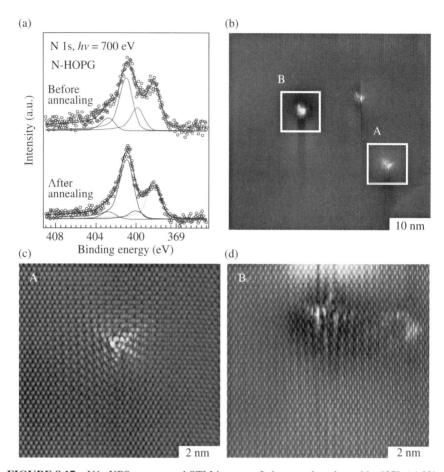

FIGURE 8.17 N1s XPS spectra and STM images of nitrogen-doped graphite [37]. (a) N1s XPS spectra of graphite surface after nitrogen ion bombardment at 200 eV. The results before and after annealing at 900 ± 50 K for 300 s are shown. Deconvoluted components are pyridinic-N (398.5 eV; before, 32.6%; after, 31.4%), pyrrolic-N (399.9 eV; before, 17.4%; after, 7.6%), graphitic-N (401.1 eV; before, 40.1%; after, 54.3%), and oxide-N (403.2 eV; before, 9.8%; after, 6.8%). (b) Typical STM topographic image of the graphite surface at about 5.3 K (scan size: 49.45×46.97 nm^2, tunneling current $I_t = 179$ pA, sample bias $V_s = 98.8$ mV). (c and d) STM topographic images of regions A and B in (b), respectively (in both case, scan size: 9.88×9.26 nm^2, $I_t = 97.5$ pA, $V_s = -109$ mV).

position for pyrrolic-N is known to be similar to that for cyanide-type-N (N connected to single C, 399.5 eV) and amine-N (N connected to single C and two H, 399.4 eV) [39], possibly also contributing to the same peak. In our samples, pyridinic-N and graphitic-N were the dominant components similar to the real nitrogen-containing carbon-based catalyst as shown in Figure 8.17a.

A typical STM image of the nitrogen-doped graphite surface is shown in Figure 8.17b. Two types of bright species were observed with an average diameter of about 4 nm: one was surrounded by a bright region (type A), and the other was surrounded by a dark region (type B). Because it is generally difficult to identify surface atom elements by STM alone, we have carried out the characterization by combining STS analysis and density functional theory (DFT) calculations. We show here that type A and type B defects can be assigned to pyridinic-N species and graphitic-N species, respectively. These indeed correspond to the two dominant components observed in the XPS spectra of Figure 8.17a. The effect of each dopant N on the electronic structure near the Fermi level at the surrounding carbon atoms are discussed in the following. Note that in Figure. 8.17c and d some of the bright regions consist of the well-known superstructure of graphite formed by the standing wave of π electrons around surface defects [42, 43]. In particular, this region does not spread isotropically around the defect but seems to propagate in three directions, as observed in the case of a single-atom vacancy defect on a graphite surface [15].

8.4.3.1 Pyridinic-N

The experimentally observed STM image and STS spectrum of the type A defect (Fig. 8.17c) are shown in Figure 8.18. The STS spectrum shown in Figure 8.18b was measured at the position indicated by the arrow in Figure 8.18a, and it consists of a large peak at about −370 mV and smaller peaks within a parabolic background. The spectrum is different from that of a single-atom vacancy with a single and large STS peak at [44] or just above the Fermi level [15]. The STS spectrum of the type A defect is again different from that of the single-atom vacancy in graphite in terms of a propagation feature [15, 45]. That is, the modified electronic states of the single vacancy propagate anisotropically to three directions, while such an anisotropic propagation was not observed for the type A defect [37]. To identify the defect species in Figure 8.18a, STM images and STS spectra were simulated for several types of structures having the pyridinic-N or the graphitic-N species because these species are the dominant species in our sample as shown by the XPS peak components in Figure 8.17a. After comparison with the results of different defect models, it was found that the simulated STM and STS features of pyridinic-N well reproduced those measured in experiments as shown in Figure 8.18. That is, the propagation of the bright region in STM and the appearance of the STS peak at about −370 mV (Fig. 8.18a and c) are common between experiment and theory. The defect was thus assigned as the pyridinic-N. Based on the DFT calculation, the large STS peak at −370 mV in Figure 8.18b and d can be assigned to localized π states (p_z orbitals) because of the localized DOS character [37].

Here, we note that the localized states appear in the occupied region for pyridinic-N. The formation of the lone pair suggests that the N atom of the pyridinic defect is negatively charged. The negatively charged N is consistent with the chemical shift in N1s binding energy to lower energy for the pyridinic-N (398.5 eV) in the XPS spectrum compared to that for the graphitic-N (401.1 eV). If N is negatively charged, the surrounding carbon atoms would be charged positively because of the screening effect. The positive charge of carbon can explain a shift of the localized π state of carbon from the Fermi level to the lower energy level. Following this picture, we have proposed that the states corresponding to the STS peak in the occupied region near the Fermi level may act as "Lewis base" toward molecular species [37].

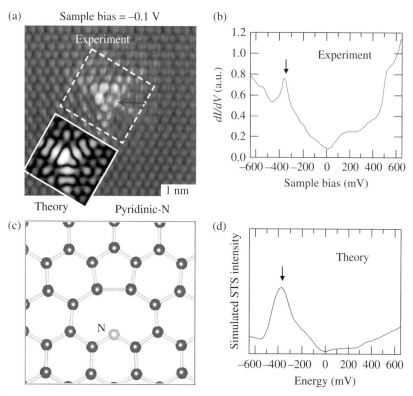

FIGURE 8.18 STM and STS of pyridinic-N [37]. (a) STM topographic image of region A in Figure 8.19b (scan size: 4.81×4.61 nm^2, $I_t = 96.9$ pA, $V_s = -108$ mV). The simulated STM image ($V = -0.1$ V) is also shown for comparison. (b) STS spectrum measured at the position indicated by the arrow in (a). (c) The equilibrium geometry of pyridinic-N defect calculated by DFT. (d) Simulated STS spectrum of pyridinic-N.

8.4.3.2 *Graphitic-N*   The experimentally observed STM image at +500 mV and STS spectrum of a type B defect (Fig. 8.17d) are shown in Figure 8.19. The STS is different from that for the pristine graphite in terms of the asymmetric LDOS with respect to 0 V (the Fermi level), where larger intensity can be recognized in the positive bias region (unoccupied region). The small shoulder peak can be seen at the positive bias voltage, which is in contrast to that at the negative voltage observed for the pyridinic-N. The STS spectra with the same shape have been measured in the vicinity of the defect independent of the lateral position as measured in the pyridinic-N [37]. To identify the defect species in Figure 8.19a, DFT simulations of STM images and STS spectra were carried out assuming several types of defect structures. After comparison with the results of different defect models, it was found that the simulated STM and STS features of graphitic-N well reproduced those measured in experiments as shown in Figure 8.19. That is, triangular bright spots in STM image and STS peak (enhanced intensity) at about +500 mV are common between experiment and theory. The defect was thus assigned to the graphitic-N. Similar STM

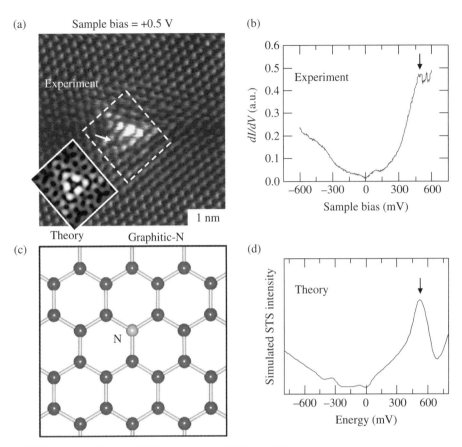

(a) Sample bias = +0.5 V

Experiment

Theory   Graphitic-N

(c)

N

(b)

Experiment

dI/dV (a.u.)

Sample bias (mV)

(d)

Simulated STS intensity

Theory

Energy (mV)

FIGURE 8.19 STM and STS of graphitic-N [37]. (a) STM topographic image corresponding to the defect shown in region B in Figure 8.19b (image was taken from different sample, scan size: $5.09 \times 5.08\,nm^2$, $I_t = 39.0\,pA$, $V_s = 500\,mV$). The simulated STM image (V = +0.5 V) is also shown for comparison. (b) STS spectrum measured at the position indicated by the arrow in (a). (c) The equilibrium geometry of graphitic-N defect calculated by DFT. (d) Simulated STS spectrum of graphitic-N.

and STS characters have also been observed recently on the nitrogen-doped graphene on copper by Zhao et al. and have also been assigned as graphitic-N characters [46]. In our DFT calculation, the position of N atom in the graphitic-N species is almost the same as that of C atom in graphite (shorter by 0.002 nm for CN bond length than for CC bond length). That is, the nitrogen species takes the sp^2 planar structure of graphite. Concerning the origin of the STS peak at 500 mV, the state is ascribed to unoccupied π states (p_z orbitals), which is similar to the edge state because of the localized character.

Here, we note that the localized states appear in the unoccupied region for graphitic-N. The positively charged N is expected by the chemical shift of N1s to higher energies for graphitic-N (401.1 eV) in contrast to the lower energies for pyridinic-N (398.5 eV).

Indeed, the difference of 2.6 eV is so large and the binding energy of 401.1 eV is comparable with that for pyridinium ion, $C_5H_5NH^+$ (401.2 eV), and ammonium ion, NH_4^+ (401.5 eV), where N is positively charged in both cases [39], suggesting the positive charge of N of the graphitic-N. The positively charged N of the graphitic-N has also been reported recently based on the theoretical results by Yu et al. [47] and Mayer et al. [48]. The positive charge can be explained by electron transfer from N atom to the π conjugated state. If N is positively charged, the surrounding carbon atoms should be charged negatively because of the screening effect. The negative charge of carbon can explain a shift of the localized π state of carbon from the Fermi level to the upper energy level. In addition, the screening of positive N charge may be the source of the dark halo around the graphitic-N defect seen in Figure. 8.17b and d, that is, the charge density rearrangement may be induced by the positive charge of N, and it may cause the modulated contrast in the STM image around N atom. Here, we have also proposed that the states with STS peak in the unoccupied region near the Fermi level may act as "Lewis acid" [37].

Here, we discuss which nitrogen species, pyridinic-N and graphitic-N, may contribute to the ORR reaction as related element to the active site. As described earlier, the carbon atoms around the pyridinic-N species has the nature of "Lewis base," while those around the graphitic-N species has the nature of "Lewis acid." The oxygen atom is neither an acid nor a base molecule but has high electronegativity, 3.44 (Pauling scale). Therefore, it can be considered that the carbon atoms around the pyridinic-N can be the oxygen adsorption site rather than those around graphitic-N, that is, oxygen molecule may adsorb on the carbon atoms around the pyridinic-N and receive the electron from the surface to form O_2^- as in the case of A-NH$_3$-900°C in Figure 8.15.

8.4.4 Summary of Active Sites for ORR

The catalytic activity of the nitrogen-containing carbon-based catalysts is high for ORR in acidic condition when the catalyst synthesized from Fe- and/or Co-containing precursors as reported by Dodelet's group [21] and Zelenay's group [23]. Therefore, M–N–C (M: Fe or Co) structure should contribute as an active site and/or to the active site formation. On the other hand, it has also been reported that the catalytic activity for ORR appears even for the catalysts synthesized from precursors without metal species or catalysts thoroughly washed by acid to remove metals such as Fe and Co as shown in Figure 8.14. Thus, the catalytic active site for ORR is probably not only M–N–C sites but also some specific N–C species. In this review, we have focused on the possible active site related to C–N species for ORR rather than M–N–C (M: Fe or Co) structure.

As shown in Figure 8.15, the nitrogen doping has been reported to create the site for reversible oxygen adsorption in the carbon-based material. Since the oxygen adsorption is an initial step of ORR, the creation of the oxygen adsorption site should be related to the active site for ORR. It has been reported that the pyridinic-N (N bonded to two C) and the graphitic-N (N bonded to three C) are the dominant nitrogen species in nitrogen-containing carbon-based catalysts (Fig. 8.16). To clarify the effect of nitrogen doping in carbon-based materials and to clarify which nitrogen contributes to the creation of the oxygen adsorption site in carbon-based materials, we have examined the local electronic structure at the atomic scale of the nitrogen-doped graphite as a model system. Based on our STM, STS, DFT, and XPS results, it has been clarified that the pyridinic-N causes the appearance of the localized p_z states at C atoms around N atom in the unoccupied region near the Fermi level (Fig. 8.18), while graphitic-N causes the localized state appearance at the occupied region near the Fermi level (Fig. 8.19). Pyridinic-N and graphitic-N are then proposed as "Lewis base" and "Lewis acid," respectively. We have then considered that the nitrogen species contributing to create the oxygen adsorption site on the carbon atoms is the pyridinic-N rather than the graphitic-N because oxygen has a higher electronegativity, 3.44 (Pauling scale), that is, oxygen molecule may adsorb on the carbon atoms around the pyridinic-N and receive the electron from the surface to form O_2^- as in the case of A-NH_3-900°C in Figure. 8.15. The adsorbed O_2^- species may react with proton to eventually form H_2O in ORR.

REFERENCES

[1]   J. Nakamura, Novel support materials for fuel cell catalysts. In T. Okada and M. Kaneko (Eds). *Molecular Catalysts for Energy Conversion*. Springer Series in Materials Science, vol. **111**. Berlin: Springer-Verlag, pp. 185–197 (2008).

[2]   T. Matsumoto, T. Komatsu, K. Arai, T. Yamazaki, M. Kijima, H. Shimizu, Y. Takasawa and J. Nakamura, Reduction of Pt usage in fuel cell electrocatalysts with carbon nanotube electrodes, *Chem. Commun.*, 840 (2004).

[3]   E. Yoo, T. Okada, T. Akita, M. Kohyama, J. Nakamura and I. Honma, Enhanced electrocatalytic activity of Pt subnanoclusters on graphene nanosheet surface, *Nano Lett.* **9**, 2255 (2009).

[4]   D. Villers, S. H. Sun, A. M. Serventi and J. P. Dodelet, Characterization of Pt nanoparticles deposited onto carbon nanotubes grown on carbon paper and evaluation of this electrode for the reduction of oxygen, *J. Phys. Chem. B* **110**, 25916 (2006).

[5]   W. Li, C. Liang, J. Qiu, W. Zhou, H. Han, Z. Wei, G. Sun and Q. Xin, Carbon nanotubes as support for cathode catalyst of a direct methanol fuel cell, *Carbon* **40**, 791 (2002).

[6]   E. Yoo, T. Okada, T. Kizuka and J. Nakamura, Effect of various carbon substrate materials on the CO tolerance of anode catalysts in polymer electrolyte fuel cells, *Electrochemistry* **75**, 146 (2007).

[7]   E. Yoo, T. Okada, T. Kizuka and J. Nakamura, Effect of carbon substrate materials as a Pt–Ru catalyst support on the performance of direct methanol fuel cells, *J. Power Sources* **180**, 221 (2008).

[8]   E. Antolini, Graphene as a new carbon support for low-temperature fuel cell catalysts, *Appl. Catal. B* **123–124**, 52 (2012).

[9]   L. Dong, R. R. S. Gari, Z. Li, M. M. Craig and S. Hou, Graphene-supported platinum and platinum–ruthenium nanoparticles with high electrocatalytic activity for methanol and ethanol oxidation, *Carbon* **48**, 781 (2010).

[10] S. Bong, Y. R. Kim, I. Kim, S. Woo, S. Uhm, J. Lee and H. Kim, Graphene supported electrocatalysts for methanol oxidation, *Electrochem. Commun.* **12**, 129 (2010).

[11] R. A. Siburian, T. Kondo and J. Nakamura, Size control to a sub-nanometer scale in platinum catalysts on graphene, *J. Phys. Chem. C* **117**, 3635 (2013).

[12] T. Kondo, K. Izumi, K. Watahiki, Y. Iwasaki, T. Suzuki and J. Nakamura, Promoted catalytic activity of a Platinum monolayer cluster on Graphite, *J. Phys. Chem. C* **112**, 15607 (2008).

[13] T. Kondo, Y. Iwasaki, Y. Honma, Y. Takagi, S. Okada and J. Nakamura, Formation of nonbonding π electronic states of graphite due to Pt-C hybridization, *Phys. Rev. B* **80**, 233408 (2009).

[14] J. Oh, T. Kondo, D. Hatake, Y. Iwasaki, Y. Honma, Y. Suda, D. Sekiba, H. Kudo and J. Nakamura, Significant reduction in adsorption energy of CO on platinum clusters on graphite, *J. Phys. Chem. Lett.* **1**, 463 (2010).

[15] T. Kondo, Y. Honma, J. Oh, T. Machida and J. Nakamura, Edge states propagating from a defect of graphite: Scanning tunneling spectroscopy measurements, *Phys. Rev. B* **82**, 153414 (2010).

[16] T. Kondo, T. Suzuki and J. Nakamura, Nitrogen doping of graphite for enhancement of durability of supported platinum clusters, *J. Phys. Chem. Lett.* **2**, 577 (2011).

[17] F. Jaouen, E. Proietti, M. Lefèvre, R. Chenitz, J. P. Dodelet, G. Wu, H. T. Chung, C. M. Johnston and P. Zelenay, Recent advances in non-precious metal catalysis for oxygen-reduction reaction in polymer electrolyte fuel cells, *Energy Environ. Sci.* **4**, 114 (2011).

[18] Y. Shao, J. Sui, G. Yin and Y. Gao, Nitrogen-doped carbon nanostructures and their composites as catalytic materials for proton exchange membrane fuel cell, *Appl. Catal. B* **79**, 89 (2008).

[19] Z. Chen, D. Higgins, A. Yu, L. Zhang and J. Zhang, A review on non-precious metal electrocatalysts for PEM fuel cells, *Energy Environ. Sci.* **4**, 3167 (2011).

[20] G. Wu and P Zelenay, Nanostructured nonprecious metal catalysts for oxygen reduction reaction, *Acc. Chem. Res.* **46**, 1878 (2013).

[21] E. Proietti, F. Jaouen, M. Lefèvre, N. Larouche, J. Tian, J. Herranz and J. P. Dodelet, Iron-based cathode catalyst with enhanced power density in polymer electrolyte membrane fuel cells, *Nat. Commun.* **2**, 1427 (2011).

[22] M. Lefevre, E. Proietti, F. Jaouen and J. P. Dodelet, Iron-based catalysts with improved oxygen reduction activity in polymer electrolyte fuel cells, *Science* **324**, 71 (2009).

[23] G. Wu, K. L. More, C. M. Johnston and P. Zelenay, High-performance electrocatalysts for oxygen reduction derived from polyaniline, iron, and cobalt, *Science* **332**, 443 (2011)

[24] T. Iwazaki, R. Obinata, W. Sugimoto and Y. Takasu, High oxygen-reduction activity of silk-derived activated carbon, *Electrochem. Commun.* **11**, 376 (2009).

[25] M. Chokai, M. Chokai, M. Taniguchi, S. Moriya, K. Matsubayashi, T. Shinoda, Y. Nabae, S. Kuroki, T. Hayakawa, M. Kakimoto, J. Ozaki and S. Miyata, Preparation of carbon alloy catalysts for polymer electrolyte fuel cells from nitrogen-containing rigid-rod polymers, *J. Power Sources* **195**, 5947 (2010).

[26] Y. Nabae, S. Kuroki, T. Hayakawa, M. Kakimoto and S. Miyata, Carbon-based cathode materials doped with a new borazine compound for electrochemical oxygen reduction, *Chem. Lett.* **41**, 923 (2012).

[27] H. Kiuchi, H. Niwa, M. Kobayashi, Y. Harada, M. Oshima, M. Chokai, Y. Nabae, S. Kuroki, M. Kakimoto, T. Ikeda, K. Terakura and S. Miyata, Study on the oxygen adsorption property of nitrogen-containing metal-free carbon-based cathode catalysts for oxygen reduction reaction, *Electrochim. Acta* **82**, 291 (2012).

[28] N. P. Subramanian, X. Li, V. Nallathambi, S. P. Kumaraguru, H. C. Mercado, G. Wu, J. W. Lee and B. N. Popov, Nitrogen-modified carbon-based catalysts for oxygen reduction reaction in polymer electrolyte membrane fuel cells, *J. Power Sources* **188**, 38 (2009).

[29] M. Chokai, Y. Nabae, S. Kuroki, T. Hayakawa, M. Kakimoto and S. Miyata, Preparation of carbon-based catalysts for PEFC from nitrogen-containing aromatic polymers, *J. Photopolym. Sci. Technol.* **24**, 241 (2011).

[30] Y. Nabae, M. Malon, S. M. Lyth, S. Moriya, K. Matsubayashi, N. Islam, S. Kuroki, M. Kakimoto, J. Ozaki and S. Miyata, The role of Fe in the preparation of carbon alloy cathode catalysts, *ECS Trans.* **25**, 463 (2009).

[31] Y. Nabae, Y. Kuang, M. Chokai, T. Ichihara, A. Isoda, T. Hayakawa and T. Aoki, High performance Pt-free cathode catalysts for polymer electrolyte membrane fuel cells prepared from widely available chemicals, *J. Mater. Chem. A* **2**, 11561 (2014).

[32] Y. Nabae, Development trends of carbon-based cathode catalysts, *J. Fuel Cell Technol.* **12**, 24 (2013) (in Japanese)

[33] B. Stohr, H. P. Boehm and R. Schlögl, Enhancement of the catalytic activity of activated carbons in oxidation reactions by thermal treatment with ammonia or hydrogen cyanide and observation of a superoxide species as a possible intermediate, *Carbon* **29**, 707 (1991).

[34] T. Ikeda, M. Boero, S.-F. Huang, K. Terakura, M. Oshima and J. Ozaki, Carbon alloy catalysts: active sites for oxygen reduction reaction, *J. Phys. Chem. C* **112**, 14706 (2008).

[35] X. Hu, Y. Wu, H. Li and Z. Zhang, Adsorption and activation of O_2 on nitrogen-doped carbon nanotubes, *J. Phys. Chem. C* **114**, 9603 (2010).

[36] C. Kozlowski and P. M. A. Sherwood, X-ray photoelectron-spectroscopic studies of carbon-fibre surfaces. Part 5.—The effect of pH on surface oxidation, *J. Chem. Soc. Faraday Trans. 1* **81**, 2745 (1985).

[37] T. Kondo, S. Casolo, T. Suzuki, T. Shikano, M. Sakurai, Y. Harada, M. Saito, M. Oshima, M. I. Trioni, G. T. Franco and J. Nakamura, Atomic-scale characterization of nitrogen-doped graphite: Effects of dopant nitrogen on the local electronic structure of the surrounding carbon atoms, *Phys. Rev. B* **86**, 035436 (2012).

[38] S. Maldonado, S. Morin and K. J. Stevenson, Structure, composition, and chemical reactivity of carbon nanotubes by selective nitrogen doping, *Carbon* **44**, 1429 (2006).

[39] J. R. Pels, F. Kapteijn, J. A. Moulijn, Q. Zhu and K. M. Thomas, Evolution of nitrogen functionalities in carbonaceous materials during pyrolysis, *Carbon* **33**, 1641 (1995).

[40] E. R. Piñero, D. C. Amorós, A. L. Solano, J. Find, U. Wild and R. Schlögl, Structural characterization of N-containing activated carbon fibers prepared from a low softening point petroleum pitch and a melamine resin, *Carbon* **40**, 597 (2002).

[41] S. Wrabetz, R. Blume, M. Lerch, J. McGregor, E. P. J. Parrott, J. A. Zeitler, L. F. Gladden, A. Knop-Gericke, R. Schlögl and D. S. Su, Tuning the acid/base properties of nanocarbons by functionalization via amination, *J. Am. Chem. Soc.* **132**, 9616 (2010).

[42] P. Ruffieux, M. Melle-Franco, O. Gröning, M. Bielmann, F. Zerbetto and P. Gröning, Charge-density oscillation on graphite induced by the interference of electron waves, *Phys. Rev. B* **71**, 153403 (2005).

[43] H. A. Mizes and J. S. Foster, Long-range electronic perturbations caused by defects using scanning tunneling microscopy, *Science* **244**, 559 (1989).

[44] M. M. Ugeda, I. Brihuega, F. Guinea and J. M. Gomez-Rodriguez, Missing atom as a source of carbon magnetism, *Phys. Rev. Lett.* **104**, 096804 (2010).

[45] J. Oh, T. Kondo, D. Hatake, Y. Honma, K. Arakawa, T. Machida and J. Nakamura, He and Ar beam scatterings from bare and defect induced graphite surfaces, *J. Phys. Condens. Matter* **22**, 304008 (2010).

[46] L. Zhao, R. He, K. T. Rim, T. Schiros, K. S. Kim, H. Zhou, C. Gutiérrez, S. P. Chockalingam, C. J. Arguello, L. Palova, D. Nordlund, M. S. Hybertsen, D. R. Reichman, T. F. Heinz, P. Kim, A. Pinczuk, G. W. Flynn and A. N. Pasupathy, Visualizing individual nitrogen dopants in monolayer graphene, *Science* **333**, 999 (2011).

[47] S.-S. Yu and W.-T. Zheng, Effect of N/B doping on the electronic and field emission properties for carbon nanotubes, carbon nanocones, and graphene nanoribbons, *Nanoscale* **2**, 1069 (2010).

[48] J. C. Meyer, S. Kurasch, H. J. Park, V. Skakalova, D. Künzel, A. Groß, A. Chuvilin, G. Algara-Siller, S. Roth, T. Iwasaki, U. Starke, J. H. Smet and U. Kaiser, Experimental analysis of charge redistribution due to chemical bonding by high-resolution transmission electron microscopy, *Nat. Mater.* **10**, 209 (2011).

PART III

CARBON NANOMATERIALS FOR ENERGY STORAGE

9

SUPERCAPACITORS BASED ON CARBON NANOMATERIALS

VASILE V.N. OBREJA

National R&D Institute for Microtechnology (IMT Bucuresti), Bucharest, Romania

9.1 INTRODUCTION

Supercapacitors known also with the terms of ultracapacitors or electrochemical double-layer capacitors (EDLCs) are the result of technology evolution from the traditional electrolytic capacitors [1]. A first supercapacitor demonstrator was reported in [2]. Two porous carbon electrodes immersed in electrolyte have been used. Fired tar lampblack electrodes were used, and significantly higher capacitance value was found in comparison with an electrolytic capacitor of comparable size at that time. By replacing the dielectric films from the anode and cathode aluminum foil surface of an electrolytic capacitor, with activated (porous) carbon-based layers, further advance in supercapacitor research and development was realized [3].

At this time, carbon nanomaterials are regarded to have a significant role for energy conversion and storage and, in particular, for performance enhancement of supercapacitors as energy storage devices [4]. In [5], an overview on carbon-based nanostructured materials and their composites for use as supercapacitor electrodes is provided. In a review on carbon-based supercapacitors [6], nanocarbons are considered as main electrode materials for such devices. Carbon materials, metal oxides, and conducting polymers for electrochemical supercapacitors are also reviewed in [7]. Carbon nanostructures such as carbon nanotubes, graphene-based materials, porous carbon-based nanostructures, and other three-dimensional (3D) carbon-based

Carbon Nanomaterials for Advanced Energy Systems: Advances in Materials Synthesis and Device Applications, First Edition. Edited by Wen Lu, Jong-Beom Baek and Liming Dai.
© 2015 John Wiley & Sons, Inc. Published 2015 by John Wiley & Sons, Inc.

nanostructures for advanced supercapacitors are reviewed in [8]. Other recent reviews related to nanostructured carbon–metal oxide composite electrodes [9] or to carbon-based supercapacitors [10, 11] have been published.

In this chapter, the role of carbon nanomaterials for development and performance improvement of supercapacitors is considered. Nanoporous carbon (activated carbon), graphene with its derivatives, carbon nanotubes, composites of these carbon nanomaterials, and mixtures of carbon nanomaterials with other materials like metal oxides or conductive polymers are used at this time to advance supercapacitor technology.

9.2 SUPERCAPACITOR TECHNOLOGY AND PERFORMANCE

Available as commercial products, supercapacitors are already used in electrical energy storage applications. Typical products are shown in Figure 9.1. Similar packages can be encountered for commercially available electrolytic capacitors. Inside of a package from Figure 9.1, a winding like in Figure 9.2 is found. The winding has two aluminum foil bands (current collectors) coated on both sides with a carbon-based porous layer (electrodes) and a thin paper band (separator) so that an electric contact between electrodes can be avoided.

The porous carbon-based electrodes and the porous separator are impregnated with liquid electrolyte, and the ions can move freely through the porous separator. An SEM image of a cross section through the current collector (aluminum foil) coated on both sides with carbon-based electrodes is shown in Figure 9.3. A piece of aluminum foil with electrodes was embedded in epoxy resin and then polished to obtain this cross section. The white color band gives the thickness of aluminum foil of about $50\,\mu m$, whereas the thickness of the carbon-based electrode is around $200\,\mu m$.

An SEM image at the surface of electrode is shown in Figure 9.4. Carbon particles with a size lower and higher than $1\,\mu m$ are visible.

FIGURE 9.1 Typical commercial supercapacitors cells (5000F, 2700F, 200F) at 2.5 V; AA size battery for comparison (Obreja V.V.N., unpublished work).

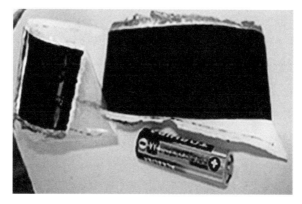

FIGURE 9.2   Winding of two aluminum foils coated with carbon-based electrode films and thin paper as separator between electrodes extracted from a sample of commercial supercapacitor package (Obreja V.V.N., unpublished work).

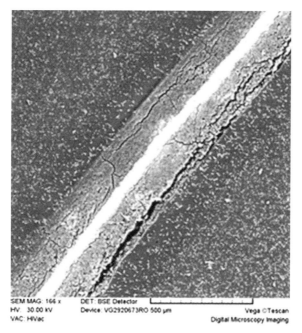

FIGURE 9.3   Cross section through carbon-based electrodes on both sides of aluminum foil substrate (Fig. 9.2); scale bar 500 μm (Obreja V.V.N., unpublished work).

Porosity at the nanoscale level is exhibited at the surface of particles. Activated carbon or nanoporous carbon, a commercially available material, is composed of such particles. A typical SEM image of activated carbon particles is shown in Figure 9.5. Like in Figure 9.4, porosity at the nanoscale level is visible in Figure 9.5, and as a consequence, a specific surface area of the order of $1000 \, m^2/g$ is exhibited. Nevertheless,

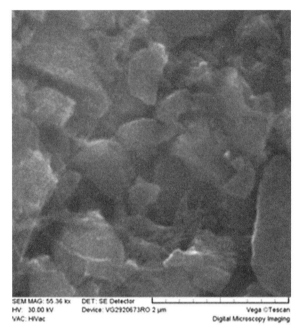

FIGURE 9.4   Typical SEM image at the porous carbon electrode surface from Figure 9.2; scale bar 2 μm (Obreja V.V.N., unpublished work).

FIGURE 9.5   Typical SEM image of activated carbon particles from a commercial material sample (Obreja V.V.N., unpublished work).

while in Figure 9.5 the nanoporous carbon particles can be moved, in Figure 9.4 the particles are bonded among them by means of binder material, so that electric contact from one to another one and further to the aluminum current collector is achieved.

Polytetrafluoroethylene (PTFE), polyvinylidene fluoride (PVDF), and N-methyl pyrrolidine are examples of binder materials used to fabricate electrodes. For electrode manufacturing, a composite slurry (nanoporous carbon powder, carbon black, polymer binder, and solvent) is prepared. As solvent, water, acetone, and alcohol are seldom used. The slurry is coated on metallic substrate (e.g., aluminum foil current collector) and then is dried and cured in oven, and then pressure is applied to the carbon-based film to improve electric conductivity. Finally, the metallic substrate with carbon-based film electrode is cut in bands to be used in the assembling of a supercapacitor cell (Fig. 9.2). As electrolytes impregnating the electrodes and separator in a cell, organic electrolytes such as 1 M $(C_2H_5)_4NBF_4$ (tetraethylammonium tetrafluoroborate) in acetonitrile or aqueous electrolytes such as 6 M KOH (potassium hydroxide) are used in commercial devices. Most of commercial supercapacitor cells available at this time are based on organic electrolyte because the maximum working voltage is 2.5 V, whereas for aqueous electrolyte, the maximum working voltage is limited to 1–1.2 V.

As in the case of batteries, for performance assessment of supercapacitors, key indicators such as the specific energy and specific power are considered. The specific energy is given by the relation

$$E_s = CV^2 / 2 \times 3600 \times \text{mass} \tag{9.1}$$

where C is the capacitance of the cell and V is the working voltage and the cell's mass is considered in kilograms. Sometimes instead of mass, volume of the cell is considered.

The specific power is given by the relation

$$P_s = V^2 / 4 \times \text{ESR} \times \text{mass(impedance match)} \tag{9.2}$$

where ESR is the equivalent series resistance of supercapacitor cell.

For commercial supercapacitor cells available at this time [12], based on organic electrolyte and nanoporous carbon electrodes (Figs. 9.2, 9.3, and 9.4), a maximum value of E_s of 6 Wh/kg is specified for maximum working voltage $V_{max} = 2.7$ V. The specific energy of common lead-acid batteries is of around 35 Wh/kg, but for other battery types, higher values are exhibited. A major disadvantage of supercapacitors in comparison with batteries is low value of specific energy. Nevertheless, supercapacitors have other advantages in comparison with batteries: significant higher cycle life, higher specific power, charging–discharging in short time, and extended operation temperature. As a consequence, most of research and development work has been directed toward specific energy increase. From relation (9.1), one can see that increase of capacitance (C) or of working voltage (V) can contribute to higher specific energy. Increase of the working voltage V is related to the electrolyte used in the cell. Increase of capacitance C is mainly related to the materials used in electrodes. In the case of nanoporous or activated carbon electrodes used in commercial supercapacitor cells based on organic electrolyte, for a value of maximum specific energy $E_{smax} = 6$ Wh/kg, a specific capacitance (C_s) can be derived from relation (9.1) with a

(a)

(b)

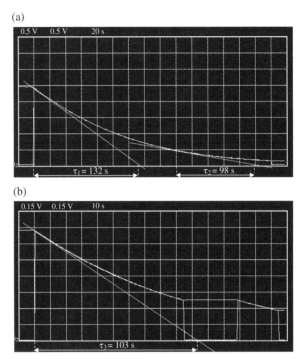

FIGURE 9.6 Discharge of a commercial supercapacitor cell on a load resistance of $0.5\,\Omega$:
(a) initially charged at about 2.5 V, 0.5 V/div, and 20 s/div and (b) initially charged at about 1 V,
0.15 V/div, and 10 s/div (Obreja V.V.N., unpublished work).

value $C_s = 5926\,\text{F/kg}$ or $5.93\,\text{F/g}$. Taking into account only the mass of nanoporous
carbon used in a commercial supercapacitor cell [13], which is around 1/5 of the total
mass of cell, then a specific capacitance $C_{sc} \sim 30\,\text{F/g}$ is derived for 1 g of carbon
nanomaterial used in the cell. For a cell with two identical electrodes, a specific
capacitance for only one electrode $C_{se} \sim 4 \times 30\,\text{F/g} \sim 120\,\text{F/g}$ can be derived. The two
electrodes are accompanied by two electric double-layer capacitances in series, and
this is the reason that the specific capacitance of the cell is multiplied by 4 to get the
specific capacitance for one electrode.

A value of a specific capacitance for cell of 25 F/g is reported in [14] for 1 M Et_4NBF_4
in acetonitrile electrolyte and a value around 100 F/g for a single electrode. The electrodes with a thickness of 500 µm were prepared by using YP17 (Kuraray Chemical,
Japan) commercial activated carbon (90%) and PTFE (10%) as composition.

Typical discharge of a commercial supercapacitor on a load resistance is shown in
Figure 9.6. The discharging curve recorded by means of an oscilloscope resembles to
discharging of a traditional capacitor for which an exponential voltage or current
dependence on time is expected. From a tangent line at the curve, a time constant can
be found and practically, after three time constants, complete discharge takes place
as in the case of a common capacitor. From the time constant $\tau = RC$ where R is the

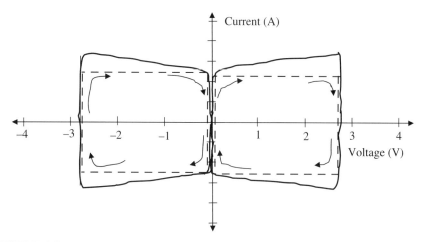

FIGURE 9.7 Schematics of cycling current voltage characteristic (voltammogram) for a commercial supercapacitor cell for low voltage scan rate (pure electric double-layer charge storage) (Obreja V.V.N., unpublished work).

discharging resistance and C is the capacitance, the supercapacitor capacitance can be derived if the resistance R is known. In the case of Figure 9.6, the load resistance is $0.5\,\Omega$, and the equivalent series resistance (ESR) of supercapacitor (lower than $0.01\,\Omega$) is negligible in comparison with the load resistance.

It is seen from Figure 9.6a that a constant value of τ is not exhibited. At higher charged voltage, higher capacitance value is exhibited. Similar behavior is reported in [14], and this is attributed to the space charge of the electric double layer from the carbon–electrolyte interface. The width of the space charge layer increases with the charging voltage, and larger capacitance is expected at higher voltage. In Figure 9.6b, it is seen that after discharge interruption for short time (the discharge current drops to 0), constant voltage is exhibited, as expected also in the case of a common capacitor.

For characterization of supercapacitors, cycling voltammetry, galvanostatic charge–discharge cycles, and electrochemical impedance spectroscopy are usually used. In Figure 9.7, schematic voltammograms (positive or negative polarization) are shown, which are usually exhibited by commercial supercapacitor cells. Almost a perfect rectangular shape (the dashed line) can be exhibited by conventional capacitors. Such voltammograms are exhibited at low scan rate or when the capacitor's ESR value is low (can be neglected).

From Figure 9.7, the capacitance value can be derived if the value of the current (I) is known. The capacitance is obtained from the relation $C = I/s$ where s is the scan rate. In many published works for the reported experimental cyclic voltammograms, a capacitance value is indicated on the ordinate of the voltammogram's plot. A schematic galvanostatic charge–discharge curve at constant current, usually, exhibited by a commercial supercapacitor cell is shown in Figure 9.8. A perfect linear dependence (the dotted line) is usually exhibited by traditional capacitors where the capacitance value does not depend on voltage.

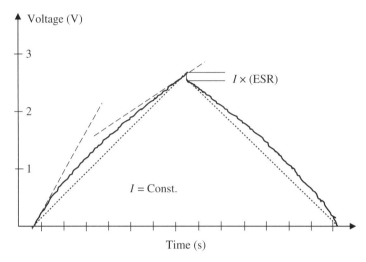

FIGURE 9.8 Schematics of galvanostatic charging–discharging curve for a supercapacitor cell with a voltammogram shape as in Figure 9.7; $I \times ESR$ is the voltage drop given by the internal equivalent series resistance (ESR) (Obreja V.V.N., unpublished work).

From Figure 9.8, capacitance value (C) can be derived from the relation $C = I/(dV/dt)$ where I is the constant charging or discharging current and (dV/dt) is the slope of curve given by the tangent at this curve (dashed lines). It is seen from Figure 9.8 that the slope at low voltage is higher and hence lower capacitance (current) in Figure 9.7 is exhibited. At higher voltage, the slope is lower and higher current (capacitance) is exhibited in Figure 9.7. Further, experimental reported voltammograms and galvanostatic charging–discharging curves will be compared with the shapes from Figures 9.7 and 9.8.

9.3 NANOPOROUS CARBON

A lot of research and development work has been dedicated to supercapacitor advancement by using porous carbon at the nanoscale level. Activated carbon is a well-known form of nanoporous porous carbon powder material (Fig. 9.5) with major contribution to supercapacitor development. Nevertheless, other forms of porous carbon—for example, ordered mesoporous carbon, carbon fibers and nanofibers, carbon cloth, carbon paper, and others—have been investigated. Aqueous and nonaqueous electrolytes were used in many reported experiments with nanoporous carbon.

9.3.1 Supercapacitors with Nonaqueous Electrolytes

As nonaqueous electrolytes, organic electrolytes and ionic liquids are usually used in the reported experiments. In [15], commercial activated carbon (RP20) was reactivated by steam in a temperature range of 950–1150°C, and physical characterization was performed. A specific surface area of up to $2240 \, m^2/g$ was obtained with pore width distribution in a range of 1–5 nm. Experimental supercapacitor cells were assembled by using electrodes with a composition of activated carbon (96%) and PTFE binder (4%) with a thickness of 100 μm. Aluminum foil attached to electrodes was used as current collectors, and Celgard 2400 separator of 25 μm thickness was used between electrodes. As electrolyte impregnating the electrodes and separator, 1.2 mol $(C_2H_5)_3CH_3NBF_4$ salt in acetonitrile was used. Displayed cyclic voltammograms up to 3.2 V resemble with what is shown in Figure 9.7 (positive polarization). For a scan rate of 1 mV/s, specific capacitances in a range of 113–117 F/g have been obtained at 3 V. Reactivation at high temperature resulted in a significant increase of specific area from 1358 to $1580 \, m^2/g$ (at 950°C), $2240 \, m^2/g$ (at 1050°C), and $1899 \, m^2/g$ (at 1150°C). Nevertheless, as shown in the same reference [15], the specific capacitance has increased from 113.4 F/g for initial RP20 untreated material ($1358 \, m^2/g$) to a maximum value of 118 F/g for $2240 \, m^2/g$ (treated at 1050°C) at a scan rate of 1 mV/s. Little change in specific capacitance occurred at higher scan rate, for example, a decrease from 118 F/g (1 mV/s) to 112 F/g (50 mV/s) and 3 V polarization voltage. Also, the voltammograms for a scan rate of 50 mV/s and lower values (5, 10, 20 mV/s) exhibit little change of the shape in comparison with that for 1 mV/s.

Carbide-derived nanoporous carbon prepared from Mo_2C was used in [16] for experimental supercapacitor cells, practically, realized in the same manner as described in [15]. 1 mol $(C_2H_5)_3CH_3NBF_4$ in acetonitrile as electrolyte was used. Cyclic voltammograms with a shape resembling to that one shown in Figure 9.7 (positive voltage) and galvanostatic charge–discharge curves with a shape resembling to that one shown in Figure 9.8 are reported. For nanoporous carbon powder synthesized at 700°C with a specific surface area of $1811 \, m^2/g$, a specific capacitance value of 140 F/g was obtained at 2.5 V and a scan rate of 1 mV/s. Comparison with values of specific capacitance from above [15] shows that for the same surface area or even lower surface area, higher value of specific capacitance is possible.

Recently [17], nanoporous carbon was prepared from paper pulp sludge by a process called hydrothermal carbonization followed by KOH activation at 700 and 800°C. Electrodes were fabricated from a mixture of 80 wt% prepared activated carbon, 10 wt% carbon black, and 10 wt% PVDF as binder in N-methylpyrrolidone solution, and a slurry was formed. The slurry composite was rolled onto aluminum foil and dried at 100°C, and disks with a diameter of approximately 1.5 cm were punched out. Coin-like symmetric supercapacitor cells were realized by using two electrode disks with a porous polymeric separator between them. The cells were assembled inside an Ar-filled glove box (<0.1 ppm of both oxygen and H_2O). The thickness of nanoporous carbon-based electrode was 60–100 μm. The size of activated carbon particles from electrode was in a range of 1–40 μm, and pore size distribution was in a range from below 1 nm to 4–5 nm. Nanoporous carbon powder resulted after activation at 700°C exhibited a specific surface area (BET) of $1470 \, m^2/g$. After activation at 800°C, $2340 \, m^2/g$ surface area at lower content of added KOH and $2980 \, m^2/g$ at higher content of KOH were determined. Commercial activated carbon powder (Norit) with BET surface area of $2050 \, m^2/g$ was also used for comparison. As electrolytes, 1.5 M tetraethylammonium tetrafluoroborate solution in acetonitrile

(TEABF$_4$/AN) and two ionic liquids were used. For these electrolytes, cyclic voltammograms for scan rates of 2 and 200 mV/s and galvanostatic charging–discharging curves at 0.1 and 10 A/cm^2 were measured. Only for the above commercial activated carbon sample used for comparison, the shape of voltammograms at 2 mV/s both in organic electrolyte and ionic electrolytes can be approximated with that from Figure 9.7 (full line). A small difference is related to the voltage dependence for the charging and discharging part. For the charging part from 1–1.5 V up to 2.25 V (organic electrolyte) and up to 3 V (ionic liquid electrolyte), slight deviation from linear uniform increase of current (capacitance) takes place with higher value than expected from linear variation. For the discharging part, under 1–1.5 V for both electrolytes, slight deviation from linear uniform decrease of the current (absolute value) takes place with higher value than expected from linear variation. The shown charging–discharging curves at low current density (0.1 A/cm^2) for both electrolytes have the shapes similar with that one shown in Figure 9.8. Values of specific capacitance in the range of 108.4–123.4 F/g from the voltammograms at 2 mV/s and from the charging–discharging curves both for organic and ionic liquid were determined for the above activated carbon commercial sample used for comparison. For the above prepared nanoporous carbon by hydrothermal carbonization, followed by KOH activation, more visible deviation of the voltammograms at 2 mV/s and of charging–discharging curves at 0.1 A/cm^2 from the shape of Figure 9.7 is manifested than those corresponding to the activated carbon commercial sample. Such a behavior indicates that the electric double layer is not the only charge storage mechanism in this experiment. For the hydrothermal carbonized and KOH activated sample at 700°C (1470 m^2/g), a specific capacitance in a range of 113.4–119.3 F/g at the same low scan rate or current density as before for the organic electrolyte and for one of the ionic liquids was determined. For the other ionic liquids, the capacitance value was 82.4–85.7 F/g. For the sample activated at 800°C (2345 m^2/g), a capacitance value in a range of 121.4–149.4 F/g for both organic electrolyte and the two ionic liquids (low scan rate and current density) was found. For the last sample activated at 800°C but with more KOH (2980 m^2/g), a capacitance range of 161.5–190.3 F/g was found, in the same conditions. From this experiment, it results that for the two samples of activated carbon processed in different conditions and with different surfaces, practically, the same value of specific capacitance is possible. The specific capacitance may increase with the surface area but without a proportional relation. In [17], voltammograms and charging–discharging curves are also shown at a scan rate of 200 mV/s and high current density of 1 A/cm^2. Significant change of the shapes takes place due to the value of ESR, which cannot be neglected in such conditions. Consequently, the values of specific capacitance determined in such conditions are significantly lower than those specified earlier. A maximum specific energy of 30 Wh/kg for organic electrolyte and 51 Wh/kg for ionic electrolyte is specified in [17].

A value of specific capacitance of 145 F/g for activated carbon of 2258 m^2/g in 1 M Et$_4$NBF$_4$ in propylene carbonate (PC) is also reported in [18]. A specific energy of 31 Wh/kg corresponds to this capacitance in a cell. Similar electrode preparation such as in [17] took place, and dominant pores with a size of 0.5–3 nm were exhibited for the activated carbon prepared from mesophase pitch with KOH activation at

800°C. No cyclic voltammograms are shown. Only discharging curves from 2.5 to 0 V at different currents are shown, and these have a linear variation with the time (dotted line in Fig. 9.8).

In [19], activated carbon has been prepared from phenolic resin precursor through physical activation in CO_2 atmosphere. A capacitance of 160 F/g and energy density of 35 Wh/kg is reported in this work for nonaqueous organic electrolyte (SBPBF$_4$/ PC) at a current density of 1 mA/cm^2. Only the abstract of the paper was available, and voltammogram or charge–discharge curve analysis, as mentioned before, was not performed.

Ordered mesoporous/microporous carbon material was prepared in [20] from chlorination of ordered mesoporous titanium carbide/carbon composites. A surface area of 1698 m^2/g was determined with a mesopore size centered at 4.4 nm. The micropore sizes were 0.52, 0.76, and 1.52 nm. A capacitance of 132 F/g at 0.5 A/g in nonaqueous electrolyte was measured. Cyclic voltammograms with a scan rate in a range of 20–2000 mV/s were measured with positive voltage up to 1.2 V and negative voltage up to −1.75 V. The shape of voltammograms at low scan rate of 20 mV/s (positive and negative direction) resembles to that one shown in Figure 9.7. At high scan rate, distortion of the shape from low scan rate is exhibited due to the influence of ESR of the cell. At 2000 mV/s, a capacitance of 79% from the value at low scan rate was determined. After 5000 charging–discharging cycles, the capacitance retention was 84%. Such behavior indicates that the capacitance originates from the electric double layer formed at the carbon–electrolyte interface.

The same pure capacitive behavior results from the work [21] for commercial carbon black SC3 from Cabot Corporation with a particle size in the range of 2–10 μm (most of particles with a size in 2–4 μm range). Electrodes with a weight composition of 70% SC3, 20% Super-P (another carbon black material), and 10% binder were prepared and were used in a three-electrode Swagelok-type cell. Organic electrolyte 1 M Et$_4$NBF$_4$ in PC and a glass microfiber filter of 675 μm in thickness and 12 mm in diameter as separator were used. The SC3 material was compared with commercial activated carbon DLC Super 30 (Norit) for supercapacitors and with commercial activated carbon BP 20 (also, supercapacitor grade) from Kuraray Chemical. All of them have a surface area of 1500–1800 m^2/g. From a cyclic voltammogram at 20 mV/s, a specific capacitance of about 115 F/g (40 mA/g specific capacity at 5 mA/ cm^2) was derived for SC3 (carbon black). The voltammogram is shown between −1 and 1.5 V. The shape of voltammogram for 0.3–1.5 V resembles well with that shown in Figure 9.7 (positive polarization). For the part with negative polarization from 0.3 to −1 V similar to the shape shown in Figure 9.7 and for the part from −1 to 0.3 V, there is some deviation.

In [22], nanoporous carbon with a tailored pore size prepared from phenol formaldehyde resin as precursor and KOH/ZnCl$_2$ as activating agent gave a specific capacitance of 141.56 F/g at 120 mA/g in 1 M Et$_3$MeNBF$_4$/PC organic electrolyte. This specific capacitance is not higher than the value of 145 F/g reported in [18] for which, as mentioned earlier, a specific energy of 31 Wh/kg corresponds. No cyclic voltammograms or constant current charging–discharging curves are presented in [22]. Nevertheless, an average specific energy of 74.13 Wh/kg is indicated.

The values of the specific energy given earlier (30–35 Wh/kg) (Refs. [17–19]) can be derived from the relation (9.1) taking into account a value of V of about 2.5 V for organic electrolyte and the specific capacitance divided to 4 to get the cell capacitance C (two electrodes and symmetric capacitor). In the case of value of 30 Wh/kg given earlier (Ref. [17]), a value for $V = 2.25$ V is considered. In Ref. [22] for the specific energy of 74.13 Wh/kg and 141.56 F/g, a value of voltage (V) toward 4 V has to be used in the relation (9.1), and this is not possible for organic electrolyte.

Activated carbon was prepared in [23] from waste coffee grounds by carbonization and activation with $ZnCl_2$ at 900°C. A BET surface area of 940–1020 m^2/g was obtained, and a comparison with Maxsorb (commercial activated carbon) of 1840 m^2/g surface area was performed. From a mixture of 90 wt% activated carbon, 5 wt% carbon black, and 5 wt% PVDF in N-methylpyrrolidone, a slurry was formed, and then this slurry was painted on aluminum 1 cm^2 area substrate (current collector). Symmetric cells were assembled by using two electrodes attached to current collectors. Between electrodes, glassy fiber paper was used. The electrodes and the separator were impregnated with 1 M $TEABF_4$ in acetonitrile electrolyte. Cyclic voltammograms at 5 and 50 mV/s, positive polarization up to 2.7 V, for four activated carbon samples including Maxsorb are presented. For the Maxsorb sample, deviation from the shape in Figure 9.7 consists that the charging part from 0 to 2.7 V exhibits nonlinear increasing variation, and for the discharging part from 2.7 to 0 V, no visible current decrease in absolute value takes place. For one of the samples activated with lower content of $ZnCl_2$ at 900°C, the current decreases above 1.5 V for the charging part, and for the discharging part, the current increases in absolute value from 2.7 V toward 1.5 V and then toward 0 V remains constant. The influence of higher ESR value on the voltammograms of these samples is exhibited. The voltammograms for two samples activated with higher content of $ZnCl_2$ at 900°C approach the shape given in Figure 9.7. A small difference is that for the charging part, slight deviation is manifested above 1.5 V and that for the discharging part, constant value of the current is exhibited. As in Ref. [17], some deviation of voltammograms at low scan rate from the shape in Figure 9.7 indicates that the exhibited capacitance is not only the electric double-layer contribution and that some contribution comes from other charge storage mechanisms. For the samples with voltammograms approaching the shape shown in Figure 9.7 (surface area of 940, 1021 m^2/g, more $ZnCl_2$ for activation), a specific capacitance at low scan rate (0.5 mV/s) or current density (0.05 A/g) of about 134 F/g is determined. The capacitance value is not lower than 100 F/g at 5 A/g. Nevertheless, for the sample with voltammogram with clear distortion from the shape in Figure 9.7 (surface area 1019 m^2/g, lower content of $ZnCl_2$), a specific capacitance of about 105 F/g in the same conditions was found. For this sample at 1 A/g current density, the capacitance is below 50 F/g. For the Maxsorb sample (1840 m^2/g, commercial product), a specific capacitance of 170 F/g was observed at 0.05 A/g and 65 F/g at 5 A/g. To understand the above results, it is mentioned that in the case of Maxsorb sample and of the sample with lower content of $ZnCl_2$, the ESR values are higher than for the other two samples. As a consequence for these samples with ESR values, the voltammograms at 50 mV/s exhibit more distortion than other two samples when compared with the shape shown in Figure 9.7. It is worth also mentioning that the sample with low content of $ZnCl_2$ and the Maxsorb sample have significantly lower volume of mesopores than the two samples with higher content of $ZnCl_2$. Nevertheless, these samples have significantly higher volume of micropores in comparison with the two samples with higher content of $ZnCl_2$. No charge–discharge curves are presented.

In [24], activated carbon was prepared from cotton stalk as precursor by carbonization and activation with phosphoric acid (H_3PO_4) at 800°C for 2 h. From a mixture of prepared activated carbon, carbon black for conductivity improvement, and PVDF (binder) in a weight ratio of 80:15:5 in distilled water, a slurry was formed. Electrodes were realized by pressing the slurry onto aluminum foil with a diameter of 1.2 cm and then were dried in a vacuum oven. Coin-shaped symmetric cells were assembled in an argon-filled glove box. 1 M Et_4NBF_4 electrolyte was used. Cyclic voltammograms are presented only at a scan rate of 50 mV/s, and a distorted shape from the expected one at low scan rate (1–5 mV/s), as shown in Figure 9.7, is exhibited. This is due to the high value of ESR measured by electrochemical impedance spectrometry. However, the galvanostatic charge–discharge curves have a symmetric shape similar to that one shown in Figure 9.8. For a sample of prepared activated carbon of 1480 m²/g surface area with major pore distribution of 1–6 nm, a specific capacitance of 114 F/g is exhibited at 0.5 A/g, but under 0.1 A/g, the specific capacitance is 140 F/g. In the paper, SEM pictures of the activated carbon prepared powder similar to that of Figure 9.5 are shown. For most carbon particles from powder, size distribution is in a range of 1–25 μm.

In [25], activated carbon was prepared from poly(vinylidene chloride) (PVDC) precursor. The material was pyrolyzed at 400°C, and then activation at 600°C with NaOH was performed. Specific surface areas in a range of 1829–2675 m²/g were obtained, depending on the ratio of PVDC char and NaOH in the activation process. The pore size distribution was in a 0.5–4 nm range. Experimental cells with aluminum current collectors and prepared activated carbon-based electrodes with a thickness of 30 μm were realized. The electrodes were realized from a slurry with a content of 85 wt% activated carbon, 10 wt% acetylene black, and 5 wt% PVDF mixed in *N*-methyl-2-pyrrolidinone solvent. A polypropylene membrane as separator and 1 mol/l Et_4NBF_4/PC as electrolyte were used. The specific capacitance measured at 50 mA/g was 108 F/g for 1829 m²/g (the sample with the lowest content of NaOH for activation and lowest volume of mesopores) and 155 F/g at 2675 m²/g (the sample with the highest content of NaOH for activation and the highest volume of mesopores) surface area. Cyclic voltammograms for the sample with 155 F/g are presented from 5 to 200 mV/s. Because of the used current scale, the shape of the voltammogram for 5 mV/s cannot be seen to be compared with the shape from Figure 9.7. Nevertheless, the galvanostatic charge–discharge curve has a shape resembling well with that one shown in Figure 9.8.

The values of specific capacitance of 120 F/g were reported in [26] for activated carbon obtained from cherry stones as precursor with a specific surface area of about 1130 m²/g in $(C_2H_5)_4NBF_4$/acetonitrile electrolyte. Only two current voltage cyclic voltammograms are presented at 50 mV/s, and these have distorted shape in comparison with the shape in Figure 9.7 due to higher ESR value.

Nanoporous carbon with a highly ordered porous structure was prepared in [27] by using mesoporous silica-templating procedure and different carbon precursors such as propylene, sucrose, and pitch. Performance investigation of the obtained material in TEABF$_4$/acetonitrile-based organic electrolyte reveals specific capacitance values not higher than reported above. The presented voltammograms have a similar shape to that one shown in Figure 9.7. Porous carbon obtained in [28] from metal–organic framework as both template and carbon precursor was further activated by using KOH. Electrochemical characterization was performed in organic electrolyte, 1.5 M Et$_4$NBF$_4$/tetraethylammonium tetrafluoroborate in acetonitrile, and in aqueous electrolyte (6 M KOH). For a specific surface area of 2222 m^2/g, a value of a specific capacitance of 156 F/g was obtained in organic electrolyte and 271 F/g in aqueous electrolyte at 2 mV/s scan rate. Cyclic voltammograms at 100 mV/s and galvanostatic charge–discharge curves at 0.25 A/g with a shape approaching the one shown in Figures 9.7 and 9.8 (some difference in shape is due to higher ESR value). Nevertheless, for the voltammograms in aqueous KOH electrolyte, no "rectangular" shape is observed, as mentioned in the paper. The shape of voltammograms and charging–discharging curves is completely different from a shape shown in Figure 9.7 or 9.8, and this behavior will be discussed in the following. For the value of 156 F/g of capacitance in organic electrolyte, a specific energy of 31.2 Wh/kg is indicated in the paper, whereas for 271 F/g in aqueous electrolyte, only 9.4 Wh/kg is specified because of lower working voltage (1 V). In [29], activated carbon obtained from polyethylene terephthalate waste gave a specific capacitance of 94 F/g in aprotic 1 M (C$_2$H$_5$)$_4$NBF$_4$/acetonitrile and 197 F/g in 2 M H$_2$SO$_4$ aqueous electrolyte. Significant difference is exhibited by the voltammograms in organic electrolyte and aqueous electrolyte as one can see also in [28].

Activated carbon fibers were prepared in [30] from polyacrylonitrile (PAN) fibers as precursor by carbonization at 600°C, followed by activation with NaOH at the same temperature. Surface area increases when more NaOH is present in the mixture with carbonized fibers for activation. A maximum value of 3250 m^2/g (pore size distribution of 1–5 nm) was obtained, and the corresponding value of specific capacitance was 188 F/g in ionic electrolyte composed of LiTFSI and OZO at a molar ratio of 1:4.5. Measurements were performed on button-type cells with electrodes of 11 mm diameter obtained by pressing into pellets a mixture of activated material (87%), acetylene black (10%), and PTFE (3%). It is mentioned that for activated carbon fiber sample with a surface area of 2085 m^2/g (pore size distribution of 0.5–3 nm), the specific capacitance was only 54 F/g. Nevertheless, for the same surface area, the specific capacitance was 180 F/g in LiClO$_4$ in PC electrolyte. A high value of 188 F/g for ionic electrolyte is obtained for a sample activated with more content of NaOH reaching a surface area of 3250 m^2/g. In this case, the percentage of mesopores is higher than for the samples activated with lower content of NaOH at 600°C. Such behavior is attributed to ion penetration in the pores. Cyclic voltammograms and charging–discharging curves (up to 2.5 V) have different shapes from those shown in Figures 9.7 and 9.8. Similar behavior is observed in [17] for ionic electrolyte. The change is mainly for the discharging part and will be discussed in the following.

In [31], activated carbon with a surface area of $1420 \, m^2/g$ was used as electrode material in cells with $PYR_{14}TFSI$ ionic liquid as electrolyte. A specific capacitance of $90 \, F/g$ for $5 \, mV/s$ scan rate and then decreasing to $60 \, F/g$ at $20 \, mV/s$ is reported. Similar difference from Figures 9.7 and 9.8 for the discharging part as in [30] is exhibited for the shown voltammogram and charging–discharging curves. Graphite fine powders [32], usually used in lithium-ion battery cathodes as conductive additives, were investigated as electrode materials in cells with $1 \, M \, LiPF_6/EC+DMC$ organic electrolyte. A working voltage of $4 \, V$ can be reached with this electrolyte, also used in lithium-ion rechargeable batteries. Specific capacitance values of $41 \, F/g$ for $10 \, m^2/g$ specific surface area, $35 \, F/g$ for $21 \, m^2/g$, and $33 \, F/g$ for $21 \, m^2/g$ are given in [32]. The cyclic voltammogram and charge–discharge curves differ from the shapes shown in Figures 9.7 and 9.8.

9.3.2 Supercapacitors with Aqueous Electrolytes

As aqueous electrolytes, acidic solutions (e.g., sulfuric acid), basic solutions (e.g., potassium hydroxide), and neutral solutions (e.g., potassium chloride) are used in many experiments. The same prepared activated carbon as described earlier [23] was used in symmetric cells in [33, 34] but this time with aqueous electrolyte, $1 \, M \, H_2SO_4$. The shown cyclic voltammograms up to 1–$1.2 \, V$ have a different shape from the shape shown for nonaqueous organic electrolyte (approximating the shape shown in Fig. 9.7). The shape of voltammograms in aqueous electrolyte for the same activated carbon, also used in organic electrolyte in the same conditions of low scan rate, resembles to that one shown in Figure 9.9 (positive direction). The shape of the charging–discharging curves at constant current resembles to that one shown in Figure 9.10, which is different from the shape shown in Figure 9.8.

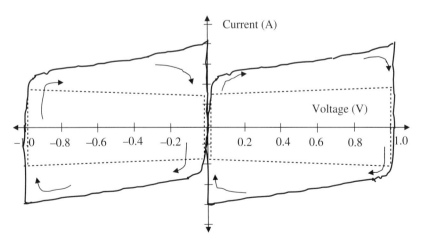

FIGURE 9.9 Schematics of cyclic voltammogram for a supercapacitor with pseudocapacitance contribution (type I) (Obreja V.V.N., unpublished work).

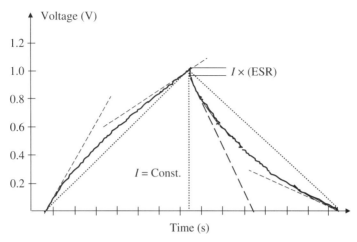

FIGURE 9.10 Schematics of galvanostatic charging–discharging curve for supercapacitor cell with voltammogram shape as in Figure 9.9 (type I) (Obreja V.V.N., unpublished work).

A comparison of voltammogram (Fig. 9.7) found sometimes for organic electrolyte with voltammogram (Fig. 9.9) found sometimes for aqueous electrolyte reveals that for the charging part, the same linear increasing tendency with voltage is manifested. Nevertheless, as was shown earlier for organic electrolytes, deviation from this linear variation may take place for real voltammograms. In Figure 9.7, decrease of current (capacitance) in absolute value from the maximum voltage toward the minimum voltage takes place at discharge, whereas in Figure 9.9, increase of current in absolute values for the discharging part from the maximum voltage toward 0V is observed. Comparison of Figure 9.8 with Figure 9.10 also reveals significant difference in the exhibited shapes for the discharging parts of curves. Whereas in Figure 9.8 symmetrical shape is exhibited for the discharging part, in Figure 9.10, the shape for the charging part is similar to the corresponding one in Figure 9.8 for the charging part, but for the discharging part, no symmetrical curve shape is observed. For aqueous electrolyte and the same nanoporous carbon material, significantly higher specific capacitance than for organic electrolyte (110–160F/g, as mentioned in [23]) was found. Thus, in [33], a specific capacitance reaches 400F/g, whereas in [34], a specific capacitance of 300F/g for organic electrolyte is reported. Such values of specific capacitance cannot be explained if only the capacitance given by the electric double layer (electrostatic attraction of opposite electric charges as in a conventional electrolytic capacitor) is taken into consideration. Cyclic voltammograms or charge–discharge curves as in Figures 9.7 and 9.8 are not exhibited more. Lower capacitance is exhibited at lower voltage, and larger capacitance at higher voltage is exhibited for the charging part in Figures 9.9 and 9.10. Nevertheless, for the discharging part, higher capacitance is exhibited at lower voltage and lower capacitance at higher voltage.

Similar voltammograms to that of Figure 9.9 are also shown in [35, 36] for aqueous electrolytes, where at voltage increase (charging) up to 1 V steady increase of current takes place and at voltage decrease toward 0V steady increase of current in absolute value is observed. Nevertheless, charge–discharge curves are not presented. Nonwoven mats of activated carbon nanofibers have been fabricated in [35] with nanofiber diameter in the range of 200–400 nm by using only a carbonization process. The nanofibers have been electrospun from blends of PAN and sacrificial Nafion at different compositions. A surface area of up to 1600 m^2/g with dominance of mesopores (2–4 nm) was obtained. A specific capacitance of up to 210 F/g at a scan rate of 100 mV/s in 1 M H_2SO_4 electrolyte was measured. At lower scan rate of 1–2 mV/s, even higher value is possible. The testing cells had electrodes realized only from carbonized nanofibers without addition of binder material. As a consequence, lower value of ESR is achieved, and this is seen from the shape of voltammograms (positive polarization), which resemble fairly well with that from Figure 9.9. Nevertheless, in the paper, it is mentioned that these voltammograms have "near rectangular" shapes, that is, approaching a shape like in Figure 9.7. In [36], highly porous binderless activated carbon monolith electrodes were produced from fibers of oil palm empty fruit bunches through nitrogen carbonization followed by activation. For a surface area of 1704 m^2/g, a specific capacitance of 150 F/g and a specific energy of 4.297 Wh/kg are reported for aqueous electrolyte.

In [37], symmetric cells with activated carbon-based electrodes and with 0.5 mol/l Na_2SO_4 aqueous electrolyte were realized. The cyclic voltammograms plotted up to 1.2 V resembles well with that one shown in Figure 9.9 for positive voltage. It is said that these voltammograms "exhibit a rectangular shape characteristic of a pure capacitive behaviour." The galvanostatic charge–discharge curves resemble to that one shown in Figure 9.10. Nevertheless, for pure capacitive behavior, the charge–discharge curves should have a shape like that one in Figure 9.8.

In [38], activated carbon with a surface area of 2176 m^2/g was oxidized in nitric acid (HNO_3) or treated with urea or melamine to create at the surface functional groups enriched with oxygen or nitrogen. The samples treated with urea or melamine were heated at 950°C. After treatment, significant decrease of surface area took place.

For a sample oxidized and treated with melamine, the surface area decreased to the lowest value of 721 m^2/g. Nevertheless, its capacitance in 1 M H_2SO_4 aqueous electrolyte was 235 F/g, whereas for the initial sample of 2176 m^2/g area, the specific capacitance was 253 F/g. For a sample treated only with melamine, the specific surface area decreased to 1435 m^2/g, but the capacitance increased to 330 F/g. The values of the specific capacitance are given at 50 mA/g current density. This sample has a cyclic voltammogram similar to that one in Figure 9.9 and a charge–discharge curve similar to that one in Figure 9.10. As a consequence, a voltammogram like in Figure 9.9 and a charge–discharge curve like in Figure 9.10 cannot exhibit "pure capacitive behavior" given by the electric double layer as mentioned in [37]. An extra capacitance called pseudocapacitance that is the result of surface functional groups from the carbon surface is manifested in such a case. Another evidence for pseudocapacitance contribution is the decrease of specific capacitance at 1 A/g current density given in [38]. Thus, from 330 F/g at 50 mA/g, a decrease to 236 F/g at 1 A/g takes place on behalf of the pseudocapacitance. The same behavior takes place at cycling voltammetry when the scan rate increases from 5 to 100 mV/s. As was shown earlier, for organic electrolyte-based supercapacitors, where electric double-layer capacitance is the dominant one and the ESR value is low, such significant reduction of specific capacitance with the current density or scan rate does not take place.

The pseudocapacitance contribution is also revealed in [39] where porous carbon was prepared from sorghum pith, firstly carbonized at 200°C and then treated with NaOH(20%) solution for 8 h. After this carbonization, further treatment at temperature of 300–500°C took place for 3 h, and three samples of porous carbon with an estimated surface area of 35 m^2/g (300°C), 27 m^2/g (400°C), and 17 m^2/g (500°C) were obtained. Cycling voltammetry in 1 M H_2SO_4 electrolyte reveals visible increase of positive charging current from 0 to 1.2 V and decrease of negative discharging current from 1.2 to 0 V. However, the current increase and decrease are not uniform with the voltage change as in Figure 9.9, and some peaks can be distinguished. A specific capacitance of 320.6 F/g was indicated for the sample with 17 m^2/g (500°C), whereas for the sample with 35 m^2/g (300°C), only 13.6 F/g was found at 10 mV/s scan rate. For a scan rate of 100 mV/s, significant decrease of specific capacitance takes place. This behavior is attributed to larger pore size for carbonization at higher temperature and to functional groups from the carbon surface.

Cyclic voltammograms for positive and negative voltage in 1 M H_2SO_4 electrolyte with the same behavior as shown in Figure 9.9 are presented in [40] for high-porosity carbon derived from mesophase pitch. Galvanostatic charging–discharging curves resembling well with that from Figure 9.10 are also presented. The mesophase pitch powder ground and sieved at average size particle of 20 μm was impregnated with KOH solution and after drying was carbonized and activated at 700°C. Activated carbon samples with a surface area in a range of 1310–2860 m^2/g depending on the KOH/pitch ratio were obtained. Symmetric cells with electrodes pressed on stainless steel foil from a slurry of activated carbon (80%) and PVDF as binder material and filter paper as separator were manufactured. It has been found that for electrode composite material, significant decrease of the specific surface area took place. Thus, in electrode composites, the surface area of a 1930 m^2/g sample decreased to 797 m^2/g and that of a 2860 m^2/g sample to 1370 m^2/g. This is due to the fact that the binder material could block some pore entrances. A specific capacitance of 103 F/g was obtained for the above low surface area and 130 F/g for the above high surface area. Significant decrease of the specific capacitance with increasing scan rate or discharging current took place.

In a study performed in [41], a schematic voltammogram similar to that one given in Figure 9.9 for positive voltage is shown. It is said that the exhibited current from V_{min} (0 V) to V_{max} (1 V) is an oxidation current, whereas the current from V_{max} (1 V) to V_{min} (0 V) is a reduction current. These currents are related to the redox reactions from the carbon–electrolyte interface, and these are seldom mentioned in the supercapacitor literature. Pseudocapacitance due to redox reactions is exemplified in [42] for a modified activated carbon (Black Pearls from Cabot Corporation) of initially 1500 m^2/g surface area. Anthraquinone groups were grafted at the carbon surface. As a consequence, the specific capacitance has increased from 100 to 195 F/g. The cyclic voltammograms in 0.1 M H_2SO_4 at a scan rate of 10 mV/s for unmodified and modified activated carbon differ. For unmodified carbon, a voltammogram from −0.5 to 0.5 V resembles to that one shown in Figure 9.7 attributed to electric double layer. For modified carbon, a voltammogram with noticeable peaks is exhibited. By increasing the mass percentage

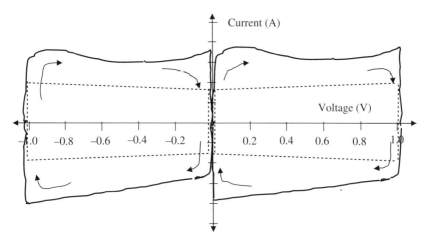

FIGURE 9.11 Schematics of cyclic voltammogram for a supercapacitor with pseudocapacitance contribution (type II) (Obreja V.V.N., unpublished work).

of grafted anthraquinone, a significant pseudocapacitance component is added to the electric double-layer component, and an increase of total capacitance is exhibited.

Cyclic voltammograms at low scan rate in H_2SO_4 electrolyte resembling to that one shown in Figure 9.9 are also shown in [43, 44] for nanoporous carbon. However, if the electrolyte is changed, for example, Nafion instead of H_2SO_4 in [43] and KOH instead of H_2SO_4, then cyclic voltammograms like in Figure 9.11 are exhibited. In this case, slight increase, then decrease, and again increase of current from 0 to 1 V take place, but steady current decrease (negative value) is manifested from 1 to 0 V. Nevertheless, in [45], similar curves with that one given in Figure 9.11 are shown—1 M H_2SO_4 aqueous electrolyte—for nanoporous carbon prepared from sucrose with a templated method. Hence, the shape of the cyclic voltammogram (Fig. 9.9 or 9.11) depends on the utilized electrolyte and the nature of nanoporous carbon. In [45], galvanostatic charge–discharge curves up to 3 V are presented for organic $LiPF_6$/PE+CE electrolyte. Cyclic voltammograms are not shown, but those presented in [30, 31] have a shape similar to that one in Fig. 9.9 or 9.11. Similar behavior is valid for a supercapacitor with polymer electrolyte based on ionic liquids [46].

Nitrogen-doped nanoporous carbon has been synthesized [47] by direct pyrolysis of a nitrogen-containing organic salt at a temperature of 600–900°C without any activation process, with a surface area of 408–1170 m^2/g with low value for 600°C and high value for 900°C. The nitrogen content decreases from 8.59% for 600°C pyrolysis to 1.02% for 900°C. A specific capacitance as high as 245 F/g in 6 mol/l KOH aqueous electrolyte was obtained for 700°C (708 m^2/g) at 7.74% nitrogen content. For 900°C (1170 m^2/g), the specific capacitance is only 133 F/g due to low content of nitrogen and hence little contribution from the pseudocapacitance component to total capacitance. Cyclic voltammogram like in Figure 9.9 and charging–discharging curve like in Figure 9.10 are exhibited for the nanoporous sample pyrolyzed at 700°C. Nonetheless, for the temperature of 900 °C, the voltammogram and the charge–discharge curve approach the shape from Figures 9.11 and 9.12, respectively. Significant decrease of the specific capacitance with current density increase or scan rate increase is exhibited at higher content of nitrogen (pseudocapacitance contribution).

By treating a phenol-based activated carbon with different concentrations of phosphoric acid [48], a specific capacitance increase to 452 F/g from 256 F/g at a scan rate of 5 mV/s takes place, if the activated carbon sample is treated with a 2 M solution of phosphoric acid. As electrolyte, 1 M H_2SO_4 aqueous solution was used. Electrochemical characterization was performed with electrodes in three-electrode system with working electrode, counter electrode, and reference electrode. The working electrode was realized from a slurry of 80 wt% treated or untreated activated carbon, 10 wt% carbon black as conductive additive, and 10 wt% PVDF as binder. The slurry was coated on titanium plate, and after drying at room temperature, a pressure of 200 bars was applied to obtain the final working electrode. Cyclic voltammograms at 5 mV/s resemble to that one in Figure 9.11. This capacitance enhancement is attributed to further development of nanopores and oxygen functional groups (particularly C=O) due to the etching reaction at the carbon surface.

Nitrogen-enriched activated carbon from waste medium density fiberboard base using K_2CO_3 as an activating agent has been prepared in [49]. Cyclic voltammograms and charge–discharge curves in 7 mol/l KOH aqueous solution are similar to that one in Figures 9.11 and 9.12, respectively. A maximum value of specific capacitance of 230 F/g was obtained at 2 mV/s scan rate and for 400°C carbonization temperature. Significant decrease of specific capacitance with the current density is exhibited. In [50], phenolic-based carbon nanofiber webs with an average diameter of 390 nm exhibited a surface area of 416 m^2/g and a capacitance of 171 F/g at 5 mV/s and remained 84% (143.64 F/g) at 100 mV/s. Cyclic voltammograms in 6M KOH electrolyte resemble to that from Figure 9.11.

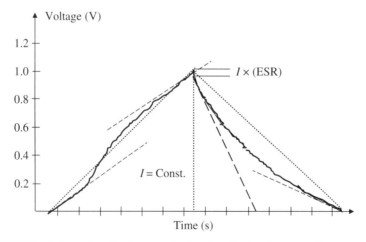

FIGURE 9.12 Schematics of galvanostatic charging–discharging curve for supercapacitor cell with voltammogram shape as in Figure 9.11 (type II) (Obreja V.V.N., unpublished work).

Mesopore carbon microspheres with a diameter of 0.5–2 μm and main mesopore sizes of 2.6–4 nm and surface area in a range of 449–1212 m²/g were prepared [51] and used in electrodes. Electrodes were realized from a paste with 80% carbon powder, 10% graphite powder, and 10% PTFE binder composition using ethanol. The paste was pressed onto nickel foil at 20 MPa for 10 s. The voltammograms in 6 M KOH electrolyte (three-electrode system) and charging–discharging curves indicate pseudocapacitive influence. These have the shape as in Figure 9.11 or 9.12. A specific capacitance of 171 F/g at 1 A/g and 157 F/g at 10 A/g current density was determined for an electrode of 1010 m²/g surface area. Higher value is expected at 0.1 A/g but is not specified in the paper. For an electrode of 1212 m²/g, the capacitance was 126 F/g. Nyquist plots of the two electrodes from electrochemical impedance spectroscopy tests indicate an internal resistance higher for the electrode of 1212 m²/g (about two times higher). Similar behavior takes place in [52] for microspheres with a diameter of 2–5 μm and hierarchical mesopores in a range of 3.5–60 nm. For a surface area of 1320 m²/g, a specific capacitance of 208 F/g at 0.5 A/g is reported. A 2 M H_2SO_4 electrolyte was used. The same pseudocapacitive influence as in [51] is exhibited by the cyclic voltammograms and the galvanostatic charging–discharging curves. For the same material of 1320 m²/g surface area, investigation in organic electrolyte, 1.0 M $(C_2H_5)_4NBF_4$ (TEABF$_4$) in PC, was performed (paper supporting information). A specific capacitance of 97 F/g was found at 0.5 A/g, and the cyclic voltammogram up to 3 V resembles well with that one shown in Figure 9.7 with a typical "rectangular shape" as it is mentioned in the paper. As a consequence, due to pseudocapacitance contribution, a double value of specific capacitance is possible in aqueous electrolyte.

Phenol–melamine–formaldehyde resin as precursor was used in [53] to obtain activated carbon. After carbonization of the polymer at 700°C, activation was performed in a mixed gas of oxygen and nitrogen at 350–450°C. A BET surface area of 674.3 m²/g was determined for a sample of activated material at 400°C, with a specific capacitance of 210 F/g, whereas before activation, the carbonized material exhibited 30.6 m²/g and 58 F/g. The electrodes were manufactured by pressing a mixture of activated carbon, acetylene black, and PVDF with a proportion of 80%:10%:10% onto nickel foil, and experimental cells were realized. As electrolyte, 6 mol/l KOH solution has been used. Cyclic voltammograms and charging–discharging curves with a shape as in Figures 9.11 and 9.12 are shown. Such behavior indicates again that the pseudocapacitance can have significant contribution to the specific capacitance. In [54], nanoporous carbon from H_2SO_4-doped polyaniline precursor was prepared by carbonization at 800°C, followed by activation with oxygen and nitrogen gas mixture. Similar preparation of the electrodes and the same electrolyte as in [53] was used. Also, similar cyclic voltammograms and charging–discharging curves are shown. Before activation, a surface area of 325 m²/g with a specific capacitance of 135 F/g was obtained, and after activation, a surface area of 514 m²/g with a capacitance of 235 F/g was determined at 1 A/g current density. As in [53, 54] [55], doped nanoporous carbons have been prepared from polyaniline for supercapacitors. Polyaniline was carbonized and etched with KOH at 700°C. A carbonized sample without etching was also prepared. Similar results as in [53, 54] have been obtained after electrochemical characterization in 6 mol/l KOH. The effect of nitrogen atomic content and functionalities on the pseudocapacitive property has been studied. Two mechanisms of energy storage—electric double-layer formation and pseudocapacitance—have been confirmed. It is concluded that the specific capacitance exhibited by doped unactivated carbon is mainly attributed to pseudocapacitance, whereas that of doped activated carbon prepared by physical activation is given by combination of double-layer capacitance and pseudocapacitance. It is also said that the specific capacitance of doped activated carbon prepared by chemical activation is mainly given by the double-layer capacitance. This is due to the fact that activation treatment at higher temperature diminishes the content in surface functionalities.

Significantly higher specific capacitance of 455 F/g but for a surface area of 1976 m²/g was obtained in [56] for activated carbon derived from polyaniline after carbonization and KOH activation. Instead of a two-electrode cell used in [53–55], a three-electrode system was used for electrochemical characterization in 6 M KOH electrolyte, and cyclic voltammograms with a shape as in Figure 9.11 were exhibited. After carbonization of rod-shaped polyaniline (a diameter of 170 nm) at 850°C for 2 h and before activation at the same temperature for additional 1.5 h, a specific surface area of only 17 m²/g with a specific capacitance of 112 F/g was determined. For another prepared sample for which graphite oxide (GO) was added into the carbonized polyaniline before activation, a surface area of 1774 m²/g with a specific capacitance of 324 F/g was determined. Introduction of graphene nanosheets into activated carbon resulted in lower performance.

In [57], binderless electrodes after carbonization at 900°C of pieces of poplar wood were prepared and treated with nitric acid. No other activation process was used. A pore structure including macropores with a width of the order of 1 μm and nanopores is exhibited by this material. For a surface area of 416 m²/g, a specific capacitance of 234 F/g at 5 mA/cm² was obtained for the porous wood carbon monolith. For untreated material with nitric acid with a surface area of 467 m²/g, the capacitance was 80 F/g. The cyclic voltammograms at low scan rate (5–10 mV/s) and charge–discharge curves in 2 M KOH electrolyte resemble to that one shown in Figures 9.11 and 9.12. Distorted cyclic voltammograms at higher scan rate are due to higher ESR value. Significant increase of capacitance after treatment is due to functional groups introduced by nitric acid. Porous carbonaceous hydrogels and aerogels by using crude biomass (watermelon) as the carbon source were prepared in [58]. The carbonaceous gels have a 3D structure and consist of both carbonaceous nanofibers and nanospheres. Electrochemical characterization reveals a capacitance of 333.1 F/g at a current density of 1 A/g in 6 M KOH solution. After 1000 cycles of charge–discharge, 96% of the capacitance is retained. Cyclic voltammograms and charge–discharge curves have a shape like in Figures 9.11 and 9.12.

Cationic starch as precursor was used in [59] to prepare activated carbon by KOH, ZnCl₂, and ZnCl₂/CO₂ activation. A high surface area of 3332 m²/g resulted from KOH activation, whereas for the other two activation methods, 1600–1900 m²/g was obtained. Electrochemical characterization in prototype capacitors by means of cycling voltammetry and galvanostatic charge–discharge characteristics in 30 wt% KOH solution with resulted voltammograms like in Figure 9.11 was performed.

For KOH activated material, a specific capacitance of 238 F/g at 370 mA/g was measured. The capacitances for the other two activation methods were 119 and 139 F/g. Activated carbon samples from a series of starch were also fabricated in [60] by KOH activation at 850°C after carbonization at 350°C, with a surface area between 1330 and 1510 m²/g. The specific capacitance was between 170 and 200 F/g at 370 mA/g. Cyclic voltammetric curves in 30 wt% KOH aqueous solution, with a shape as in Figure 9.11, are shown at low and higher scan rate and are said to have an "approximate rectangular shape." Nevertheless, significant decrease of the specific capacitance with the current density is exhibited. Other experiments with potato starch-based activated carbon [61, 62] obtained by carbonization and activation with KOH or CO_2 of the precursor were reported. A capacitance of 335 F/g is obtained at a current density of 50 mA/g for a specific surface area of 2342 m²/g (KOH activation) and 326 F/g for a surface area of 2018 m²/g (CO_2 activation).

In [63], resorcinol and formaldehyde as precursors were used to obtain activated carbon by KOH activation at 700°C. By changing the mass ratio of KOH to resorcinol, the surface area increased from 522 to 2760 m²/g, but the average pore diameter decreased from 4.4 to 2.5 nm. A specific capacitance as high as 294 F/g for a surface area of 1543 m²/g in 30% KOH aqueous solution electrolyte was determined. At 522 m²/g, the capacitance was 186 F/g and at 2760 m²/g was 229 F/g. Cycling voltammetry and galvanostatic charge–discharge curves resemble to that one shown in Figure 9.11 (negative polarization) and Figure 9.12 (charge–discharge). Waste newspaper precursor was used in [64] to realize activated carbon by simple carbonization and KOH activation at 500°C. For a surface area of 417 m²/g (average pore diameter of 5.9 nm), a specific capacitance of 180 F/g at 2 mV/s was determined. Cycling voltammetry curve (negative polarization) and charging–discharging curve (6 M KOH electrolyte) have shapes resembling to that one shown in Figures 9.11 and 9.12 (very well for the charging–discharging curve).

Ordered mesoporous carbon with a moderate amount of nitrogen functionality by using mesoporous silica and sucrose as carbon sources was prepared in [65] by carbonization at 850°C. Surface areas in a range of 1100–1300 m²/g were obtained and capacitances in a range of 119–180 F/g. Cycling voltammetry curves for different samples in 1 M H_2SO_4 have shapes resembling with that one shown in Figure 9.9 or 9.11. Sucrose as precursor material was also used in [66] where by pyrolysis and activation with CO_2 at 900°C resulted in activated carbon with a surface area of up to 3000 m²/g and a capacitance of up to 160 F/g in 1 M H_2SO_4.

Nanoporous carbon was obtained in [67] by ball milling of high-purity flake natural graphite powder, and the surface area was increased from 7 to 580 m²/g. For a surface area of 313 m²/g, a specific capacitance of 205 F/g was obtained. For 580 m²/g, the capacitance was 150 F/g. Cycling voltammetry curves in a three-electrode system (6 M KOH electrolyte) have a shape similar to that one in Figure 9.11 (negative polarization).

In [68], an approach is proposed to evaluate the contribution of electric double layer and pseudocapacitance to the total capacitance of supercapacitors. Dependence of specific capacitance with the operating voltage window is taken into consideration to find the contribution from each energy storage mechanism. It has been found that in the case of KOH activated carbon, the contribution of pseudocapacitance (faradaic origin) can be up to 40%. In [69, 70], a microelectrode technique was used to get insight into faradaic phenomena. Cycling voltammetry curves as shown in Figure 9.11, with some peaks, are revealed by this technique. The effect of thermal treatment at 1000°C of activated carbon on specific capacitance is investigated in [71, 72]. Cycling voltammetry and charging–discharging curves with a shape as in Figures 9.11 and 9.12 are shown. After thermal treatment at 1000°C, a reduction of specific capacitance value takes place on behalf of pseudocapacitance component, but the same shape of voltammograms is exhibited. This means that further pseudo-capacitance contribution takes place on the total capacitance.

Contribution of pseudocapacitance is also possible in the case of nonaqueous electrolytes [30–32]. It was shown earlier that deviation of the cyclic voltammo-grams or charging–discharging curves from the trapezoidal shape in Figure 9.7 or from the symmetric shape in Figure 9.8 is exhibited. The shapes approach those shown in Figures 9.9 and 9.11 or 9.10 and 9.12 because of pseudocapacitance contribution.

It is not easy to separate the electric double-layer contribution and the pseudoca-pacitance contribution to the total exhibited capacitance. In Figures 9.9 and 9.11, the area given by the dashed line of approximately rectangular form (a form like in Fig. 9.7) would correspond to electric double-layer capacitance. The pseudocapacitance contribution corresponds to the area outside of the rectangular form, and variable current voltage dependence due to pseudocapacitance is obtained. Pseudocapacitance contribution is also visible in Figures 9.10 and 9.12 for the discharging part of the curve (voltage time dependence from 1 to 0V). In this case, the instantaneous capacitance ($I/(dV/dt)$) at the start of discharge is different from that at the end of charge (Fig. 9.7). I is the constant charging–discharging current, and dV/dt is the slope given by the tangent at the curve, the dashed line. Because the discharge of electric double layer is a faster process than the discharge from pseudocapacitance (slower process, chemical reactions), it is expected that at discharge start, the instan-taneous capacitance corresponds in good part to the electric double layer. As a consequence, the energy stored by the electric double layer would be given approxi-mately by the area delineated by the thicker dashed line and by dotted line, multiplied by I. To the energy stored by pseudocapacitance, the area outside the thicker dashed line, multiplied by I, approximately would correspond.

It has been said that usually, in practice, discharge does not take place at constant current and a constant load resistance is used. In this case, the discharging current is not constant. An example of such discharge is shown in Figure 9.13 for a symmetric experimental cell with commercial activated carbon electrodes and 1.5 M KOH elec-trolyte. This is a better illustration of pseudocapacitance contribution. As in the case of Figure 9.6, the cell is connected to a load resistance by means of a switch. The time voltage dependence at the cell output and the current (voltage) on the load are recorded. At the first discharge start, after the voltage drop given by the ESR of the cell, discharge begins and an instantaneous time constant, τ_{i1}, can be determined, given by the intersection of the tangent with the time axis. Also, an instantaneous capacitance C_{i1} could be derived taking into account the value of instantaneous

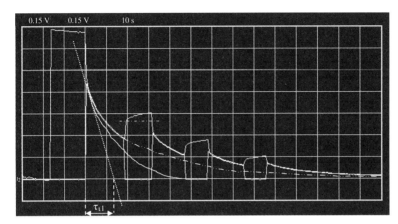

FIGURE 9.13 Discharge of an experimental supercapacitor cell with pseudocapacitance contribution (Obreja V.V.N., unpublished work).

current and the slope (dV/dt) given by the tangent at the curve. C_{il} could be also derived from the relation $\tau_{il} = (R_L + ESR) \times C_{il}$ where R_L is the load resistance and ESR is the cell's internal equivalent resistance given by the voltage drop in Figure 9.13. Evaluation of τ_{il} from Figure 9.13 enables evaluation of C_{il}. From Figure 9.13, it is seen that in the case of no interruptions, the discharge will approximately follow the white dashed line. At the start of first discharge, τ_{il} of approximate 12 s corresponds roughly to the electric double-layer capacitance to which, with approximation, an exponential time dependence corresponds (the full white line), and this ends after about three time constants. Nevertheless, from Figure 9.13, it is seen that the cell's discharge even in the case of no interruptions lasts more than 3 time constants ($\sim$40 s). During the discharge interruption, one can see that the current reaches 0 value, but in contrast with Figure 9.6b, the voltage at the cell's output does not remain at constant value after the voltage drop, and some transient increase is exhibited (in the case of Fig. 9.6b, the voltage drop is practically 0 because ESR of very low value is negligible). If no discharge would be further repeated, the voltage would increase at more than 3×0.15 V after more time. Nevertheless, at the second discharge start, the voltage is higher than at the end of the first discharge. To our best knowledge, this is the first time that a voltage increase at the end of discharge is revealed for a supercapacitor. This effect of voltage increase in very short time after discharging interruption toward the initial value before discharge is encountered in the case of batteries. This effect of voltage increase after discharge interruption could be a means to check the presence of pseudocapacitance contribution.

9.4 GRAPHENE AND CARBON NANOTUBES

An overview of graphene-based materials and their applications in supercapacitors is given in [73]. Graphene oxide is the main precursor for the preparation of different graphene derivative materials [74]. Graphene is a newer form of carbon material (first time demonstrated in 2004) than carbon nanotubes. Due to rich oxygen functional groups attached to both sides of a graphene oxide nanoplatelet, the measured thickness (around 1 nm) is higher than the expected thickness of a pure graphene single layer (about 0.35 nm). A single-walled carbon nanotube (SWCNT) can be imagined as a turned over graphene sheet into a cylindrical form and multi-walled carbon nanotube as more graphene sheets into a cylindrical form.

In [75] for functionalized exfoliated graphene oxide, a specific capacitance of 146 F/g and an energy density of 20 Wh/kg are reported. This value of energy density is too high for aqueous capacitor, and it is not clear from the paper how it was determined. The electrodes for electrochemical characterization in a three-electrode system with 1 M KOH electrolyte were manufactured from a paste composed of functionalized exfoliated graphene oxide and PTFE as binder material. The paste was deposited onto a nickel mesh serving as current collector. Cyclic voltammograms and constant current charging–discharging curves with a shape as in Figures 9.11 and 9.12 exhibit significant pseudocapacitance contribution due to functional groups.

In [76] for novel borane-reduced graphene oxide used in two- and three-electrode electrochemical measurements (aqueous electrolyte), a specific capacitance of 200 F/g for a surface area of 466 m^2/g is reported. It is shown that energy storage in such supercapacitors was contributed by ion adsorption on the surface of the nano-platelets (electric double layer) in addition to electrochemical redox reactions (pseudocapacitance). Cyclic voltammograms (6 M KOH electrolyte) and charge–discharge curves similar to that one in Figures 9.11 and 9.12 are presented.

Graphene surface functionalities have significant impact on the specific capacitance [77]. Some functional groups may contribute more than others to the specific capacitance. It is shown, for example, that the specific capacitance of graphene oxide is lower than that of reduced graphene oxide. Nitrogen-doped reduced graphene oxide has lower capacitance than reduced graphene oxide, but amine-modified reduced graphene oxide has the highest capacitance. Cyclic voltammetry curve in 1 M H$_2$SO$_4$ electrolyte for graphene oxide is similar to that one in Figure 9.9 (positive polarization), whereas for reduced graphene oxide and functionalized graphene, the voltammograms resemble to that one from Figure 9.11 (positive polarization). Charge–discharge curves with a shape approaching the shape given in Figure 9.12 are shown. Significant pseudocapacitance contribution to specific capacitance is manifested.

A developed acid-assisted method to achieve specifically functionalized graphene by tailoring the structure of graphene sheets is reported in [78]. Acid incorporating graphene oxide was treated at 900°C. A coupling between pseudocapacitance and the electrical double-layer capacitance is realized, and a specific capacitance as high as 505 F/g at 0.1 A/g in 6 M KOH electrolyte for a surface area of 391 m^2/g is achieved. Cyclic voltammograms and charging–discharging curves with a shape like in Figures 9.11 and 9.12 are shown.

Electrochemically reduced graphene oxide was obtained in [79] from graphene oxide by using an electrochemical method of repetitive cathodic potential cycling. A good part of the oxygen functional groups from graphene oxide were removed, improving the charge transfer. The material gave a specific capacitance of 223.6 F/g at a scan rate of 5 mV/s in a three-electrode system with 1 M H_2SO_4 aqueous electrolyte. No specific surface area is indicated. The voltammograms and galvanostatic charging–discharging curves have shapes approaching the shapes shown in Figures 9.9 and 9.10. The difference is that visible peaks are exhibited due to significant pseudo-capacitance contribution.

An electrochemical reduction process [80] of graphene oxide sheets resulted in a specific capacitance of 246 F/g in 1 M H_2SO_4 electrolyte. The presented cyclic voltammograms exhibit symmetric redox peaks for acid electrolyte but not for aqueous neutral electrolyte (0.2 M KCl). In this case, the voltammogram resembles well with that from Figure 9.11 (positive polarization). In the paper, it is said that the voltammogram for neutral electrolyte has a rectangular shape although this is not visible. An ideal double-layer capacitor behavior is attributed to such a voltammogram. Similar charging–discharging behavior like in [79] is manifested by the constant current charging–discharging curves.

Vertically oriented graphene uniformly grown on a metallic current collector by one-step atmospheric plasma-enhanced chemical vapor deposition (PECVD) was electrochemically characterized in a two-electrode system using aqueous 6 M KOH electrolyte and 1 M $TEABF_4$/AN electrolyte [81]. For the presented cyclic voltammograms, the scan rate is higher than 10 mV/s, reaching 1000 mV/s. Below 100 mV/s, the shape of voltammograms is not distinguishable. Only voltammograms at low scan rate can give indication about the nature of charge storage mechanism (electric double layer or pseudocapacitance). At higher scan rate, the ESR of the cell can distort the voltammogram. Nevertheless, in the paper, it is said that the voltammograms keep "quasi-rectangular shape even at a scan rate as high as 1000 mV/s, which indicates the predominant EDL capacitive behavior." Values of 140 F/g specific capacitance in aqueous electrolyte and 132 F/g in organic electrolyte at a current density of 1 A/g are reported.

For high-quality graphene scrolls, a capacitance of 162.2 F/g is reported in [82] at a current density of 1 A/g in 6 M KOH aqueous electrolyte. For graphene sheets, a value of 110 F/g was measured.

Activated microwave-exfoliated GO electrodes [83] were tested in a eutectic mixture of ionic liquids. A high specific surface area of around 2000 m^2/g was obtained by KOH activation of exfoliated GO. A specific capacitance of up to 180 F/g with an electrochemical window of up to 3.5 V and a wide temperature range from −50 to 80°C for the first time was demonstrated. The cyclic voltammograms at low scan rate (1 mV/s) and up to ±1.5 V have a shape similar to that one shown in Figure 9.7. Similar behavior is exhibited in [84] up to ±2 V for activated carbon at 1 mV/s with ionic liquid as electrolyte. Such behavior indicates that the capacitance is given by electric double layer without any pseudocapacitance contribution. Nevertheless, above +2 or −2 V, a slight distortion of voltammogram is manifested in [84], and this is also manifested for activated carbon in ionic electrolyte. It is supposed that this distortion manifested with peaks around 2 V for voltammograms explored up to +3.5 or −3.5 V could be the result of some functional groups from the carbon surface (pseudocapacitive behavior).

A simple process to prepare nitrogen-modified few-layer graphene for a supercapacitor electrode is described in [85]. Few-layer graphene is directly prepared from graphite flakes with the aid of melamine and ultrasonication in acetone. The melamine adsorbed on the surface of graphene is transformed by an annealing process into a condensation product of melamine, which dopes the graphene with nitrogen. The nitrogen-modified few-layer graphene shows a specific capacitance of 227 F/g at 1 A/g significantly higher than that of reduced graphene oxide (133 F/g). Cyclic voltammograms in 6 M NaOH at 20 mV/s and up to 0.4 V and charging–discharging curves have a shape approaching what is shown in Figures 9.11 and 9.12.

In [86], a maximum specific capacitance of 348 F/g in 1 M H_2SO_4 at 0.2 A/g current density was obtained for partially reduced graphene oxide using hydrobromic acid, which is a weak reduction agent. In ionic liquid (BMIPF$_6$), the specific capacitance was 158 F/g. The shape of cyclic voltammogram in 1 M H_2SO_4 aqueous electrolyte at 1 mV/s cannot be clearly seen at the scale where voltammograms for 50 and 100 mV/s with visible redox peaks are shown. Nevertheless, the shape of charging–discharging curves corresponds to that of Figure 9.10 or 9.12. The shape of voltammogram at 1 mV/s in ionic liquid is not clear because of the scale. As a consequence, pseudocapacitance contribution in the case of ionic liquid is manifested.

In [87], functionalized graphene was prepared by using a solvothermal method. Graphene oxide dispersed in dimethylformamide (DMF) was treated at 150°C for a fine control of the density of functionalities. Functionalities enable high pseudocapacitance, good wetting of the surface, and acceptable electric conductivity. A specific capacitance of up to 276 F/g at 0.1 A/g in 1 M H_2SO_4 electrolyte was obtained. In a schematic voltammogram at 10 mV/s in a three-electrode system, its area is shown to include a central rectangle corresponding to double-layer capacitance and the rest of the area corresponding to pseudocapacitance. Similar delimitation is shown in Figures 9.9 and 9.11.

Use of chemically modified or functionalized graphene in supercapacitor electrodes [88] resulted in specific capacitance of 135 F/g in 5.5 M KOH electrolyte and 99 F/g in TEABF$_4$/AN organic electrolyte. Cyclic voltammograms at 20 and 40 mV/s in aqueous electrolyte approach the shape given in Figure 9.11 (positive polarization). Voltammograms at 20 and 40 mV/s in organic electrolyte approach the shape shown in Figure 9.7 (positive polarization). There is slight deviation from the expected linear variation. A value of 117 F/g specific capacitance in aqueous 1 M H_2SO_4 electrolyte has been reported in [89] for graphene-based electrodes. Exfoliation of graphitic oxide at 1050°C was the material source with 925 m^2/g surface area, and the cyclic voltammogram has a shape as in Figure 9.11. Decrease of specific capacitance with the scan rate or current density supports pseudocapacitance influence for this material. For graphene material obtained from nanodiamond heating at 1650°C, the surface area was 520 m^2/g, with a specific capacitance of about 30 F/g, and the cyclic voltammogram of up to 1 V had a shape like in Figure 9.7 (dashed line). No dependence of specific capacitance of the scan rate or current density was exhibited. In ionic liquid electrolyte, the specific capacitance was 75 F/g corresponding to an energy density of 31.9 Wh/kg taking into account a working voltage of 3.5 V. Distorted cyclic voltammograms at 100 mV/s due to ESR cannot be compared with the expected shape given in Figure 9.7, 9.9, or 9.11.

A maximum specific capacitance of 205 F/g has been reported in [90] for graphene-based electrodes (320 m²/g surface area) in aqueous solution (30 wt% KOH). The cyclic voltammograms do not have a rectangular shape as said in the paper (significant contribution from pseudocapacitance is included in 205 F/g), and the specific energy value of 28.5 Wh/kg is too high for a supercapacitor cell with the indicated specific capacitance. As in the case of carbon nanotubes, functionalized graphene sheets can provide higher specific capacitance. A maximum value of 230 F/g in aqueous KOH electrolyte and in nonaqueous EC/DEC electrolyte 73 F/g has been reported in [91]. The functionalized graphene was prepared from GO by thermal exfoliation at 250, 300, 350, and 400°C. From the sample exfoliated at 300°C, other two samples were obtained after treatment at 700 and 900°C. It has been found that the surface area increases with the exfoliation temperature or temperature treatment from 328 m²/g (250°C) to 418 m²/g (400°C) and to 574 m²/g (700°C) and to 737 m²/g (900°C). For exfoliation in the temperature range of 250–400°C, the specific capacitances were in the range of 221–233 F/g. For thermal treatment at 700°C, the capacitance was 114 F/g and at 900°C was 95 F/g. Lower capacitance at higher surface area is due to a decrease in the surface functionalities with the temperature treatment and hence pseudocapacitance contribution reduction. This is also supported by the cyclic voltammograms in 2 mol/l KOH at 5 mV/s scan rate, with a shape like in Figure 9.11 (negative polarization). Significant restricted area of voltammograms for 700–900°C is exhibited in comparison with 250–400°C. Also, the discharge curves with a shape like in Figure 9.12 indicate pseudocapacitance reduction due to treatment at 700 and 900°C. Other evidence related to pseudocapacitance contribution is given by the dependence of specific capacitance with increasing scan rate. The voltammograms in ionic liquid electrolyte (1.5–4 V) indicate a shape approaching the shape shown in Figure 9.7 in the case of sample treated at 900°C (low pseudocapacitance contribution) and a shape like that in Figure 9.11 (high pseudocapacitance contribution) in a case of sample exfoliated at 300°C without treatment at 900°C. The charge–discharge curves in ionic electrolyte have a shape like in Figure 9.12 (for 300°C) and approaching the shape given in Figure 9.8 (for 900°C).

A comparison of SWCNT electrochemical capacitor electrode fabrication methods is performed in [92]. Four solution-based electrode fabrication methods— for example, drop casting, airbrushing, filtration, and electrospraying—were used. Specific capacitances in a range of 25–45 F/g have been obtained. No specific surface area is given. A cyclic voltammogram for 25 mV/s in an aqueous electrolyte and three-electrode system has a shape approaching the shape shown in Figure 9.11 (negative polarization).

In [93], functionalization of SWCNT by treatment with nitric acid resulted in an important increase of the specific capacitance. Carboxylic groups were found for major contribution to capacitance increase. By varying the treatment time in nitric acid, an increase of specific capacitance from 32 F/g (untreated sample) to 142 F/g at 4 A/g current density took place. Electrode films with a thickness of about 10 μm and assembled capacitors were realized. It is interested that the BET surface area decreased from 326 m²/g to 11–20 m²/g depending on the acid treatment time. It is suggested that the oxygenated functional groups and nitrate radicals left in SWCNT samples after acid treatment may prevent nitrogen access for most surfaces of SWCNTs.

In published work until 2006–2007 [94], for carbon nanotube-based electrodes, the reported specific capacitance is limited to 150–180 F/g.

Carbon nanotube film electrodes have exhibited 108 F/g in 1 M H_2SO_4 electrolyte in a three-electrode system [95]. The film electrodes were realized by electrostatic spray deposition (ESD) of a carbon nanotube solution onto a metallic substrate. In [96], electrodes have been fabricated by direct grow of aligned carbon nanotubes on conductive substrates. The specific capacitance of the carbon nanotube electrodes increased with decreasing nanotube lengths from 10.75 to 21.57 F/g. The maximum energy and power density have ranged from 2.3 to 5.4 Wh/kg and 19.6 to 35.4 kW/kg, respectively.

A supercapacitor assembled with SWCNT film electrodes and an organic electrolyte was investigated in [97]. Values of specific capacitance of 25–35 F/g were found. Surface functionalization of SWCNTs by chemical and electrochemical oxidation [98] provides significantly higher specific capacitance of 209 F/g in 1 M H_2SO_4 electrolyte at a current density of 50 mA/g due to pseudocapacitance contribution. A specific capacitance of 202 F/g has been obtained for 1 M Et_3MeNBF_4 in polycarbonate electrolyte at 50 mA/g. In [99], cyclic voltammograms for onion-like carbon electrodes in supercapacitor cells with ionic liquid electrolyte with approximate shape like in Figure 9.7 are shown up to 3–3.5 V. However, for carbon nanotube electrodes in ionic liquid electrolyte, cyclic voltammograms with approximate shape like in Figure 9.9 are shown. Cyclic voltammograms for carbon nanotube electrodes in a three-electrode system in ionic liquid electrolyte, displayed up to 1.5 V [100], have similar shape like in Figure 9.7. Similar shape (up to 1.5 V) like in Figure 9.7 for voltammograms of aligned carbon nanotube-based electrodes in a three-electrode system with ionic electrolyte is shown in [101]. A value of 440 F/g specific capacitance was reported.

9.5   NANOSTRUCTURED CARBON COMPOSITES

In [102], a mixture of porous carbon at the nanoscale level and graphene was prepared by simple carbonization of a GO/PVDF film. No activation process was performed. The specific surface area was 737 m^2/g. The presence of graphene in the mixture enables a specific capacitance of 311 F/g in comparison with 249 F/g for only nanoporous carbon obtained from PVDF (686 m^2/g specific surface area). Cyclic voltammograms and charging–discharging curves in aqueous electrolyte have approximate shapes with those from Figures 9.9 and 9.10 supporting pseudocapacitance contribution.

High surface area carbon nanoparticle/graphene composite was prepared in [103]. The carbon nanoparticles between the graphene sheets prevent their restacking during drying, and a specific surface area of 1256 m^2/g was obtained. This composite used in a supercapacitor achieved a specific capacitance of 326.4 F/g at 0.3 A/g current density. Similar results are reported in [104] where 1% functionalized carbon black was used for the spacer nanoparticles between the graphene sheets.

A composite consisting of hollow carbon spheres and carbon nanotubes was synthesized by a chemical vapor deposition process and used for supercapacitor electrode [105]. A maximum specific capacitance of 201.5 F/g at 0.5 A/g in 6 mol/l KOH electrolyte was reported. Symmetric supercapacitors with these materials exhibited a maximum energy density of 11.3 Wh/kg and a power density of 11.8 kW/kg in 1 mol/l Na_2SO_4 electrolyte solution.

A mesoporous carbon/graphene composite [106] exhibited a specific capacitance of 242 F/g in 6 M KOH electrolyte at a current density of 0.5 A/g. Good cycling stability was displayed by this composite, and the final capacitance was higher than the initial capacitance after 2000 cycles.

Binder-free activated carbon/carbon nanotube paper electrodes for use in supercapacitors were fabricated in [107]. The specific capacitance reached a maximum value of 267.6 F/g for a loading of 5 wt% of carbon nanotubes. The energy density was 22.5 Wh/kg and the power density was 7.3 kW/kg at high current density of 20 A/g. The composite electrode showed good performance after 5000 cycles at a scan rate of 200 mV/s with 97.5% capacitance retention.

Electrodes from graphene oxide/carbon nanotube sandwich papers with a content of 12.5 wt% of carbon nanotubes exhibited specific capacitance of 151 F/g at 0.5 A/g [108]. This value was obtained after a moderate reduction of graphene oxide by using hydrazine or annealing at 220°C in air.

In [109] by means of a solution-casting method, graphene oxide/carbon nanotube films with different carbon nanotube contents were prepared. The films were annealed at 200°C so that the oxygen-containing groups can be removed. Due to good mechanical and electrical properties of these hybrid films, these could be used as supercapacitor electrodes without a need of binders or current collectors. The specific capacitance was in a range of 70–110 F/g at 1 mV/s scan rate.

A nanostructured composite of exfoliated graphite nanosheets and carbon nanotubes was prepared by a chemical reduction method [110]. By electrochemical characterization, a specific capacitance of 266 F/g at a current density of 0.1 A/g was obtained. For exfoliated graphite nanosheets, the specific capacitance was 185 F/g. At higher current density, the specific capacitance was 220 F/g for the composite. Nanocomposites using activated carbon, carbon nanotubes, and ionic liquid for supercapacitor electrodes were prepared and characterized in [111]. A specific capacitance of 188 F/g was obtained, and a specific energy of 50 Wh/kg could be reached in prototype supercapacitors, taking into account a voltage window of 4 V of the ionic electrolyte.

9.6 OTHER COMPOSITES WITH CARBON NANOMATERIALS

Use of ruthenium dioxide in composite electrodes for supercapacitors has been investigated [112]. A composite including $RuO_2 \cdot nH_2O$, aligned multiwalled carbon nanotubes, and titanium gave a specific capacitance of up to 1652 F/g in 1.0 M H_2SO_4 electrolyte solution. The carbon nanotubes were coated with ruthenium dioxide. A ruthenium dioxide/graphene sheet composite was prepared in [113] by hydrothermal and low-temperature annealing process. The Ru content in the composite was varied to study influence on the electrochemical properties. A composite with a Ru 40 wt% showed a specific capacitance of 551 F/g at a current density of 1 A/g in 1 M H_2SO_4 electrolyte solution. The specific capacitance of the composite is significantly higher than that of graphene sheets. In [114], an activated carbon/ruthenium dioxide composite was prepared and characterized. It was found that the specific capacitance increases with the content of RuO_2 in the composite. For pure ruthenium dioxide, the specific capacitance was 705.6 F/g, whereas for activated carbon, a value of 119.3 F/g was obtained. A ruthenium oxide/graphene sheet composite was prepared in [115]. The fine RuO_2 particles with a size of 5–20 nm separate the graphene sheets. The composite-based supercapacitors exhibited a specific capacitance of 570 F/g for 38.3 wt% Ru content and a current density of 0.1 A/g. The specific energy is 20.1 Wh/kg and 97.9% capacitance retention after 1000 cycles was determined.

It has been said that ruthenium oxide is an expensive material, and for this reason, other cost-effective metal oxides, for example, manganese oxide and ferrous oxide, have been investigated for supercapacitor electrodes. The theoretical specific capacitance of MnO_2 is 1370 F/g [116]. Nevertheless, manganese oxide and other metal oxides have poor conductivity, and lower performance is achieved for supercapacitors with metal oxide electrodes. For this reason, composites including metal oxides have been investigated for improved performance.

A ternary composite of carbon nanotube/MnO_2/graphene oxide was synthesized in [117]. The crystals of manganese oxide with a needlelike shape of 100 nm thickness were coated on the carbon nanotube surface, and then on the MnO_2 surface, graphene oxide was coated. A specific capacitance of 473 F/g at 5 mV/s and a good stability after 500 charging–discharging cycles were obtained for the composite.

In [118], the specific capacitance of manganese dioxide deposited on carbon paper substrate was 345 F/g but increases to 749 F/g if the carbon paper substrate undergoes an optimum thermal treatment.

Conducting polymer/reduced graphene oxide composites were investigated in [119]. Polyaniline, polypyrrole, and poly(3,4-ethylenedioxythiophene) (PEDOT) were deposited on the surface of reduced graphene sheets with different concentrations of these conducting polymers. A specific capacitance of 361 F/g was exhibited by the polyaniline/reduced graphene composite, whereas the polypyrrole/reduced graphene composite exhibited 248 F/g, and for the PEDOT/reduced graphene oxide, the specific capacitance was only 108 F/g. These values were determined for the same current density of 0.3 A/g. After 1000 charging–discharging cycles, the specific capacitance was higher than 80% of its initial value. On the other hand, in [120], a specific capacitance of 526 F/g at 0.2 A/g for polyaniline nanofiber/graphene composite was obtained. The composite was a homogeneous dispersion of individual graphene sheets in the polymer matrix.

For a polypyrrole/reduced graphene oxide composite, a 424 F/g specific capacitance is reported in [121] at 1 A/g current density in 1 M H_2SO_4 electrolyte solution. This value is higher than that of a pristine polypyrrole film.

Multiwalled carbon nanotube/polyaniline composite prepared in [122] exhibited specific capacitance of 560 F/g at 66 wt% polyaniline content. After 700 cycles, a capacitance decay to 70% of the initial value took place.

For nanoporous carbon/polyaniline composite [123], a specific capacitance of 587 F/g was exhibited. The polyaniline film was uniformly deposited on the surface of nanoporous carbon. The cycling stability of composite is better than that of polyaniline. For the polyaniline, a capacitance decay from 513 to 334 F/g after 50 cycles took place.

A nanocomposite containing graphene oxide nanosheets and a novel conjugated polymer, poly(3,4-propylenedioxythiophene), was prepared in [124]. A specific capacitance of composite at 10 mV/s scan rate of 158 F/g was determined. The cyclic voltammograms at low scan rate and the charging–discharging curves at 5 A/g have a shape resembling to that one in Figures 9.9 and 9.10, respectively.

In a review for hybrid nanostructured materials [125], the role of binary and ternary carbon-based/conducting polymers/metal oxides for supercapacitor performance improvement was analyzed.

9.7 CONCLUSIONS

Different nanostructured carbon-based materials and related composites are used for the manufacture of electrodes used in supercapacitor cells with aqueous or non-aqueous electrolytes. Nanoporous carbon powder or activated carbon, which is commercially available, has been extensively used in commercial and experimental supercapacitor cell. Ordered microporous (pore size under 2 nm) and mesoporous (pore size higher than 2 nm) carbons, porous carbon at the nanoscale level, and carbon nanofibers and fibers have been investigated for supercapacitor performance improvement. Organic-based matter as precursor was used for processing at high temperature to obtain such carbon nanomaterials. New forms of nanostructured carbon like carbon nanotubes and graphenes have been used to manufacture electrodes used in experiments with supercapacitor cells aiming at performance improvement. Different mixtures of the aforementioned carbon nanomaterials have been used in composite electrodes to investigate their performance for supercapacitors. Binary mixtures of carbon nanomaterials and metal oxides or conducting polymers and even ternary mixtures have been investigated. A specific capacitance with related specific energy and specific power (related to ESR) is seldom mentioned as material performance result of experimental supercapacitor investigations. Cycling voltammetry, galvanostatic charging–discharging, and electrochemical impedance spectrometry are usually used for supercapacitor characterization. For nanoporous carbon-based electrodes, a specific capacitance of up to 150–200 F/g is mentioned for organic electrolyte (maximum working voltage of 2.5–3.0 V) or ionic electrolyte (maximum working voltage of 3.5–4 V), whereas for aqueous electrolyte (maximum working voltage of 1–1.2 V), the value of specific capacitance is limited to 450–500 F/g. By taking into account the value of the specific capacitance, specific energy values in a range of 10–50 Wh/kg (depending on the electrolyte type and electrochemical cell) have been reported. The specific capacitance values are given only for one electrode and for the carbon material used in the electrode. For a supercapacitor cell, where two electrodes and also other materials for cell assembling and packaging are used, the aforementioned values have to be divided by a factor higher than two. As a consequence, for nanoporous carbon that is an affordable and cost-effective material, specific energy higher than 10 Wh/kg for a practical supercapacitor symmetrical cell hardly could be reached in the present state of the art.

Carbon nanotubes, graphene material derivatives, and different related composites have been used in many experiments reported last time. Nevertheless, in spite of the outstanding properties of these materials, significant increase of the specific capacitance or of the specific energy in comparison with nanoporous carbon is not achieved up to the present. These materials are significantly expensive than nanoporous carbon, and commercial cost-effective supercapacitors have not been realized yet.

Pure capacitive behavior given by the electric double layer is encountered in the case of supercapacitors with nonaqueous electrolytes, especially with organic electrolytes. For supercapacitors with aqueous electrolytes, significant contribution to the total capacitance is given by the pseudocapacitance component related to the functional groups from the carbon surface inducing redox or faradaic reactions, a behavior that is similar to battery. Due to this component, severe distortion of the cyclic voltammograms from a shape approaching a rectangular form takes place. The presence of the pseudocapacitive component causes lack of proportionality between the specific capacitance and the specific surface area. Absence of proportionality is also exhibited, even when the pseudocapacitance influence is low for specific surface area higher than $1000\,m^2/g$ toward $2500–3000\,m^2/g$. The electrode preparation method has a significant influence. A sample of carbon nanomaterial with significantly lower specific area than other sample, when used in an electrode, can have specific capacitance significantly higher than the sample with higher surface area. Sometimes, it is difficult to correlate the results from one paper to another one, because different conditions for measurements were used. An experimental cycling voltammetry curve and galvanostatic charge–discharge curve for a cell with dominant electric double-layer charge storage differ from the corresponding curve, measured in the same conditions for a cell where significant pseudocapacitance contribution takes place. Discharging supercapacitor voltage time dependence at constant current or variable current on a load resistance is a means to reveal the presence of double-layer and pseudocapacitance contribution. In some papers, significant pseudocapacitance contribution is attributed to electric double-layer contribution. Better understanding of the resulting experimental results can lead to better carbon nanomaterial utilization for higher-performance supercapacitors.

REFERENCES

[1] Kenmochi, K. Aluminum electrolytic capacitor and a manufacturing method therefor. U.S. Patent No. 4,663,824, 1987.

[2] Becker, H.J. Low voltage electrolytic capacitor. U.S. Patent No. 2,800,616, 1957.

[3] Yoshida, A.; Nonaka, S.; Aoki, I.; Nishino, A. Electric double-layer capacitors with sheet-type polarizable electrodes and application of the capacitors. *Journal of Power Sources* 1996, **60**, 213–218.

[4] Dai, L.; Chang, D.W.; Baek, J.B.; Lu, W. Carbon nanomaterials for advanced energy conversion and storage. *Small* 2012, **8**, 1130–1166.

[5] Bose, S.; Kuila, T.; Mishra, A.K.; Rajasekar, R.; Kim, N.H.; Lee, J.H. Carbon-based nanostructured materials and their composites as supercapacitor electrodes. *Journal of Materials Chemistry* 2012, **22**, 767–784.

[6] Ghosh, A.; Lee, Y.H. Carbon-based electrochemical capacitors. *ChemSusChem* 2012, 480–499.

[7] Wang, G.; Zhang, L.; Zhang, J. A review of electrode materials for electrochemical supercapacitors. *Chemical Society Reviews* 2012, **41**, 797–828.

[8] Jiang, H.; Lee, P.S.; Li, C. 3D carbon based nanostructures for advanced supercapacitors. *Energy & Environmental Science* 2013, **7**, 41–53.

[9] Zhi, M.; Xiang, C.; Li, J.; Li, M.; Wu, N. Nanostructured carbon–metal oxide composite electrodes for supercapacitors: a review. *Nanoscale* 2013, **5**, 72–88.

[10] Frackowiak, E.; Abbas, Q.; Beguin, F. Carbon/carbon supercapacitors. *Journal of Energy Chemistry* 2013, **22**, 226–240.

[11] Beguin, F.; Raymundo-Pinero, E. Nanocarbons for supercapacitors, in *Batteries for Sustainability* (Brodd R.J., ed., Springer, New York) 2013, 393–421.

[12] Maxwell Technologies. Products & Solutions, Ultracapacitors, 2013, http://www.maxwell.com (accessed March 3, 2015).

[13] Obreja, V.V.N.; Dinescu, A.; Obreja, A.C. Activated carbon based electrodes in commercial supercapacitors and their performance. *International Review of Electrical Engineering-IREE* 2010, **5**, 272–282.

[14] Ruch, P.W.; Cericola, D.; Foelske-Schmitz, A.; Kotz, R.; Wokaun, A. Aging of electrochemical double layer capacitors with acetonitrile-based electrolyte at elevated voltages. *Electrochimica Acta* 2010, **55**, 4412–4420.

[15] Janes, A.; Kurig, H.; Lust, E. Characterisation of activated nanoporous carbon for supercapacitor electrode materials. *Carbon* 2007 ,**45**, 1226–1233.

[16] Thomberg, T.; Janes, A.; Lust, E. Energy and power performance of electrochemical double-layer capacitors based on molybdenum carbide derived carbon. *Electrochimica Acta* 2010, **55**, 3138–3143.

[17] Wang, H.; et al. Supercapacitors based on carbons with tuned porosity derived from paper pulp mill sludge biowaste. *Carbon* 2013, **57**, 317–328.

[18] Zhai, D.; Li, B.; Kang, F.; Du, H.; Xu, C. Preparation of mesophase-pitch-based activated carbons for electric double layer capacitors with high energy density. *Microporous and Mesoporous Materials* 2010, **130**, 224–228.

[19] Lei, C.; Amini, N.; Markoulidis, F.; Wilson, P.; Tennison, S.R.; Lekakou, C. Activated carbon from phenolic resin with controlled mesoporosity for an electric double-layer capacitor (EDLC). *Journal of Materials Chemistry A* 2013, **1**, 6037–6042.

[20] Zhou, D.D.; Du, Y.J.; Song, Y.F.; Wang, Y.G.; Wang, C.X.; Xia, Y.Y. Ordered hierarchical mesoporous/microporous carbon with optimized pore structure for supercapacitors. *Journal of Materials Chemistry A* 2013, **1**, 1192–1200.

[21] Krause, A.; Kossyrev, P.; Oljaca, M.; Passerini, S.; Winter, M.; Balducci, A. Electrochemical double layer capacitor and lithium-ion capacitor based on carbon black. *Journal of Power Sources* 2011, **196**, 8836–8842.

[22] Chen, H.; Wang, F.; Tong, S.; Guo, S.; Pan, X. Porous carbon with tailored pore size for electric double layer capacitors application. *Applied Surface Science* 2012, **258**, 6097–6102.

[23] Rufford, T.E.; Hulicova-Jurcakova, D.; Fiset, E.; Zhu, Z.; Lu, G.Q. Double-layer capacitance of waste coffee ground activated carbons in an organic electrolyte. *Electrochemistry Communications* 2009, **11**, 974–977.

[24] Chen, M.; Kang, X.; Wumaier, T.; Dou, J.; Gao, B.; Han, Y.; Xu, G.; Liu, Z.;Zhang, L. Preparation of activated carbon from cotton stalk and its application in supercapacitor. *Journal of Solid State Electrochemistry* 2013, **17**, 1005–1012.

[25] Xu, B.; Wu, F.; Mu, D.; Dai, L.; Cao, G., Zhang, H., Chen, S. Activated carbon prepared from PVDC by NaOH activation as electrode materials for high performance EDLCs with non-aqueous electrolyte. *International Journal of Hydrogen Energy* 2010, **35**, 632–637.

[26] Olivares-Marin, M.; Fernandez, J.A.; Lazaro, M.J.; Fernandez-Gonzalez, C.; Macias-Garcia, A.; Gomez, V. Cherry stones as precursor of activated carbons for supercapacitors. *Materials Chemistry and Physics* 2009, **114**, 223–227.

[27] Vix-Guterl, C.; Saadallah, S.; Jurewicz, K.; Frackowiak, E.; Reda, M.;Parmentier, J.; Patarin, J.; Beguin, F. Supercapacitor electrodes from new ordered porous carbon materials obtained by a templating procedure. *Materials Science and Engineering* 2004, **B108**, 148–155.

[28] Hu, J.; Wang, H.; Gao, Q.; Guo, H. Porous carbons prepared by using metal–organic framework as the precursor for supercapacitors. *Carbon* 2010, **48**, 3599–3606.

[29] Domingo-García, M.; Fernández, J.A.; Almazán-Almazán, M.C.; López-Garzón, F.J.; Stoeckli, F.; Centeno, T.A. Poly (ethylene terephthalate)-based carbons as electrode material in supercapacitors. *Journal of Power Sources* 2010, **195**, 3810–3813.

[30] Xu, B.; Wu, F.; Chen, R.; Cao, G.; Chen, S.; Yang, Y. Mesoporous activated carbon fiber as electrode material for high-performance electrochemical double layer capacitors with ionic liquid electrolyte. *Journal of Power Sources* 2010, **195**, 2118–2124.

[31] Balducci, A.; Dugas, R.; Taberna, P.L.; Simon, P.; Plee, D.;Mastragostino, M.; Passerini, S. High temperature carbon–carbon supercapacitor using ionic liquid as electrolyte. *Journal of Power Sources* 2007, **165**, 922–927.

[32] Michael, M.S.; Prabaharan, S.R.S. High voltage electrochemical double layer capacitors using conductive carbons as additives. *Journal of Power Sources* 2004, **136**, 250–256.

[33] Rufford, T.E.; Hulicova-Jurcakova, D.; Zhu, Z.; Lu, G.Q. Nanoporous carbon electrode from waste coffee beans for high performance supercapacitors. *Electrochemistry Communications* 2008, **10**, 1594–1597.

[34] Rufford, T.E.; Hulicova-Jurcakova D.; Khosla, K.; Zhu, Z.; Lu, G.Q. Microstructure and electrochemical double-layer capacitance of carbon electrodes prepared by zinc chloride activation of sugar cane bagasse. *Journal of Power Sources* 2010, **193**, 912–918.

[35] Tran, C.; Kalra, V. Fabrication of porous carbon nanofibers with adjustable pore sizes as electrodes for supercapacitors. *Journal of Power Sources* 2013, **235**, 289–296.

[36] Farma, R.; et al. Preparation of highly porous binderless activated carbon electrodes from fibres of oil palm empty fruit bunches for application in supercapacitors. *Bioresource Technology* 2013, **132**, 254–261.

[37] Demarconnay, L.; Raymundo-Piñero, E.; Béguin, F. A symmetric carbon/carbon supercapacitor operating at 1.6 V by using a neutral aqueous solution. *Electrochemistry Communications* 2010, **12**, 1275–1278.

[38] Seredych, M.; Hulicova-Jurcakova, D.; Lu, G.Q.; Bandosz, T.J. Surface functional groups of carbons and the effects of their chemical character, density and accessibility to ions on electrochemical performance. *Carbon* 2008, **46**, 1475–1488.

[39] Senthilkumar, S.T.; Senthilkumar, B.; Balaji, S.; Sanjeeviraja, C.; Kalai Selvan, R. Preparation of activated carbon from sorghum pith and its structural and electrochemical properties. *Materials Research Bulletin* 2011, **46**, 413–419.

[40] Weng, T.C.; Teng, H. Characterization of high porosity carbon electrodes derived from mesophase pitch for electric double-layer capacitors. *Journal of the Electrochemical Society* 2001, **148**, A368–A373.

[41] Wang, X.; Zheng, J.P. The optimal energy density of electrochemical capacitors using two different electrodes. *Journal of the Electrochemical Society* 2004, **151**, A 1683–A 1689.

[42] Pognon, G.; Brousse, T.; Demarconnay, L.; Belanger, D. Performance and stability of electrochemical capacitor based on anthraquinone modified activated carbon. *Journal of Power Sources* 2011, **196**, 4117–4122.

[43] Lufrano, F.; Staiti, P. Performance improvement of Nafion based solid state electrochemical supercapacitor. *Electrochimica Acta* 2004, **49**, 2683–2689.

[44] Toupin, M.; Belanger, D.; Hill, I.R.; Quinn, D. Performance of experimental carbon blacks in aqueous supercapacitors. *Journal of Power Sources* 2005, **140**, 203–210.

[45] Liu, G.; Kang, F.; Li, B.; Huang, Z.; Chuan, X. Characterization of the porous carbon prepared by using halloysite as template and its application to EDLC. *Journal of Physics and Chemistry of Solids* 2006, **67**, 1186–1189.

[46] Lewandowski, A.; Swiderska, A. Electrochemical capacitors with polymer electrolytes based on ionic liquids. *Solid State Ionics* 2003, **161**, 243–249.

[47] Xu, B.D.; Zheng, D.; Jia, M.; Cao, G.; Yang, Y. Nitrogen-doped porous carbon simply prepared by pyrolyzing a nitrogen-containing organic salt for supercapacitors. *Electrochimica Acta* 2013, **98**, 176–182.

[48] Yu, H.R.; Cho, S.; Jung, M.J.; Lee, Y.S. Electrochemical and structural characteristics of activated carbon-based electrodes modified via phosphoric acid. *Microporous and Mesoporous Materials* 2013, **172**, 131–135.

[49] Jin, X.J.; Zhang, M.Y.; Wu, Y.; Zhang, J.; Mu, J. Nitrogen-enriched waste medium density fiberboard-based activated carbons as materials for supercapacitors. *Industrial Crops and Products* 2013, **43**, 617–622.

[50] Ma, C.; Song, Y.; Shi, J.; Zhang, D.; Zhong, M.; Guo, Q.; Liu, L. Phenolic-based carbon nanofiber webs prepared by electrospinning for supercapacitors. *Materials Letters* 2012, **76**, 211–214.

[51] Xiong, W.; et al. A novel synthesis of mesoporous carbon microspheres for supercapacitor electrodes. *Journal of Power Sources* 2011, **196**, 10461–10464.

[52] Li, Q.; et al. Synthesis of mesoporous carbon spheres with a hierarchical pore structure for the electrochemical double-layer capacitor. *Carbon* 2011, **49**, 1248–1257.

[53] Shen, H.; et al. A novel activated carbon for supercapacitors. *Materials Research Bulletin* 2012, **47**, 662–666.

[54] Li, L.; et al. A doped activated carbon prepared from polyaniline for high performance supercapacitors. *Journal of Power Sources* 2010, **195**, 1516–1521.

[55] Li, L.; Enhui Liu, E.; Shen, H.; Yang, Y.; Huang, Z.; Xiaoxia Xiang, X.; Tian, Y. Charge storage performance of doped carbons prepared from polyaniline for supercapacitors. *Journal of Solid State Electrochemistry* 2011, **15**, 175–182.

[56]　Yan, J.; et al. A high-performance carbon derived from polyaniline for supercapacitors. *Electrochemistry Communications* 2010, **12**, 1279–1282.

[57]　Liu, M.C.; Kong, L.B.; Zhang, P.; Luo, Y.C.; Kang, L. Porous wood carbon monolith for high-performance supercapacitors. *Electrochimica Acta* 2012, **60**, 443–448.

[58]　Wu, X.L.; Wen, T.; Guo, H.L.; Yang, S.; Wang, X.; Xu, A.W. Biomass-derived sponge-like carbonaceous hydrogels and aerogels for supercapacitors. *ACS Nano* 2013, **7**, 3589–3597.

[59]　Wang, H.; Zhong, Y.; Li, Q.; Yang, J.; Dai, Q. Cationic starch as a precursor to prepare porous activated carbon for application in supercapacitor electrodes. *Journal of Physics and Chemistry of Solids* 2008, **69**, 2420–2425.

[60]　Li, Q.Y.; Wang, H.Q.; Dai, F.D.; Yang, J.H.; Zhong, Y.L. Novel activated carbons as electrode materials for electrochemical capacitors from a series of starch. *Solid State Ionics* 2008, **179**, 269–273.

[61]　Zhao, S.; Wang, C.Y.; Chen, M.M.; Zhi, J.W.; Shi, Z.Q. Potato starch-based activated carbon spheres as electrode material for electrochemical capacitor. *Journal of Physics and Chemistry of Solids* 2009, **70**, 1256–1260.

[62]　Zhao, S.; Xiang, J.; Wang, C.Y.; Chen, M.M. Characterization and electrochemical performance of activated carbon spheres prepared from potato starch by CO_2 activation. *Journal of Porous Materials* 2013, **20**, 15–20.

[63]　Zhu, Y.; Hu, H.; Li, W.; Zhang, X. Resorcinol-formaldehyde based porous carbon as an electrode material for supercapacitors. *Carbon* 2007, **45**, 160–165.

[64]　Kalpana, D.; Cho, S.H.; Lee, S.B.; Lee, Y.S.; Misra, R.; Renganathan, N.G. Recycled waste paper: a new source of raw material for electric double-layer capacitors. *Journal of Power Sources* 2009, **190**, 587–591.

[65]　Kim, N.D.; Kim, W.; Joo, J.B.; Oh, S.; Kim, P.; Kim, Y.; Yi, J. Electrochemical capacitor performance of N-doped mesoporous carbons prepared by ammoxidation. *Journal of Power Sources* 2008, **180**, 671–675.

[66]　Wei, L.; Yushin, G. Electrical double layer capacitors with activated sucrose-derived carbon electrodes. *Carbon* 2011, **49**, 4830–4838.

[67]　Li, H.Q.; Wang, Y.G.; Wang, C.X.; Xia, Y.Y. A competitive candidate material for aqueous supercapacitors: high surface-area graphite. *Journal of Power Sources* 2008, **185**, 1557–1562.

[68]　Ruiz, V.; Roldan, S.; Villar, I.; Blanco, C.; Santamaria, R. Voltage dependence of carbon-based supercapacitors for pseudocapacitance quantification. *Electrochimica Acta* 2013, **95**, 225–229.

[69]　Ruiz, V.; Malmberg, H.; Blanco, C.; Lundblad, A.; Björnbom, P.; Santamaría, R. A study of Faradaic phenomena in activated carbon by means of macroelectrodes and single particle electrodes. *Journal of Electroanalytical Chemistry* 2008, **618**, 33–38.

[70]　Malmberg, H.; Ruiz, V.; Blanco, C.; Santamaria, R.; Lundblada, A.; Bjomborn, P. An insight into Faradaic phenomena in activated carbon investigated by means of the microelectrode technique. *Electrochemistry Communications* 2007, **9**, 2320–2324.

[71]　Ruiz, V.; Blanco, C.; Raymundo-Pi˜nero, C.; Khomenko, V.; Beguin, F.; Santamarıa, R. Effects of thermal treatment of activated carbon on the electrochemical behaviour in supercapacitors. *Electrochimica Acta* 2007, **52**, 4969–4973.

[72]　Ruiz, V.; Blanco, C.; Granda, M.; Santamaria, R. Enhanced life-cycle supercapacitors by thermal treatment of mesophase-derived activated carbons. *Electrochimica Acta* 2008, **54**, 305–310.

[73] Huang, Y.; Liang, J.; Chen, Y. An overview of the applications of graphene-based materials in supercapacitors. *Small* 2012, **8**, 1805–1834.

[74] Obreja, A.C.; Cristea, D.; Gavrila, R.; Schiopu, V.; Dinescu, A.; Danila, M.; Comanescu, F. Isocyanate functionalized graphene/P3HT based nanocomposites. *Applied Surface Science* 2013, **276**, 458–467.

[75] Karthika, P.; Rajalakshmi, N.; Dhathathreyan, K.S. Functionalized exfoliated graphene oxide as supercapacitor electrodes. *Soft Nanoscience Letters* 2012, **2**, 59–66.

[76] Han, J.; et al. Generation of B-doped graphene nanoplatelets using a solution process and their supercapacitor applications. *ACS Nano* 2013, **7**, 19–26.

[77] Lai, L.; et al. Preparation of supercapacitor electrodes through selection of graphene surface functionalities. *ACS Nano* 2012, **6**, 5941–5951.

[78] Fang, Y.; et al. Renewing functionalized graphene as electrodes for high-performance supercapacitors. *Advanced Materials* 2012, **24**, 6348–6355

[79] Yang, J.; Gunasekaran, S. Electrochemically reduced graphene oxide sheets for use in high performance supercapacitors. *Carbon* 2013, **51**, 36–44.

[80] Yu, H.; He, J.; Sun, L.; Tanaka, S.; Fugetsu, B. Influence of the electrochemical reduction process on the performance of graphene-based capacitors. *Carbon* 2013, **51**, 94–101.

[81] Bo, Z.; Wen, Z.; Kim, H.; Lu, G.; Yu, K.; Chen, J. One-step fabrication and capacitive behavior of electrochemical double layer capacitor electrodes using vertically-oriented graphene directly grown on metal. *Carbon* 2012, **50**, 4379–4387.

[82] Zeng, F.; Kuang, Y.; Liu, G.; Liu, R.; Huang, Z.; Fu, C.; Zhou, H. Supercapacitors based on high-quality graphene scrolls. *Nanoscale* 2012, **4**, 3997–4001.

[83] Tsai, W.Y.; et al. Outstanding performance of activated graphene based supercapacitors in ionic liquid electrolyte from −50 to 80°C. *Nano Energy* 2013, **2**, 403–411.

[84] Wei, L.; Yushin, G. Electrical double layer capacitors with sucrose derived carbon electrodes in ionic liquid electrolytes. *Journal of Power Sources* 2011, **196**, 4072–4079.

[85] Xiao, N.; Lau, D.; Shi, W.; Zhu, J.; Dong, X.; Hng, H.; Yan, Q. A simple process to prepare nitrogen-modified few-layer graphene for a supercapacitor electrode. *Carbon* 2013, **57**, 184–190.

[86] Chen, Y.; Zhang, X.; Zhang, D.; Yu, P.; Ma, Y. High performance supercapacitors based on reduced graphene oxide in aqueous and ionic liquid electrolytes. *Carbon* 2011, **49**, 573–580.

[87] Lin, Z.; et al. Superior capacitance of functionalized graphene. *The Journal of Physical Chemistry C* 2011, **115**, 7120–7125

[88] Stoller, M.D.; Park, S.; Zhu, Y.; An, J.; Ruoff, R.S. Graphene-based ultracapacitors. *Nano Letters* 2008, **8**, 3498–3502.

[89] Vivekchand, S.R.C.; Rout, C.S.; Subrahmanyam, K.S.; Govindaraj, A.; Rao, C.N.R. Graphene-based electrochemical supercapacitors. *Journal of Chemical Sciences* 2008, **120**, 9–13.

[90] Wang, Y.; Shi, Z.; Huang, Y.; Ma, Y.; Wang, C.; Chen, M.; Chen, Y. Supercapacitor devices based on graphene materials. *The Journal of Physical Chemistry C* 2009, **113**, 13103–13107.

[91] Du, Q.; Zheng, M.; Zhang, L.; Wang, Y.; Chen, J.; Xue, L.; Dai, W. Preparation of functionalized graphene sheets by a low-temperature thermal exfoliation approach and their electrochemical supercapacitive behaviors. *Electrochimica Acta* 2010, **55**, 3897–3903.

[92] Ervin, M.H.; Miller, B.S.; Hanrahan, B.; Mailly, B.; Palacios, T. A comparison of single-wall carbon nanotube electrochemical capacitor electrode fabrication methods. *Electrochimica Acta* 2012, **65**, 37–43.

[93] Shen, J.; et al. How carboxylic groups improve the performance of single-walled carbon nanotube electrochemical capacitors? *Energy & Environmental Science* 2011, **4**, 4220–4229.

[94] Obreja, V.V.N. On the performance of supercapacitors with electrodes based on carbon nanotubes and carbon activated material: a review. *Physica E* 2008, **40**, 2596–2605.

[95] Kim, J.H.; Nam, K.W.; Ma, S.B.; Kim, K.B. Fabrication and electrochemical properties of carbon nanotube film electrodes. *Carbon* 2006, **44**, 1963–1968.

[96] Shah, R.; Zhang, X., Talapatra, S. Electrochemical double layer capacitor electrodes using aligned carbon nanotubes grown directly on metals. *Nanotechnology* 2009, **20**, 395202.

[97] Masarapu, C.; Zeng, H.F.; Hung, K.H.; Wei, B. Effect of temperature on the capacitance of carbon nanotube supercapacitors. *ACS Nano* 2009, **3**, 2199–2206.

[98] Mukhopadhyay, I.; Suzuki, Y.; Kawashita, T.; Yoshida, Y.; Kawasaki, S. Studies on surface functionalized single wall carbon nanotube for electrochemical double layer capacitor application. *Journal of Nanoscience and Nanotechnology* 2010, **10**, 4089–4094.

[99] Lin, R.; et al. Capacitive energy storage from −50 to 100°C using an ionic liquid electrolyte. *The Journal of Physical Chemistry Letters* 2011, **2**, 2396–2401.

[100] Cao, G.; Yang, Y.; Gu, Z. Capacitive performance of an ultralong aligned carbon nanotube electrode in an ionic liquid at 60°C. *Carbon* 2008, **46**, 30–34.

[101] Lu, W.; Qu, L.; Henry, K.; Dai, L. High performance electrochemical capacitors from aligned carbon nanotube electrodes and ionic liquid electrolytes. *Journal of Power Sources* 2009, **189**, 1270–1277.

[102] Kim, K.S.; Park, S.J. Electrochemical performance of graphene/carbon electrode contained well-balanced micro-and mesopores by activation-free method. *Electrochimica Acta* 2012, **65**, 50–56.

[103] Wang, M.; Liu, Q.; Sun, H.; Stach, E.A.; Zhang, H.; Stanciu, L.; Xie, J. Preparation of high-surface-area carbon nanoparticle/graphene composites. *Carbon* 2012, **50**, 3845–3853.

[104] Wang, M.; Liu, Q.; Sun, H.; Stach, E.A.; Xie, J. Preparation of high surface area nanostructured graphene composites. *ECS Transactions* 2012, **41**, 95–105.

[105] Wang, Q.; et al. Hollow luminescent carbon dots for drug delivery. *Carbon* 2013, **52**, 209–218.

[106] Wang, L.; Sun, L.; Tian, C.; Tan, T.; Mu, G.; Zhang, H.; Fu, H. A novel soft template strategy to fabricate mesoporous carbon/graphene composites as high-performance supercapacitor electrodes. *RSC Advances* 2012, **2**, 8359–8367.

[107] Xu, G.; Zheng, C.; Zhang, Q.; Huang, J.; Zhao, M.; Nie, J.; Wang, X.; Wei, F. Binder-free activated carbon/carbon nanotube paper electrodes for use in supercapacitors. *Nano Research* 2011, **4**, 870–881.

[108] Huang, Z.D.; et al. Effects of reduction process and carbon nanotube content on the supercapacitive performance of flexible graphene oxide papers. *Carbon* 2012, **50**, 4239–4251.

[109] Li, J.J.; Ma, Y.W.; Jiang, X.; Feng, X.M.; Fan, Q.L.; Huang, W. Graphene/carbon nanotube films prepared by solution casting for electrochemical energy storage. *IEEE Transactions on Nanotechnology* 2012, **11**, 3–7.

[110] Wang, H.; Liu, Z.; Chen, X.; Han, P.; Dong, S.; Cui, G. Exfoliated graphite nanosheets/carbon nanotubes hybrid materials for superior performance supercapacitors. *Journal of Solid State Electrochemistry* 2011, **15**, 1179–1184.

[111] Lu, W.; Hartman, R.; Qu, L.; Dai, L. Nanocomposite electrodes for high-performance supercapacitors. *The Journal of Physical Chemistry Letters* 2011, **2**, 655–660.

[112] Hsieh, T.; Chuang, C.; Chen, W.; Huang, J.; Chen, W.; Shu, C. Hydrous ruthenium dioxide/multi-walled carbon-nanotube/titanium electrodes for supercapacitors. *Carbon* 2012, **50**, 1740–1747.

[113] Lin, N.; Tian, J.; Shan, Z.; Chen, K.; Liao, W. Hydrothermal synthesis of hydrous ruthenium oxide/graphene sheets for high-performance supercapacitors. *Electrochimica Acta* 2013 **99**, 219–224.

[114] Wang, K.; Li, L.; Wu, X. Synthesis of graphene and electrochemical performance. *International Journal of Electrochemical Science* 2013, **8**, 6574–6578.

[115] Wu, Z.; Wang, D.; Ren, W.; Zhao, J.; Zhou, G.; Li, F. Anchoring hydrous RuO_2 on graphene sheets for high-performance electrochemical capacitors. *Advanced Functional Materials* 2010, **20**, 3595–3602.

[116] Kang, J.; et al. Enhanced supercapacitor performance of MnO_2 by atomic doping. *Angewandte Chemie* 2013, **125**, 1708–1711.

[117] Im, C.; Yun, Y.S.; Kim, B.; Park, H.H.; Jin, H.J. Amorphous carbon nanotube/MnO_2/graphene oxide ternary composite electrodes for electrochemical capacitors. *Journal of Nanoscience and Nanotechnology* 2013, **13**, 1765–1768.

[118] Zhao, H.; Han, G.; Chang, Y.; Li, M.; Li, Y. The capacitive properties of amorphous manganese dioxide electrodeposited on different thermally-treated carbon papers. *Electrochimica Acta* 2013, **91**, 50–57.

[119] Zhang, J.; Zhao, X.S. Conducting polymers directly coated on reduced graphene oxide sheets as high-performance supercapacitor electrodes. *Journal of Physical Chemistry C* 2012, **116**, 5420–5426.

[120] Mao, L.; Zhang, K.; Chan, H.; Wu, J. Nanostructured MnO_2/graphene composites for supercapacitor electrodes: the effect of morphology, crystallinity and composition. *Journal of Materials Chemistry* 2012, **22**, 80–85.

[121] Chang, H.; Chang, C.; Tsai, Y.; Liao, C. Electrochemically synthesized graphene/polypyrrole composites and their use in supercapacitor. *Carbon* 2012, **50**, 2331–2336.

[122] Zhou, Y.; Qin, Z.; Li, L.; Zhang, Y.; Wei, Y.; Wang, L.; Zhu, M. Polyaniline/multi-walled carbon nanotube composites with core–shell structures as supercapacitor electrode materials. *Electrochimica Acta* 2010, **55**, 3904–3908.

[123] Wang, Q.; Li, J.; Gao, F.; Li, W.; Wu, K.; Wang, X. Activated carbon coated with polyaniline as an electrode material in supercapacitors. *New Carbon Materials* 2008, **23**, 275–280.

[124] Kumar, N.A.; Choi, H.J.; Bund, A.; Baek, J.B.; Jeong, Y.T. Electrochemical supercapacitors based on a novel graphene/conjugated polymer composite system. *Journal of Materials Chemistry* 2012, **22**, 12268–12274.

[125] Yu, G.; Xie, X.; Pan, L.; Bao, Z.; Cui, Y. Hybrid nanostructured materials for high-performance electrochemical capacitors. *Nano Energy* 2013, **2**, 213–234.

10

LITHIUM-ION BATTERIES BASED ON CARBON NANOMATERIALS

BRIAN J. LANDI[1,2], REGINALD E. ROGERS[1,2], MICHAEL W. FORNEY[1] AND MATTHEW J. GANTER[1]

[1]*NanoPower Research Laboratory, Rochester Institute of Technology, Rochester, NY, USA*
[2]*Department of Chemical Engineering, Rochester Institute of Technology, Rochester, NY, USA*

10.1 INTRODUCTION

There is an ever-increasing demand for improved energy storage to meet the needs of a rapidly growing technological society. Rechargeable lithium-ion batteries are emerging as the primary energy storage medium due to higher-power and higher-energy options than legacy technologies like Pb acid, Ni–Cd, or Ni metal hydrogen [1]. However, as shown in the Ragone plot (Fig. 10.1), current lithium-ion battery technology is designed for either high power OR high energy, but not both. Therefore, recent efforts have been made to develop advanced materials that will meet the requirement of high power AND high energy, moving toward the upper right of the Ragone plot. Achievement of this ultimate goal will open up new opportunities with high-demand applications (e.g., power tools, electric vehicles, and portable electronics) and increase adoption.

Lithium-ion batteries are composed of four primary components: a cathode, an anode, a separator, and electrolyte. The cathode and anode are comprised of a current collector (aluminum for the cathode and copper for the anode), an active material, a binder, and conductive additive (typically carbon). The separator is made of a

Carbon Nanomaterials for Advanced Energy Systems: Advances in Materials Synthesis and Device Applications, First Edition. Edited by Wen Lu, Jong-Beom Baek and Liming Dai.
© 2015 John Wiley & Sons, Inc. Published 2015 by John Wiley & Sons, Inc.

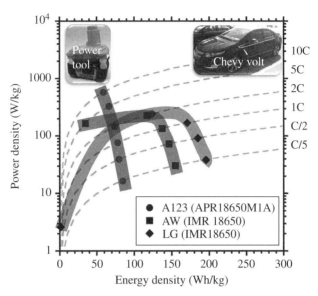

FIGURE 10.1   Ragone plot showing power and energy capabilities for current Li-ion batteries.

polypropylene/polyethylene/polypropylene porous membrane that permits Li ions to pass through to either the anode or cathode. Traditional electrolytes are composed of a mixture of a salt (e.g., $LiPF_6$) and organic solvents. Figure 10.2a shows a schematic for how the Li-ion battery is constructed. Li is typically integrated within the crystal structure of the cathode active material prior to operation. During operation, the battery is charged when Li ions move from the cathode through the separator via the conductive electrolyte to the anode. During this process, electrons flow along the outside via the current collector to the anode where a reduction reaction takes place to form Li metal compound. When the current polarity is reversed, discharge occurs with Li ions moving from the anode to the cathode in a similar manner as was done on charge. During the discharge process, electrolyte breakdown leads to the buildup of a passivation layer called the solid electrolyte interface (SEI). This layer leads to a significant loss in the capacity retained on the first cycle. The coulombic efficiency (CE) is an important value that relates to the percentage of charge that is reversibly cycled. This value quantifies how efficient the cell is and is typically lower on the first cycle. The CE of the anode will also affect capacity matching where excess loss of lithium due to SEI formation, typically due to high surface area, has to be accounted for through excess cathode loading (decreasing cell energy density) or through anode prelithiation techniques.

   In developing a Li-ion battery, there are a suite of active material options one can consider for both the cathode and anode. Figure 10.2b presents the cathode and anode active materials that have been or are currently in use by researchers and commercial entities. On the cathode, currently, oxide materials such as $LiCoO_2$ and $LiNiCoAlO_2$ (NCA) are used in most applications. However, safety concerns associated with

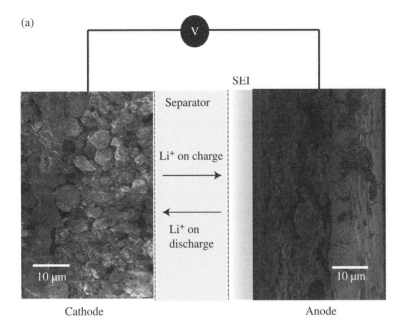

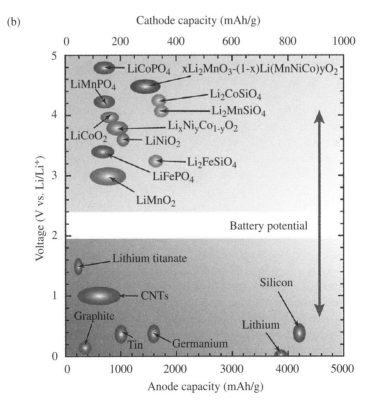

FIGURE 10.2 (a) Schematic of the inside of a Li-ion battery. (b) Active material options for the cathode and anode as a function of voltage versus Li.

thermal runaway potential have prompted alternative cathode materials to be considered. Such materials need to promote enhanced safety during abuse conditions (overcharge, internal shorts, temperature changes) to mitigate the potential of hazards such as fires from occurring. For example, lithium transition metal phosphates, such as $LiFePO_4$, are now being commercialized for consumer use in applications such as power tools due to the potential for achieving fast charge–discharge rates and avoiding issues associated with overcharge of the cathode [2]. $LiCoPO_4$ has also been extensively studied given the high voltage plateau, making it desirable for high-voltage applications [3]. In addition to these chemistries, researchers are also investigating alternative chemistries that will enhance the overall energy density of the battery, especially when considering transportation needs. Both lithium-rich metal oxides and lithium transition metal orthosilicates (Li_2MSiO_4, where M = transition metal) have the potential to double the theoretical capacity and energy density if they can be practically integrated into a full battery [4–9]. On the anode, most batteries have utilized graphite as the active material. Alternatives to graphite that have been studied include tin, germanium, and silicon; however, issues associated with material stability during cycling of the battery have slowed large-scale adoption. Instead, researchers have looked into combining carbonaceous materials with tin, germanium, and/or silicon into hybrid composites to maximize the capacity and energy potential of the individual components [10–15].

In the early stages of development, a cathodic or anodic material of interest is typically tested opposite a lithium metal electrode. By using a lithium electrode, the reversible capacity of the candidate electrode is able to be assessed before pairing against the actual cathode or anode in a full battery. The cathode discharge voltage profiles with changing chemistries are shown in Figure 10.3a. The average voltage and discharge capacity are directly related to the cathode chemistry. As can be seen from Figure 10.3a, a higher voltage and capacity would provide the largest energy. However, current electrolytes become unstable above 4.5 V and must be addressed to utilize higher-voltage cathodes [3, 16]. When testing an anode against lithium, it is important to understand the charge and discharge capacities are reversed and the anode is acting as the cathode in the cell. Looking back at Figure 10.2b, the reference material is lithium, which has a potential of zero. All other anode and cathode active material potentials are determined against this reference. Since lithium has the lower potential versus the other anode active material, it will function as the anode when paired against the other anode materials, causing them to function as cathodes in such systems. Therefore, for clarity, when testing anodes versus lithium, the axes are labeled insertion and extraction capacity based on whether lithium ions are being inserted into or extracted from the anode. An example of anode voltage profiles versus lithium can be seen in Figure 10.3b. The extraction capacity is the reversible capacity that is reported as the active material's specific capacity. The average voltage plateau will affect the battery energy when paired versus a cathode. A lower anodic voltage increases the energy by increasing the overall battery potential difference.

When the cathode and anode are paired in a typical battery configuration, the resulting voltage profile is lowered by the potential difference of the anode versus

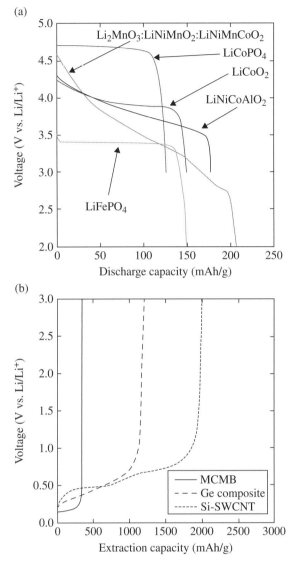

FIGURE 10.3 Voltage profiles of (a) cathode active materials and (b) anode active materials. Curves shown tested against Li foil at C/10 rate using 1.2M LiPF$_6$ EC:EMC (3:7 v/v).

lithium. Figure 10.4 demonstrates the changing voltage profiles when pairing NCA cathode versus different active anode materials. The change in voltage directly correlates with the change in anode voltage versus lithium as shown. The cathode and anode capacities must be matched by areal capacity (mAh/cm^2) in order to maximize energy and cyclability. A slightly higher anode areal loading is typically utilized to minimize the chance of lithium plating and maximize the use of cathode capacity.

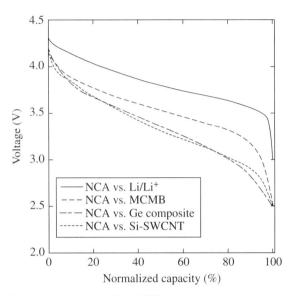

FIGURE 10.4 Discharge voltage profiles of NCA cathode versus varying anode chemistries. Cells testing performed at a C/10 rate and with 1.2M LiPF$_6$ EC:EMC 3:7 in all cases.

The cathode active material capacity is currently significantly lower (150–180 mAh/g) than the standard graphitic anode material capacity (300 mAh/g). Therefore, the anode composite has a lower areal mass loading (mg/cm^2) to match the cathode, resulting in a thinner coating. This has an effect on energy density especially in high-power designs where the electrodes are thinned to enable faster transport. In that case, the mass of the inactive metal current collector becomes a greater percentage and reduces overall energy density. The lower cathode capacity also reduces the benefits of a higher-capacity anode such as Si and Ge, because the anode layers have to be thinned to match the cathode.

10.2 IMPROVING LI-ION BATTERY ENERGY DENSITY

There exist a number of ways to improve the energy density of a lithium-ion battery. Specific energy is typically stated in terms of Wh/kg, which is the energy provided by the active materials (the materials shuttling lithium ions) divided by the entire mass of the battery that includes packaging, electrolyte, current collectors, binders, and conductive additives. Increases in energy density can be made by decreasing component masses and/or increasing the energy of the battery by changing active material loading or structure. The majority of research to improve energy density has focused on increasing the active material capacity and/or voltage.

The use of higher-capacity anode materials, such as silicon (Si) and germanium (Ge), has been shown to improve the energy density of the battery with specific capacities reaching 3000 mAh/g for Si [17, 18]. The full battery energy density increase is typically only around 10% due to the higher density of the copper current collector and the low-capacity cathodes [19]. The full benefit of the high-capacity anode is not realized above 1000 mAh/g because the cathode specific capacity is not high enough to match the higher anode capacity in standard designs. The typical Si-based anodes are restricted to being thinly coated on much heavier copper current collectors, which leads to the marginal gains in energy density.

10.3 IMPROVEMENTS TO LITHIUM-ION BATTERIES USING CARBON NANOMATERIALS

10.3.1 Carbon Nanomaterials as Active Materials

In addition to Si and Ge, carbon nanomaterials such as single-walled carbon nanotubes (SWCNTs) and multi-walled carbon nanotubes (MWCNTs) have also been considered for use as anode materials [1, 20, 21]. Both SWCNT and MWCNT are limited in their use not only due to capacity matching with the cathode but also the formation of an SEI that severely reduces their reversible capacity performance or CE. Table 10.1 summarizes the electrochemical performance of select carbon nanomaterials that are currently being investigated as anode active materials in comparison to conventional MCMB.

Figure 10.5 compares a pure SWCNT anode against a Si–SWCNT anode. While the SWCNT is able to achieve an insertion capacity of over 2500 mAh/g, its extraction capacity is limited to around 600 mAh/g. The reason for this large loss in capacity is the formation of the SEI layer, which is a major challenge for capacity matching. Research, however, has shown that combining silicon with the SWCNT not only reduces the SEI layer but improves the CE fourfold [13]. In addition to their pure forms, SWCNT and Si (along with Ge) are currently being investigated in hybrid forms that provide new pathways toward improving the CE in a full battery and overall cycle life [10–15]. Such approaches make the use of carbon nanotubes (CNTs) more plausible as anode active materials and warrant further study into their use in battery applications.

TABLE 10.1 Comparison of Carbon Nanomaterial-Based Anode Active Materials versus Conventional Ones

Active Material	Voltage (V)	Capacity (mAh/g)	First Cycle Coulombic Efficiency (%)
SWCNT [1, 21, 22]	1.0	600–1000	21
MWCNT [1, 20, 21]	0.8	220	50
Si [14, 15]	0.4	1170	35
Si–SWCNT [14, 15, 21]	0.4	1000–2200	90
MCMB [1]	0.2	300	80

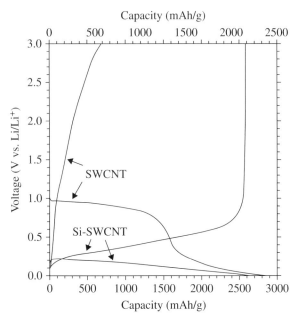

FIGURE 10.5 Charge–discharge profile comparison of pure SWCNT and Si–SWCNT hybrid for use as an anode in lithium-ion battery.

10.4 CARBON NANOMATERIALS AS CONDUCTIVE ADDITIVES

10.4.1 Current and SOA Conductive Additives

Within traditional lithium-ion battery electrode composites, the cathode and anode active materials are a primary interest due to the direct relationship to lithium-ion electrochemical performance. However, the optimization of electrode design (e.g., thickness, porosity, etc.), including the selection and concentration of binder and conductive additive, is crucial to realizing the potential performance of the active materials [23, 24]. The materials within the composite and the processing of the electrode will influence electrical and ionic conductivity, electrode density, and flexibility. Conductive additives are especially important in cathode composites to provide sufficient electronic conductivity between the active cathode particles and the metal foil current collector [25]. Electrodes are engineered to provide a balance between electrical and ionic conductivity, by altering composite porosity and density, where sufficient electrode conductivity is achieved to meet the desired power capabilities without sacrificing active mass loading. Typical conductive additives are carbon based (e.g., carbon black or graphite powders) and are employed at sufficient weight loadings to form an electrical percolation network (the concentration in a composite that sustains long-range connectivity or contiguous pathways for electrons to move) in the composite. Conductive additives have evolved commercially from acetylene black and graphite to more sophisticated long-chain conductive carbon blacks for

NCA–carbon black   NCA–VGCF/MWCNT   NCA–SWCNT

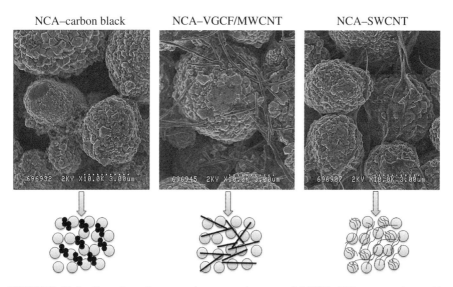

FIGURE 10.6 Scanning electron microscopy images of LiNiCoAlO$_2$ composites with varying conductive additives. Representative images demonstrating the interaction between active particle and conductive additive are shown below SEM images.

improved rate performance [26]. Recent research has investigated the use of high aspect ratio one-dimensional (1D) carbon nanomaterial conductive additives to meet the demand for higher conductivity at lower weight loadings. Examples include vapor-grown carbon fiber (VGCF) and CNTs, which have been shown to provide an enhanced percolation network by enabling long-range connectivity and particle connections across void spaces between active particles [27]. SEM images of composites with standard carbon black and 1D carbon nanomaterial additives are shown in Figure 10.6. The difference between additive interactions can be seen qualitatively from the SEM images and the diagrams. Carbon black additives are around 30–40 nanometers and can agglomerate to form longer-range connection. Due to the spherical nature, a larger amount of additive is needed to provide connection throughout the composite and meet the percolation threshold. 1D carbon nanomaterial additives have larger aspect ratios (ratio of length to diameter ranging from 10,000 to 1) with lengths from 1 to 30 μm and can form an equivalent electrical connection throughout the composite at a lower weight loading. VGCF and MWCNT additives exhibit rodlike structures that have lower flexibility compared to the SWCNT additives in the images. The flexibility of the SWCNT additives can provide the advantage of improved composite adhesion and mechanical properties that can prevent active particle separation with cycling.

The structure of the conductive additive influences how the particle interacts with the active materials and the tortuosity of the conduction path. However, conductive additives must possess other inherent properties such as high electronic and thermal conductivity and must be inert to eliminate any potential side reactions. CNTs possess excellent inherent properties such as a high theoretical electrical conductivity of 1×10^8 S/m, exceeding copper (5.8×10^7 S/m), and a large thermal conductivity of SWCNTs (3500 W/m·K) surpassing graphite (~300 W/m·K) and carbon black (<1 W/m·K) by more than an order of magnitude [28–30]. MWCNTs have been predominantly investigated in cathodes, showing approximately 10% improvement in the reversible capacity of the electrodes up to an equivalent of a 3C rate for $LiCoO_2$ compared to carbon black counterparts [1, 31, 32]. $LiFePO_4$ composites with a direct replacement of carbon black additives with MWCNTs improved cycling and increased the specific capacity by approximately 20% at a 2C rate and a weight loading of 3% w/w [33]. Graphene has also been investigated as a conductive additive and initially demonstrated improved capacity at modest discharge rates [34]. However, a recent study has shown that graphene additives can reduce capacity retention at higher discharge rates (>2C), suggesting that a large planar structure causes polarization and limits ionic conductivity [35]. SWCNTs have only recently been investigated as a conductive additive in Li-ion battery cathode composites [36–39]. SWCNT additives can not only bridge gaps between active particles but can also wrap particles to provide improved connection between particles in contact as evident by SEMs. The use of SWCNTs as a conductive additive in a typical $LiCoO_2$ cathode coating found that 0.5% commercially prepared high-pressure carbon monoxide (HiPco) SWCNTs can effectively replace 10% carbon black and realize nearly equivalent electrode conductivity, cell impedance, and rate capability [36]. SWCNT additives at a 1% w/w loading with $LiNi_{0.8}Co_{0.2}O_2$ were found to have 3 times the capacity retention at a 10C rate compared to a 4% w/w loading of carbon black additives [39]. That study also demonstrated that SWCNT additives can improve the thermal stability of electrodes and the SWCNT additive composite had a 40% reduction in exothermic energy release upon overcharge compared to the carbon black additive sample. It should be emphasized that proper dispersion of carbon nanomaterials is critical in the utilization as additives and advanced mixing procedures for SWCNTs to prevent excessive SWCNT bundling are what is attributed to the performance. Also, SWCNTs can vary in purity, diameter, length, and defect content that can affect the efficacy of SWCNTs as conductive additives.

10.5 SWCNT ADDITIVES TO INCREASE ENERGY DENSITY

Another advantage of SWCNT additives that is gaining attention is the ability to improve energy density by enabling higher composite areal loading. At higher composite loadings, the electronic and ionic conductivity of the composite plays a larger role as electrons travel through a thicker composite to the current collector and electrolyte must be present throughout the composite to enable lithium-ion conduction. Typical areal loading capacities for commercial lithium-ion battery electrodes are around 2 mAh/cm^2 with a thickness in the range of 60 μm and go as high as 4 mAh/cm^2 and 100 μm thick. Batteries containing higher areal loadings are considered for energy density only and are used for applications that allow for slow charge–discharge rates. At faster discharge rates, a higher voltage drop and a lower capacity are found due to resistive losses caused by a thicker low conductivity

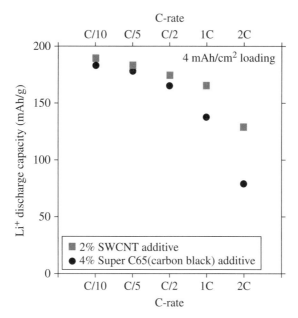

FIGURE 10.7 Comparison of $4\,\text{mAh/cm}^2$ LiNiCoAlO_2 (NCA) cathode areal loadings with 4.0% w/w Super C65 (carbon black) and 2.0% w/w SWCNT additives at increasing discharge rates.

composite. SWCNTs have been shown to improve rate capability and conductivity of composites previously at a low loading of 1% w/w. However, a higher percentage of SWCNTs may be necessary to improve the rate performance of thicker and higher areal loading electrodes. An example of this is shown in Figure 10.7, where a LiNiCoAlO_2 (NCA) cathode composite at a loading of $4\,\text{mAh/cm}^2$ ($100\,\mu\text{m}$ thick) is compared for a cathode with 4% carbon black additive versus 2% SWCNT additive. Previous work compared 1% w/w SWCNTs with 4% w/w carbon black at a low loading of approximately $1.6\,\text{mAh/cm}^2$ and demonstrated similar capacities up to a 2C rate with the SWCNT composite having higher capacity and voltage retention at 5C and 10C rates [39]. However, at a higher areal loading of $4\,\text{mAh/cm}^2$ and higher SWCNT additive loading, a higher capacity was found starting at a C/2 rate. This suggests that the high conductivity of the SWCNTs is necessary to enhance performance of thicker composites. However, other factors may be affected as well such as ionic conductivity and composite porosity.

A significant increase in energy density can be realized by going to higher areal loadings, such as $8\,\text{mAh/cm}^2$, to reduce the relative mass of the inactive current collector. However, there can be limitations as seen in Figure 10.8, where a composite with 4% w/w carbon black additives suffered from poor adhesion and composite flexibility. A composite with that high of active material areal loading with carbon black additives on standard aluminum foil is not practical, and it wasn't possible to test the electrode. The composite containing 1% w/w SWCNTs was flexible and well

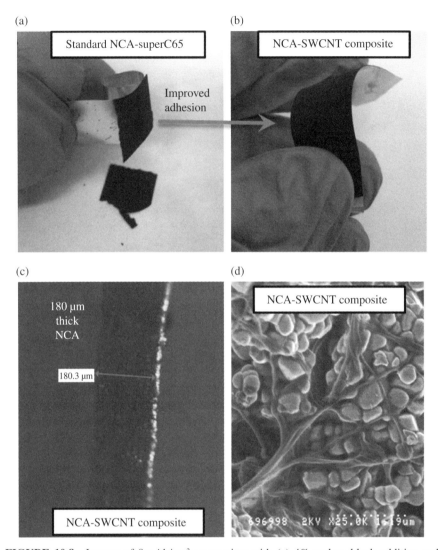

FIGURE 10.8 Images of 8 mAh/cm^2 composites with (a) 4% carbon black additive and (b) 1% SWCNT additive. (c) Cross section of 8 mAh/cm^2 composite and (d) SEM of NCA composite with 1% SWCNTs.

adhered to the current collector owing to the ability of the SWCNTs to entangle particles, uptake the polymer binder, and create long-range connectivity as shown in the SEM image. The photographs in Figure 10.8 demonstrate the improved flexibility of electrodes using SWCNT additives, which could also lead to improved cycling performance.

Cathodes containing NCA at areal loadings of 4 and 8 mAh/cm^2 with 2% w/w SWCNT were compared to monitor discharge rate limit for maintaining capacity (Fig. 10.9). The results at increasing discharge rates demonstrate a drop in capacity

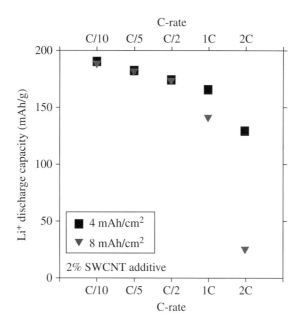

FIGURE 10.9 Comparison of 4 and 8 mAh/cm² NCA cathode areal loadings with 2% w/w SWCNT additives at increasing discharge rates.

starting at 1C; however, similar discharge capacities were achieved up to a C/2 rate, which is a common discharge rate for higher-energy applications. These results demonstrate that increased weight loadings of SWCNT additives can significantly improve rate performance of higher areal capacity NCA composites compared to traditional composites with carbon black additives. At extremely high loadings (8 mAh/cm²), SWCNT additives are necessary to provide composite mechanical stability and adhesion to the Al current collector. At higher loadings, the rate performance may be further enhanced by optimizing ionic conductivity through controlled electrode porosity.

10.6 CARBON NANOMATERIALS AS CURRENT COLLECTORS

10.6.1 Current Collector Options

Traditionally, metal foils are used for current collectors due to their high electrical conductivity and mechanical robustness. Foil thicknesses are typically 10–20 μm (lower thickness translates to higher energy density), and the metal selection can affect the operational voltage window of the battery. An additional method to increase the energy density of lithium-ion batteries is to use a free-standing graphene or CNT paper as a current collector replacement, which can reduce the current collector density by an order of magnitude (Table 10.2) while maintaining acceptable conductivity. Such an approach can use composite slurries coated onto CNT papers, free-standing hybrid

TABLE 10.2   Current Collector Options and Their Properties

Material	Density (g/cm^3)	Conductivity (S/m)
Aluminum	2.7	3.5×10^7
Copper	8.9	6.0×10^7
Titanium	4.5	2.4×10^6
SWCNT	0.5–0.75 [40, 41]	1–5×10^6 [41, 42]
MWCNT	0.5–0.75 [40, 41]	1–5×10^6 [41, 42]
Graphene	<1	$\sim 8 \times 10^3$ [43, 44]

composites (active materials and SWCNTs are mixed and coated or filtered to form a three-dimensional composite), or deposition of active materials (e.g., Si or Ge) onto CNT papers by e-beam evaporation or chemical vapor deposition (CVD) techniques. In all cases, the increase in energy is due to the replacement of the metal foil current collector, which is particularly beneficial on the anode side by replacing the dense copper foil.

The total electrode specific capacity gives a much more accurate indication of the actual effect on energy density of a full battery because all of the components of the electrode are included (i.e., the current collector, binder, and conductive additive). Figure 10.10a compares the specific capacity of NCA active material to electrode specific capacity at different areal loadings when coated on Al foil or CNT papers. The specific capacity of the intrinsic NCA active material is 185 mAh/g regardless of the current collector used. As the areal electrode loading is increased from 2 to 8 mAh/cm^2, the relative mass of the current collector is decreased, thus increasing electrode energy density. Since a CNT paper has a density of 0.5–0.75 g/cm^3 [40, 41], the change in electrode capacity with loading isn't as significant as with a metal foil. When the CNT paper is the current collector for the cathode, a 28% improvement in electrode capacity is found at the standard loading of 2 mAh/cm^2 and a modest improvement of 8% at higher 8 mAh/cm^2 loadings. Equivalent electrode specific capacities can be reached with a 2 mAh/cm^2 NCA coating on SWCNTs versus an 8 mAh/cm^2 coating on Al foil, where the thinner coating allows for a simultaneous improvement in power capability.

In the case of the anode, the comparison of electrode specific capacity using MCMB active material composites coated on Cu or CNTs is shown in Figure 10.10b. There is a dramatic impact on electrode specific capacity at a standard 2 mAh/cm^2 loading when including the copper current collector mass, which decreases the electrode specific capacity from 330 mAh/g for MCMB active materials to 86 mAh/g for MCMB composite (91.5% active:0.5% carbon black:8% PVDF binder) coated on Cu. In comparison, the electrode specific capacity is only reduced to 247 mAh/g for MCMB composite on CNT current collectors. As shown, replacing the Cu current collector with CNTs can nearly quadruple the electrode specific capacity at 2 mAh/cm^2 and double the electrode capacity at 4 mAh/cm^2 and still have a 50% improvement at 8 mAh/cm^2. Thus, there will be a considerable energy density gain that can be garnered by using CNT current collectors as a replacement for conventional copper foil. In addition, there is benefit from CNTs exhibiting increased depth

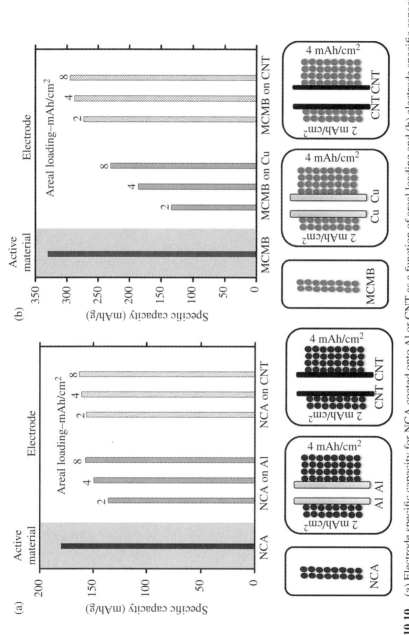

FIGURE 10.10 (a) Electrode specific capacity for NCA coated onto Al or CNT as a function of areal loading and (b) electrode specific capacity for MCMB coated onto Cu or CNT as a function of areal loading.

of discharge below 2.5 V compared to traditional copper current collector [41] and the potential for having a zero volt state of charge, which can improve safety concerns during transport and storage.

The effect of CNT paper properties such as purity, conductivity, and CNT type on current collector performance has not yet been fully documented in the literature. The purity of CNT materials can typically be related back to the performance of the material in a variety of devices and is dependent on synthesis method and postprocessing. The presence of metallic impurities may cause undesirable side reactions or be corroded when used as a current collector [45]. An example study with commercially available CNTs as current collectors for traditional lithium-ion battery electrode composites shows how purification procedures can influence the current collector performance.

10.7 IMPLEMENTATION OF CARBON NANOMATERIAL CURRENT COLLECTORS FOR STANDARD ELECTRODE COMPOSITES

10.7.1 Anode: MCMB Active Material

CNT current collectors were used to replace copper current collectors on the anode by coating a standard composite of MCMB active material, PVDF binder, and carbon black onto commercial CNT paper from Nanocomp Technologies. The coating method was the same as it would be with a copper current collector, which is important for transitioning this technology to industrial scale, and the composite qualitatively appeared to adhere well to the CNT paper. Cross-sectional SEM images of a cleaved MCMB composite coated on purified CNTs can be seen in Figure 10.11c and d at 1,000× and 10,000× magnification, respectively. The images show a uniform CNT paper that is compliant with the composite and can compress to provide enhanced contacting. Even in areas that became delaminated due to electrode cleaving, the SWCNTs can be seen bridging the gap between the CNT paper and the composite in the higher magnification image. Therefore, it is evident that a CNT paper may be able to provide electrical contact even after partial composite delamination.

The first cycle voltage profiles of MCMB anode composites on as-received and purified CNT and copper current collectors are shown in Figure 10.11a. The extraction capacities of the MCMB composites coated on Cu and purified CNTs are equivalent up to typical cutoff voltages of 1.0 V versus Li/Li^+ equal to approximately 300 mAh/g. The composite on purified CNTs has a slightly higher capacity above 1.0 V versus Li/Li^+ due to a small amount of lithium-ion storage in the CNT paper. However, the extraction capacity of the composite coated on the as-received (unpurified) CNT paper had a lower capacity of approximately 260 mAh/g. This result demonstrates the importance of CNT paper purity for its use as a current collector in lithium-ion batteries. However, there can be a drawback with certain CNT current collectors, namely, the first cycle loss due to SEI formation. As shown in Figure 10.11a, the first cycle CE of the anode composites on CNTs (~65%) was measured to be lower

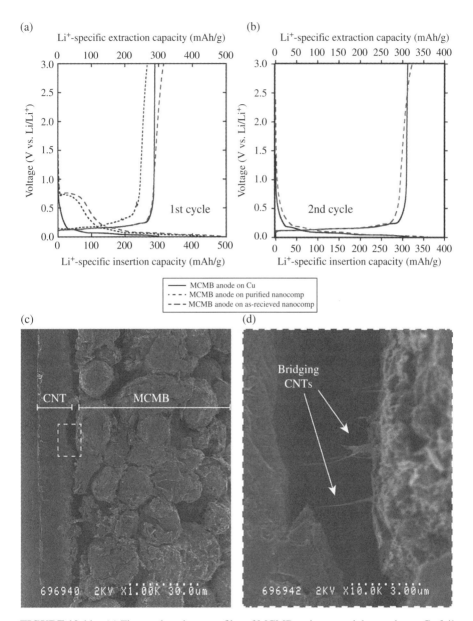

FIGURE 10.11 (a) First cycle voltage profiles of MCMB active material coated onto Cu foil and Nanocomp CNT sheets, (b) second cycle voltage profiles of MCMB coated onto Cu and purified Nanocomp CNT sheets, (c) SEM images of an MCMB slurry coated onto a CNT current collector, and (d) SEM image showing CNTs bridging between the MCMB coating and the CNT current collector.

compared to MCMB coatings on Cu (~90%). The additional SEI formation can be seen in the insertion voltage curves where a plateau is present around 0.75 V versus Li/Li$^+$ for the composite on CNT current collectors. This voltage plateau is consistent with free-standing CNT anodes and is due to the high surface area CNTs. A reduction in SEI formation when using CNTs as an anode current collector can be achieved through prelithiation or surface area reduction, which has recently been studied [46]. The CE improves in later cycles, as shown by the second cycle data in Figure 10.11b, where the CE for MCMB on CNT has increased to 90%.

10.7.2 Cathode: NCA Active Material

Purified CNT sheets have been coated with a standard cathode composite consisting of $LiNi_{0.8}Co_{0.15}Al_{0.05}O_2$ (NCA), PVDF binder, and carbon black to assess the viability of replacing standard Al foil with CNT sheets. Although the mass savings is less significant than for replacing Cu foil, as shown in Figure 10.10a, the CNTs are still less dense than Al current collectors and improve energy density. The use of CNTs as a cathode current collector should not have the issue of excess SEI formation and first cycle loss as the CNTs are anodic in nature and SEI formation on the cathode isn't as significant. The electrochemical results shown in Figure 10.12 demonstrate the ability to use CNT current collectors on the cathode and still achieve approximately 185 mAh/g, which is standard for NCA composites. The CE is demonstrated to be around 90%, which is comparable to composites coated on Al foil. The rate capability of NCA composites on Al and CNT current collectors was measured at increasing discharge rates. The cycling and capacities were similar with the composite on CNTs having a slightly higher capacity demonstrating that CNT current collectors can provide equivalent rate performance to Al.

Overall, these results demonstrate the ability to coat standard cathode and anode slurries onto lightweight CNT papers to enhance energy density, since the active composites had similar performance compared to coatings on metal current collectors. The resulting mass decrease from using CNT current collectors results in a 5–10% increase in energy density for a battery containing cathode composites on CNTs and a 20–30% increase for battery containing anode composites coated on CNTs based on previous fuel cell calculations [21].

10.8 IMPLEMENTATION OF CARBON NANOMATERIAL CURRENT COLLECTORS FOR ALLOYING ACTIVE MATERIALS

Another approach to increase battery energy density is to replace standard MCMB anode composites with alloying lithium storage materials like silicon or germanium, which have theoretical specific capacities of 4200 mAh/g [47] and 1600 mAh/g [48], respectively. In practice, more typical specific capacity values for silicon and germanium are 1500–3000 mAh/g [14] and 800–1000 mAh/g [11, 12], respectively. By combining lightweight carbon nanomaterial current collectors with high specific capacity active materials like silicon and germanium, there is the potential to

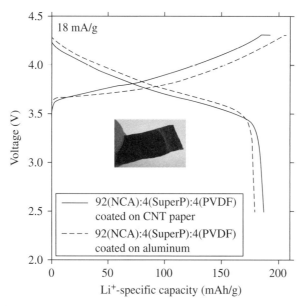

FIGURE 10.12   Voltage profiles of NCA coated onto Al and Nanocomp CNT paper with inset photograph of the NCA–CNT cathode.

significantly improve battery energy density. Consequently, this type of free-standing carbon electrode (copper-free) has been gaining attention in the literature [1, 11–14, 19, 49–53], and a variety of fabrication techniques are being evaluated.

An example of a free-standing silicon–carbon nanotube (Si–CNT) anode is pictured in Figure 10.13a. This example shows an electrode that was fabricated by plasma-enhanced chemical vapor deposition (PECVD) of silicon onto a commercially available CNT sheet (Nanocomp Technologies, Inc.). Figure 10.13b shows an SEM image of the CNT sheet before silicon deposition, and Figure 10.13c shows an SEM image of the Si–CNT anode that results from PECVD of silicon onto the CNT sheet. The PECVD method for fabricating Si–CNT electrodes has been successfully demonstrated with SWCNT [14] and MWCNT [15] current collectors, with electrode specific capacities as high as 2500 mAh/g. Alternative approaches to free-standing Si–CNT anodes typically rely on low-pressure CVD (LPCVD) and have lower electrochemical performance.

Figure 10.13d shows a photograph of a Ge–SWCNT anode that was fabricated by e-beam deposition of germanium onto a SWCNT sheet. SEM cross-sectional analysis of this type of Ge–SWCNT anode is shown in Figure 10.13e, demonstrating the discrete Ge and SWCNT layers. Ge–SWCNT anodes of this type have demonstrated electrode specific capacities up to 800 mAh/g [12]. Alternative approaches for Ge-based anodes start with Ge nanoparticles, which are then mixed with CNTs [11, 13] or graphene [54] to form free-standing anodes. An example of this type of free-standing Ge-SWCNT anode is shown in Figure 10.13f, and have demonstrated improved electrode specific capacities up to 1000 mAh/g [11].

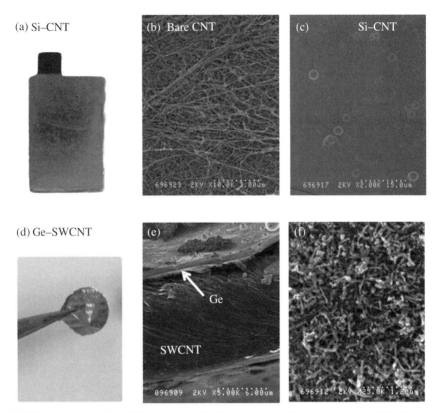

FIGURE 10.13   (a) Si–CNT anode, (b) SEM image of bare CNT paper before CVD Si deposition, (c) Si–CNT anode after Si deposition, (d) Ge–SWCNT anode formed by e-beam deposition of Ge onto SWCNT paper, (e) SEM image of e-beam deposited Ge–SWCNT anode, and (f) Ge–NP–SWCNT hybrid anode.

10.9   ULTRASONIC BONDING FOR POUCH CELL DEVELOPMENT

A challenge when working with carbon nanomaterials for bulk applications is the ability to interface (e.g., electrically or mechanically) with macroscale metal wires or foils. A specific challenge when working with carbon nanomaterial-based electrodes for useful battery form factors, like pouch cells, is that they are often porous, which can lead to leakage of electrolyte solution from pouch cells through the current collector contact tab. One approach to solving this problem is to ultrasonically bond a metal foil, for example, copper, to the carbon nanomaterial current collector, which "welds" the two materials together with ultrasonic acoustic energy. Figure 10.14a shows a photograph of a Si–CNT anode and a copper tab extension before ultrasonic bonding. The bonded Cu–Si–CNT electrode is pictured in Figure 10.14b resting between the ultrasonic head and the anvil immediately after ultrasonic bonding. This Cu–CNT bond provides both an electrical connection and mechanical strength.

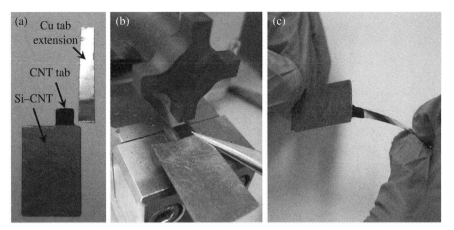

FIGURE 10.14 (a) Photograph of copper tab extension and Si–CNT anode before ultrasonic bonding, (b) photograph ultrasonic bonder with a copper tab extension bonded to a Si–CNT anode, and (c) a flexed Si–CNT anode ultrasonically bonded copper tab extension.

The Cu–Si–CNT electrode is being torqued as shown in Figure 10.14c to demonstrate the mechanical integrity of the ultrasonic bond. Thus, ultrasonic bonding is a technique that enables the scale-up of free-standing carbon nanomaterial-based electrodes for use in pouch cell batteries.

An example of electrochemical data from a pouch cell battery containing a Si–CNT anode paired with an NCA cathode is presented in Figure 10.15. The charge and discharge voltage profiles from the first two cycles are plotted, and a discharge capacity of 13.2 mAh (target of 14 mAh) was achieved with nearly identical discharge voltage profiles. The difference in the charge voltage profiles results from the use of Stabilized Lithium Metal Powder (SLMP®) to prelithiate the Si–CNT anodes [46], thus overcoming the first cycle loss that is typically associated with high surface area nanomaterial electrodes for lithium-ion batteries.

10.10 CONCLUSION

Carbon nanomaterial current collectors offer great potential to increase the gravimetric energy density of lithium-ion batteries, especially by replacing the copper for the anode. As described earlier, two approaches can be taken to increase the electrode specific capacity: (i) replacing metal foils with carbon nanomaterial current collectors and (ii) combining high specific capacity active materials with carbon nanomaterial current collectors. Computational modeling has been performed to demonstrate the magnitude of the gains that can be achieved through these two approaches. In the examples shown in Figure 10.16a, the predicted gravimetric energy density of an 18650 cylindrical cell is plotted as a function of areal capacity (mAh/cm^2) of the electrodes. The four curves plotted are (i) standard MCMB coated on copper, (ii) standard MCMB coated on a CNT current collector, (iii) Si coated on

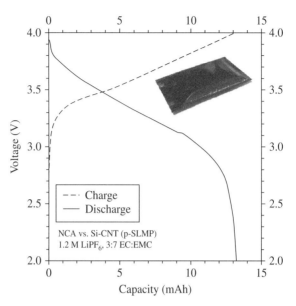

FIGURE 10.15   Voltage profile of a pouch cell (inset photograph) with an NCA cathode paired with a free-standing Si–CNT anode.

copper, and (iv) a Si–CNT anode. In all four cases, the cathode is assumed to be a standard NCA cathode coated onto aluminum. When using a standard MCMB composite, replacing the standard copper current collector with a CNT current collector can yield a gravimetric energy density improvement of 10–20%, depending on the areal capacity of the electrodes. In fact, at areal capacities below $4\,mAh/cm^2$, this replacement results in a greater energy density improvement than replacing the MCMB active material with Si, but the reverse is true at higher areal capacities. It should also be reiterated that SWCNT conductive additives are necessary to achieve areal capacities greater than $4\,mAh/cm^2$ with MCMB composites. The greatest gain in gravimetric energy density is achieved with the replacement of both MCMB and copper with Si and CNT, respectively, where a gain of approximately 50% is possible. Predicted volumetric energy density values as a function of areal capacity, for the same four cases, are plotted in Figure 10.16b. Due to similar thicknesses between metal foils and CNT papers, the active material becomes the key determining factor for volumetric energy density. Regardless of current collector, the Si active material batteries should have 25–45% higher volumetric energy density than MCMB active material batteries, depending on the areal capacity of the electrodes.

   As demonstrated throughout this chapter, novel carbon nanomaterials have the potential to improve the performance and energy density of next-generation lithium-ion batteries. The use of CNT additives represents a near-term drop-in replacement for traditional additives that can lead to increased power and energy densities. The predicted gains with a more disruptive approach by replacing materials such as MCMB (active material), carbon black (conductive additive), and aluminum/copper (current collectors) with carbon nanomaterials warrant continued development efforts to

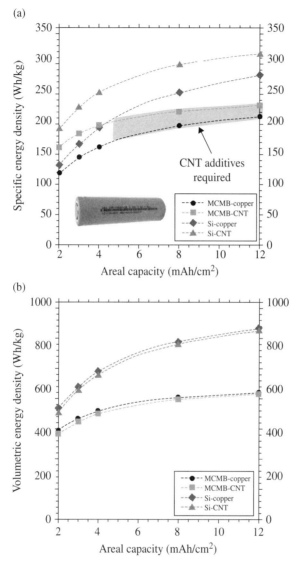

FIGURE 10.16 (a) Gravimetric energy density predictions and (b) volumetric energy density predictions for the improvements that can be achieved by using Si active material, CNT current collectors, or both.

integrate these advanced materials to utilize their full potential benefits. Additionally, carbon nanomaterials provide new opportunities for electrode engineering to develop lightweight systems, which can lead to further improvements in overall battery energy densities for commercial batteries. Understanding the full capabilities of carbon nanomaterials and developing methods for transferring the technology beyond the lab scale will help transition from fundamental science to practical applications.

REFERENCES

[1] Landi, B.J., et al., Carbon nanotubes for lithium ion batteries. *Energy & Environmental Science*, 2009a. **2**(6): pp. 638–654.

[2] Ben Mayza, A., et al., Thermal characterization of $LiFePO_4$ cathode in lithium ion cells. *Nanostructured Materials for Energy Storage and Conversion*, 2011. **35**(34): pp. 177–183.

[3] Rogers, R.E., et al., Impact of microwave synthesis conditions on the rechargeable capacity of $LiCoPO_4$ for lithium ion batteries. *Journal of Applied Electrochemistry*, 2013. **43**(3): pp. 271–278.

[4] Johnson, C.S., et al., Anomalous capacity and cycling stability of xLi(2)MnO(3)center dot(1-x)LiMO$_2$ electrodes (M = Mn, Ni, Co) in lithium batteries at 50 degrees C. *Electrochemistry Communications*, 2007. **9**(4): pp. 787–795.

[5] Johnson, C.S., et al., Synthesis, characterization and electrochemistry of lithium battery electrodes: xLi(2)MnO(3)center dot(1-x)LiMn0.333Ni0.333Co0.333O2 (0 <= x <= 0.7). *Chemistry of Materials*, 2008. **20**(19): pp. 6095–6106.

[6] Dominko, R., et al., In-situ XAS study on Li_2MnSiO_4 and Li_2FeSiO_4 cathode materials. *Journal of Power Sources*, 2009. **189**(1): pp. 51–58.

[7] Nyten, A., et al., Electrochemical performance of Li_2FeSiO_4 as a new Li-battery cathode material. *Electrochemistry Communications*, 2005. **7**(2): pp. 156–160.

[8] Nyten, A., et al., The lithium extraction/insertion mechanism in Li_2FeSiO_4. *Journal of Materials Chemistry*, 2006a. **16**(23): pp. 2266–2272.

[9] Nyten, A., et al., Surface characterization and stability phenomena in Li_2FeSiO_4 studied by PES/XPS. *Journal of Materials Chemistry*, 2006b. **16**(34): pp. 3483–3488.

[10] DiLeo, R.A., et al., Enhanced capacity and rate capability of carbon nanotube based anodes with titanium contacts for lithium ion batteries. *ACS Nano*, 2010. **4**(10): pp. 6121–6131.

[11] DiLeo, R.A., et al., Hybrid Germanium nanoparticle-single-wall carbon nanotube free-standing anodes for lithium ion batteries. *Journal of Physical Chemistry C*, 2011. **115**(45): pp. 22609–22614.

[12] DiLeo, R.A., et al., Germanium-single-wall carbon nanotube anodes for lithium ion batteries. *Journal of Materials Research*, 2010. **25**(8): pp. 1441–1446.

[13] DiLeo, R.A., et al., Balanced approach to safety of high capacity silicon–germanium–carbon nanotube free-standing lithium ion battery anodes. *Nano Energy*, 2012. **2**(2): pp. 268–275.

[14] Forney, M.W., et al., High performance silicon free-standing anodes fabricated by low-pressure and plasma-enhanced chemical vapor deposition onto carbon nanotube electrodes. *Journal of Power Sources*, 2013a. **228**: pp. 270–280.

[15] Forney, M.W., et al., High capacity Si-carbon nanotube anodes for lithium-ion batteries. *Journal of Intelligence Community Research and Development*, 2013b.

[16] Abraham, D.P., et al., Effect of electrolyte composition on initial cycling and impedance characteristics of lithium-ion cells. *Journal of Power Sources*, 2008. **180**(1): pp. 612–620.

[17] Chan, C.K., X.F. Zhang, and Y. Cui, High capacity Li ion battery anodes using Ge nanowires. *Nano Letters*, 2007. **8**(1): pp. 307–309.

[18] Teki, R., et al., Nanostructured silicon anodes for lithium ion rechargeable batteries. *Small*, 2009. **5**(20): pp. 2236–2242.

[19] Cui, L.-F., et al., Light-weight free-standing carbon nanotube-silicon films for anodes of lithium ion batteries. *ACS Nano*, 2010. **4**(7): pp. 3671–3678.

[20] Landi, B.J., et al., Multi-walled carbon nanotube paper anodes for lithium ion batteries. *Journal of Nanoscience and Nanotechnology*, 2009b. **9**(6): pp. 3406–3410.

[21] Landi, B.J., C.D. Cress, and R.P. Raffaelle, High energy density lithium-ion batteries with carbon nanotube anodes. *Journal of Materials Research*, 2010. **25**(8): pp. 1636–1644.

[22] Landi, B.J., et al., Lithium ion capacity of single wall carbon nanotube paper electrodes. *Journal of Physical Chemistry C*, 2008. **112**(19): pp. 7509–7515.

[23] Tran, H.Y., et al., Influence of electrode preparation on the electrochemical performance of $LiNi_{0.8}Co_{0.15}Al_{0.05}O_2$ composite electrodes for lithium-ion batteries. *Journal of Power Sources*, 2012. **210**: pp. 276–285.

[24] Zhang, Y., et al., Composite anode material of silicon/graphite/carbon nanotubes for Li-ion batteries. *Electrochimica Acta*, 2006. **51**(23): pp. 4994–5000.

[25] Chen, Y.H., et al., Selection of conductive additives in Li-Ion battery cathodes. *Journal of the Electrochemical Society*, 2007. **154**(10): pp. A978–A986.

[26] Spahr, M.E., et al., Development of carbon conductive additives for advanced lithium ion batteries. *Journal of Power Sources*, 2011. **196**(7): pp. 3404–3413.

[27] Lestriez, B., et al., Hierarchical and resilient conductive network of bridged carbon nanotubes and nanofibers for high-energy Si negative electrodes. *Electrochemical and Solid-State Letters*, 2009. **12**(4): pp. A76–A80.

[28] Berber, S., Y.-K. Kwon, and D. Tomanek, Unusually high thermal conductivity of carbon nanotubes. *Physical Review Letters*, 2000. **84**(20): pp. 4613–4616.

[29] Dai, H., Carbon nanotubes: Synthesis, integration, and properties. *Accounts of Chemical Research*, 2002. **35**(12): pp. 1035–1044.

[30] Jarosz, P.R., et al., Carbon nanotube wires and cables: Near-term applications and future perspectives. *Nanoscale*, 2011. **3**(11): pp. 4542–4553.

[31] Stura, E. and C. Nicolini, New nanomaterials for light weight lithium batteries. *Analytica Chimica Acta*, 2006. **568**: pp. 57–64.

[32] Li, X., F. Kang, and W. Shen, A comparative investigation on MWCNTs and carbon black as conducting additive in $LiNi_{0.7}Co_{0.3}O_2$. *Electrochemical and Solid-State Letters*, 2006. **9**: pp. A126–A129.

[33] Huang, W., Q. Cheng, and X. Qin, Carbon nanotubes as a conductive additive in $LiFePO_4$ cathode material for lithium-ion batteries. *Russian Journal of Electrochemistry*, 2010. **46**(2): pp. 175–179.

[34] Su, F.-Y., et al., Flexible and planar graphene conductive additives for lithium-ion batteries. *Journal of Materials Chemistry*, 2010. **20**(43): pp. 9644–9650.

[35] Su, F.-Y., et al., Could graphene construct an effective conducting network in a high-power lithium ion battery? *Nano Energy*, 2012. **1**(3): pp. 429–439.

[36] Dettlaff-Weglikowska, U., et al., Effect of single-walled carbon nanotubes as conductive additives on the performance of LiCoO[sub 2]-based electrodes. *Journal of the Electrochemical Society*, 2011. **158**(2): pp. A174–A179.

[37] Ban, C., et al., Extremely durable high-rate capability of a $LiNi_{0.4}Mn_{0.4}Co_{0.2}O_2$ cathode enabled with single-walled carbon nanotubes. *Advanced Energy Materials*, 2011. **1**(1): pp. 58–62.

[38] Ganter, M.J., et al., Electrochemical Performance and Safety of Lithium Ion Battery Anodes Incorporating Single Wall Carbon Nanotubes, in *2012 MRS Spring Meeting*. 2012: San Francisco, CA. Cambridge University Press, Boston, MA.

[39] Ganter, M.J., et al., Differential scanning calorimetry analysis of an enhanced $LiNi_{0.8}Co_{0.2}O_2$ cathode with single wall carbon nanotube conductive additives. *Electrochimica Acta*, 2011. **56**(21): pp. 7272–7277.

[40] Jarosz, P.R., et al., Coaxial cables with single-wall carbon nanotube outer conductors exhibiting attenuation/length within specification. *Micro & Nano Letters*, 2012. **7**(9): pp. 959–961.

[41] Alvarenga, J., et al., High conductivity carbon nanotube wires from radial densification and ionic doping. *Applied Physics Letters*, 2010. **97**(18): p. 182106.

[42] Behabtu, N., et al., Strong, light, multifunctional fibers of carbon nanotubes with ultrahigh conductivity. *Science*, 2013. **339**(6116): pp. 182–186.

[43] Li, D., et al., Processable aqueous dispersions of graphene nanosheets. *Nat Nano*, 2008. **3**(2): pp. 101–105.

[44] Gwon, H., et al., Flexible energy storage devices based on graphene paper. *Energy & Environmental Science*, 2011. **4**(4): pp. 1277–1283.

[45] Braithwaite, J.W., et al., Corrosion of lithium-ion battery current collectors. *Journal of the Electrochemical Society*, 1999. **146**(2): pp. 448–456.

[46] Forney, M.W., et al., Pre-lithiation of Silicon-Carbon Nanotube Anodes using Stabilized Lithium Metal Powder, in *2013 ACS Spring Meeting*. 2013c: New Orleans, LA.

[47] Kasavajjula, U., C.S. Wang, and A.J. Appleby, Nano- and bulk-silicon-based insertion anodes for lithium-ion secondary cells. *Journal of Power Sources*, 2007. **163**(2): pp. 1003–1039.

[48] Graetz, J., et al., Nanocrystalline and thin film germanium electrodes with high lithium capacity and high rate capabilities. *Journal of the Electrochemical Society*, 2004. **151**(5): pp. A698–A702.

[49] Hu, L., et al., Silicon–carbon nanotube coaxial sponge as Li-Ion anodes with high areal capacity. *Advanced Energy Materials*, 2011. **1**(4): pp. 523–527.

[50] Wang, R.H., et al., Flexible free-standing hollow Fe_3O_4/graphene hybrid films for lithium-ion batteries. *Journal of Materials Chemistry A*, 2013. **1**(5): pp. 1794–1800.

[51] Liang, J.F., et al., Flexible free-standing graphene/SnO_2 nanocomposites paper for Li-Ion battery. *ACS Applied Materials & Interfaces*, 2012. **4**(11): pp. 5742–5748.

[52] Yu, A.P., et al., Free-standing layer-by-layer hybrid thin film of graphene-MnO_2 nanotube as anode for lithium ion batteries. *Journal of Physical Chemistry Letters*, 2011. **2**(15): pp. 1855–1860.

[53] Wang, J.Z., et al., Flexible free-standing graphene-silicon composite film for lithium-ion batteries. *Electrochemistry Communications*, 2010. **12**(11): pp. 1467–1470.

[54] Kim, C.H., et al., High-yield gas-phase laser photolysis synthesis of germanium nanocrystals for high-performance photodetectors and lithium ion batteries. *Journal of Physical Chemistry C*, 2012. **116**(50): pp. 26190–26196.

11

LITHIUM/SULFUR BATTERIES BASED ON CARBON NANOMATERIALS

YUEGANG ZHANG[1] AND XIAOWEI YANG[2]

[1]Suzhou Institute of Nano-Tech and Nano-Bionics, Chinese Academy of Sciences, Suzhou, China
[2]School of Material Science and Engineering, Tongji University, Shanghai, China

11.1 INTRODUCTION

The energy and power characteristics of energy storage systems will critically impact the commercial viability of emerging advanced technologies. Among different electrical energy storage systems, lithium-ion batteries represent the state-of-the-art technology and play an increasingly important role because of high energy density for many applications, especially portable electronics [1]. New generations of such batteries are expected to electrify transportation and find use in stationary electricity storage. However, even when fully developed, the performance of current lithium-ion batteries is not satisfactory for tomorrow's energy storage requirements such as electric vehicles and grid-level energy storage due to their relatively low specific energy and high cost [2]. The obtainable specific capacities of current electrode materials remain insufficient to meet the ever-increasing requirements of the rapidly progressing emerging technologies [3, 4]. Therefore, explorations of new materials and new chemistries are urgently needed to go beyond the incremental improvements in the specific energy of existing batteries. With a high theoretical specific capacity (1675 mAh/g), sulfur has been considered to be one highly promising alternative to replace existing cathode materials used in lithium batteries [5]. Since their introduction by Herbet and Ulam in 1962 and by Argonne National Laboratory in 1967,

Carbon Nanomaterials for Advanced Energy Systems: Advances in Materials Synthesis
and Device Applications, First Edition. Edited by Wen Lu, Jong-Beom Baek and Liming Dai.
© 2015 John Wiley & Sons, Inc. Published 2015 by John Wiley & Sons, Inc.

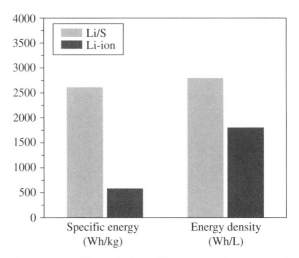

FIGURE 11.1 Comparison of theoretical specific energy and energy density of the lithium/ sulfur cell with those of current lithium-ion cells. Reproduced with permission from Ref. 5. © 2013 Royal Society of Chemistry.

lithium/sulfur cells were first investigated as high-temperature cells with a molten salt electrolyte in the late 1960s [6]. With the average voltage of 2.15 V, the theoretical specific energy and energy density of a lithium/sulfur cell are often stated as 2600 Wh/ kg and 2800 Wh/L, respectively, under the assumption of complete Li_2S formation, which is far greater than that of current lithium-ion batteries as shown in Figure 11.1 [7]. In addition to its high theoretical specific capacity, sulfur is inexpensive, abundant on earth, and environmentally benign.

With these appealing features, lithium/sulfur cells are considered to be the next-generation power source in emerging applications such as advanced portable electronics where the energy density and the cost are critical factors. However, in order to realize this potential, numerous scientific and technical challenges have to be overcome; there is still a long way to go for the development of lithium/sulfur cells with high specific energy and acceptable cycling performance.

In this chapter, we begin with a brief discussion on the operating principles and scientific challenges in the development of lithium/sulfur cells. We then introduce the advantages to improve sulfur cathode by use of carbon nanomaterials and showcase some recent progress with sulfur–carbon-based composite for lithium/sulfur cells. Finally, the opportunities and perspectives for future research directions will be discussed.

11.2 FUNDAMENTALS OF LITHIUM/SULFUR CELLS

11.2.1 Operating Principles

In the most common configuration, as schematically shown in Figure 11.2, sulfur serves as the cathode and is electrochemically coupled with the lithium metal anode [5]. During discharge, lithium ions move spontaneously through the electrolyte from the anode to

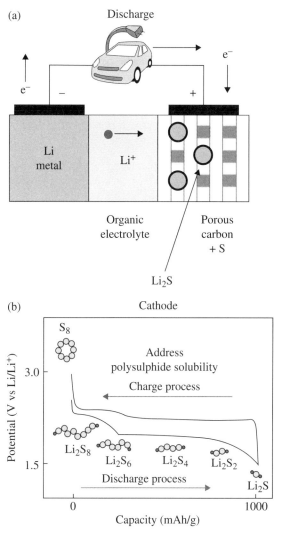

FIGURE 11.2 Introduction to the Li/S cell. (a) Schematic diagram of common configurations of lithium/sulfur cells. (b) The voltage profile and chemistry of sulfur cathode in the organic electrolyte. Reproduced with permission from Ref. 7. © 2012 Nature Publishing Group.

the cathode while electrons flow through the external circuit, delivering electrical energy. When charging, both lithium ions and electrons flow in the opposite direction under an external voltage, storing electrical energy as chemical energy in the cell.

There are alternative configurations of lithium/sulfur cells, depending on where lithium sources are located and what materials are used as anodes. When prelithiated sulfur (e.g., Li$_2$S) is used as the cathode, other high-capacity electrode materials (e.g., silicon- or tin-based compounds, which can form alloys with lithium) are often used as the anodes with improved safety [8, 9]. As the cathode is in the fully discharged state (Li_2S), the cell needs to be charged first, similar to the conventional lithium ion batteries. After the first charge, the cell (especially the reactions in cathode side) behaves more or less the same as the aforementioned lithium/sulfur configuration. For this reason, even though some of these cells do not use lithium metal anode, we sometimes still refer them as lithium/sulfur cells.

In general, the reversible electrochemical reaction in a lithium/sulfur cell can be described as in the following, assuming the formation of Li_2S from elemental sulfur during discharge, and vice versa:

$$S_8 + 16Li^+ + 16e^- \leftrightarrow Li_2S \,(cathode)$$

$$16\,Li \leftrightarrow 16Li^+ + 16e^- \,(anode)$$

In the organic electrolyte, the solid sulfur (S_8 is the most stable form at standard temperature and pressure) is typically reduced to form polysulfides during the initial process of discharge. These polysulfides are soluble in many organic solvent-based electrolytes. Typical charge–discharge profiles of lithium/sulfur cells are shown in Figure 11.2b; several plateaus appear in the discharge curve, indicating the discharge of sulfur cell proceeds through multiple steps:

$$S_8 + 2Li^- + 2e^- \rightarrow Li_2S_8$$

$$Li_2S_8 + 2Li^- + 2e^- \rightarrow 2Li_2S_4$$

$$Li_2S_4 + 2Li^- + 2e^- \rightarrow 2Li_2S_2$$

$$Li_2S_2 + 2Li^- + 2e^- \rightarrow 2Li_2S$$

The high-voltage plateau (2.4–2.1 V) is related to the reduction of elemental sulfur to the higher-order lithium polysulfides (Li_2S_n, $n \geq 4$). Li_2S_8 is unstable in many aprotic electrolytes and undergoes disproportionation to form Li_2S_n ($n \geq 4$) and further reduction of high-order polysulfides to low-order polysulfides (Li_2S_n, $n < 4$), and lithium sulfide occurs at the low-voltage plateau (<2.1 V). Because of the low solubility and the slow kinetics of Li_2S_2 in typical electrolytes, further reduction to Li_2S may not be completed. The mechanistic details of the electrochemical reduction process of lithium polysulfides are very complex and need deeper understanding. Reaction pathways may be also quite different depending on the composition of the electrolyte.

11.2.2 Scientific Problems

Although significant progress has been made during the recent years, some existing issues are still obvious, such as capacity fading during cycling. We should understand the mechanistic details of the complex and interrelated processes in order to find solutions to further extend the cycle life of lithium/sulfur cells.

11.2.2.1 Dissolution and Shuttle Effect of Lithium Polysulfides In the early stages of the discharge process, the elemental sulfur is reduced to form soluble high-order polysulfides, Li_2S_n ($n \geq 4$), at the cathode and create a concentration gradient inside the cell. These high-order polysulfides can diffuse through the separator to the anode because of their higher concentration at the cathode and undergo chemical reactions with lithium to form low-order polysulfides and even insoluble and insulated Li_2S or Li_2S_2 on the surface of lithium anode. Once insulating Li_2S forms on the lithium surface by chemical reaction of polysulfide anions with lithium metal, this passivation layer can make the cycling efficiency of the lithium electrode worse and increase the cell resistance. These lower-order polysulfides can diffuse back to the sulfur electrode and can be oxidized to higher-order polysulfides. This phenomenon, the diffusion back and forth of polysulfides between the two electrodes, is known as the "polysulfide shuttle." The solubility and dissolved properties of lithium polysulfides in the electrolyte can lead to the loss of sulfur from the positive electrode, reduction of the charge–discharge efficiency especially at the high-voltage plateau, and corrosion of the lithium metal and self-discharge [10, 11].

11.2.2.2 Insulating Nature of Sulfur and Li_2S The poor intrinsic conductivity (5×10^{-30} S/cm at 25°C) of sulfur often leads to low electrochemical utilization and limited rate capability, which necessitates intimate contact with conductors (e.g., porous carbons) [12]. Furthermore, stoichiometric Li_2S has an electronic resistivity greater than 10^{-14} Ω/cm, and the Li^+ diffusivity in Li_2S is as low as 10^{-15} cm^2/s [13]. Once a thin Li_2S layer completely covers the whole electrode, further lithiation will be largely impeded and the voltage decreases rapidly. Consequently, complete conversion of sulfur to Li_2S is difficult, and most reports show discharge capacity less than 80% of the theoretical limit.

11.2.2.3 Volume Change of the Sulfur Electrode during Cycling Sulfur has a density of 2.03 g/cm^3, while Li_2S is much lighter (1.66 g/cm^3). As a result, the volume expansion/contraction of the active electrode material during cycling may be as large as 76%, which leads to the deterioration of microstructure or architecture of the electrodes and the pulverization of sulfur electrodes.

11.2.3 Research Strategy

To enhance the electrode kinetics, improve the cyclability, and reduce the energy loss associated with the discharge–charge overpotentials, sulfur-based cathodes with high conductivity and porous nanostructures/nanoarchitectures were introduced. As discussed before, the major capacity loss results from the dissolution of lithium polysulfides and their high mobility in organic electrolytes [10, 14]. The ideal sulfur electrode must restrain the migration of the polysulfide during cell operation. When dissolved polysulfides are reduced to solid Li_2S and/or Li_2S_2, these solid phases should be uniformly deposited inside the cathode in order to allow for complete reaction of all the sulfur. Additionally, the ideal electrode has to maintain a porously structural integrity to provide pathways for lithium ions while retaining good electrical connectivity to facilitate electron transport. During charging, solid sulfur should deposit inside the porous electrode to maintain close contact with the conductive carbon.

Fortunately, recent advances in carbon nanoscience and nanotechnology have offered exciting opportunities for the development of lithium/sulfur cells [5, 15–17]. The capability to synthesize nanostructured carbon materials with tailored pores and high surface area architectures holds a great potential to maximize sulfur loading and constrain dissolved polysulfides within the cathodes, thereby dramatically improving specific energy and cycling performance. Their advantages are briefly summarized as follows:

1. Porous structures with readily tunable pore sizes and shapes offer reservoir to constrain liquid polysulfides inside the cathode. Carbon materials provide possibilities for surface modification to obtain multifunctionality in order to enhance surface reactivity. Nanocarbon with large surface area can increase the number of active sites for electrode reactions by enlarging the contact area and bonding between the carbon and the sulfur.

2. The electronic conductivity of insulted sulfur can be enhanced by the formation of sulfur–carbon nanocomposites, in which the extended interfaces between the phases ensure fast transport of electronic species. A large surface-to-volume ratio can increase the contact area between the active material and the distributed current collectors to provide for efficient utilization of the active materials, which in turn increases usable specific energy and reduces electrode polarization. Furthermore, nanostructured sulfur electrodes with short diffusion lengths associated with the nanocarbon can effectively reduce the distance that lithium and electrons must travel in the solid state through electrode materials during operation.

3. The pores can also be used to buffer the strain as a result of volume change inside the structure if an empty pore space is included in the design.

In the following section, the latest advancements in design, synthesis, and optimization of various nanostructured sulfur–carbon electrodes will be discussed.

11.3 NANOSTRUCTURE CARBON–SULFUR

The development of nanostructured carbon–sulfur cathodes can be divided into several categories based on the structure of carbon materials: (i) porous carbon–sulfur composites, (ii) one-dimensional carbon–sulfur composites, (iii) two-dimensional carbon (graphene)–sulfur composites, (iv) free-standing films. The advances in these materials systems have dramatically improved the performance of sulfur cathodes by use of nanocarbon and have broadened people's understanding of the lithium/sulfur electrochemical system.

11.3.1 Porous Carbon–Sulfur Composite

The porous carbon–sulfur composite is composed of porous carbon and sulfur where the majority of sulfur is loaded inside the pores. The porous carbon matrix helps to confine the dissolved polysulfides and improves the electronic conductivity of the sulfur cathode. Earlier work performed by Wang et al. [18, 19] used activated carbon with a pore size of around 2.5 nm as the conductive matrix and achieved a reversible capacity of 400 mAh/g.

A mesoporous material is a material containing pores with diameters between 2 and 50 nm. Since its development in the 1970s, the research activities in this field have grown steadily inspired by its potential applications in catalysis, gas sensing, and ion exchange. The use of mesoporous carbon also brings about a huge performance leap in lithium/sulfur cell. Recently, Nazar and coworkers reported that carbon–sulfur cathodes based on highly ordered composites of sulfur and mesoporous carbon (CMK-3), prepared by a simple "melt–diffusion strategy" at 155°C, can exhibit high reversible capacity with good cycling performance and efficiency, as shown in Figure 11.3 [20]. The use of a conductive mesoporous carbon framework could constrain sulfur within its pores/channels, suppress polysulfide diffusion, and reduce the polysulfide shuttle. To further constrain polysulfides, the surface properties of the carbon–sulfur composite were modified by coating with polyethylene glycol (PEG) after sulfur infiltration. The PEG-modified material could effectively trap highly polar polysulfide species by providing a hydrophilic chemical gradient on the surface, thus limiting the concentration of polysulfide anions in the electrolyte. By linking the carbon surface with PEG to further trap polysulfides, the initial discharge capacity of the CMK-3/sulfur composite electrode was raised to 1320 mAh/g from 1005 mAh/g

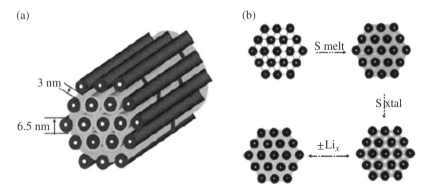

FIGURE 11.3 (a) A schematic diagram of the sulfur (gray) confined in the interconnected pore structure of mesoporous carbon (black), CMK-3, formed from carbon tubes that are propped apart by carbon nanofibers. (b) Schematic diagram of composite synthesis by impregnation of molten sulfur, followed by its densification on crystallization. The lower diagram represents subsequent discharging–charging with Li, illustrating the strategy of pore filling to tune for volume expansion/contraction. Reproduced with permission from Ref. 20. © 2009 Nature Publishing Group.

without PEG. These values are much higher than earlier reports, indicating the importance of incorporating sulfur into mesopores of carbon nanostructures. The deposition of solid Li_2S on the cathode surface was successfully inhibited, indicating the importance of "polymer protection" to obtain good cyclability of sulfur electrodes. However, the authors reported cycling performance up to only 20 cycles, still insufficient for lithium/sulfur cells to be considered as a practically viable option. Additional strategies to trap or immobilize lithium polysulfides during cycling are needed.

A porous hollow carbon sphere–sulfur composite has also been demonstrated to improve the cycle life [21]. The carbon sphere has a hollow core with a diameter of about 200 nm and a mesoporous carbon shell with a thickness of approximately 40 nm. Sulfur is stored in the porous shell and can expand toward the inside of the sphere during lithiation. A discharge capacity of over 950 mAh/g was demonstrated with a capacity decay of only approximately 10% per [22] 100 cycles.

By screening over various mesoporous carbons with different mesopore sizes and pore volumes, Li et al. concluded that (i) large pore volume is good for increasing sulfur percentage and capacity of the composite; (ii) empty pore space, that is, partial sulfur loading, is crucial for facile supply of lithium ions and adsorption of polysulfides; and (iii) polymer coating of mesoporous carbon is important for better cathode performance [23]. It is important to emphasize that the empty space in carbon–sulfur composites is necessary for buffering the volume change during lithiation–delithiation of the cathode.

Compared to mesoporous carbon, a microporous material mainly contains pores with diameters less than 2 nm. When used in lithium/sulfur cells, the composite with microporous carbon spheres and 42 wt% sulfur presented a long electrochemical stability up to 500 cycles by Gao et al. [24]. The electrochemical reaction constrained inside the narrow micropores due to strong adsorption would be the dominant factor for the enhanced long stability of the sulfur cathode. Wang et al. demonstrated better stability of sulfur cycling performance by using a hierarchical microporous–mesoporous carbon as a matrix [25]. The micropores could act as solvent-restricted reactors for sulfur lithiation that promise long cycle stability, while the mesopores provide an ion migration pathway for electrolyte. The cathode is able to operate reversibly over 800 cycles with a 1.8C discharge–recharge rate.

A desolvation effect could be used to interpret the unusual stability of sulfur in microporous carbon [15]. As generally accepted, the sulfur loss is a solid (S)–liquid (polysulfides)–solid (Li_2S_2/Li_2S) process, and the mesopores stabilize sulfur through the adsorption of dissolved polysulfide ions (Fig. 11.4a). According to the ion-desolvation theory, the polysulfides in micropores that are devoid of solvent react with desolvated lithium ions (Fig. 11.4b). This suggests a quasisolid-state reaction of the sulfur under solvent-deficient conditions. The low Li^+ conductivity in S and solid sulfides could also explain the retarded lithiation of sulfur in micropores.

It is also noteworthy that pore size affects the voltage profile and electrochemistry of the sulfur cathode. Reports on mesoporous and macroporous carbon–sulfur composites show the typical two plateau behavior of the sulfur electrode [22, 26]. In contrast, microporous carbon/sulfur composites have a good cycle life but a different

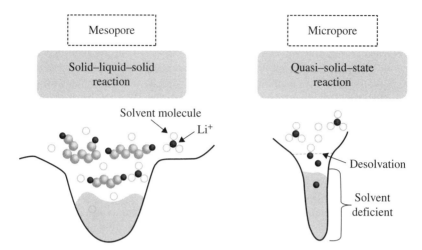

FIGURE 11.4 Illustration of the different lithiation mechanisms of sulfur confined in meso-pores and micropores. Reproduced with permission from Ref. 15. © 2013 Royal Society of Chemistry.

voltage profile, with a large capacity between 1.5 and 2 V versus Li/Li+ [24]. Wan et al. proposed that cyclo-S_{5-8} molecules with at least two dimensions larger than 0.5 nm cannot exist inside the micropores, while small S_{2-4} molecules with at least one dimension less than 0.5 nm can be hosted [27]. Consequently, the corresponding Gibbs free energy of the reaction is different.

In summary, porous carbon–sulfur composites represent a simple method to store the easily dissolved sulfur within the pores of carbon, as shown in the examples earlier. To achieve better performance, the following aspects should be carefully taken into account: (i) reaching an optimized combination of different pore sizes and shapes to trap polysulfides and to facilitate the lithium ion transport and (ii) increasing pore volume to maximize the content of sulfur for practical applications.

11.3.2 One-Dimensional Carbon–Sulfur Composite

Carbon nanotubes (CNTs) are allotropes of carbon with a long, hollow, cylindrical nanostructure. CNTs usually have a very large length-to-diameter ratio. These cylindrical carbon molecules have unusual properties, which are valuable for electronics, optics, and other fields of materials science and technology. In particular, owing to their extraordinary thermal conductivity and mechanical and electrical properties, CNTs find applications as additives to various composites.

Recently, Han et al. reported that the use of multiwalled carbon nanotubes (MWCNTs) instead of acetylene black in the cathode can increase not only the rate capability but also the cycle life of their sulfur electrode [28]. MWCNTs should be an attractive choice as the conductive additives for sulfur cathodes because they can provide a more effective electronically conductive network than the traditional conductive additives based on carbon blacks.

The interior of CNTs is hollow and favorable for holding sulfur inside the narrow channel. Wang et al. prepared hollow CNTs by a template wetting technique using commercial anodic aluminum oxide (AAO) membranes as the template and then incorporated sulfur into graphitized carbon layers by the "sulfur vapor infusion" technique. Through sulfur vapor infusion, S_6 or S_2 small molecules at high temperature were capable of diffusing through a narrow pore channel in the shell into a hollow core and enable sulfur–carbon bonding so that the conventional Li–S_8 reaction route with dissolvable polysulfide intermediate products may be altered. The small size of the narrow pore channel in a carbon shell can effectively prevent the liquid electrolyte penetration and avoid polysulfide dissolution [29].

Carbon nanofibers (CNFs) are also cylindrical nanostructures with graphene layers arranged orderly or randomly as stacked cones, cups, or plates. They have been utilized in different material systems like composites, thanks to their exceptional properties and low cost. Zhang et al. reported the synthesis of sulfur encapsulated in porous CNFs via electrospinning, carbonization, and solution-based chemical reaction–deposition method [30]. These novel porous carbon nanofiber–sulfur (CNF–S) composites with various S loadings showed high reversible capacity, good discharge capacity retention, and enhanced rate capability when they were used as cathodes in rechargeable Li/S cells. They demonstrated that an electrode prepared from a porous CNF–S nanocomposite with 42 wt% S maintains a stable discharge capacity of about 1400 mAh/g at 0.05C, 1100 mAh/g at 0.1C, and 900 mAh/g at 0.2C. The good electrochemical performance is attributed to the high electrical conductivity and the extremely high surface area of the CNFs that homogeneously disperse and immobilize S in their porous structures, alleviating the polysulfide shuttle phenomenon. SEM measurements showed that the porous CNF structures remained nearly unchanged even after 30 cycles' discharging–charging at 0.05C, as shown in Figure 11.5.

Hollow CNFs were also reported as a desirable host for a sulfur cathode mainly because of the limited diffusion pathways of polysulfides. This hollow design was

(a)   (b)

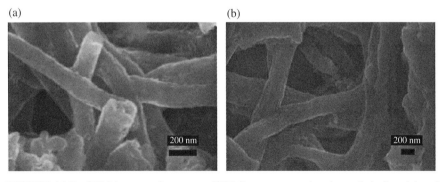

FIGURE 11.5 SEM images of a porous CNF–S nanocomposite electrode before (a) and after (b) 30 charge–discharge cycles at a constant rate of 0.05C. Reproduced with permission from Ref. 30. © 2011 Royal Society of Chemistry.

believed to contribute the following characteristics: (i) a closed structure for efficient polysulfides containment, (ii) a limited surface area for sulfur–electrolyte contact, (iii) a sufficient space to accommodate sulfur volumetric expansion/shrinkage, and (iv) a short electron and Li ion transport pathway [31].

11.3.3 Two-Dimensional Carbon (Graphene)–Sulfur

Graphene is a single-atomic thick planar sheet of carbon atoms that are packed in a honeycomb crystal lattice [32]. Graphene oxide is an oxidized form of graphene, where oxygen atoms bond to the lattice in forms such as hydroxyl and carbonyl groups. Graphene (oxide) has high surface area, high mechanical strength, and high flexibility, making it an ideal support material to immobilize active materials for lithium battery applications [22, 33]. The strong electronic coupling with insulating active materials renders the composites conducting, which significantly increases the available specific capacity and rate capability of the composite electrode materials [34].

Recently, a simple and environmentally benign chemical reaction–deposition–melt–diffusion strategy to anchor sulfur on quasi-two-dimensional graphene oxide (GO) was reported [33]. This approach provides a close contact between sulfur and carbon, which would not necessarily be the case for materials made by other methods such as ball milling. After a nanoscale-thickness sulfur coating was deposited onto GO sheets by chemical reaction in a microemulsion system, the as-prepared samples were heat treated in order to melt some bulk sulfur particles, so the bulk sulfur can diffuse into the pores of GO due to strong adsorption effects derived from the high surface area and the electrostatic interaction from the functional groups on the surface of the GO. The unique structure of the GO–S nanocomposite can significantly improve the overall electrochemical performance. The low-temperature heat-treated GO still contains various kinds of functional groups, which can chemically absorb and anchor sulfur atoms and effectively prevent the subsequently formed lithium polysulfides from dissolving into the electrolyte during cycling. This chemical absorption mechanism was confirmed by ab initio calculations and X-ray absorption spectroscopy (XAS) measurement as shown in Figure 11.6. Hydroxyl and epoxy groups can enhance the binding of S to the C atoms due to the induced lattice ripples by functional groups, showing the important roles of functional groups in extending the cycle life of sulfur electrodes. Indeed, novel GO–S nanocomposite electrodes displayed good reversibility and excellent cycling stability up to 50 cycles in the early report. To further confine sulfur to the cathode, the GO–S nanocomposite was coated with a surfactant called cetyltrimethylammonium bromide, which has a strong affinity for sulfur. This improvement, together with a new mixed ionic liquid–organic electrolyte and a new elastomeric binder, has resulted in a cycle life as long as 1500 cycles and a rate capability as high as 6C (Fig. 11.7) [35].

Graphene can also be coated onto the surface of sulfur particles or form a closed structure to wrap sulfur nanoparticles. Solution processing is an effective method to form a uniform sulfur–graphene (oxide) composite with a closed structure [33, 36]. Wang et al. demonstrated the synthesis of submicron-sized sulfur particles by reacting sodium thiosulfate with hydrochloric acid in an aqueous solution along with Triton

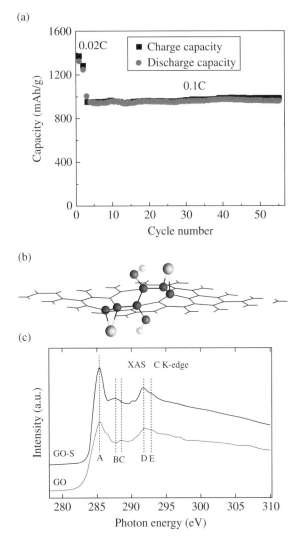

FIGURE 11.6   (a) Cycling performance of GO–S electrodes. (b) Representative drawing of GO immobilizing sulfur. Yellow, red, and white balls represent S, O, and H atoms, while others are C atoms. (c) C K-edge XAS spectra of GO and GO–S nanocomposites after heat treatment. Reproduced with permission from Ref. 33. © 2011 American Chemical Society.

X-100 surfactant, which was used as a capping agent to limit the size of the sulfur particles [36]. During the synthesis, well-dispersed mildly oxidized graphene (mGO) solution was added to the reactor. Since the graphene oxide solution is not stable under acidic conditions, a layer of graphene oxide precipitates out and coats the sulfur particles. The as-synthesized sulfur particles are less than 1 mm in size, and the surface is coated with Triton X-100 and mGO in sequence. The advantage of this structure is that during expansion of sulfur, the graphene oxide layers can adjust their position and configuration to accommodate the volumetric strain. This rational approach leads to stable specific capacities of around 600 mAh/g and decay of only 10–15% per 100 cycles. However, the capacity of this electrode is not as high as the porous carbon–sulfur composite. A possible reason is that lithium ions cannot directly penetrate the hexagonal carbon lattice but need to find boundaries between mGO sheets to diffuse toward the sulfur particles, which impedes fast charge and discharge.

In general, graphene (oxide)–sulfur composites have similar advantages as other porous carbon–sulfur composites. Furthermore, graphene (oxide) can also be readily functionalized to enhance its affinity for sulfur.

11.3.4 Three-Dimensional Carbon Paper–Sulfur

In recent years, it has been demonstrated that the improved battery performance can be achieved by reconfiguring the electrode materials of granular form into three-dimensional (3D) carbon architectures without any polymeric binder [37, 38]. The use of binder-free 3D carbon will allow for a high load of sulfur while maintaining good electronic conductivity owing to its intrinsic carbon network, as shown in Figure 11.7 [39]. Obviously, a higher sulfur load should be favorable to a higher specific capacity on account of the whole mass of cathode [40].

Note that one failure mechanism of the sulfur composite cathodes fabricated as mixture of granular sulfur and carbon power is the breakdown of the electrodes' matrices due to the stresses as a result of volume change [41]. It causes internal resistance increase, rapid capacity fading, and low coulombic efficiency of Li/S cells. The use of 3D carbon papers with the inherent mechanical strength and flexible property can accommodate the stress to some extent, leading to the improved cycling life. Recently, activated carbon fiber (ACF) cloth has been demonstrated as a binder-free cathode support for sulfur. This ACF cloth is monolithic and holds the sulfur in the micropores. The structural integrity of the monolithic ACF cloth is sufficient to withstand the volume variation during lithiation and delithiation. Meanwhile, the abundant micropores that can accommodate sulfur promised excellent cathode stability [40]. Sulfur-coated CNT paper [42] and graphene films [43] were also assembled into papers for use as flexible cathodes.

Zhou et al. developed a highly conducting carbon–sulfur nanotube membrane with good mechanical resilience. This self-supporting composite electrode can tolerate a 10 MPa stress with a 9% strain over 12,000 times while maintaining unchanged conductivity (800 S/m) (Fig. 11.8b) [39]. The unique mechanical–electrical properties arise from the molecular level mixing of carbon with sulfur where the sulfur is intercalated between the graphitic local domains, that is, residing in the tube walls.

11.3.5 Preparation Method of Sulfur–Carbon Composite

The earliest method of sulfur–carbon composite preparation is ball milling sulfur powder together with carbon materials [28]. The poor contact between sulfur and carbon resulted in a low utilization of sulfur and a specific capacity of

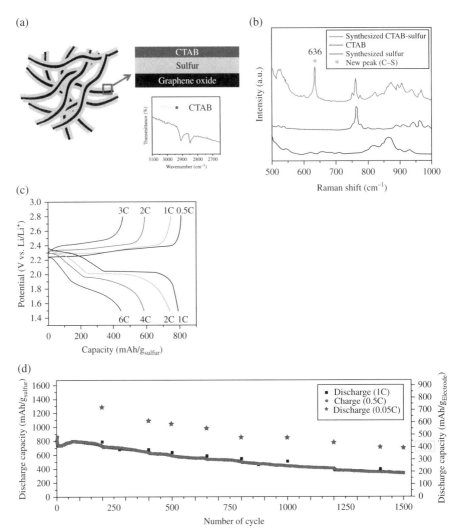

FIGURE 11.7   (a) Schematic of the S–GO nanocomposite structure. The presence of CTAB on the S–GO surface was confirmed by Fourier transform infrared spectroscopy (FTIR) and shown to be critical for achieving improved cycling performance by minimizing the loss of sulfur. (b) Enlarged view of Raman spectra on CTAB, synthesized sulfur, and CTAB-modified sulfur from 500 to 1000 cm⁻¹. It clearly shows the formation of a new peak, which can be assigned as a C–S bond (600–700 cm⁻¹), confirming that there is strong interaction between CTAB and sulfur. (c) Voltage profiles of CTAB-modified S–GO composite cathodes at different rates. (d) Long-term cycling test results of the Li/S cell with CTAB-modified S–GO composite cathodes. This result represents the longest cycle life (exceeding 1500 cycles) with an extremely low decay rate (0.039% per cycle) demonstrated so far for a Li/S cell. The S–GO composite contained 80% S, and elastomeric SBR/CMC binder was used. 1 M LiTFSI in PYR₁₄TFSI/DOL/DME mixture (2:1:1 by volume) with 0.1 M LiNO₃ was used as the electrolyte (total 60 μl). Reproduced with permission from Ref. 35. © 2013 American Chemical Society. (*See insert for color representation of the figure.*)

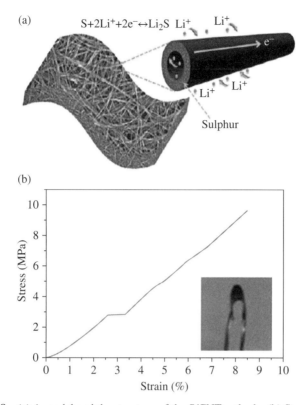

FIGURE 11.8 (a) A model and the structure of the S/CNT cathode. (b) Stress–strain curve of a flexible S/CNT membrane cathode. Inset shows a bent S/CNT membrane. Reproduced with permission from Ref. 39. © 2012 Royal Society of Chemistry.

300–500 mAh/g. As an improved method to increase the bonding of sulfur, Wang et al. introduced a two-step sulfur melting route [18], involving a thermal treatment at 200°C to melt sulfur that can diffuse into the pores of active carbon followed by a second heating at 300°C to vaporize sulfur covering the outer surface of carbon. The discharge capacity with this method was increased to 800 mAh/g during the first cycle. This heating process has been further modified with a suitable temperature and dwell time for different carbon materials to obtain high-performance sulfur–carbon composites [44].

Taking advantage of the relatively low sublimation temperature of sulfur, Jayaprakash et al. utilized a sulfur vaporizing route to infuse gaseous sulfur into porous hollow carbon and obtained outstanding cyclability, as the sulfur vapor led to molecular contacts with the carbon framework [21]. This method facilitates fast, efficient, and controlled infusion of elemental sulfur into the host carbon porous structure and yields particles with tap density of around 0.82 g/cm^3. Thermal gravimetric analysis shows that nearly 70% of the mass of the porous particles is comprised of sulfur by three passes and 35% sulfur can be incorporated in the particles in a single pass (i.e., repeat exposures of the porous carbon particles to sulfur vapor).

Solution processing of sulfur–carbon composite in an aqueous solution has become a favorable method not only because of the low cost but also due to the reaction being free of toxic solvents. Sulfur nucleates on dispersed carbon substrates after dropping acid into $Na_2S_2O_3$ solution or a polysulfide solution prepared by dissoluting of sulfur into a sodium sulfide (Na_2S) solution [30, 36]. High-purity sulfur heterogeneously nucleates on highly dispersed carbon nanoparticles in an aqueous solution as the thiosulfate ions are reduced or polysulfide ions are oxidized to sulfur. These synthesis methods involve a simple water-based reaction to bind sulfur and carbon strongly and are applicable to most carbon substrates.

11.4   CARBON LAYER AS A POLYSULFIDE SEPARATOR

Besides the fact that the porous carbon can play a role both as an electron conductor and as a trap for absorption of dissolved polysulfide species, the microporous carbon paper (MCP) can be used as the electrolyte-permeable but polysulfide-limiting membrane between the polymer separator and cathode (Fig. 11.9). This carbon interlayer can be treated as a second current collector for accommodating the migrating polysulfide from sulfur cathodes. On one side, the micropores in carbon interlayer used in this cell configuration localize the soluble polysulfide in cathodes and inhibit polysulfide shuttle in the electrolyte for long cycles; on the other side, this design can also decrease the resistance of sulfur cathodes, resulting in an enhancement of active material utilization. This approach could not only simplify the battery processing without elaborate synthesis of composites and surface chemistry modification but also improves the capacity and cycle life, thereby promoting the practical use of Li/S cells.

Recently, Manthiram et al. inserted an interlayer between the sulfur cathode and separator, such as an MCP [45] and free-standing MWCNT paper [46]. As exhibited in Figure 11.9, the cyclability at 1C rate retains over 1000 mAh/g after 100 cycles, converting to a retention rate of 85%. At a higher rate of 2C for long cycles, the cell with the MCP exhibits 846 mAh/g with an average coulombic efficiency greater than 98% after 150 cycles; it also shows a more considerable capacity increase during the first 40 cycles compared with the cell at a 1C rate.

Reduced graphene oxide (rGO) film is an ideal candidate of interlayer separator, where the oxygen functional groups on its basal planes and edges and the structural defects provide strong anchoring points for polysulfide, while its two-dimensional sheetlike structure promotes the formation of self-assembled films. Thus, rGO-based films are easily sandwiched between a sulfur cathode and the separator as a shuttle inhibitor. The work of Wang et al. indicated that the rGO-based interlayer enhances the cycling performance of the Li/S cell [47]. The cell with an rGO interlayer has an initial discharge capacity of 1260 mAh/g and remains 895 mAh/g after 100 cycles.

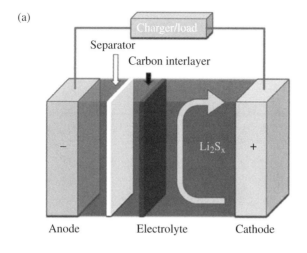

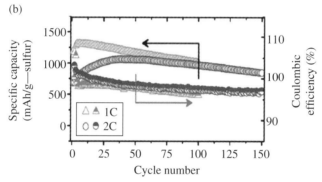

FIGURE 11.9 (a) Schematic configuration of a Li/S cell with a bifunctional microporous carbon interlayer inserted between the sulfur cathode and the separator. (b) Cycle life and coulombic efficiency of the cell with MCP at 1C and 2C for long cycles. Reproduced with permission from Ref. 45. © 2012 Nature Publishing Group.

11.5 OPPORTUNITIES AND PERSPECTIVES

During the past several years, many promising approaches that involve various nanostructured carbon materials have been explored to develop high-performance Li/S cells. The rational designs help to solve the interrelated challenges of volume expansion, poor ionic and electronic transport, and polysulfide dissolution in the electrolyte and thus dramatically enhance the capacity, cycle life, and power capability of the Li/S cells.

We also acknowledge that the improvement of carbon–sulfur cathode materials alone cannot solve all the problems in the Li/S technology. Problems with the lithium anode, the electrolyte, as well as the engineering difficulties in fabricating lithium metal batteries all need to be addressed before the Li/S batteries can be widely adopted in applications. To meet this goal, more fundamental studies are required at the material, component, and system levels.

REFERENCES

[1] Armand, M.; Tarascon, J. M. Building better batteries, *Nature* 2008, **451**, 652–657.

[2] Dunn, B.; Kamath, H.; Tarascon, J.-M. Electrical energy storage for the grid: a battery of choices, *Science* 2011, **334**, 928–935.

[3] Ellis, B. L.; Lee, K. T.; Nazar, L. F. Positive electrode materials for Li-ion and Li-batteries, *Chem. Mater.* 2010, **22**, 691–714.

[4] Whittingham, M. S. Materials challenges facing electrical energy storage, *MRS Bull.* 2008, **33**, 411–419.

[5] Song, M.-K.; Cairns, E. J.; Zhang, Y. Lithium/sulfur batteries with high specific energy: old challenges and new opportunities, *Nanoscale* 2013, **5**, 2186–2224.

[6] Cairns, E. J.; Shimotake, H. High-temperature batteries, *Science* 1969, **164**, 1347–1355.

[7] Bruce, P. G.; Freunberger, S. A.; Hardwick, L. J.; Tarascon, J.-M. Li–O$_2$ and Li–S batteries with high energy storage, *Nat. Mater.* 2012, **11**, 172.

[8] Yang, Y.; McDowell, M. T.; Jackson, A.; Cha, J. J.; Hong, S. S.; Cui, Y. New nanostructured Li$_2$S/silicon rechargeable battery with high specific energy, *Nano Lett.* 2010, **10**, 1486–1491.

[9] Cai, K. P.; Song, M. K.; Cairns, E. J.; Zhang, Y. G. Nanostructured Li$_2$S–C composites as cathode material for high-energy lithium/sulfur batteries, *Nano Lett.* 2012, **12**, 6474–6479.

[10] Mikhaylik, Y. V.; Akridge, J. R. Polysulfide shuttle study in the Li/S battery system, *J. Electrochem. Soc.* 2004, **151**, A1969–A1976.

[11] Shim, J.; Striebel, K. A.; Cairns, E. J. The lithium/sulfur rechargeable cell effects of electrode composition and solvent on cell performance, *J. Electrochem. Soc.* 2002, **149**, A1321–A1325.

[12] Cheon, S.-E.; Ko, K.-S.; Cho, J.-H.; Kim, S.-W.; Chin, E.-Y.; Kim, H.-T. Rechargeable lithium sulfur battery II. Rate capability and cycle characteristics, *J. Electrochem. Soc.* 2003, **150**, A800–A805.

[13] Yang, Y.; Zheng, G.; Misra, S.; Nelson, J.; Toney, M. F.; Cui, Y. High-capacity micrometer-sized Li$_2$S particles as cathode materials for advanced rechargeable lithium-ion batteries, *J. Am. Chem. Soc.* 2012, **134**, 15387–15394.

[14] Aurbach, D.; Pollak, E.; Elazari, R.; Salitra, G.; Kelley, C. S.; Affinito, J. On the surface chemical aspects of very high energy density, rechargeable Li–sulfur batteries, *J. Electrochem. Soc.* 2009, **156**, A694–A702.

[15] Wang, D.-W.; Zeng, Q.; Zhou, G.; Yin, L.; Li, F.; Cheng, H.-M.; Gentle, I. R.; Lu, G. Q. M. Carbon–sulfur composites for Li–S batteries: status and prospects, *J. Mater. Chem. A* 2013, **1**, 9382–9394.

[16] Bruce, P. G.; Freunberger, S. A.; Hardwick, L. J.; Tarascon, J.-M. Li–O$_2$ and Li–S batteries with high energy storage, *Nat. Mater.* 2012, **11**, 19–29.

[17] Yang, Y.; Zheng, G.; Cui, Y. Nanostructured sulfur cathodes, *Chem. Soc. Rev.* 2013, **42**, 3018–3032.

[18] Wang, J. L.; Yang, J.; Xie, J. Y.; Xu, N. X.; Li, Y. Sulfur–carbon nano-composite as cathode for rechargeable lithium battery based on gel electrolyte, *Electrochem. Commun.* 2002, **4**, 499–502.

[19] Wang, J.; Liu, L.; Ling, Z.; Yang, J.; Wan, C.; Jiang, C. Polymer lithium cells with sulfur composites as cathode materials, *Electrochim. Acta* 2003, **48**, 1861–1867.

[20] Ji, X.; Lee, K. T.; Nazar, L. F. A highly ordered nanostructured carbon–sulphur cathode for lithium–sulphur batteries, *Nat. Mater.* 2009, **8**, 500–506.

[21] Jayaprakash, N.; Shen, J.; Moganty, S. S.; Corona, A.; Archer, L. A. Porous hollow carbon@sulfur composites for high-power lithium–sulfur batteries, *Angew. Chem. Int. Ed.* 2011, **50**, 5904–5908.

[22] Schuster, J.; He, G.; Mandlmeier, B.; Yim, T.; Lee, K. T.; Bein, T.; Nazar, L. F. Spherical ordered mesoporous carbon nanoparticles with high porosity for lithium–sulfur batteries, *Angew. Chem. Int. Ed.* 2012, **51**, 3591–3595.

[23] Li, X.; Cao, Y.; Qi, W.; Saraf, L. V.; Xiao, J.; Nie, Z.; Mietek, J.; Zhang, J.-G.; Schwenzer, B.; Liu, J. Optimization of mesoporous carbon structures for lithium–sulfur battery applications, *J. Mater. Chem.* 2011, **21**, 16603–16610.

[24] Zhang, B.; Qin, X.; Li, G. R.; Gao, X. P. Enhancement of long stability of sulfur cathode by encapsulating sulfur into micropores of carbon spheres, *Energy Environ. Sci.* 2010, **3**, 1531–1537.

[25] Wang, D.-W.; Zhou, G.; Li, F.; Wu, K.-H.; Lu, G. Q.; Cheng, H.-M.; Gentle, I. R. A microporous–mesoporous carbon with graphitic structure for a high-rate stable sulfur cathode in carbonate solvent-based Li–S batteries, *Phys. Chem. Chem. Phys.* 2012, **14**, 8703–8710.

[26] He, G.; Ji, X.; Nazar, L. High "C" rate Li-S cathodes: sulfur imbibed bimodal porous carbons, *Energy Environ. Sci.* 2011, **4**, 2878–2883.

[27] Xin, S.; Gu, L.; Zhao, N.-H.; Yin, Y.-X.; Zhou, L.-J.; Guo, Y.-G.; Wan, L.-J. Smaller sulfur molecules promise better lithium–sulfur batteries, *J. Am. Chem. Soc.* 2012, **134**, 18510–18513.

[28] Han, S.-C.; Song, M.-S.; Lee, H.; Kim, H.-S.; Ahn, H.-J.; Lee, J.-Y. Effect of multi-walled carbon nanotubes on electrochemical properties of lithium/sulfur rechargeable batteries, *J. Electrochem.l Soc.* 2003, **150**, A889–A893.

[29] Guo, J.; Xu, Y.; Wang, C. Sulfur-impregnated disordered carbon nanotubes cathode for lithium–sulfur batteries, *Nano Lett.* 2011, **11**, 4288–4294.

[30] Ji, L. W.; Rao, M. M.; Aloni, S.; Wang, L.; Cairns, E. J.; Zhang, Y. G. Porous carbon nanofiber–sulfur composite electrodes for lithium/sulfur cells, *Energy Environ. Sci.* 2011, **4**, 5053–5059.

[31] Zheng, G. Y.; Yang, Y.; Cha, J. J.; Hong, S. S.; Cui, Y. Hollow carbon nanofiber-encapsulated sulfur cathodes for high specific capacity rechargeable lithium batteries, *Nano Lett.* 2011, **11**, 4462–4467.

[32] Geim, A. K.; Novoselov, K. S. The rise of graphene, *Nat. Mater.* 2007, **6**, 183–191.

[33] Ji, L. W.; Rao, M. M.; Zheng, H. M.; Zhang, L.; Li, Y. C.; Duan, W. H.; Guo, J. H.; Cairns, E. J.; Zhang, Y. G. Graphene oxide as a sulfur immobilizer in high performance lithium/sulfur cells, *J. Am. Chem. Soc.* 2011, **133**, 18522–18525.

[34] Wang, H.; Yang, Y.; Liang, Y.; Cui, L.-F.; Sanchez Casalongue, H.; Li, Y.; Hong, G.; Cui, Y.; Dai, H. $LiMn_{1-x}Fe_xPO_4$ nanorods grown on graphene sheets for ultrahigh-rate-performance lithium ion batteries, *Angew. Chem. Int. Ed.* 2011, **50**, 7364–7368.

[35] Song, M.-K.; Zhang, Y.; Cairns, E. J. A long-life, high-rate lithium/sulfur cell: a multifaceted approach to enhancing cell performance, *Nano Lett.* 2013, **13**, 5891–5899.

[36] Wang, H. L.; Yang, Y.; Liang, Y. Y.; Robinson, J. T.; Li, Y. G.; Jackson, A.; Cui, Y.; Dai, H. J. Graphene-wrapped sulfur particles as a rechargeable lithium–sulfur battery cathode material with high capacity and cycling stability, *Nano Lett.* 2011, **11**, 2644–2647.

[37] Long, J. W.; Dunn, B.; Rolison, D. R.; White, H. S. Three-dimensional battery architectures, *Chem. Rev.* 2004, **104**, 4463–4492.

[38] Zhang, H.; Yu, X.; Braun, P. V. Three-dimensional bicontinuous ultrafast-charge and -discharge bulk battery electrodes, *Nat. Nanotechnol.* 2011, **6**, 277–281.

[39] Zhou, G.; Wang, D.-W.; Li, F.; Hou, P.-X.; Yin, L.; Liu, C.; Lu, G. Q.; Gentle, I. R.; Cheng, H.-M. A flexible nanostructured sulphur–carbon nanotube cathode with high rate performance for Li-S batteries, *Energy Environ. Sci.* 2012, **5**, 8901–8906.

[40] Elazari, R.; Salitra, G.; Garsuch, A.; Panchenko, A.; Aurbach, D. Sulfur-impregnated activated carbon fiber cloth as a binder-free cathode for rechargeable Li-S batteries, *Adv. Mater.* 2011, **23**, 5641–5645.

[41] Elazari, R.; Salitra, G.; Talyosef, Y.; Grinblat, J.; Scordilis-Kelley, C.; Xiao, A.; Affinito, J.; Aurbach, D. Morphological and structural studies of composite sulfur electrodes upon cycling by HRTEM, AFM and Raman spectroscopy, *J. Electrochem. Soc.* 2010, **157**, A1131–A1138.

[42] Fu, Y.; Su, Y.-S.; Manthiram, A. Highly reversible lithium/dissolved polysulfide batteries with carbon nanotube electrodes, *Angew. Chem. Int. Ed.* 2013, **52**, 6930–6935.

[43] Jin, J.; Wen, Z.; Ma, G.; Lu, Y.; Cui, Y.; Wu, M.; Liang, X.; Wu, X. Flexible self-supporting graphene–sulfur paper for lithium sulfur batteries, *RSC Adv.* 2013, **3**, 2558–2560.

[44] Manthiram, A.; Fu, Y.; Su, Y.-S. Challenges and prospects of lithium-sulfur batteries, *Acc. Chem. Res.* 2012, **46**, 1125–1134.

[45] Su, Y.-S.; Manthiram, A. Lithium–sulphur batteries with a microporous carbon paper as a bifunctional interlayer, *Nat. Commun.* 2012, **3**, 1166.

[46] Su, Y.-S.; Manthiram, A. A new approach to improve cycle performance of rechargeable lithium–sulfur batteries by inserting a free-standing MWCNT interlayer, *Chem. Commun.* 2012, **48**, 8817–8819.

[47] Wang, X.; Wang, Z.; Chen, L. Reduced graphene oxide film as a shuttle-inhibiting interlayer in a lithium–sulfur battery, *J. Power Sources* 2013, **242**, 65–69.

12

LITHIUM–AIR BATTERIES BASED ON CARBON NANOMATERIALS

DORETTA CAPSONI, MARCELLA BINI, STEFANIA FERRARI
AND PIERCARLO MUSTARELLI

Department of Chemistry, University of Pavia, Pavia, Italy

12.1   METAL–AIR BATTERIES

Metal–air batteries are based on the electrochemical coupling of a reactive anode to an electrode open to the air (i.e., oxygen), which is the cathode-active material. In principle, this provides a battery with inexhaustible cathode reactant and, in some cases, a very high specific energy. This concept is absolutely not new and was reported already more than 30 years ago [1]. A very general and naïve scheme of such a cell is reported in Figure 12.1.

The capacity limits of metal–air batteries are determined by the anode and by the handling of the cathode reactants. From a historical point of view, the first systems under development were based on zinc, which is the most electropositive metal that can be electroplated in aqueous solution [2], by using neutral or alkaline electrolytes. The oxygen reduction half-cell reaction may be written as

$$O_2 - 2H_2O + 4e^- \rightarrow 4OH^- \qquad E^0 = 0.40V\,(vs.SHE) \qquad (12.1)$$

Some theoretical and practical specifics of metal–air cells, with reference to the cathodic half reaction (12.1), are reported in Table 12.1.

Both primary (i.e., nonrechargeable) and secondary (rechargeable) metal–air batteries were developed during the last 30 years. Among the primary ones, the

Carbon Nanomaterials for Advanced Energy Systems: Advances in Materials Synthesis and Device Applications, First Edition. Edited by Wen Lu, Jong-Beom Baek and Liming Dai.
© 2015 John Wiley & Sons, Inc. Published 2015 by John Wiley & Sons, Inc.

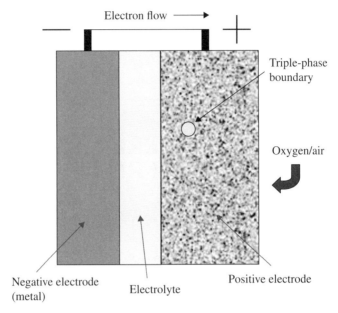

FIGURE 12.1   Scheme of a metal–air cell.

TABLE 12.1   **Characteristics of Metal–Air Cells with Oxygen Cathode and Aqueous Electrolyte**

Metal Anode	Metal Electrochemical Equivalent (Ah g⁻¹)	Theoretical Cell Voltage (V)	Practical Cell Voltage[a] (V)	Theoretical Specific Energy (Wh g⁻¹)
Li	3.86	3.4	2.4	13.0
Na	1.17	2.6	n.a.	3.0
Ca	1.34	3.4	2.0	4.6
Mg	2.20	3.1	1.4	6.8
Al	2.98	2.7	1.6	8.1
Zn	0.82	1.6	1.2	1.3
Fe	0.96	1.3	1.0	1.2

[a] Due to polarization effects.

Zn–air battery is used as a hearing-aid battery. Mg–air and Al–air cells found military applications in underwater propulsion and are under development as saline systems. Most commercial metal–air batteries, including Zn–air, Al–air, and Mg–air, may be recharged mechanically by replacing the discharged anode and the spent/degraded electrolyte with new components. Moreover, the spent electrolyte slurry containing metal oxides and oxide hydroxides can be recycled and regenerated as a metal using electrolysis or thermal decomposition [3]. The electrical recharging of metal–air batteries is a more challenging task, since it requires an electrode capable of both oxygen reduction reaction (ORR) during discharge and oxygen evolution reaction (OER) during charge. From the electrochemical point of view, it should be noted that the involved reactions, as well as the overall products (and by-products), may differ from each other among various types of metal–air batteries, depending on the metal, the electrolyte, and the catalytic materials. For more detailed information on metal–air batteries, see Ref. 4.

Generally speaking, metal–air batteries suffer from a series of technological problems involving both the anode and the cathode compartment. As far as the anode is concern, the major issues are passivation and corrosion due to the electrolyte, which decrease the overall capacity and the coulombic efficiency. Concerning the cathode, the most important problems are due to the high overpotentials and to the complexity of the oxygen chemistry, which may give origin to poorly reversible reactions whose mechanisms are often difficult to understand.

12.2 Li–AIR CHEMISTRY

Among the metal–air chemistries, rechargeable Li–air has been indeed the most investigated one during the last 10 years, it being the most promising alternative to Li-ion devices for massive applications in automotive and large-scale deployment of renewable energy [5]. The major appeal of the Li–air battery is given by its expected very high theoretical specific energy, that is, gravimetric energy density ($Wh\,kg^{-1}$), or energy density, that is, volumetric energy density ($Wh\,l^{-1}$). As a matter of fact, a value of $11,586\,Wh\,kg^{-1}$ is often quoted, which strongly exceeds that of methanol/air fuel cell ($5,524\,Wh\,kg^{-1}$), and is very near to that of gasoline engine ($11,860\,Wh\,kg^{-1}$) [6]. On the other hand, this value does not consider the addition of O_2 mass during cell discharge, and most reasonable values of 3505 and $3582\,Wh\,kg^{-1}$ have been recently reported in the case of nonaqueous and aqueous electrolytes, respectively [5] (see following text). However, Li–air batteries are expected to have the potential of about 10 times the energy density of Li-ion devices. To date, four different approaches have been reported in the literature, which substantially differ on the electrolyte nature, namely: (i) aqueous, (ii) nonaqueous aprotic, (iii) mixed aqueous/aprotic, and (iv) all solid state. In the following, the attention will be focused on the approaches (i) and (ii), which are indeed the most promising ones. It must be stressed, in any case, that significant improvements must be obtained in a variety of fields before commercial rechargeable products can be obtained.

12.2.1 Aqueous Electrolyte Cell

The aqueous Li–air cell consists of a lithium metal anode, an aqueous electrolyte, and a porous carbon cathode, in the case of hosting a metal-based catalyst. The aqueous electrolyte contains soluble lithium salts. This generally eliminates the problem of cathode clogging since the reaction products are also soluble in water. A further advantage of the aqueous design with respect to the aprotic one is given by the higher practical discharge potential than its aprotic counterpart. However, lithium metal strongly reacts with water, which requires a solid electrolyte interface

(SEI) between the lithium and electrolyte. Commonly, a lithium-conducting ceramic or glass is used [7], but the conductivity values are generally lower than $10^{-5}\,\text{S}\,\text{cm}^{-1}$ at room temperature (see Fig. 12.2). Another limitation of the aqueous cells is that the working potential window is limited to the potential where water is stable against hydrogen evolution reaction (HER) and OER, which are related to the cathode materials.

In principle, the electrolyte could be either acidic or alkaline in nature. In the case of acidic electrolyte, the cathode process can be written as

$$2Li + \tfrac{1}{2}O_2 + 2H^+ \rightarrow 2Li^+ + H_2O \qquad E^0 = 4.27V \qquad (12.2)$$

where a conjugate base is involved in the reaction. However, acidic electrolytes cause severe Li anode corrosion, and the related exothermic reactions make the overall thermal management very difficult. Moreover, it is difficult to find electrocatalysts that are stable under both the electrolyte acidic conditions and the oxidizing environment of the cathode. For these reasons, acidic configurations do not seem well suited for practical applications.

The aqueous alkaline approach has indeed the advantage of a more efficient ORR due to better kinetics and lower overpotentials with respect to the acidic one [8]. Moreover, the practical current densities are about one order of magnitude higher, which makes it possible to use nonnoble metal catalysts. On the other hand, the presence of CO_2 leads to carbonate precipitation, which can determine pore clogging and blocking of the electrode channels:

$$CO_2 + 2OH^- \rightarrow CO_3^{2-} + H_2O \qquad (12.3)$$

In order to avoid this problem, purified air and/or a selective anion-exchange membrane at the electrolyte/cathode interface are needed (see Fig. 12.2). The selective

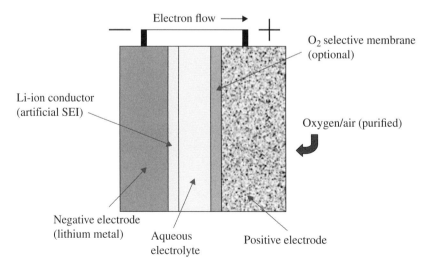

FIGURE 12.2 Scheme of a lithium–air cell with aqueous electrolyte.

membrane can also allow OH⁻ to be transported out of the cathode and to avoid Li⁺ from entering the electrode, so forcing LiOH formation outside the cathode itself.

In the case of alkaline aqueous electrolyte, the anode and the overall reactions are respectively given by

$$Li + OH^- \rightarrow LiOH + e^- \qquad E^0 = -2.95V \qquad (12.4)$$

$$4Li + O_2 + 2H_2O \rightarrow 4LiOH \qquad E^0 = 3.35V \qquad (12.5)$$

The ORR at the cathode, on the other hand, cannot be simply described by Equation (12.1), but is a complex process that includes multistep electron transfer processes and several oxygen-based species (12.4):

$$O_2 + H_2O + 2e^- \rightarrow HO_2^- + OH^- \qquad E^0 = -0.07V \qquad (12.6)$$

$$HO_2^- + H_2O + 2e^- \rightarrow 3OH^- \qquad E^0 = 0.87V \qquad (12.7)$$

$$2HO_2^- \rightarrow 2OH^- + O_2 \qquad (12.8)$$

As evidenced by reactions (12.1, 12.6–12.8), the ORR can proceed either by a four-electron (4e⁻) or a two-electron (2e⁻) pathway. Both the nature of the pathway and the related mechanisms are determined by the metal catalyst, by its crystallographic structure (surface) and by the binding energy. For insights, see Ref. 9. The abovementioned 4e⁻ and 2e⁻ reaction schemes may be concomitant and somehow competing. When searching for proper carbon cathode/catalysts combinations, the direct 4e⁻ path should be preferred due to a high energy efficiency, whereas the 2e⁻ path is less desirable both because of its lower efficiency and because the peroxide species are corrosive and may determine fast cell degradation [10]. It is now generally recognized that the 4e⁻ pathway predominates when noble metals are used as the catalyst, whereas the 2e⁻ one is more important with carbonaceous materials [11].

12.2.2 Nonaqueous Aprotic Electrolyte Cell

The aprotic cell consists of a lithium metal anode, a liquid organic electrolyte, and a porous carbon cathode (generally with a metal catalyst). The practical feasibility of a rechargeable cell with organic electrolyte was proven by Abraham and Jiang [12]. The obtained capacity was about 1300 mAh g⁻¹, and the cell was cycled several times. The organic electrolytes, very similar to those used for Li-ion cells, are based on ethers, esters, or carbonates, which can dissolve lithium salts such as $LiPF_6$, $LiAsF_6$, $LiSO_3CF_3$, and $LiN(SO_2CF_3)_2$ [13]. Recently, more performing electrolytes based on boron esters or ionic liquids (IL) have been proposed [14]. The carbon cathode is usually made of a high surface area (e.g., mesoporous) carbon-based material with a nanosized metal catalyst (e.g., MnO_2). A major advantage of this approach with respect to the aqueous cell is the spontaneous formation of a barrier between the anode and electrolyte (similar to the SEI formed between electrolyte and intercalation anode in Li-ion cells) that protects the lithium metal from further reaction with the

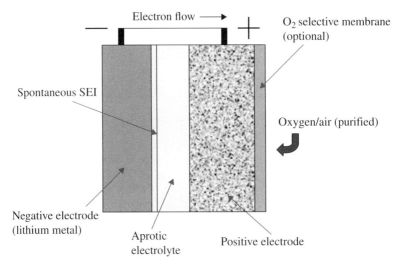

FIGURE 12.3   Scheme of a lithium–air cell with nonaqueous aprotic electrolyte.

electrolyte. A cathode-protected design similar to that discussed for aqueous cells can also be planned, for example, by employing polysiloxanes or perfluorocarbon-based membranes (see Fig. 12.3).

This O_2-specific membrane can be a film, a gel, or a liquid embedded in the pores of the carbon-based cathode or even applied to a porous Teflon® layer, which is laminated to the cathode. The membrane must (i) exhibit high O_2 solubility, (ii) retard the diffusion of moisture into the cell, and (iii) retard the diffusion of the organic solvents out of the cell.

The possible overall reactions for a nonaqueous cell can be summarized as follows, with two or four electrons transferred:

$$4Li + O_2 \rightarrow 2Li_2O \quad \left(E^0 = 2.91V \text{ vs. } Li^+/Li\right) \tag{12.9}$$

$$2Li + O_2 \rightarrow Li_2O_2 \quad \left(E^0 = 2.96 \text{ or } 3.10V \text{ vs. } Li^+/Li\right)^1 \tag{12.10}$$

Because of the very near standard potentials, from the thermodynamic point of view, both lithium oxide and lithium peroxide could be generated during cell discharge. Whereas the Equations (12.9) and (12.10) sound simple, however, the detailed mechanisms of cathode reactions are very complex, involving several steps and intermediate products, and are far to be fully understood. There is a certain consensus on the reduction pathway proposed by Laoire et al. [15]:

$$O_2 + e^- \rightarrow O_2^- \tag{12.11}$$

$$O_2^- + Li^+ \rightarrow LiO_2 \tag{12.12}$$

$$2LiO_2 \rightarrow Li_2O_2 + O_2 \tag{12.13}$$

---

[1] Depending on different tabulated values of the Gibbs energy.

$$LiO_2 + Li^+ + e^- \rightarrow Li_2O_2 \quad \left(E^0 = 3.1V \text{ vs. } Li^+ / Li\right) \quad (12.14)$$

Alternative direct reduction schemes, which are likely irreversible in nature, were proposed by McCloskey et al. [16]:

$$4Li^+ + O_2 + 4e^- \rightarrow 2Li_2O \quad \left(E^0 = 2.91V \text{ vs. } Li^+ / Li\right) \quad (12.15)$$

$$Li_2O_2 + 2Li^+ + 2e^- \rightarrow 2Li_2O \quad \left(E^0 = 2.72V \text{ vs. } Li^+ / Li\right) \quad (12.16)$$

Both Li_2O and Li_2O_2 produced at the cathode are generally insoluble in the organic electrolyte, leading to build up along the cathode/electrolyte interface. Therefore, the cathodes in nonaqueous aprotic batteries are subjected to clogging, pore blockage, and volume expansion, which reduce conductivity and degrade cell performances with time. So far, however, there has been only a limited evidence of the formation of Li_2O, and Li_2O_2 seems the most common product. The electrochemical decomposition of Li_2O_2 to Li and O_2, which leads to oxygen evolution, is given by the reaction

$$Li_2O_2 \rightarrow 2Li^+ + 2e^- + O_2 \quad (12.17)$$

which is not the reverse of the discharge reaction scheme (12.11–12.13) since it does not pass from the O_2^- intermediate. This difference is considered one of the reasons for the approximately $1V$ overpotential on charge [5]. Another reason to explain the charge overpotential can be found on the different O_2 states (singlet vs. triplet) obtained during oxidation and reduction, respectively [17]. A proper choice of electrolyte and catalyst is needed to try to reduce the overpotential. Moreover, the catalyst nature may affect both the reduction mechanisms and the discharge products.

12.2.3 Mixed Aqueous/Aprotic Electrolyte Cell

The aqueous/aprotic (or mixed) Li–air design tries to join the advantages of the aprotic and the aqueous designs. The common feature of these hybrid designs is an electrolyte divided in two parts, one part aqueous and one part aprotic, connected by a lithium-conducting layer (see Fig. 12.4). The lithium metal anode is in contact with the aprotic side of the electrolyte, in order to avoid contact with water, whereas the porous cathode is in contact with the aqueous side. This gives the further advantage of LiOH formation, which is soluble in water. A lithium-conducting ceramic membrane is employed to put in contact the two liquid electrolytes [18]. Despite the aforementioned interesting features, many serious design problems must be solved to deserve industrial development.

12.2.4 All Solid-State Cell

A solid-state Li–air cell is based on a lithium anode, a glass, ceramic, or glass–ceramic electrolyte, and a porous carbon cathode. The anode and the cathode should be separated from the electrolyte by polymer–ceramic composite layers (see Fig. 12.5), with

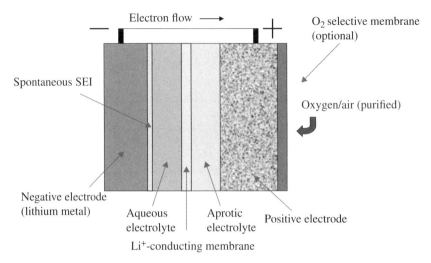

FIGURE 12.4   Scheme of a lithium–air cell with mixed aqueous/aprotic electrolyte.

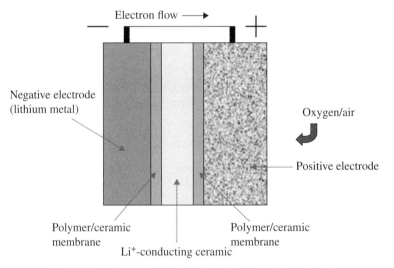

FIGURE 12.5   Scheme of an all solid-state lithium–air cell.

the aim to reduce the overall impedance by enhancing charge transfer and to compensate for the different thermal expansion coefficients that can lead to mechanical detachment of the compartments on thermal cycling. The all solid-state design is chiefly attractive from a safety point of view, since it eliminates the possibility of cracks and ignition. The main drawback of the solid-state design is the low conductivity of most glass–ceramic electrolytes [7].

12.3 CARBON NANOMATERIALS FOR LI–AIR CELLS CATHODE

As previously stated, in Li–air cells, carbonaceous (nano-)materials are at the basis of the cathode structure and can also host metal (or mixed metal/nonmetal) catalysts. In the following, the attention will be focused on the carbon component of the cathode.

Carbonaceous materials can roughly be divided into two main categories, amorphous and graphitic carbons, which have different physicochemical properties, including electrical conductivity, surface reactivity and mechanical properties, structural disorder, and porosity. All of these features should be properly tailored to meet the requirements of the Li–air cathode, which can be summarized as follows:

1. High surface area
2. Micro- and mesoporosity
3. Electrocatalytic properties
4. Wettability

One has to keep in mind that the particles morphology of the discharge products strictly influences the OER and so the reversibility of the electrochemical process. At the same time, carbon features play a role in determining the Li_2O_2 particle shape. Therefore, other characteristics of the cathode that should be optimized in order to maximize the electrochemical performances of the cell are particles morphology, number of unsaturated C atoms, site defects, functionalized C atoms, and self-standing cathode structure without the need of binders. Moreover, it has also to be taken into account that carbon can decompose due to the presence of the electrolyte, and this fact is well discussed in the case of aqueous electrolytes, for which electrode potentials of carbon corrosion during oxygen evolution/reduction reactions are known [19]. For the nonaqueous Li–air battery, it has been shown that on charge up to 4 V, an oxidative decomposition occurs and several Li-based by-products (e.g., Li_2CO_3 and Li carboxylates) can be formed [20].

In the following, different carbonaceous (nano-)materials belonging to the amorphous and graphitic carbons will be discussed, highlighting advantages and limitations for their application in Li–air batteries. Emphasis will be put on the application in nonaqueous Li–air or Li–O$_2$ cells, which at present are indeed the most investigated ones. In fact, for Li–air aqueous systems, the carbon cathode is less or not subjected to the pore clogging (see earlier discussion) and commercial carbons are typically used. The research in this field is more focused on the protecting layers for anode and cathode and on the electrolytes.

12.4 AMORPHOUS CARBONS

12.4.1 Porous Carbons

The first attempts of designing air electrodes for Li–air batteries were done by simply transferring in this field the knowledge acquired on the gas diffusion electrode (GDE) of polymer electrolyte membrane fuel cells (PEMFCs). Commercial carbons applied

TABLE 12.2   Physical Properties and Discharge Capacity of Several Carbon Nanomaterials

Carbon Material	Surface Area $(m^2 g^{-1})$	Particle Size (nm)	Pore Size (nm)	Pore Volume $(cm^3 g^{-1})$	Discharge Capacity $(mAh g^{-1})$
Ketjenblack EC600JD	1325	~40 (primary) ~125 (aggregate)	—	2.47	2600[a]
Ketjenblack EC300JD	890	~30 (primary) >125 (aggregate)	—	1.98	~2000[a]
Super P	62	—	—	0.32	956[a]
DENKA BLACK	60	—	—	0.23	757[a]
ENSACO 250G	62	—	—	0.18	579[a]
VULCAN XC-72	250	2	2	—	762[b]
AC	2100	2	2	—	414[b]

[a] Data taken from Ref. 21.
[b] Data taken from Ref. 22.

in PEMFCs were at first tested, including VULCAN XC-72, Super P, Ketjenblack, DENKA BLACK, ENSACO, and acetylene black (AC). The electrochemical results put into evidence significant differences in discharge capacity and reversibility, and a thorough characterization of these carbons was made by several research groups, in order to understand the relationships between the physical parameters of carbons and the capacity of the cells. Some important factors were identified, namely, surface area, pore volume, overall dimension of cathode, and morphology. The relevant physical properties of carbons and the initial discharge capacity are reported in Table 12.2. It is evident that the surface area is neither the only nor likely the most effective parameter to be taken into account (compare, e.g., Ketjenblack and AC). Super P shows a good capacity also, thanks to the presence of mesopores that can contain a large amount of discharge products, so preventing the pore clogging.

Some attempts were also made to increase the electrocatalytic activity of carbon, so improving ORR and consequently the electrochemical performances. To reach this goal, N doping of commercial carbon was used, and higher surface area and pore volumes were obtained, and also, the C surface defectivity was increased [23].

Anyway, to prepare an air electrode by using amorphous carbons, a support is needed, such as nickel foam or carbon paper, as well as a polymeric binder. This last one can negatively affect the overall performance of the Li–air battery forming a layer on the carbon surface that decreases the amount of sites available for the Li and O_2 reaction [24]. In Figure 12.6, the carbon morphology changes with increasing binder amount are shown.

Different cathode architectures were developed to favor the presence of both large tunnels and mesopores, in order to optimize the rapid O_2 diffusion and to provide an efficient triple-phase region for O_2 reduction. In this regard, mesocellular carbon foam and hierarchically porous honeycomb-like carbon (HCC) have been recently obtained starting from silica templates [22, 25]. To obtain HCC, two different silica were used as templates, with 100 and 400 nm pore diameter, obtaining the HCC-100 and

(a)

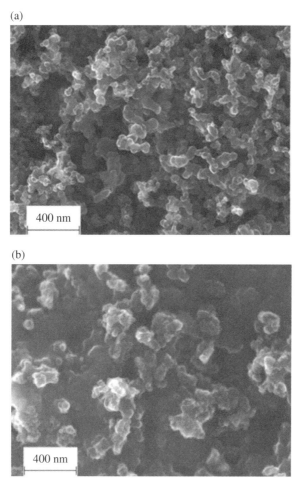

(b)

FIGURE 12.6   SEM micrographs of the cathode films with (a) 80:20 and (b) 20:80 carbon–Kynar ratio, respectively. Reproduced with permission from Ref. 24. © 2011 Elsevier.

HCC-400 samples, respectively. In Figure 12.7a, the discharge performances of these samples at different current densities are compared. The HCC-100 carbon cathode shows the best performance due to its hierarchically porous structure (see Fig. 12.7b).

12.5   GRAPHITIC CARBONS

12.5.1   Carbon Nanotubes

Carbon nanotubes (CNTs) were used to prepare the air electrodes chiefly by means of two different approaches. One of these makes use of proper supports such as buckypapers [26], nickel foam [27], or others, in the same way as for porous carbons

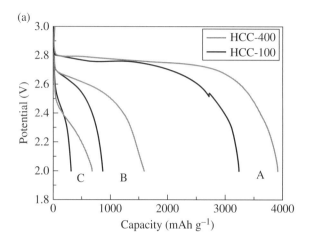

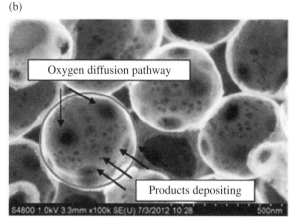

FIGURE 12.7 (a) Discharge characteristics of the HCC-400 and HCC-100 electrodes at various current densities of (A) $0.05\,mA\,cm^{-2}$, (B) $0.2\,mA\,cm^{-2}$, and (C) $0.5\,mA\,cm^{-2}$. (b) FESEM micrograph of HCC-100 carbon. Reproduced with permission from Ref. 25. © 2013 RSC. (*See insert for color representation of the figure.*)

but taking advantage of the unique ability of CNTs to form large bundles or fibers, so avoiding the use of polymeric binders that are detrimental for the mass transport during battery operation. Freestanding air electrodes have been prepared by means of different preparation routes [28, 29]. Regardless of the preparation methodology adopted for the air cathode, however, a high (hierarchical) porosity, which guarantees high O_2 diffusion in micro- and mesopores and also in voids present among the adjacent CNT walls, is a common feature. SEM micrographs reported in Figure 12.8a, b clearly reveal the morphology of an electrode-based cup-stacked-type carbon nanotubes (CS-CNTs), which was constituted by a tubular surface, which is retained also in the discharged cathode (Fig. 12.8c–e) [28].

FIGURE 12.8 SEM images of as-prepared CS-CNT-based cathodes (a and b) before and (c–e) after discharge. Reproduced with permission from Ref. 28. © 2012 Elsevier.

In the last panel of the figure, small particles of less than 50 nm, which represent the discharge products, are arranged along the nanotubes. The CNTs used in that work possessed numerous activated reaction sites derived from edges and defects on the carbon surface, besides other amorphous-like and graphitic components. In contrast, heat-treated CNTs are more graphitized, and a completely different morphology of the Li_2O_2 discharge products, forming agglomerated particles, was observed (see Fig. 12.9). Also, the cathode morphology is different, and the tubular surface is not seen anymore. So, the surface structure of carbon strongly affected the morphology

FIGURE 12.9   Heat-treated CNTs after discharge. Reproduced with permission from Ref. 28. © 2012 Elsevier.

of the discharge products and consequently the electrochemical features: a better cycling was obtained for the electrode based on CS-CNTs, with a coulombic efficiency of 71% against 40% obtained in the second case (heat-treated CNTs).

A similar Li_2O_2 segregation on CNTs was observed in the case of hierarchical-fibril CNT electrode formed by well-aligned CNTs [29]. This kind of cathode architecture allowed to obtain very interesting discharge capacities at a current density as high as $5000 \, mAh \, g^{-1}$ after 20 cycles. The reversibility of the electrochemical reactions is related to the Li_2O_2 decomposition during the OER, which is influenced by the Li_2O_2 morphology as well as by the electrocatalytic properties of the CNTs. This last property can be enhanced by the unique structure of CNTs in which sites and edge defects and high surface (due to inner and outer surface sites) are dominant. A further increase of the catalytic properties of CNTs can be obtained by doping with N [30] and F [31] atoms or by preparing composites with, for example, Co_3O_4 [32] and α-MnO_2 [33]. Finally, partially cracked CNTs offer the advantage of mixing exfoliated and uncracked tubes, so improving porosity and the amount of active edge sites, making the materials more interesting for the electrochemical application and approaching graphene performance in O_2 absorption [34].

12.5.2   Graphene

In these last years, graphene has undoubtedly emerged as one of the most interesting materials for applications in energy production, conversion, and storage. Indeed, it is gaining a lot of success also in the batteries field. Graphene nanosheets (GNS) were initially tested as an anode material for Li-ion batteries, even though with no impressive results, and it is only in the last few years that their potential as a cathode material in Li–air batteries has been explored [35]. The main reasons for which graphene is so attractive in this field are high electrocatalytic activity superior to that of acetylene carbon black, easiness of obtaining freestanding 2D or 3D films with high

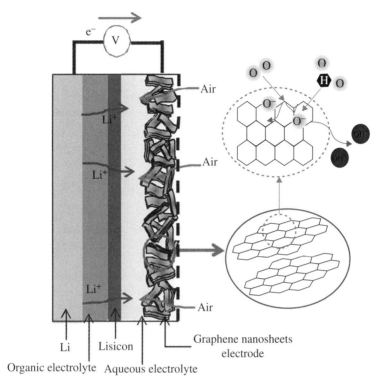

FIGURE 12.10 Structure of the rechargeable Li–air battery based on GNS as an air electrode. Reproduced with permission from Ref. 36. © 2011 ACS.

porosity for oxygen diffusion, and very high surface area of the order of $2600\,\mathrm{m^2\,g^{-1}}$. In particular, this last feature is due to the unique 2D-layered structure in which the presence of numerous sites, defects, and unsaturated carbon atoms plays a major role. Actually, the 2D structure seems not so suitable for Li–air applications, whereas a 3D air electrode can be a better choice to obtain a bimodal porous structure that can host the reaction by-products and, at the same time, assure a high Li$^+$ and O$_2$ diffusion rate and maximize the three-phase electrochemical area. As a matter of fact, hierarchical, functionalized, and doped GNS seem to offer a solution to the major problems of the air electrode. The first innovative application of GNS was as a metal-free catalyst for oxygen reduction [36] (see Fig. 12.10), so introducing the new idea for which GNS are not simply a support host for catalysts as in fuel cells, but are themselves the electrocatalytic material.

In this regard, both SEM and XRPD analysis showed that the reaction product Li$_2$O$_2$ is easily decomposed during the charge process. In addition, as stated previously, graphene can be used to obtain gas diffusion layers (GDL) with wrinkled structure (see Fig. 12.11) [37] that showed superior capacity ($8706\,\mathrm{mAh\,g^{-1}}$) with respect to commercial carbons such as VULCAN XC-72 ($1054\,\mathrm{mAh\,g^{-1}}$) and BP-2000 ($1909\,\mathrm{mAh\,g^{-1}}$). A further gain in capacity ($15,000\,\mathrm{mAh\,g^{-1}}$) can be obtained

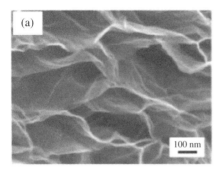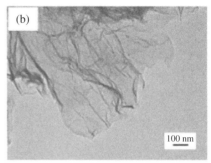

FIGURE 12.11   SEM (a) and TEM (b) images of GNS electrodes. Reproduced with permission from Ref. 37. © 2011 RSC.

passing from 2D to 3D porous air electrodes containing functional groups and lattice defect sites energetically favored for pinning Li_2O_2 as isolated particles, so preventing the clogging of pores [38].

Once again, this result highlights the major importance of pore structure, size, and volume with respect to the surface area in obtaining high discharge capacities. In fact, air electrodes with high meso- and microporosity with interconnected channels for mass diffusion may be considered the best choice. Besides, the addition of a catalyst to GNS could be an alternative approach to improve ORR and the reversibility of the electrochemical process. Noble metals (Pt, Rh) typically applied as catalysts in electrodes for fuel cells could be in principle used also for the air electrode, but their high cost is indeed a drawback. As an alternative to noble metals, catalysts such as ruthenium [39] and, less expensive, light elements such as nitrogen and sulfur have been preferred to functionalize graphene and provide an electrocatalytic activity [40, 41]. For example, a 3% N doping leads to the formation of pyridinic-N, pyrrolic-N, and graphitic-N, which increase the amount of unsaturated C atoms, activating the reaction with O_2 to form oxygen-containing groups [42]. The structure and morphology of the resulting air electrode and of the discharge products, which are influenced by the doping elements, are the key factors to improve the cathode performance at high current densities. The role of the dopants is not yet really understood, and further investigations are needed to this aim. In addition, novel 3D architectures could be designed to optimize long-term cycling and round-trip efficiency that are really scarce. In fact, high discharge capacity values are shown only for a few cycles, and a detailed study of the graphene stability at OER potential for prolonged cycling is still missing.

12.5.3   Composite Air Electrodes

Recently, new concepts for the air electrode preparation have been developed. In particular composites, fibrous and other types of cathodes have been proposed to overcome some specific limitations of porous carbons. For example, a composite of nitrogen-doped carbon (N–C) blend with lithium aluminum germanium phosphate (LAGP) was investigated, and two main advantages were identified: (i) the N–C blend is mesoporous and favors the access of O_2 to the reaction sites, and (ii) the LAGP showed an electrocatalytic activity toward ORR [43]. The addition of the ceramic material increases the discharge capacity but, at the same time, lowers the gravimetric energy density of the cell, so a good compromise between LAGP percentage and cell performance has to be studied.

Morphology and pore size have been also optimized by using hierarchical activated carbon microfiber [44] (see Fig. 12.12) prepared without any binder.

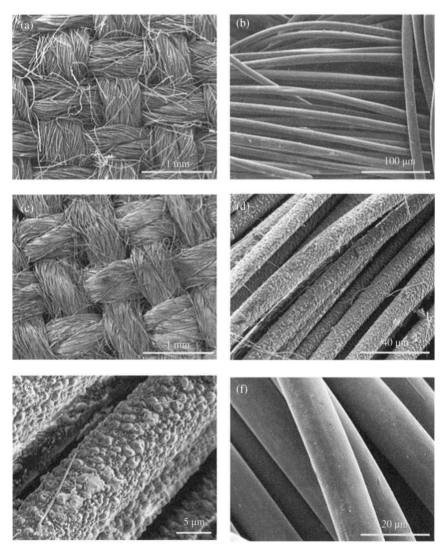

FIGURE 12.12 SEM images of (a and b) pristine electrodes, (c–e) electrodes discharged to 2 V in triglyme–LiTFSI (1 M) electrolyte solution, and (f) electrodes charged to 4.3 V in triglyme–LiTFSI (1 M) electrolyte solution. Reproduced with permission from Ref. 44. © 2013 RSC.

This air electrode showed superior electrochemical performances ($4116\,\text{mAh}\,\text{g}^{-1}$) due to its high surface area and macrostructure, which includes interwoven fibers and hierarchical porosity with diffusion channels not blocked during ORR. Moreover, the electrodes charged at 4.3 V are completely Li_2O_2 free, thus suggesting also the efficient electrocatalytic activity of these microfibers.

Microfibers and nanofibers can also be used as support for catalysts [44, 45] such as α-MnO_2 or Fe, so obtaining, in the former case, a relevant increase in capacity ($9000\,\text{mAh}\,\text{g}^{-1}$) and, in the latter one, performances comparable to those of commercial Pt-based catalysts.

Focusing on the wettability issue of the carbon cathode, a recent approach is based on the combined use of IL and CNTs to prepare a gel. This approach offers several advantages: the conventional three-phase electrochemical interface could be expanded into a 3D network (see Fig. 12.13) [46], and the use of CNTs allows the efficient

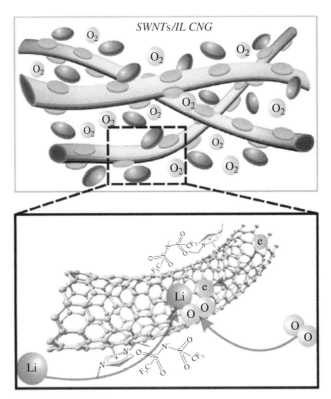

FIGURE 12.13 Three-dimensional continuous passage of electrons, ions, and oxygen. Electrons conduct along the carbon nanotubes. Lithium ions transferred from the ionic liquid electrolyte outside into the cross-linked network gel become coordinated by the inside-anchored [NTf2] ion. Oxygen in the cross-linked network incorporates with the lithium ions and electrons along the CNTs, thereby turning into the discharge products. Reproduced with permission from Ref. 46. © 2012 John Wiley & Sons, Inc.

passage of electrons. IL is then responsible for the high conduction of Li^+ into the air electrode and O_2 easily diffuse through the numerous channels present in the gel cathode. The high porosity is particularly suitable to avoid the electrolyte permeation in the cathode and prevents the clogging of pores by the discharge products.

12.6 CONCLUSIONS

Li–air rechargeable batteries are one of the most promising storage systems under development, and the nonaqueous aprotic scheme will be likely the architecture of choice. On the other hand, their US DoD Technology Readiness Level is still between three and four (Research to Prove Feasibility/Technology Development), and the expected time to market for demanding applications such as in automotive may be estimated in the order of 30 years [47]. Among the critical issues to be solved, one of the most important is indeed the development of a carbon cathode with an optimal mesoporous/hierarchical structure, possibly coupled to a protecting membrane to prevent CO_2 and moisture migration into the air electrode while allowing fast O_2 diffusion. Further investigations are also required to understand the role played by the carbon materials as catalysts (or cocatalysts) in the charge–discharge processes.

As we have shown, numerous attempts were done to improve porosity, surface area, and morphology of the air electrode, but some issues related to the complex behavior of carbonaceous materials in nonaqueous Li–air batteries remain still open. Although we showed herein that different cathode carbons and architectures strongly improve the initial capacity of the cell, the cycling efficiency and long-term cycling are still very poor and remain major issues for the future research.

REFERENCES

[1] D.A.J. Rand, *J. Power Sources* **4**, 101–143 (1979).

[2] D. Linden, T.B. Reddy (eds.), *Handbook of Batteries and Fuel Cells*, McGraw-Hill, New York (2002).

[3] Y.H. Wen, J. Cheng, S.Q. Ning, Y.S. Yang, *J. Power Sources* **188**, 301–307 (2009).

[4] F. Cheng, J. Chen, *Chem. Soc. Rev.* **41**, 2172–2192 (2012).

[5] P.G. Bruce, S.A. Freunberger, L.J. Hardwick, J.-M. Tarascon, *Nat. Mater.* **11**, 19–29 (2012).

[6] DOE, Properties of Fuels (2006). http://www.afdc.energy.gov/fuels/fuel_comparison_chart.pdf (accessed May 8, 2015).

[7] E. Quartarone, P. Mustarelli, *Chem. Soc. Rev.* **40**, 2525–2540 (2011).

[8] K. Kinoshita, *Electrochemical Oxygen Technology*, Wiley, New York (1992).

[9] P.A. Christensen, A. Hamnett, D. Linares-Moya, *Phys. Chem. Chem. Phys.* **13**, 5206–5214 (2011).

[10] W. Vielstich, A. Lamm, H.A. Gasteiger, *Handbook of Fuel Cells: Fundamentals, Technology and Applications*, Wiley & Sons, Ltd, Chichester (2003).

[11] A. Morozan, B. Jousselme, S. Palacin, *Energy Environ. Sci.* **4**, 1238–1254 (2011).

[12]   K.M. Abraham, Z. Jiang, *J. Electrochem. Soc.* **143**, N01–N05 (1996).

[13]   K. Xu, *Chem. Rev.* **104**, 4303–4418 (2004).

[14]   D. Capsoni, M. Bini, S. Ferrari, E. Quartarone, P. Mustarelli, *J. Power Sources* **220**, 253–263 (2012).

[15]   C.O. Laoire, S. Mukerjee, K.M. Abraham, E.J. Plichta, M.A. Hendrickson, *J. Phys. Chem. C* **113**, 20127–20134 (2009).

[16]   B.D. McCloskey, D.S. Bethune, R.M. Shelby, G. Girishkumar, A.C. Luntz, *J. Phys. Chem. Lett.* **2**, 1161–1166 (2011).

[17]   J. Hassoun, F. Croce, M. Armand, B. Scrosati, *Angew. Chem. Int. Ed.* **50**, 2999–3002 (2011).

[18]   H. Zhou, Y. Wang, H. Li, P. He, *ChemSusChem* **3**, 1009–1019 (2010).

[19]   K. Kinoshita, *Handbook of Battery Materials*, John Wiley and Sons, Inc., Hoboken (2011), pp. 274–275.

[20]   M.M.O. Thotiyl, S.A. Freunberger, Z. Peng, P.G. Bruce, *J. Am. Chem. Soc.* **135**, 494–500 (2013).

[21]   C.K. Park, S.B. Park, S.Y. Lee, H. Lee, H. Jang, W.I. Cho, *Bull. Korean Chem. Soc.* **31**, 3221–3224 (2010).

[22]   X.-H. Yang, P. He, Y.Y. Xia, *Electrochem. Commun.* **11**, 1127–1130 (2009).

[23]   P. Kichambare, J. Kumar, S. Rodrigues, B. Kumar, *J. Power Sources* **196**, 3310–3316 (2011).

[24]   S.R. Younesi, S. Urbonaite, F. Bjorefors, K. Edstrom, *J. Power Sources* **196**, 9835–9838 (2011).

[25]   X. Lin, L. Zhou, T. Huang, A. Yu, *J. Mater. Chem. A* **1**, 1239–1245 (2013).

[26]   Y. Li, K. Huang, Y. Xing, *Electrochim. Acta* **81**, 20–24 (2012).

[27]   H. Wang, K. Xie, L. Wang, Y. Han, *RSC Adv.* **3**, 8236–8241 (2013).

[28]   S. Nakanishi, F. Mizuno, K. Nobuhara, T. Abe, H. Iba, *Carbon* **50**, 4794–4803 (2012).

[29]   H.-D. Lim, K.-Y. Park, H. Song, E.Y. Jang, H. Gwon, J. Kim, Y.H. Kim, M.D. Lima, R.O. Robles, X. Lepró, R.H. Baughman, K. Kang, *Adv. Mater.* **25**, 1348–1352 (2013).

[30]   Y. Li, J. Wang, X. Li, J. Liu, D. Geng, J. Yang, R. Li, X. Sun, *Electrochem. Commun.* **13**, 668–672 (2011)

[31]   Y. Tian, H. Yue, Z. Gong, Y. Yang, *Electrochim. Acta* **90**, 186–193 (2013).

[32]   T.H. Yoon, Y.J. Park, *Nanoscale Res. Lett.* **7**, 28 (2012).

[33]   G.Q. Zhang, J.P. Zheng, R. Liang, C. Zhang, B. Wang, M. Au, M. Hendrickson, E.J. Plichta, *J. Electrochem. Soc.* **158**, A822–A827 (2011).

[34]   J. Li, B. Peng, G. Zhou, Z. Zhang, Y. Lai, M. Jia, *ECS Electrochem. Lett.* **2**, A25–A27 (2013).

[35]   S.L. Candelaria, Y. Shao, W. Zhouc, X. Li, J. Xiao, J.-G. Zhang, Y. Wang, J. Liu, J. Li, G. Cao, *Nano Energy* **1**, 195–220 (2012).

[36]   E. Yoo, H. Zhou, *ACS Nano* **5**, 3020–3026 (2011).

[37]   Y. Li, J. Wang, X. Li, D. Geng, R. Li, X. Sun, *Chem. Commun.* **47**, 9438–9440 (2011).

[38]   J. Xiao, D. Mei, X. Li, W. Xu, D. Wang, G.L. Graff, W.D. Bennett, Z. Nie, L.V. Saraf, I.A. Aksay, J. Liu, J.-G. Zhang, *Nano Lett.* **11**, 5071–5078 (2011).

[39]   H-G. Yung, Y.S. Jeong, J.B. Park, Y.K. Sun, B. Scrosati, Y.J. Lee, *ACS Nano* **7**, 3532–3539 (2013).

[40] G. Wu, N.H. Mack, W. Gao, S. Ma, R. Zhong, J. Han, J.K. Baldwin, P. Zelenay, *ACS Nano* **6**, 9764–9776 (2012).

[41] Y. Li, J. Wang, X. Li, D. Geng, M.N. Banis, Y. Tang, D. Wang, R. Li, T.-K. Sham, X. Sun, *J. Mater. Chem.* **22**, 20170–20174 (2012).

[42] Y. Li, J. Wang, X. Li, D. Jeng, M.N. Banis, R. Li, X. Sun, *Electrochem. Commun.* **18**, 12–15 (2012).

[43] P. Kichamabare, S. Rodrigues, J. Kumar, *ACS Appl. Mater. Interfaces* **4**, 49–52 (2012).

[44] V. Etacheri, D. Sharon, A. Garsuch, M. Afri, A.A. Frimer, D. Aurbach, *J. Mater. Chem. A* **1**, 5021–5030 (2013).

[45] J. Wu, H.W. Park, A. Yu, D. Higgins, Z. Chen, *J. Phys. Chem. C* **116**, 9427–9432 (2012).

[46] T. Zhang H. Zhou, *Angew. Chem. Int. Ed.* **51**, 11062–11067 (2012).

[47] G. Girishkumar, B. McCloskey, A.C. Luntz, S. Swanson, W. Wilcke, *J. Phys. Chem. Lett.* **1**, 2193–2203 (2010).

13

CARBON-BASED NANOMATERIALS FOR H$_2$ STORAGE

Fen Li[1], Jijun Zhao[1] and Zhongfang Chen[2]

[1] *Laboratory of Materials Modification by Laser, Electron, and Ion Beams and College of Advanced Science and Technology, Dalian University of Technology, Dalian, China*
[2] *Department of Chemistry, Institute for Functional Nanomaterials, University of Puerto Rico, San Juan, PR, USA*

13.1 INTRODUCTION

Energy and environment are two of the most critical problems in our modern society. The demand of industrialization and vehicles is still increasing, while the world supply of fossil fuels is going down, and the concern for global warming, climate change, and air pollution is uprising. Obviously, the petroleum reserve is limited, and the rising cost of petroleum is deteriorating everyone's life quality.

To solve the energy and environment crisis, an attractive strategy is to develop an alternative environment-friendly energy fuel, like hydrogen. With the superior merits such as high energy content of $142\,MJ\,kg^{-1}$ (three times of petroleum) and clean combustion (producing only water), hydrogen is currently considered as one of the most potential fuels. However, it is not an energy source but only an energy carrier. Therefore, before the application of hydrogen energy, we have to overcome several significant technical steps including hydrogen production, storage, and conversion. Among these, hydrogen storage is the most critical step for hydrogen energy commercialization.

Carbon Nanomaterials for Advanced Energy Systems: Advances in Materials Synthesis and Device Applications, First Edition. Edited by Wen Lu, Jong-Beom Baek and Liming Dai.
© 2015 John Wiley & Sons, Inc. Published 2015 by John Wiley & Sons, Inc.

The development of hydrogen economy requires an economic, safe, lightweight, and high-capacity storage medium. Recently, the US Department of Energy (DOE) has lowered the hydrogen storage target to 5.5 wt% and 40 g l^{-1} at 2017 [1] due to the technical difficulties for hydrogen storage. So far, no available hydrogen storage materials can satisfy all the requirements for utility use.

Several strategies have been considered for hydrogen storage. High-pressure and cryogenic hydrogen storage systems are not ideal for civilian applications due to the safety concerns and the energies consumed in pressurizing and cooling the systems. Currently, the actively investigated hydrogen storage procedures are storing hydrogen in the complex hydrides [2–5], metal hydrides [6–8], ammonia borane [9–11], porous sorbent [12–15], graphene-based materials [16], molecular clathrates [17–19], and so on.

Due to the high surface areas and potentially high energy capacity, sorbents have been distinguished from other hydrogen storage media. To effectively store hydrogen, the sorbents need to fulfill several criteria. Generally, to fulfill the hydrogen storage at room temperature, the ideal H_2 binding energy should lie between the physisorption and chemisorption (0.21–0.42 eV) [20]. In addition, the host sorbents should have large surface areas and be light in weight; thus, carbon-based nanomaterials are of choice. However, the interaction between molecular hydrogen and pristine carbon nanomaterials, such as fullerenes, nanotube, and graphene, is too weak, which leads to low storage capacity at ambient circumstance. Herein, to achieve reversible hydrogen storage on carbon-based sorbents under practical operating conditions, it is highly desirable to enhance the H_2 binding energy on these carbon-based nanomaterials.

In this chapter, we mainly focus on the theoretical design of carbon-based sorbents for hydrogen storage. Unless specified otherwise, the computations are performed by density functional theory (DFT). Various low-dimensional nanomaterials, such as fullerenes, nanotubes, and graphenes, as well as their derivatives, have been considered. Different strategies adopted to increase the hydrogen uptake are summarized; both the advantages and disadvantages of these strategies are also discussed.

13.2 HYDROGEN STORAGE IN FULLERENES

For the pure carbon-based sorbents, the H_2 binding energy was theoretically evaluated to be 0.01–0.06 eV [21], which is too weak to keep the H_2 molecules at ambient conditions. Therefore, effective strategies to enhance the interaction between the H_2 molecule and the sorbent materials have been intensively searched. In a pioneering work, Zhang and coworkers proposed to use the transition metal (TM) doping on the fullerenes for hydrogen storage [22]. In such complexes, the metal atoms can donate electrons to fullerenes, thus leaving the metal atom in the cationic form. Therefore, the H_2 molecules can be adsorbed stronger on the fullerene surface by the charge polarization mechanism. TM atoms tend to bond on the pentagonal (P) faces of fullerenes by charge transfer interactions to produce stable organometallic buckyballs and to adsorb many H_2 molecules by the Kubas interaction [23]. Hydrogen storage densities can approach to 9 wt% when the $C_{48}B_{12}$ [ScH(H_2)$_5$]$_{12}$ OBB is fully charged (Fig. 13.1), with the ideal H_2 binding energy range of 0.24–0.35 eV. The high

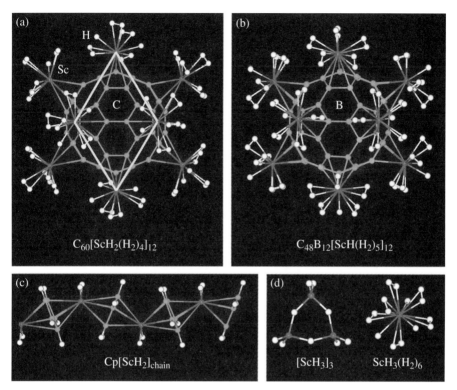

FIGURE 13.1 (a) $C_{60}[ScH_2(H_2)_4]_{12}$, (b) $C_{48}B_{12}[ScH(H_2)_5]_{12}$, (c) $Cp[ScH_2]chain$, and (d) $[ScH_3]_3$ (left) and $ScH_3(H_2)_6$ (right). Reproduced with permission from Ref. 22. © 2005 American Physical Society.

hydrogen capacity of fullerenes implies that the operation condition should be within a relatively small pressure range at room temperature.

Shortly after, Yildirim *et al.*'s DFT investigation of TM doping on the C_{60} showed similar conclusions: Ti atom on C_{60} could bind up to four H_2 molecules with an average binding energy of 0.3–0.5 eV, and the fully Ti-covered fullerenes ($C_{60}Ti_{14}$) could hold 56 H_2 molecules, corresponding to 7.5 wt% hydrogen [24]. Note that their calculations predicted that the Sc and Ti prefer the hexagonal (H) site rather than the former reported pentagonal (P) faces of C_{60} fullerenes, since the first adsorbed H_2 molecule does not dissociate at H site, unlike in the case of P site.

These two delightful studies predicted a promising prospect of carbon nanostructure for hydrogen storage by TM doping. However, their conclusions were wholly based on the assumption that the TM atoms coated on the fullerene surface remain isolated from each other. Unfortunately, Sun *et al.* revealed that the TM atoms tend to cluster on the C_{60} surface [25]. In the case of single Ti atom-coated C_{60}, they obtained similar results as others; however, at the high Ti coverage, Ti atoms prefer to form clusters on the C_{60} surface, and the nature of H bonding switches from molecular adsorption to atomic adsorption. When 12 Ti atoms are coated on C_{60}, all the

(a)   (b)

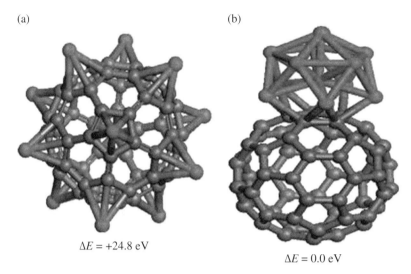

$\Delta E = +24.8$ eV

$\Delta E = 0.0$ eV

FIGURE 13.2   Two configurations of Ti$_{12}$C$_{60}$: (a) 12 individual Ti atoms are located above 12 pentagons, (b) a cluster of 12 Ti atoms is attached to C$_{60}$. Reproduced with permission from Ref. 25. © 2005 American Chemical Society.

Ti atoms prefer to aggregate together. The newly formed Ti cluster has lost most of its activities for trapping H$_2$ molecules since only those directly bonded on the C$_{60}$ surface are positively charged and the Ti atoms below the cluster surface are not capable to adsorb H$_2$ (Fig. 13.2). These results suggested that to achieve and maintain higher hydrogen capacity, we have to avoid metal clustering on support materials.

In a later investigation using Ni-dispersed fullerenes for hydrogen storage, it was found that a single Ni coated on the C$_{60}$ fullerene surface could store up to three H$_2$ molecules [26]. Consequently, at high Ni coverage, Ni-dispersed fullerenes could store up to 6.8 wt% H$_2$ and thus were considered to be the novel hydrogen storage media. However, the metal aggregation problem on the fullerene surface has not been examined.

Since the TM atoms tend to cluster on the C$_{60}$ fullerene, decorating other metals on the fullerene for hydrogen storage has been considered. One way to avoid metal clustering is to reduce the interactions between metal atoms. Along this line, it was suggested that Li-coated C$_{60}$ fullerene can bind H$_2$ without the aggregation problem [27]. The Li$_{12}$C$_{60}$ could store 60 H$_2$ molecules (9 wt%) with a binding energy of 0.075 eV/H$_2$, which is much lower than the case of TM doping. The small H$_2$ binding energy of H$_2$ molecules to Li$_{12}$C$_{60}$ clusters means that the storage system should be operated at lower temperature than ambient condition. Another shortcoming is that though the Li atoms on the C$_{60}$ surface do not tend to aggregate, the Li$_{12}$C$_{60}$ clusters prefer to interact with each other. Consequently, the linking Li atoms are unable to trap H$_2$ molecules, which lowers the hydrogen storage capacity. These theoretical predictions were confirmed by the recent experiments of Teprovich *et al.* [28]. The lithium-doped fullerene (Li$_x$-C$_{60}$-H$_y$) they synthesized can reversibly store hydrogen

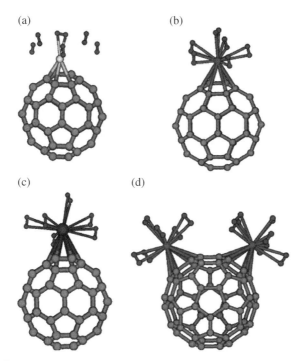

FIGURE 13.3 Adsorption of hydrogen molecules on alkali metal-doped fullerenes. (a) $C_{60}Li(H_2)_2$, (b) $C_{60}Na(H_2)_6$, (c) $C_{60}K(H_2)_6$, and (d) $C_{60}Na_2(H_2)_{12}$. Reproduced with permission from Ref. 29. © 2008 American Chemical Society.

through chemisorption with a capacity up to 5 wt% at approximately 270°C, which directly demonstrated the feasibility of metal decoration on fullerene for practical hydrogen storage.

Besides Li, other alkali metals have also been explored for hydrogen storage [29–32]. Doping Na on C_{60} fullerene has shown more pronounced molecular hydrogen storage capacity than doping of Li and K (see Fig. 13.3) [29]. The stronger hydrogen adsorption is attributed to the electrostatic interaction between the induced dipole and the quadrupolar interaction of the molecules with the field due to the charge transfer from the sodium atom to the C_{60} molecule. The sodium-doped fullerene molecule, Na_8C_{60}, can be stably formed, and the corresponding hydrogenated species, $[Na(H_2)_6]_8C_{60}$, with 48 hydrogen molecules surrounded, can reach a hydrogen adsorption density of approximately 9.5 wt% with the average H_2 binding energy of 0.09 eV. Additionally, hydrogen storage in the alkali metal functionalized fullerenol $M^+_{12}(C_{60}O_{12})_{12}$ (M = Li and Na) clusters have been investigated [31], in which both metal and oxygen sites can contribute to the H_2 adsorption, whereas H_2 molecules on the oxygen site only concerns with the polarization mechanism. Promising hydrogen storage capacities of 9.78 and 8.33 wt% were obtained for $C_{60}(OLi)_{12}$·54 H_2 (binding energy of 0.115 eV/H_2) and for $C_{60}(ONa)_{12}$·54 H_2 (0.122 eV/H_2), respectively. Obviously, if well dispersed on the fullerene surface, the alkali metal doping on the

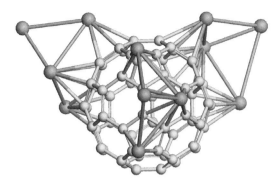

FIGURE 13.4   Na-coated fullerene $C_{60}Na_{12}$. Reproduced with permission from Ref. 32. © 2010 American Chemical Society.

fullerenes could dramatically enhance the H_2 adsorption. However, further studies of the structural and electronic properties of both Na_nC_{60} and Li_nC_{60} ($n \leq 12$) clusters revealed that though the Li atoms could coat homogeneously on the surface of C_{60} fullerene via pentagonal sites, the sodium atoms prefer to form 4-atom islands on the surface that is unfavorable to the H_2 adsorption (Fig. 13.4) [32].

Apart from the TM and alkali metal atoms, other metal atoms doping on fullerenes have also been considered for hydrogen storage [33–35]. Calcium doping stands out as a promising strategy to improve the H_2 adsorption [33, 34]. Yoon *et al.* predicted that Ca atoms prefer to coat on the C_{60} fullerene as a monolayer by an intriguing charge transfer mechanism involving the empty *d* levels of the metal elements [33]. The highly Ca-coated $Ca_{32}C_{60}$ cluster could trap 92 H_2 molecules, corresponding to a hydrogen uptake of 8.4 wt%, with the binding energy range of 0.2–0.4 eV/H_2. However, a further study by Wang *et al.* showed that the $Ca_{32}C_{60}$ cluster could only bind 62 H_2 molecules [34]. Even when increasing the fullerene size, the fully Ca-coated fullerene cannot store more than 6.25 wt% hydrogen. Moreover, the first 30 H_2 molecules would dissociate and bind atomically on the 60 triangular faces made by Ca and carbon atoms in C_{60} fullerene. The remaining 32 H_2 molecules (3.2 wt%) bound on the second layer quasimolecularly have only an average binding energy of 0.11 eV/H_2.

Charge polarization is the underlying mechanism for hydrogen storage in the aforementioned systems. An alternative strategy for hydrogen storage is to directly introduce the charge on the fullerenes. Yoon *et al.*'s computations demonstrated that the H_2 molecules could be effectively polarized on positively or negatively charged fullerenes C_n ($20 \leq n \leq 82$) and the binding energy per H_2 could be dramatically enhanced to 0.18–0.32 eV [36]. Note that the enhanced binding is attributed to the polarization of the hydrogen molecules by the high electric field near the surface of the charged fullerene. At the full hydrogen coverage, the charged fullerenes could uptake up to 8.0 wt% of hydrogen, which is comparable to the metal-coated fullerenes.

There was also an endeavor of doping C_{60} fullerene by other atoms such as Si, B, N, and Ti. For Si doping, only one H_2 molecule could be bound on the Si atom, which is far lower than the case of metal-doped fullerene [37]. In the Li-decorated

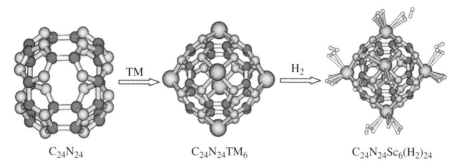

$C_{24}N_{24}$ $C_{24}N_{24}TM_6$ $C_{24}N_{24}Sc_6(H_2)_{24}$

FIGURE 13.5 Schematic of TM-decorated $C_{24}N_{24}$ for hydrogen storage. Reproduced with permission from Ref. 39. © 2012 American Chemical Society.

B-substituted heterofullerene ($L_{12}C_{48}B_{12}$), the Li atoms could stay isolated on the surface, and the positively charged Li ion could bind up to three H_2 molecules with the corresponding binding energy of 0.135–0.172 eV/H_2 [38]. These hydrogen storage behaviors are quite similar with the case of Li coated on the pristine C_{60} fullerene [27]. Additionally, Srinivasu *et al.* found that the porphyrin-like porous fullerene, $C_{24}N_{24}$, is capable of strongly binding the TM atoms, therefore inhibiting the aggregation problem of TM [39]. The TM atoms adsorb H_2 molecules through the well-known Kubas-type interactions, and the $C_{24}N_{24}Sc_6$ cluster could trap up to 24 H_2 molecules (~5.1 wt%) with the binding energy of 0.11 eV (Fig. 13.5). It was also predicted that substitutionally doping C_{60} with Ti atoms could maintain the Ti atoms isolated on the C_{60} surface [40]. The first two H_2 molecules dissociate, while the remaining four H_2 molecules could be trapped in molecular form with suitable binding energy. The Ti_6C_{48} cluster exhibits the hydrogen storage capacity of 7.7 wt%. Very recently, the effect of C vacancy on physisorption of hydrogen onto Ti-functionalized C_{60} was investigated [41]. It was found that creating a C vacancy in the fullerene can improve the adsorption energy of Ti at the cost of reducing the H_2 adsorption energy. The defect-free (no C vacancy) $C_{60}TiH_2$ complex shows more favorable hydrogen adsorption energies (−0.23 eV along *x*-axes, −0.21 eV along *y*-axes of C_{60}) than the defect-containing $C_{59}TiH_2$ complex, which does not exhibit adsorption energies within the targeted energy range.

Decorating the inner surface of fullerenes has also been examined. Unfortunately, the metal encapsulated fullerenes only display negligible enhancement of hydrogen adsorption since the charge donated by the encapsulated metal atoms is mainly confined inside the fullerene cage [36, 42]. Yoon *et al.* investigated the interaction between molecular hydrogen with pristine and La-encapsulated carbon fullerenes C_n ($20 \leq n \leq 82$) [42]. Though three electrons are transferred from the La atom to the fullerene cage, there is no significant enhancement in the hydrogen binding energy, since the transferred charge is mainly delocalized inside the carbon cage. Despite the so-called screen effect, the interaction of hydrogen anions (H_3^-, H_5^-, and H_7^-) with $Li^+@C_{60}$ has been explored for hydrogen storage [43]. The H_3^--introduced $Li^+@C_{60}$ cluster is able to trap H_2 molecules. With the existence of H_3^-, the metallofullerene

$Li^+@C_{60}$ has a high hydrogen capacity of 9 wt% with a binding energy of 0.05 eV/H_2, which could be practical for hydrogen storage at reduced temperatures.

Fullerenes can be combined with other materials for hydrogen storage [44–46]. Yamada *et al.* experimentally examined the hydrogen adsorption onto an electron-doped C_{60} monolayer on Cu (111) at room temperature [44]. They found that hydrogen molecules randomly adsorb on the surface area of the electron-doped C_{60} monolayer, while the ordered structure of the C_{60} monolayer remains unchanged. Some parts of the hydrogen are dissociatively chemisorbed, suggesting that the interaction between the electron-doped C_{60} monolayer and hydrogen is stronger than that expected from theoretical studies. In addition, a series of new metal-doped fullerene-intercalated phthalocyanine (Pc) covalent organic frameworks (COFs) and metal organic frameworks were designed for hydrogen storage [45, 46]. First-principle calculations showed that Li atoms could stably locate on the surface of C_{30}-, C_{36}-, C_{60}-, and C_{70}-Pc-PBBA COFs. At room temperature and 100 bar, the Li-doped C_n-Pc-PBBA COFs could uptake 4.2 wt% of H_2 [45].

13.3   HYDROGEN STORAGE IN CARBON NANOTUBES

The carbon nanotubes (CNTs) have been considered as promising hydrogen storage media since their synthesis [47]. Some early experiments argued that the CNTs have superior hydrogen storage, but later experiments proved otherwise [48–53]. Critical experimental studies demonstrated that the pure CNTs can only uptake hydrogen below 1 wt% [54–56]. Theoretical calculations also confirmed that the physisorption of H_2 molecules on the pure CNTs cannot lead to high hydrogen capacity [57, 58].

Again, one of the most promising strategies to improve hydrogen adsorption capacity is metal doping, as already demonstrated in the case of fullerenes. Among metal atoms, the highest storage capacities reported experimentally are in Li- and K-doped CNTs (20 and 14 wt%, respectively) [59]. However, later investigation demonstrated that the reported high hydrogen storage capacity was due to the presence of water impurities in their experiments [60, 61].

To improve the H_2 adsorption, a pioneering approach is to decorate the CNTs with TM atoms as proposed by Yildrim *et al.* [62]. Their first-principle calculations predicted that a single Ti atom could trap up to four H_2 molecules, leading to 8 wt% of hydrogen storage capacity at high Ti coverage for practical applications (see Fig. 13.6). Subsequently, more efforts have been ignited to further investigate TM decoration on CNTs for hydrogen storage by first-principle calculations [63–74]. Later investigations showed that the weak physisorption of H_2 cannot be enhanced significantly by increasing the curvature of the surface through radial deformation [63]. To promote H_2 uptake, Dag *et al.* utilized the Pt to dope the single-walled carbon nanotube (SWCNT) and found that the bonding character changes dramatically when SWCNT is functionalized by Pt atoms. The first H_2 molecule is adsorbed dissociatively and the second one is adsorbed molecularly, while the nature of bonding is a very weak physisorption for the third adsorbed H_2. The mechanism of H_2 binding on the TM-decorated SWCNTs is electrostatic Coulomb attraction caused by the

(a) C_8TiH_8 (5.3 wt%)   (b) C_4TiH_8 (7.7 wt%)

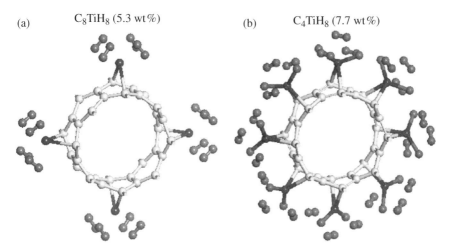

FIGURE 13.6 Two high-density hydrogen coverages on a Ti-coated (8,0) nanotube. (a) and (b) have different Ti modification ratios. Reproduced with permission from Ref. 62. © 2005 American Physical Society.

charge transfer from metal 4s orbital to 3d ones [73]. However, note that all these results were based on the hypothesis that the Pt atoms are isolated on the surface of SWCNT.

Similar to the case of metal-decorated fullerenes, clustering problem persists in the metal-coated SWCNT systems. For example, in the Ni-mediated CNTs [64], each Ni atom dispersed on the surface of CNTs can bind up to five H_2 molecules with an enthalpy change of 0.26 eV/H_2 in hydrogen adsorption, and at a high Ni coverage, the hydrogen capacity can reach 10 wt%. However, the experimental study demonstrated that only about 3 wt% of hydrogen capacity was achieved since the Ni atoms prefer to form clusters of 1 nm sizes [75]. The Sc atoms were also demonstrated to prefer to cluster on SWCNT due to the lower migration barrier and strong metal–metal attraction [66]. Although the well-separated Sc atoms on SWCNT could improve the H_2 adsorption, the hydrogen storage capacity of Sc cluster-doped SWCNT is significantly decreased, for example, a Sc_4 cluster has analogous hydrogen uptake as a single Sc atom.

In order to avoid the aggregation problem of TM atoms on the SWCNT surface, the interaction between TM and C atoms of CNTs has to be enhanced and/or the interaction between the TM and TM has to be weakened [76–81]. To improve the TM–sorbent interaction, the first strategy is to directly substitute the C atoms on the CNTs with TM atoms. Ciraci and coworkers proposed to substitute the C with Be atoms on SWCNT for hydrogen storage [76]. Nevertheless, the substituted Be has lost most of its activities to attract H_2 molecules; each Be atom could only hold one H_2 molecule with the binding energy of 0.31 eV. Four Be atoms could be substituted, leading to 2.4 wt% hydrogen storage capacity. The second strategy is to dope TM on the defective sites of CNTs. Shevlin *et al.* attempted to pin the TM atoms on the native point defects (Stone–Wales defects and vacancies) of CNT [77] and found that the vacancies can strongly bind Ti atoms and prevent their coalescence. According to their DFT computations, the defect-modulated Ti doping on the CNTs is stable at room temperature, each defect-modulated Ti atom can adsorb up to five H$_2$ molecules with the binding energy range of 0.2–0.7 eV/H$_2$, and at high Ti coverage, the Ti-doped CNT can uptake the hydrogen capacity of 7.1 wt% (C$_{112}$Ti$_{16}$H$_{160}$). Similarly, Corral et al. showed that both Pd and H$_2$ prefer to adsorb on the C vacancies of (5,5) SWCNT [78]. However, note that the presence of vacancies affects the stability of adsorbed Pd, leaving less availability to interact with the hydrogen molecule. The third strategy is to introduce some functional groups to stabilize the TM adsorption on the CNTs. For example, Singha et al. experimentally demonstrated that polyvinylpyrrolidone is capable of acting as a stabilizing agent to keep the uniform distribution of Pd nanoparticles on the multiwalled carbon nanotube (MWCNT) [79]. The hydrogen storage capacity of polyvinylpyrrolidone-capped and Pd-doped MWCNT has increased to 4 wt%, compared to 1.7 wt% for the merely Pd-functionalized MWCNT.

Alkali metal-dispersed CNTs were also predicted to be an efficient medium for hydrogen storage [63, 82–86]. Early investigation suggested that doping Li atoms could slightly strengthen the H$_2$ adsorption, even though the bonding nature is still the physisorption [63]. Obviously, the H$_2$ physisorption on Li-doped CNTs is not strong enough to store H$_2$ for practical applications. Chen et al. suggested that encapsulating fullerene molecule inside the Li-doped CNTs could further enhance the H$_2$ binding [82] and rationalized that the charge transfer between the nanotube and C$_{60}$ further facilitates the charge transfer from Li to the nanotube, thus facilitating hydrogen bonding.

The hydrogen storage capability can be improved by optimizing the number and position of dopants, as demonstrated by Liu et al. [83]. In the best structure they obtained (Fig. 13.7), eight Li atoms are dispersed at the hollow sites above the hexagonal carbon rings. Besides the H$_2$ molecules directly bound around Li atoms, each carbon atom can be fully ionized to store one H$_2$ molecule, leading to an extremely high H$_2$ storage capacity of 13.45 wt%. The average adsorption energy (0.17 eV/H$_2$) is only slightly lower than the ideal range for hydrogen storage. The authors pinpoint the two key factors to achieve high hydrogen storage capacity: first, the bands of the dopants should strongly overlap with those of H$_2$ and nanotube simultaneously; second, all the carbon atoms in the nanotube are fully ionized.

Similarly, Na doping on the CNTs for hydrogen storage were investigated. Nagare et al. found that a single Na atom always prefers to occupy the hollow site of a hexagonal carbon ring in SWCNTs considered [85]. Each Na atom could further bind up to six hydrogen molecules with a binding energy of 0.26 eV/H$_2$. The dipole interaction induced by the charge transfer from Na atoms to CNT is responsible for the high hydrogen uptake of 9.2–11.28 wt% for the Na-coated SWCNTs. They also indicated that the Na-doped SWCNTs are highly stable without the tendency for metal clustering with Na–Na distance of 5.9 Å.

Calcium doping on the CNTs exhibits superior hydrogen storage behaviors [87–90]. Lee et al.'s computations showed that the Ca atoms can be uniformly

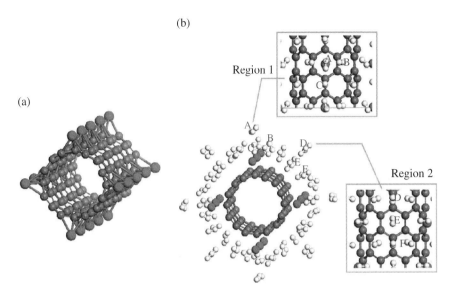

FIGURE 13.7 Schematic plots for the storage material: (a) the Li-dispersed carbon nano-tube with Li:C (1:8), (b) the relaxed $(H_2)_{64}/Li_8/C_{64}$ system. Reproduced with permission from Ref. 83. © 2009 American Chemical Society.

dispersed on the doped boron and defect sites on CNTs, thus effectively inhibiting the aggregation problem [87]. Each Ca atom is able to trap up to six H_2 molecules with a desirable binding energy of approximately $0.2\,eV/H_2$. The gravimetric capacity of approximately $5\,wt\%$ hydrogen can be achieved by the Ca-decorated B-doped (5,5) CNTs ($C_{75}B_5\cdot5Ca\cdot30H_2$). Moreover, Yang *et al.* reported a higher gravimetric capacity of approximately $9\,wt\%$ hydrogen ($8Ca\cdot32H_2$) for the narrow (4,0) CNT with the average binding energy of $0.12\,eV/H_2$ [88]. They also predicted that the Ca adsorbates on CNTs are highly stable without the tendency for clustering. A recent study found that the Ca atoms could be adsorbed stably on the acetylenic ring of the graphyne nanotube (GNT) without Ca clustering (Fig. 13.8); such a system could uptake $7.44–8.96\,wt\%$ hydrogen ($4H_2/Ca$) with the average adsorption energy in the range of $0.13–0.33\,eV/H_2$ [90]. Both the polar interactions and the orbital hybridizations contributed to the adsorption of H_2 molecules.

Apart from metal doping, N atom and some hydrides have also been employed to decorate the CNTs for hydrogen storage [76, 91–94]. Rangel *et al.* found that nanotubes decorated with atomic nitrogen could improve the strength of H_2 adsorption energy [91]. The N-decorated (8,0) SWCNTs can uptake up to $6.0\,wt\%$ hydrogen at 300 K and ambient pressure, with an average adsorption energy of $0.08\,eV/H_2$. The high hydrogen capacity is attributed to the charge transfer from the vicinity of the nanotube to the N atom, which facilitated the H_2 adsorption on CNT, though the binding energy is still lower than the ideal range. Mousavipour *et al.* explored the possibility of using hydride-functionalized (BH_3, AlH_3, NH_3, NiH_2, and LiH) SWCNTs for hydrogen storage; their computations showed that CNT–BH_3 has the

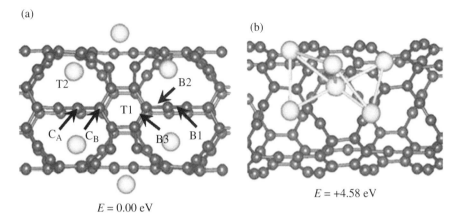

FIGURE 13.8 (a) Isolated and (b) clustered configurations of the Ca$_6$/GNT complexes. Reproduced with permission from Ref. 90. © 2012 Elsevier Inc.

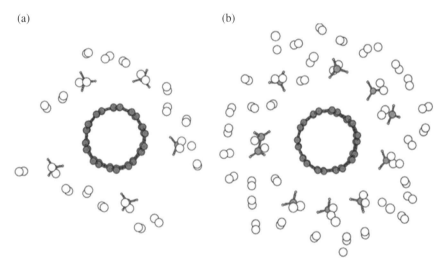

FIGURE 13.9 The optimized structures of (a) CNT–5(BH$_3$+4H$_2$) and (b) CNT–10(BH$_3$+4H$_2$). Reproduced with permission from Ref. 94. © 2011 WILEY-VCH Verlag GmbH & Co. KGaA, Weinheim.

best hydrogen storage behaviors (followed by LiH) among these hydrides and the ultimate hydrogen storage can reach 6.12–11.5 wt% with the binding energy of 0.18–0.24 eV/H$_2$ under the full coverage (CNT–10BH$_3$; Fig. 13.9) [94]. The charge transfer and the induced electrostatic interactions are responsible for the synergetic action of SWCNT and hydrides on H$_2$ adsorption at ambient conditions.

Substitutional doping by B or N atoms can also affect the hydrogen storage capability of nanotubes, which was demonstrated by Zhou *et al.* [95]. For the atomic hydrogen adsorption, B doping forms an electron-deficient six-membered ring

(a)   (b)

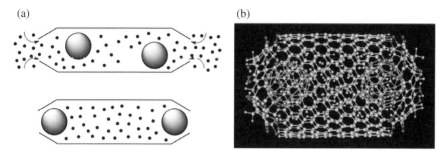

FIGURE 13.10 (a) The design for a nanocontainer with the cap and the ball (C_{60}) together serving as a molecular valve that traps the hydrogen after the release of the external pressure; (b) the side view of the atomistic model of a nanocontainer, in which two C_{60} molecules are attached to a (20,0) SWCNT. Reproduced with permission from Ref. 97. © 2007 Elsevier.

structure, which enhances hydrogen binding due to the formation of B–H bond. However, N doping forms an electron-rich six-membered ring structure, which decreases the atomic hydrogen adsorption energies in both zigzag and armchair nanotubes. For hydrogen molecular adsorption, both B and N doping decrease the adsorption energies in SWCNTs.

Another study showed that axial relaxation favors the H_2 adsorption and dominates the effect of charge transfer from Ti 3d to C 2p in Ti-decorated armchair (5,5) SWCNT, in contrast to axial compression, which lead to a new way of modulating the H_2 adsorption by strain loading [96].

The weak van der Waals potential is the main reason that the physisorption-based methods are not effective to store molecular hydrogen. To conquer this, Ye *et al.* explored the other side of such a weak potential, the well-known compressibility of hydrogen, and proposed a new method to store hydrogen through a fill-and-lock mechanism and the correlation between pressure and diffusion barrier through the molecular valve [97]. They designed a (20,0) SWCNT-based nanocontainer, in which a C_{60} "peapod" at the cap section of the nanotube serves as a molecular valve (Fig. 13.10). H_2 could be firstly filled into the container upon compression at low temperature and then be locked inside it after the release of external pressure. At 2.5 GPa, the storage capacity approaches a promising 7.7 wt%. Though its actual synthesis is going to be a challenge, as the authors also recognized, the new concept, that is, taking advantage of the unique and remarkable compressibility of H_2, may inspire further studies along this direction.

13.4 HYDROGEN STORAGE IN GRAPHENE-BASED MATERIALS

Recently, graphene has emerged as a promising candidate for hydrogen storage media due to its superior properties, such as high surface area and light weight. Similar to the cases of fullerenes and CNTs, the pristine graphene exhibits weak affinity to H_2 molecules; thus, we have to make some measures to improve the hydrogen storage capability.

Compressing the H$_2$ molecules into the graphene layers is a potential approach to use graphene for hydrogen storage. In principle, hydrogen could be stored between graphene layers. Patchkovskii et al. computationally demonstrated that one mono-layer of H$_2$ could be accommodated within the intergraphene structure (2–3 wt% storage capacity at 5 MPa) when graphene layers are separated by a distance of 6 Å [21]. By separating two graphene layers to 8 Å, the hydrogen uptake can be up to 5.0–6.5 wt%. These theoretical predictions inspired many experimental efforts, for example, Jin et al. developed an efficient method to produce the carbon scaffolds for hydrogen storage [98]. The functionalization and cross-linking of thermally exfoli-ated graphene exhibited enhanced hydrogen storage capacity (~1.9 wt% at 77 K and 2 bar) compared to the original thermally exfoliated graphene material.

Decorating graphene nanosheets with metal atoms, such as TM, alkali metal, and alkali-earth metal, serves as another promising approach to enhance the H$_2$ adsorp-tion on graphene. The same as for fullerenes and nanotubes, the clustering problem has to be avoided.

TM decoration on graphene surface has attracted great attention of enhancing the hydrogen adsorption [99–111]. Introducing defects and substitutional doping and taking advantage of functional groups on graphene are well-used strategies to con-quer the clustering problem.

Ti atoms embedded in double-vacancy graphene (Ti@DVG) could hold up to eight H$_2$ per unit in molecular form; the Ti@DVG with the interlayer distance of 8.5 Å can store hydrogen up to 6.3 wt% [99].

Kim et al. showed that boron substitutional doping could provide the acceptor-like states in the absorbents, which is essential for enhancing the metal adsorption strength and for increasing the hydrogen storage capacity [100], The Sc + B-doped graphene complex could store hydrogen up to 7 wt% with the average binding energy of 0.5 eV. Nevertheless, their later investigation revealed that nitrogen defects are able to create the highly localized stated near the Fermi level, thus inducing stronger TM bindings and more favorable hydrogen adsorption in the pyridine-like nitrogen-doped graphene than those in pure or B-doped graphenes [101]. The strong TM binding prevents the metal aggregation and improves the material stability. The Sc-dispersed graphene complex (Sc + C$_{34}$N$_{12}$) could uptake hydrogen up to 5 wt%. Furthermore, the hydrogen storage capacity of Ti-decorated pyridinic nitrogen-doped graphene can be improved by the compressive strain [102]. By applying compressive strain, the system could adsorb four H$_2$ molecules per Ti atom. The variation of binding energy for the second and third adsorbed H$_2$ molecule according to the strain was large, that is, 0.217 and 0.254 eV, respectively.

Apart from introducing defects and atom substitution, Wang et al. proposed to use graphene oxide (GO) to anchor the Ti atoms without clustering [105]. Their computations showed that the Ti atoms can be stably anchored by the hydroxyl groups on GO surface and at the same time retain the activity to adsorb the H$_2$ mole-cules. As shown in Figure 13.11, each Ti is able to bind multiple H$_2$ with the desired binding energies (0.14–0.43 eV/H$_2$), corresponding to the theoretical gravimetric and volumetric densities of 4.9 wt% and 64 g l^{-1}, respectively.

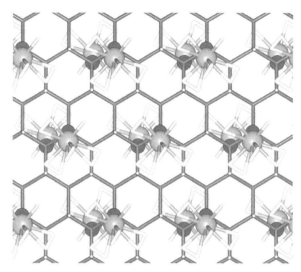

FIGURE 13.11 The structure of Ti-GO fully loaded with H_2. It has a 2×2 Ti periodicity, and the O–O separation on the same side is $d_{O-O} = 5.1$ Å. Reproduced with permission from Ref. 105. © 2009 American Chemical Society.

Alkali metal doping on graphene has also been predicted to be an effective sorbent with enhanced hydrogen adsorption [112–128]. Besides pristine graphene nanosheets, porous graphene and substituted graphene were also considered. Zhao *et al.* proposed cointercalating graphite with lithium and organic molecules as a practical strategy to improve the hydrogen storage on graphene [112]. The cointercalated species can expand the interlayer graphene distance and thus create more free space to accommodate multiple H_2 species around Li cations with a binding energy of 0.1–0.23 eV.

Li atoms can be bound strongly in porous graphene, in which the hydrogen uptake can reach 12 wt% [119]. The oxidized porous graphene has even better performance to bind the Li atoms due to the strong interaction between Li and O atoms, thus avoiding the clustering problem even better [120]. Each doped Li atom can bind five H_2 molecules, corresponding to a gravimetric density of 9.43 wt%.

Additionally, with its porous nature, graphyne was proposed to bind Li atoms for hydrogen adsorption [121]. Each Li is able to trap four H_2 molecules, resulting in 9.26 wt% hydrogen storage for one-sided Li decoration and 15.15 wt% for double-sided Li decoration, respectively.

Boron-substituted graphene is effective to well disperse Li atoms for hydrogen adsorption [123–125]. B atoms in graphene can provide available frontier orbitals to interact with the 2p orbital of Li, thus enhancing the binding strength of Li. Each dispersed Li adatom is capable of trapping four H_2 molecules for one-sided adsorption and eight H_2 molecules for double-sided adsorption on B-doped graphene. The binding energy (0.13 eV/H_2) allows the system to operate under ambient thermodynamic conditions [123].

Similarly, nitrogen-doped graphene with pyridinic and pyrrolic defects can also well disperse the Li atoms without clustering problem [126]. Each Li atom on the N-doped graphene can bind three H$_2$ molecules with the binding energy range of 0.12–0.2 eV/H$_2$. Further introduction of external electric field on Li-functionalized N-doped graphene can improve the hydrogen binding energy to the range of 0.14–0.27 eV/H$_2$ [127].

Very recently, Li *et al.* proposed a feasible strategy of combining the graphene oxide and lithium amidoborane (LiAB) to bridge the chemical hydrogen storage and hydrogen storage by physisorption (Fig. 13.12) [128]. The dehydrogenation process of GO–LiAB may offer a feasible way of uniform Li doping on the GO surface. Both of the two dominant groups, –O– and OH–, contribute to the facile combination between GO and LiAB. The possible dehydrogenated products of GO–Li$_{(n)}$ can store up to 5 wt% of H$_2$, and the GO–(Li$_3$N$_3$B$_3$)$_{(n)}$ can still store 5 wt% of H$_2$.

Alkali-earth metal decoration on graphene also has superior ability for hydrogen storage [129–140]. Available studies suggest that Ca has the best performance.

Kim *et al.* proposed that the Ca-intercalated pillared graphite can obtain higher volumetric capacity than the liquid hydrogen [129]. Ca-decorated porous three-dimensional (3-D) graphene exhibits even better Ca dispersion than the planar graphene sites [130]. The s–d level exchange in Ca chain facilitates the H$_2$ adsorption on the porous graphene, leading to the gravimetric capacity of about 5–6 wt% (Fig. 13.13). Boron-doped graphene is effective for dispersing alkali-earth metals (especially Ca) and also for improving the hydrogen adsorption (up to four H$_2$ molecules can be stably bound to a Ca atom) [131, 132]. Vacancy defects also efficiently enhance the alkali-earth metal binding and thus prevent the metal aggregation [133]. Ca-vacancy complexes exhibit the most favorable hydrogen adsorption characteristics, 6 wt% with the adsorption energy of 0.1 eV/H$_2$, among the considered metals [135].

Mg atom can be strongly bound to graphene oxide and also stay active to adsorb H$_2$ molecules [137]. The Mg and O atoms together could jointly produce a stronger electric field to polarize H$_2$ molecules, leading to a gravimetric hydrogen capacity of 5.6 wt% (four H$_2$ per Mg atom, with the average binding energy of 0.38 eV/H$_2$) at 200 K without any hydrogen pressure.

Aluminum-decorated graphene was also investigated for hydrogen storage [141–145]. Al decoration alters the electronic structures of both C and H$_2$ [141]. Ao *et al.* predicted that the 5.13 wt% hydrogen storage capacity can be achieved at 300 K and 0.1 GPa with the enhanced binding energy of 0.26 eV/H$_2$; they further predicted that when both sides can store hydrogen in the Al-decorated graphene, up to 13.79 wt% (in excess of 6 wt%) with an average adsorption energy of 0.19 eV/H$_2$ can be obtained [142]. The Al-doped bulk graphite with wide layer distance of 4.5 Å was also predicted to be ideal for H$_2$ adsorption, with 3.48 wt% hydrogen storage capability at 300 K and 0.1 GPa and the binding energy of 0.26 eV/H$_2$ [143]. The dispersion of Al (also Ti) can be enhanced by boron substitution on graphene due to the repulsive Coulomb interaction between metal adatoms and strong bonding force between dispersed metal atom and B-substituted graphene [145]. The Al (Ti)-decorated B-substituted graphene can uptake eight H$_2$ molecules per metal atom for double-sided adsorption, corresponding to 9.9 wt% (7.9 wt%) hydrogen storage.

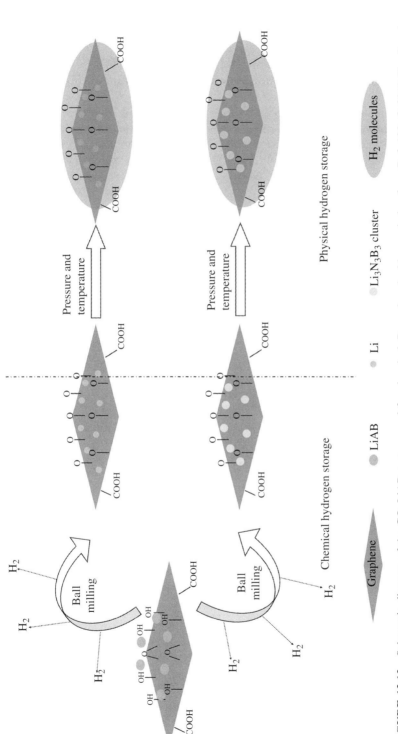

FIGURE 13.12 Schematic diagram of the GO–LiAB system used for chemical. Reproduced with permission from Ref. 128. © 2013 The Royal Society of Chemistry.

(a)

(b)

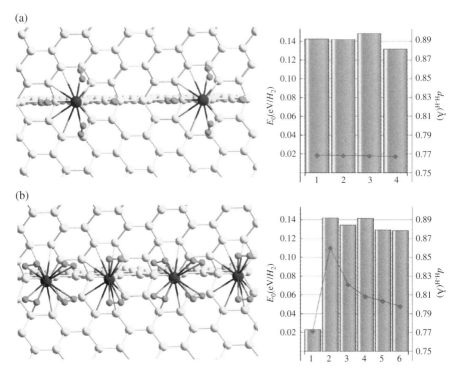

FIGURE 13.13 The optimized atomic structure of (a) atomic Ca in CBG(3) with four H$_2$ molecules adsorbed [Ca(H$_2$)$_4$+CBG(3)] and (b) Ca chain in CBG(3) with six H$_2$ molecules adsorbed in total [Ca(H$_2$)$_4$+Ca(H$_2$)$_2$+CBG(3)]. Reproduced with permission from Ref. 131. © 2009 American Physical Society.

Moreover, substitutionally doping graphene nanoribbons by Al atoms for hydrogen storage was also considered. Though the substitutionally Al-doped graphene nanoribbons with an Al atom occupying a central, lateral, or subedge site cannot satisfy the DOE binding energy criterion, the graphene nanoribbons with Al doping at an edge site and also by zigzag graphene nanoribbons with adsorbed Al at lateral hole sites are indeed promising [144].

Despite the theoretical predictions of the high hydrogen capacity in metal-decorated carbon materials, there are few experimental supports for the expectations. One possible reason is the interference of other air components in the gas phase. Sigal *et al.* systematically investigated the effect of air components on the hydrogen storage properties of metal-decorated (Li, Na, K, Al, Ti, V, Ni, Cu, Pd, Pt) graphene [146] and found that oxygen interferences would block the adsorption site or cause the irreversible oxidation of metal decoration. Among the metals considered, Ni, Pd, and Pt are the most promising decorations with minimal oxygen interference. This work raised a serious doubt about the straightforward application of the theoretically predicted systems for hydrogen application and suggested that even in the most promising cases, we have to suppress oxygen access to allow good hydrogen storage.

(a)

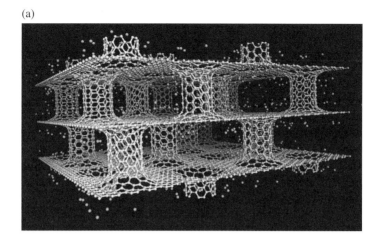

(b)

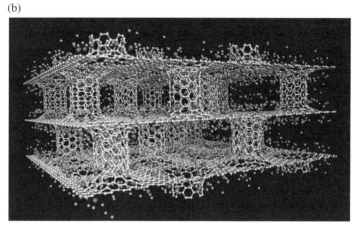

FIGURE 13.14   (a) Snapshot from the GCMC simulations of pure pillared structure at 77 K and 3 bar; (b) snapshot from the GCMC simulations of lithium-doped pillared structure at 77 K and 3 bar. Hydrogen molecules are represented in green, while lithium atoms are in purple. Reproduced with permission from Ref. 147. © 2008 American Chemical Society. (*See insert for color representation of the figure.*)

Interestingly, combining CNTs and graphene sheets can produce a novel 3-D pillared graphene (CNTs support the graphene layers like pillars and assemble a 3-D building block), which possesses large surface area and tunable pore size for hydrogen storage. The bare surface of pillared graphene still exhibits weak hydrogen adsorption and low hydrogen storage capacity at ambient conditions [147, 148]. Therefore, lithium decoration has been applied on this system to improve hydrogen adsorption [147]. As shown in the snapshots of the grand canonical Monte Carlo (GCMC) simulation (Fig. 13.14), Li-decorated pillared graphene can uptake plenty of H_2 molecules. However, the Li aggregation problem has not been considered, which would significantly decrease the hydrogen adsorption capacity. Moreover, Tylianakis *et al.* proposed replacing the graphene sheets by GO and then substituting the OH groups with the O–Li groups, in which the resulting material can effectively prevent the Li atoms from clustering [149]. Li-decorated pillared GO with pore dimensions of $d = 23$ Å and an O/C ratio of 1/8 can reach a gravimetric H$_2$ capacity greater than 10 wt% and a volumetric H$_2$ capacity of 55 g l^{-1} at 77 K and 100 bar. Kim *et al.* pointed out that it is important to identify the optimal GO interlayer distance since the hydrogen storage capacity is dependent on the distance of GO interlayer [150]. An optimal GO interlayer of 6.3 Å has led to the maximum hydrogen storage capacity in three GO–amine composites. Wu *et al.* profoundly investigated the effects of pressure, temperature, and geometric structure on hydrogen storage properties of pillared graphene by means of molecular dynamics simulations [151]. They found that in the pillared structure, the graphene sheets present better hydrogen storage capacity than the CNTs pillars, especially at the favorable conditions of low temperature, high pressure, and large spacing between graphene sheets. Later, Aboutalebi *et al.* experimentally demonstrated that processability of GO dispersions could be further exploited to fabricate self-aligned GO–multiwalled carbon nanotube (GO-MWCNT) hybrid frameworks, which offers a simple way to designing GO-based hybrid frameworks [152]. The GO-MWCNT with proper interlayer distance exhibited high hydrogen adsorption ability of 2.6 wt% at room temperature.

Recently, graphene oxide frameworks (GOFs), another new class of 3-D pillared porous materials, have drawn much attention for hydrogen storage due to its tunable porosity, accessible surface area, and versatile electronic properties [153–155]. Yildirim and coworkers proposed to intercalate the boronic acid into the GO layers; in this way, they successfully synthesized the porous 3-D materials (Fig. 13.15) with the cheap and environmentally friendly GO [153]. GCMC simulation of hydrogen adsorption isotherms at 77 K for several representative GOF structures predicted that GOF-32 with one linker per 32 graphene carbon atoms possesses an H$_2$ adsorption capacity of 6.1 wt% at 77 K and 1 bar, and such predictions were confirmed by the initial experiments. Yildirim and coworkers further synthesized a range of porous GOFs with strong boronate ester bonds between GO layers [154]. With an optimum interlayer spacing between graphene planes, the hydrogen molecules can interact with both surfaces; thus, high isosteric heat of adsorption and hydrogen adsorption capacity (twice of typical porous carbon material and comparable to metal organic frameworks) were observed experimentally. Chan *et al.* investigated the hydrogen storage properties of GOFs (GOF-120, GOF-66, GOF-28, and GOF-6) with a mathematic model and found that GOF-28 has the highest hydrogen uptake of 6.33 wt% [155], which was attributed to the mechanical support, porous spaces, and most importantly the enhanced hydrogen adsorption provided by the benzenediboronic acid pillars between graphene sheets. The remarkable synergy between GO and other materials in a hybrid structure offers a novel yet simple way of designing GO-based hybrid frameworks with extraordinary hydrogen storage capacities by incorporating two different materials neither of which alone might be completely perfect for the required application. Very recently, Liu *et al.* synthesized the complexes of Cu-MOF and GO for hydrogen storage [156]. The nanosized Cu-BTC (copper benzene-1,3,5-tricarboxylate) is well dispersed by the incorporation of GO, and the composite

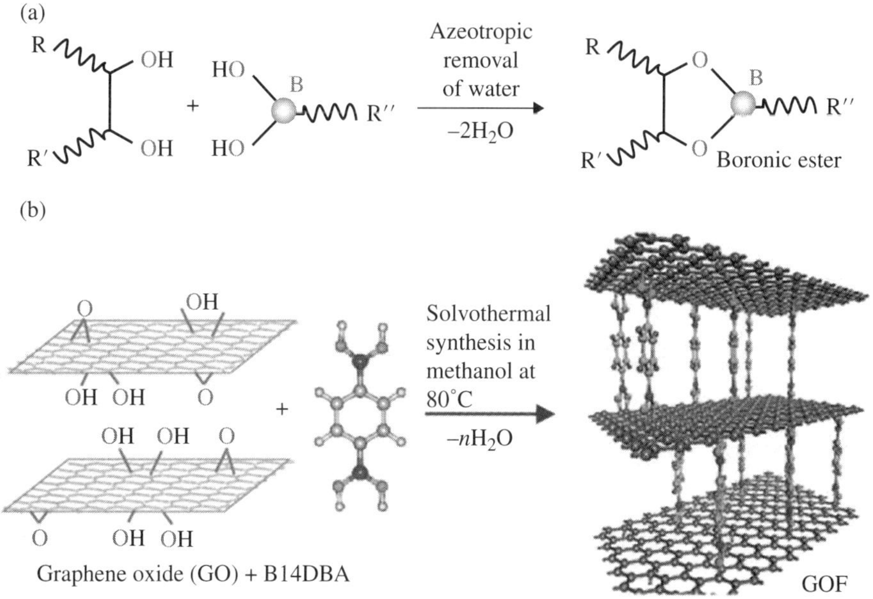

(a)

Azeotropic removal of water

$-2H_2O$

Boronic ester

(b)

Solvothermal synthesis in methanol at 80°C

$-nH_2O$

Graphene oxide (GO) + B14DBA

GOF

FIGURE 13.15 Representations of (a) boronic ester and (b) GOF formation. Idealized graphene oxide framework (GOF) materials proposed in this study are formed from layers of graphene oxide connected by benzenediboronic acid pillars. Reproduced with permission from Ref. 153. © 2010 Wiley-VCH Verlag GmbH & Co. KGaA, Weinheim.

exhibits great improvement of hydrogen storage capacity compared to the pristine Cu-BTC (from 2.81 wt% of Cu-BTC to 3.58 wt% of CG-9 at 77 K and 42 atm).

13.5 CONCLUSIONS

Herein, we briefly reviewed the H_2 storage behaviors in carbon-based nanomaterials, such as fullerene, nanotube, graphene, and their derivatives. Both state-of-the-art theoretical computations and critical experimental evaluations confirmed that the physisorption of H_2 molecules on the pristine bare carbon-based sorbents cannot lead to sufficient hydrogen capacity. Therefore, many effective strategies were proposed to enhance the hydrogen adsorption and uptake, including introduction of defects, substitutional doping, metal decoration, construction of hybrid complex, etc.

Encouragingly, TM like Ti, Sc, and Ca and alkaline metal like Li have shown significant enhancement of H_2 adsorption and storage capacity. However, it is still very challenging to realize the well-dispersed metal decoration on the surface of carbon-based sorbents. More feasible approaches to decorate the carbon nanostructures by uniformly dispersed single atoms will be continually pursued in the future.

The interlayer space between graphene (or GO) layers can store H_2 molecules, and the storage capability can be enhanced by optimizing the interlayer distance.

Theoretical studies predicted that intercalating CNTs into graphene (or GO) layers could form a new class of 3D porous materials with tunable porosity and accessible surface area for hydrogen storage, and now, we have the ability to synthesize such 3D porous frameworks.

The progresses in the last few years are exciting, and we are so proud of the endeavors and achievements of our peers. Modern computations offer us a wonderful tool for material design and provide many insights and ideas for us to explore hydrogen storage materials. We strongly believe that the joint efforts by theoreticians and experiments will lead to even better hydrogen storage materials and realize our dreamed hydrogen society soon.

ACKNOWLEDGMENTS

This work was supported in China by the National Natural Science Foundation of China (no.51101145) and in the United States by the DoD (Grant W911NF-12-1-0083). We want to thank the professor for offering the structure of Figure 13.4.

REFERENCES

[1]   US Department of Energy. "Targets for onboard hydrogen storage systems for light-duty vehicles," http://www1.eere.energy.gov/hydrogenandfuelcells/mypp/pdfs/storage.pdf (accessed May 8, 2015).

[2]   Grochala, W.; Edwards, P. P. Thermal decomposition of the non-interstitial hydrides for the storage and production of hydrogen. *Chem. Rev.* 2004, **104**, 1283–1315.

[3]   Orimo, S.; Nakamori, Y.; Eliseo, J. R.; Züttel, A.; Jensen, C. M. Complex hydrides for hydrogen storage. *Chem. Rev.* 2007, **107**, 4111–4132.

[4]   Graetz, J. New approaches to hydrogen storage. *Chem. Soc. Rev.* 2009, **38**, 73–82.

[5]   Hamilton, C. W.; Baker, R. T.; Staubitz, A.; Manners, I. B-N compounds for chemical hydrogen storage. *Chem. Soc. Rev.* 2009, **38**, 279–293.

[6]   Chandra, D.; Reilly, J. J.; Chellappa, R. Metal hydrides for vehicular applications: The state of the art. *JOM* 2006, **58**, 26–32.

[7]   Sakintuna, B.; Lamari-Darkrim, F.; Hirscher, M. Metal hydride materials for solid hydrogen storage: A review. *Int. J. Hydrogen Energy* 2007, **32**, 1121–1140.

[8]   Harder, S.; Spielmann, J.; Intermann, J.; Bandmann, H. Hydrogen storage in magnesium hydride: The molecular approach. *Angew. Chem. Int. Ed.* 2011, **50**, 4156–4160.

[9]   Marder, T. B. Will we soon be fueling our automobiles with ammonia–borane? *Angew. Chem. Int. Ed.* 2007, **46**, 8116–8118.

[10] Demirci, U. B.; Miele, P. Sodium borohydride versus ammonia borane, in hydrogen storage and direct fuel cell applications. *Energy Environ. Sci.* 2009, **2**, 627–637.

[11] Staubitz, A.; Robertson, A. P. M.; Sloan, M. E.; Manners, I. Amine- and phosphine-borane adducts: New interest in old molecules. *Chem. Rev.* 2010, **110**, 4023–4078.

[12] Meek, B. S. T.; Greathouse, J. A.; Allendorf, M. D. Metal-organic frameworks: A rapidly growing class of versatile nanoporous materials. *Adv. Mater.* 2011, **23**, 249–267.

[13] Xuan, W. M.; Zhu, C. F.; Liu, Y.; Cui Y. Mesoporous metal-organic framework materials. *Chem. Soc. Rev.* 2012, **41**, 1677–1695.

[14] Getman, R. B.; Bae, Y. S.; Wilmer, C. E.; Snurr, R. Q. Review and analysis of molecular simulations of methane, hydrogen, and acetylene storage in metal–organic frameworks. *Chem. Rev.* 2012, **112**, 703–723.

[15] Feng, X.; Ding X. S.; Jiang D. L. Covalent organic frameworks. *Chem. Soc. Rev.* 2012, **41**, 6010–6022.

[16] Tozzini, V.; Pellegrini, V. Prospects for hydrogen storage in graphene. *Phys. Chem. Chem. Phys.* 2013, **15**, 80–89.

[17] Hu, Y. H.; Ruckenstein, E. Clathrate hydrogen hydrate–a promising material for hydrogen storage. *Angew. Chem. Int. Ed.* 2006, **45**, 2011–2013.

[18] Struzhkin, V. V.; Militzer, B.; Mao, W. L.; Mao, H.; Hemley, R. J. Hydrogen storage in molecular clathrates. *Chem. Rev.* 2007, **107**, 4133–4151.

[19] Sugahara, T.; Haag, J. C.; Prasad, P. S. R.; Warntjes, A. A.; Sloan, E. D.; Sum, A. K.; Koh, C. A. Increasing hydrogen storage capacity using tetrahydrofuran. *J. Am. Chem. Soc.* 2009, **131**, 14616–14617.

[20] Lochan, R. C.; Head-Gordon, M. Computational studies of molecular hydrogen binding affinities: The role of dispersion forces, electrostatics, and orbital interactions. *Phys. Chem. Chem. Phys.* 2006, **8**, 1357–1370.

[21] Patchkovskii, S.; Tse, J. S.; Yurchenko, S. N.; Zhechkov, L.; Heine, T.; Seifer, G. Graphene nanostructures as tunable storage media for molecular hydrogen. *Proc. Natl. Acad. Sci. U. S. A.* 2005, **102**, 10439–10444.

[22] Zhao, Y. F.; Kim, Y. H.; Dillon, A. C.; Heben, M. J.; Zhang, S. B. Hydrogen storage in novel organometallic buckyballs. *Phys. Rev. Lett.* 2005, **94**, 15504–15507.

[23] Kubas, G. J. Metal–dihydrogen and σ-bond coordination: The consummate extension of the Dewar–Chatt–Duncanson model for metal–olefin π bonding. *J. Organomet. Chem.* 2001, **635**, 37–68.

[24] Yildirim, Y.; Íñiguez, J.; Ciraci, S. Molecular and dissociative adsorption of multiple hydrogen molecules on transition metal decorated C_{60}. *Phys. Rev. B* 2005, **72**, 153403–153406.

[25] Sun, Q.; Wang, Q.; Jena, P.; Kawazoe, Y. Clustering of Ti on a C_{60} surface and its effect on hydrogen storage. *J. Am. Chem. Soc.* 2005, **127**, 14582–14583.

[26] Shin, W. H.; Yang, S. H.; Goddard III, W. A.; Kang, J. K. Ni-dispersed fullerenes: Hydrogen storage and desorption properties. *Appl. Phys. Lett.* 2006, **88**, 053111–153113.

[27] Sun, Q.; Jena, P.; Wang, Q.; Marquez, M. First-principles study of hydrogen storage on $Li_{12}C_{60}$. *J. Am. Chem. Soc.* 2006, **128**, 9741–9745.

[28] Teprovich, Jr., J. A.; Wellons, M. S.; Lascola, R.; Hwang, S. J.; Ward, P. A.; Compton, R. N.; Zidan, R. Synthesis and characterization of a lithium-doped fullerane (Li_x-C_{60}-H_y) for reversible hydrogen storage. *Nano Lett.* 2012, **12**, 582–589.

[29] Chandrakumar, K. R. S.; Ghosh, S. K. Alkali-metal-induced enhancement of hydrogen adsorption in C_{60} fullerene: An ab initio study. *Nano Lett.* 2008, **8**, 13–19.

[30] Hu, X.; Trudeau, M.; Antonelli, D. M. Hydrogen storage in mesoporous titanium oxide–alkali fulleride composites. *Inorg. Chem.* 2008, **47**, 2477–2484.

[31] Peng, Q.; Chen, G.; Mizuseki, H.; Kawazoe, Y. Hydrogen storage capacity of $C_{60}(OM)_{12}$ (M = Li and Na) clusters. *J. Chem. Phys.* 2009, **131**, 214505–214512.

[32] Rabilloud, F. Structure and electronic properties of Alkali–C60 nanoclusters. *J. Phys. Chem. A* 2010, **114**, 7241–7247.

[33] Yoon, M.; Yang, S. Y.; Hicke, C.; Wang, E.; Geohegan, D.; Zhang, Z. Y. Calcium as the superior coating metal in functionalization of carbon fullerenes for high-capacity hydrogen storage. *Phys. Rev. Lett.* 2008, **100**, 206806–206809.

[34] Wang, Q.; Sun, Q.; Jena, P.; Kawazoe, Y. Theoretical study of hydrogen storage in Ca-coated fullerenes. *J. Chem. Theory Comput.* 2009, **5**, 374–379.

[35] Lee, H.; Huang, B.; Duan, W. H.; Ihm, J. Ab initio study of beryllium-decorated fullerenes for hydrogen storage. *J. Appl. Phys.* 2010, **107**, 084304–084307.

[36] Yoon, M.; Yang, S. Y.; Wang, E.; Zhang, Z. Y. Charged fullerenes as high-capacity hydrogen storage media. *Nano Lett.* 2008, **7**, 2578–2583.

[37] Kaiser, A.; Leidlmair, C.; Bartl, P.; Zöttl, S.; Denifl, S.; Mauracher, A.; Probst, M.; Scheier, P.; Echt1, O. Adsorption of hydrogen on neutral and charged fullerene experiment and theory. *J. Chem. Phys.* 2013, **138**, 074311–074323.

[38] Sun, Q.; Wang, Q.; Jena, P. Functionalized heterofullerenes for hydrogen storage. *Appl. Phys. Lett.* 2009, **94**, 013111–013113.

[39] Srinivasu, K.; Ghosh, S. K. Transition metal decorated porphyrin-like porous fullerene: Promising materials for molecular hydrogen adsorption. *J. Phys. Chem. C* 2012, **116**, 25184–25189.

[40] Guo, J.; Liu, Z. G.; Liu, S. Q.; Zhao, X. H.; Huang, K. L. High-capacity hydrogen storage medium: Ti doped fullerene. *Appl. Phys. Lett.* 2011, **98**, 023107–023109.

[41] Shalabi, A. S.; Mahdy, A. M. E.; Taha, H. O. The effect of C-vacancy on hydrogen storage and characterization of H$_2$ modes on Ti functionalized C$_{60}$ fullerene a first principles study. *J. Mol. Model.* 2013, **19**, 1211–1225.

[42] Yoon, M.; Yang, S. Y.; Zhang, Z. Y. Interaction between hydrogen molecules and metallofullerenes. *J. Chem. Phys.* 2009, **131**, 064707–064711.

[43] Liu, Z. Clustering of molecular hydrogen anion (H$_3^-$) on a Li$^+$ @C$_{60}$ surface. *Int. J. Hydrogen Energy* 2007, **32**, 3987–3989.

[44] Yamada, Y.; Satake, Y.; Watanabe, K.; Yokoyama, Y.; Okada, R.; Sasaki, M. Hydrogen adsorption on electron-doped C$_{60}$ monolayer on Cu(111) studied by He atom scattering. *Phys. Rev. B* 2011, **84**, 235425–235430.

[45] Guo, J. H.; Zhang, H.; Miyamoto, Y. New Li-doped fullerene-intercalated phthalocyanine covalent organic frameworks designed for hydrogen storage. *Phys. Chem. Chem. Phys.* 2013, **15**, 8199–8207.

[46] Thornton, A. W.; Nairn, K. M.; Hill, J. M.; Hill, A. J.; Hill, M. R. Metal-organic frameworks impregnated with magnesium-decorated fullerenes for methane and hydrogen storage. *J. Am. Chem. Soc.* 2009, **131**, 10662–10669.

[47] Iijima, S. Helical microtubules of graphitic carbon. *Nature* 1991, **354**, 56–58.

[48] Dillon, A. C.; Jones, K. M.; Bekkedahl, T. A.; Kiang, C. H.; Bethune, D. S.; Heben, M. J. Storage of hydrogen in single-walled carbon nanotubes. *Nature* 1997, **386**, 377–379.

[49] Ye, Y.; Ahn, C. C.; Whitam, C.; Fultz, B.; Liu, L.; Rinzler, A. G.; Colbert, D.; Smith, K. A.; Smalley, R. E. Hydrogen adsorption and cohesive energy of single-walled carbon nanotubes. *J. Appl. Phys. Lett.* 1999, **74**, 2307–2309.

[50] Liu, C.; Fan, Y. Y.; Liu, M.; Cong, H. T.; Cheng, H. M.; Dresselhaus, M. Hydrogen storage in single-walled carbon nanotubes at room temperature. *Science* 1999, **286**, 1127–1129.

[51] Liu, C.; Yang, Q. H.; Tong, Y.; Cong, H. T.; Cheng, H. M. Volumetric hydrogen storage in single-walled carbon nanotubes. *Appl. Phys. Lett.* 2002, **80**, 2389–2391.

[52] Chen, X.; Detlaff-Weglikowska, U.; Haluska, M.; Hulman, M.; Roth, S.; Hirscher, M.; Becher, M. Pressure isotherms of hydrogen adsorption in carbon nanostructures. *Mater. Res. Soc. Symp. Proc.* 2002, **706**, 295–300.

[53] Smith, M. R.; Bittner, E. W.; Shi, W.; Johnson, J. K.; Bockrath, B. C. Chemical activation of single-walled carbon nanotubes for hydrogen adsorption. *J. Phys. Chem. B* 2003, **107**, 3752–3760.

[54] Tibbetts, G. G.; Meisner, C. P.; Olk, C. H. Hydrogen storage capacity of carbon nanotubes, filaments, and vapor-grown fibers. *Carbon* 2001, **39**, 2291–2301.

[55] Shiraishi, M.; Takenobu, T.; Ata, M. Gas–solid interactions in the hydrogen/single-walled carbon nanotube system. *Chem. Phys. Lett.* 2003, **367**, 633–636.

[56] Kajiura, H.; Tsutsui, S.; Kadono, K.; Kakuta, M.; Ata, M.; Murakami, Y. Hydrogen storage capacity of commercially available carbon materials at room temperature. *Appl. Phys. Lett.* 2003, **82**, 1105–1107.

[57] Dodziuk, H.; Dolgonos, G. Molecular modeling study of hydrogen storage in carbon nanotubes. *Chem. Phys. Lett.* 2002, **356**, 79–83.

[58] Zhou, Z.; Zhao, J.; Chen, Z.; Gao, X.; Yan, T.; Schleyer, P. v. R. Comparative study of hydrogen adsorption on carbon and BN nanotubes. *J. Phys. Chem. B* 2006, **110**, 13363–13369.

[59] Chen, P.; Wu, X.; Lin, J.; Tan, K. L. High H_2 uptake by alkali-doped carbon nanotubes under ambient pressure and moderate temperatures. *Science* 1999, **285**, 91–93.

[60] Yang, R. T. Hydrogen storage by alkali-doped carbon nanotubes-revisited. *Carbon* 2000, **38**, 623–641.

[61] Pinkerton, F. E.; Wicke, B. G.; Olk, C. H.; Tibbetts, G. G.; Meisner, G. P.; Meyer, M. S.; Herbst, J. F. Thermogravimetric measurement of hydrogen absorption in alkali-modified carbon materials. *J. Phys. Chem. B* 2000, **104**, 9460–9467.

[62] Yildirim, T.; Ciraci, S. Titanium-decorated carbon canotubes as a cotential high-capacity hydrogen storage medium. *Phys. Rev. Lett.* 2005, **94**, 175501–175504.

[63] Dag, S.; Ozturk, Y.; Ciraci, S.; Yildirim, T. Adsorption and dissociation of hydrogen molecules on bare and functionalized carbon nanotubes. *Phys. Rev. B* 2005, **72**, 155404–15411.

[64] Lee, J. W.; Kim, H. S.; Lee, J. Y.; Kang, J. K. Hydrogen storage and desorption properties of Ni-dispersed carbon nanotubes. *Appl. Phys. Lett.* 2006, **88**, 143126–143128.

[65] Krasnov, P. O.; Ding, F.; Singh, A. K.; Yakobson, B. I. Clustering of Sc on SWNT and reduction of hydrogen uptake: Ab-initio all-electron calculations. *J. Phys. Chem. C* 2007, **111**, 17977–17980.

[66] Durgun, E.; Ciraci, S.; Yildirim, T. Functionalization of carbon-based nanostructures with light transition-metal atoms for hydrogen storage. *Phys. Rev. B* 2008, **77**, 085405–085413.

[67] Xiao, H.; Li, S. H.; Cao, J. X. First-principles study of Pd-decorated carbon nanotube for hydrogen storage. *Chem. Phys. Lett.* 2009, **483**, 111–114.

[68] López-Corral, I.; Germán, E.; Volpe, M. A.; Brizuela, G. P.; Juan, A. Tight-binding study of hydrogen adsorption on palladium decorated graphene and carbon nanotubes. *Int. J. Hydrogen Energy* 2010, **35**, 2377–2384.

[69] Zhang, Z. W.; Zheng W. T.; Jiang, Q. Hydrogen adsorption on Ce/SWCNT systems: A DFT study. *Phys. Chem. Chem. Phys.* 2011, **13**, 9483–9489.

[70]   Reyhani, A.; Mortazavi, S. Z.; Mirershadi, S.; Moshfegh, A. Z.; Parvin, P.; Nozad Golikand, A. Hydrogen storage in decorated multiwalled carbon nanotubes by Ca, Co, Fe, Ni, and Pd nanoparticles under ambient conditions. *J. Phys. Chem. C* 2011, **115**, 6994–7001.

[71]   Gialampouki, M. A.; Lekka, Ch. E. TiN decoration of single-wall carbon nanotubes and graphene by density functional theory computations. *J. Phys. Chem. C* 2011, **115**, 15172–15181.

[72]   Tabtimsai, C.; Keawwangchai, S.; Nunthaboot, N.; Ruangpornvisuti, V.; Wanno, B. Density functional investigation of hydrogen gas adsorption on Fe-doped pristine and Stone-Wales defected single-walled carbon nanotubes. *J. Mol. Model.* 2012, **18**, 3941–3949.

[73]   Lu, J. L.; Xiao, H.; Cao, J. X. Mechanism for high hydrogen storage capacity on metal-coated carbon nanotubes: A first principle analysis. *J. Solid State Chem.* 2012, **196**, 367–371.

[74]   Shalabi, A. S.; Abdel Aal, S.; Assem, M. M.; Abdel Halim, W. S. Ab initio characterization of Ti decorated SWCNT for hydrogen storage. *Int. J. Hydrogen Energy* 2013, **38**, 140–152.

[75]   Kim, H. S.; Lee, H.; Han, H. K. S.; Kim, J. H.; Song, M. S.; Park, M. S.; Lee, J. Y.; Kang, J. K. Hydrogen storage in Ni nanoparticle-dispersed multiwalled carbon nanotubes. *J. Phys. Chem. B* 2005, **109**, 8983–8986.

[76]   Durgun, E.; Jang, Y. R.; Ciraci, S. Hydrogen storage capacity of Ti-doped boron-nitride and B/Be-substituted carbon nanotubes. *Phys. Rev. B* 2007, **76**, 073413–073416.

[77]   Shevlin, S. A.; Guo, Z. X. High-capacity room-temperature hydrogen storage in carbon nanotubes via defect-modulated titanium doping. *J. Phys. Chem. C* 2008, **112**, 17456–17464.

[78]   López-Corral, I.; Celis, J. D.; Juan, A.; Irigoyen, B. DFT study of H$_2$ adsorption on Pd-decorated single walled carbon nanotubes with C-vacancies. *Int. J. Hydrogen Energy* 2012, **37**, 10156–10164.

[79]   Singha, P.; Kulkarni, M. V.; Gokhale, S. P.; Chikkali, S. H.; Kulkarni, C. V. Enhancing the hydrogen storage capacity of Pd-functionalized multi-walled carbon nanotubes. *Appl. Surf. Sci.* 2012, **258**, 3405–3409.

[80]   Pérez-Bueno, J. J.; Mendoza López, M. L.; Brieño Enriquez, K. M.; Ledesma García, J.; Godínez Mora–Tovar L. A.; Angeles Chavez, C. Hydrogen storage enhancement attained by fixation of Ti on MWNTs. *Adv. Mater. Sci. Eng.* 2012, 801230–801237.

[81]   Chakraborty, B.; Modak, P.; Banerjee, S. Hydrogen storage in yttrium-decorated single walled carbon nanotube. *J. Phys. Chem. C* 2012, **116**, 22502–22508.

[82]   Chen, L.; Zhang, Y. M.; Koratkar, N.; Jena, P.; Nayak, S. K. First-principles study of interaction of molecular hydrogen with Li-doped carbon nanotube peapod structures. *Phys. Rev. B* 2008, **77**, 033405–033408.

[83]   Liu, W.; Zhao, Y. H.; Li, Y.; Jiang, Q.; Lavernia, E. J. Enhanced hydrogen storage on Li-dispersed carbon nanotubes. *J. Phys. Chem. C* 2009, **113**, 2028–2033.

[84]   Liu, W. H.; Xu, S. F.; Li, C.; Yuan, G. Alkali-earth metal adsorption behaviors on capped single-walled carbon nanotubes based on first-principle calculations. *Diam. Rel. Mater.* 2012, **29**, 59–62.

[85]   Nagare, B. J.; Habale, D.; Chacko S.; Ghos, S. Hydrogen adsorption on Na-SWCNT systems. *J. Mater. Chem.* 2012, **22**, 22013–22021.

[86] Wang, Y. S.; Li, M.; Wang, F.; Sun, Q.; Jia, Y. Li and Na Co-decorated carbon nitride nanotubes as promising new hydrogen storage media. *Phys. Lett. A* 2012, **376**, 631–636.

[87] Lee, H.; Ihm, J.; Cohen, M. L.; Louie, S. G. Calcium-decorated carbon nanotubes for high-capacity hydrogen storage: First-principles calculations. *Phys. Rev. B* 2009, **80**, 115412–115416.

[88] Yang, X. B.; Zhang, R. Q.; Ni, J. Stable calcium adsorbates on carbon nanostructures: Applications for high-capacity hydrogen storage. *Phys. Rev. B* 2009, **79**, 075431–075434.

[89] Cazorla, C.; Shevlin, S. A.; Guo, Z. X. First-principles study of the stability of calcium-decorated carbon nanostructures. *Phys. Rev. B* 2010, **82**, 155454–155465.

[90] Wang, Y. S.; Yuan, P. F.; Li, M.; Jiang, W. F.; Sun, Q.; Jia, Y. Calcium-decorated graphyne nanotubes as promising hydrogen storage media: A first-principles study. *J. Solid State Chem.* 2013, **197**, 323–328.

[91] Rangel, E.; Ruiz-Chavarria, G.; Magana, L. F.; Arellano, J. S. Hydrogen adsorption on N-decorated single wall carbon nanotubes. *Phys. Lett. A* 2009, **373**, 2588–2591.

[92] Mousavipour S. H.; Chitsazi, R. A theoretical study on the effect of intercalating sulfur atom and doping boron atom on the adsorption of hydrogen molecule on (10,0) single-walled carbon nanotubes. *J. Iran. Chem. Soc.* 2010, **7**, 92–102.

[93] Surya, V. J.; Iyakutti, K.; Rajarajeswari, M.; Kawazoe, Y. Functionalization of single-walled carbon nanotube with borane for hydrogen storage. *Phys. E* 2009, **41**, 1340–1346.

[94] Surya, V. J.; Iyakutti, K.; Venkataramanan, N. S.; Mizuseki, H.; Kawazoe, Y. Single walled carbon nanotubes functionalized with hydrides as potential hydrogen storage media: A survey of intermolecular interactions. *Phys. Status Solidi B* 2011, **248**, 2147–2158.

[95] Zhou, Z.; Gao, X. P.; Yan, J.; Song, D. Y. Doping effects of B and N on hydrogen adsorption in single-walled carbon nanotubes through density functional calculations. *Carbon* 2006, **44**, 939–947.

[96] Shalabi, A. S.; El Mahdy, A. M.; Taha, H. O. Theoretical characterization of axial deformation effects on hydrogen storage of Ti decorated armchair (5,5) SWCNT. *Mol. Phys.* 2013, **111**, 661–671.

[97] Ye, X.; Gu, X.; Gong, X. G.; Shing, T. K. M.; Liu, Z. F. A nanocontainer for the storage of hydrogen. *Carbon* 2007, **45**, 315–320.

[98] Jin, Z.; Lu, W.; O'Neill, K. J.; Parilla, P. A.; Simpson, L. J.; Kittrell, C.; Tour, J. M. Nano-engineered spacing in graphene sheets for hydrogen storage. *Chem. Mater.* 2011, **23**, 923–925.

[99] Chu, S. B.; Hu, L. B.; Hu, X. R.; Yang, M. K.; Deng J. B. Titanium-embedded graphene as high-capacity hydrogen-storage media. *Int. J. Hydrogen Energy* 2011, **36**, 12324–12328.

[100] Kim, G.; Jhi, S. H.; Park, N.; Louie, S. G.; Cohen, M. L. Optimization of metal dispersion in doped graphitic materials for hydrogen storage. *Phys. Rev. B* 2008, **78**, 085408–085412.

[101] Kim, G.; Jhi, S. H.; Park, N. Effective metal dispersion in pyridinelike nitrogen doped graphenes for hydrogen storage. *Appl. Phys. Lett.* 2008, **92**, 013106–013108.

[102] Kim, D.; Lee, S.; Jo, S.; Chung, Y. C. Strain effects on hydrogen storage in Ti decorated pyridinic N-doped graphene. *Phys. Chem. Chem. Phys.* 2013, **15**, 12757–12761.

[103] Choi, W. I.; Jhi, S. H.; Kim, K.; Kim, Y. H. Divacancy-nitrogen-assisted transition metal dispersion and hydrogen adsorption in defective graphene: A first-principles study. *Phys. Rev. B* 2010, **81**, 085441–085445.

[104] Mashoff, T.; Takamura, M.; Tanabe, S.; Hibino, H.; Beltram, F.; Heun, S. Hydrogen storage with titanium-functionalized graphene. *Appl. Phys. Lett.* 2013, **103**, 013903–013906.

[105] Wang, L.; Lee, K.; Sun, Y. Y.; Lucking, M.; Chen, Z. F.; Zhao, J. J.; Zhang, S. B. Graphene oxide as an ideal substrate for hydrogen storage. *ACS Nano* 2009, **3**, 2995–3000.

[106] Hong, W. G.; Kim, B. H.; Lee, S. M.; Yu, H. Y.; Yun, Y. J.; Jun, Y.; Lee, J. B.; Kim, H. J. Agent-free synthesis of graphene oxide/transition metal oxide composites and its application for hydrogen storage. *Int. J. Hydrogen Energy* 2012, **37**, 7594–7599.

[107] López-Corral, I.; Germán, E.; Juan, A.; Volpe, M. A.; Brizuela, G. P. DFT study of hydrogen adsorption on palladium decorated graphene. *J. Phys. Chem. C* 2011, **115**, 4315–4323.

[108] López-Corral, I.; Germán, E.; Juan, A.; Volpe, M. A.; Brizuela, G. P. Hydrogen adsorption on palladium dimer decorated graphene: A bonding study. *Int. J. Hydrogen Energy* 2012, **37**, 6653–6665.

[109] Sen, D.; Thapa, R.; Chattopadhyay, K. K. Small Pd cluster adsorbed double vacancy defect graphene sheet for hydrogen storage: A first-principles study. *Int. J. Hydrogen Energy* 2013, **38**, 3041–3049.

[110] Wang, Y.; Liu, J. H.; Wang, K.; Chen, T.; Tan, X.; Li, C. M. Hydrogen storage in Ni-B nanoalloy-doped 2D graphene. *Int. J. Hydrogen Energy* 2011, **36**, 12950–12954.

[111] Zhou, M.; Lu, Y. H.; Zhang, C.; Feng, Y. P. Strain effects on hydrogen storage capability of metal-decorated graphene: A first-principles study. *Appl. Phys. Lett.* 2010, **97**, 103109–103111.

[112] Zhao, Y. F.; Kim, Y. H.; Simpson, L. J.; Dillon, A. C.; Wei, S. H.; Heben, M. J. Opening space for H$_2$ storage: Cointercalation of graphite with lithium and small organic molecules. *Phys. Rev. B* 2008, **78**, 144102–144106.

[113] Ataca, C.; Aktürk, E.; Ciraci, S.; Ustunel, H. High-capacity hydrogen storage by metallized graphene. *Appl. Phys. Lett.* 2008, **93**, 043123–043125.

[114] Hussain, T.; Pathak, B.; Maark, T. A.; Araujo, C. M.; Scheicher R. H.; Ahuja, R. Ab initio study of lithium-doped graphane for hydrogen storage. *EPL* 2011, **96**, 27013–27016.

[115] Li, Y.; Zhao, G. F.; Liu, C. S. Wang, Y. L.; Sun, J. M.; Gu, Y. Z.; Wang, Y. X.; Zeng, Z. The structural and electronic properties of Li-doped fluorinated graphene and its application to hydrogen storage. *Int. J. Hydrogen Energy* 2012, **37**, 5754–5761.

[116] Wang, F. D. ; Wang, F.; Zhang, N. N.; Li, Y. H.; Tang, S. W.; Sun, H.; Chang, Y. F.; Wang, R. S. High-capacity hydrogen storage of Na-decorated graphene with boron substitution: First-principles calculations. *Chem. Phys. Lett.* 2013, **555**, 212–216.

[117] Antipina, Y. L.; Avramov, P. V.; Sakai, S.; Naramoto, H.; Ohtomo, M.; Entani, S.; Matsumoto, Y.; Sorokin, P. B. High hydrogen-adsorption-rate material based on graphane decorated with alkali metals. *Phys. Rev. B* 2012, **86**, 085435–085441.

[118] Zhou, W. W.; Zhou, J. J.; Shen, J. Q.; Ouyang, C. Y.; Shi, S. Q. First-principles study of high-capacity hydrogen storage on graphene with Li atoms. *J. Phys. Chem. Solids* 2012, **73**, 245–251.

[119] Du, A. J.; Zhu, Z. H.; Smith, S. C. Multifunctional porous graphene for nanoelectron-
       ics and hydrogen storage: New properties revealed by first principle calculations.
       J. Am. Chem. Soc. 2010, **132**, 2876–2877.

[120] Huang, S. H.; Miao, L.; Xiu, Y. J.; Wen, M.; Li, C.; Zhang, L.; Jiang, J. J. Lithium-
       decorated oxidized porous graphene for hydrogen storage by first principles study.
       J. Appl. Phys. 2012, **112**, 124312–124316.

[121] Zhang, H. Y.; Zhao, M. W.; Bu, H. X.; He, X. J.; Zhang, M.; Zhao, L. X.; Luo, Y. H.
       Ultra-high hydrogen storage capacity of Li-decorated graphyne: A first principles
       prediction. *J. Appl. Phys.* 2012, **112**, 084305–084309.

[122] Rangel, E.; Ramírez-Arellano, J. M.; Carrillo, I.; Magana, L. F. Hydrogen adsorption
       around lithium atoms anchored on graphene vacancies. *Int. J. Hydrogen Energy* 2011,
       36, 13657–13662.

[123] Liu, C. S.; Zeng, Z. Boron-tuned bonding mechanism of Li-graphene complex for
       reversible hydrogen storage. *Appl. Phys. Lett.* 2010, **96**, 123101–123103.

[124] Park, H. L.; Yi, S. C.; Chung, Y. C. Hydrogen adsorption on Li metal in boron-
       substituted graphene: An ab initio approach. *Int. J. Hydrogen Energy* 2010, **35**,
       3583–3587.

[125] Liu, C. S.; Zeng, Z.; Fan, C.; Ju, X. Li-doped B$_2$C graphene as potential hydrogen
       storage medium. *Appl. Phys. Lett.* 2011, **98**, 173101–173103.

[126] Lee, S.; Lee, M.; Choi, H.; Yoo, D. S.; Chung, Y. C. Effect of nitrogen induced defects in
       Li dispersed graphene on hydrogen storage. *Int. J. Hydrogen Energy* 2013, **38**, 4611–4617.

[127] Lee, S.; Lee, M.; Chung, Y. C. Enhanced hydrogen storage properties under external
       electric fields of N-doped graphene with Li decoration. *Phys. Chem. Chem. Phys.*
       2013, **15**, 3243–3248.

[128] Li, F.; Gao, J. F.; Zhang, J.; Xu, F.; Zhao, J. J.; Sun, L. X. Graphene oxide and lithium
       amidoborane: a new way to bridge chemical and physical approaches for hydrogen
       storage. *J. Mater. Chem. A* 2013, **1**, 8016–8022.

[129] Kim, Y. H.; Sun, Y. Y.; Zhang, S. B. Ab initio calculations predicting the existence of
       an oxidized calcium dihydrogen complex to store molecular hydrogen in densities up
       to 100 g/L. *Phys. Rev. B* 2009, **79**, 115424–115428.

[130] Kim, G.; Jhi, S. H. Ca-decorated graphene-based three-dimensional structures for
       high-capacity hydrogen storage. *J. Phys. Chem. C* 2009, **113**, 20499–20503.

[131] Kim, G.; Jhi, S. H.; Lim, S.; Park, N. Crossover between multipole Coulomb and
       Kubas interactions in hydrogen adsorption on metal-graphene complexes. *Phys. Rev. B*
       2009, **79**, 155437–155443.

[132] Beheshti, E.; Nojeh, A.; Servati, P. A first-principles study of calcium-decorated, boron-
       doped graphene for high capacity hydrogen storage. *Carbon* 2011, **49**, 1561–1567.

[133] Kim, G.; Jhi, S. H.; Lim, S.; Park, N. Effect of vacancy defects in graphene on metal
       anchoring and hydrogen adsorption. *Appl. Phys. Lett.* 2009, **94**, 173102–173104.

[134] Lee, H.; Ihm, J.; Cohen, M. L.; Louie, S. G. Calcium-decorated graphene-based
       nanostructures for hydrogen storage. *Nano Lett.* 2010, **10**, 793–798.

[135] Hussain, T.; Pathak, B.; Ramzan, M.; Maark, T. A. Ahuja, R. Calcium doped graphane
       as a hydrogen storage material. *Appl. Phys. Lett.* 2012, **100**, 183902–183906.

[136] Li, C.; Li, J. B.; Wu, F. M.; Li, S. S.; Xia, J. B.; Wang, L. W. High capacity hydrogen
       storage in Ca decorated graphyne: A first-principles study. *J. Phys. Chem. C* 2011,
       115, 23221–23225.

[137] Chen, C.; Zhang, J.; Zhang, B.; Duan, H. M. Hydrogen adsorption of Mg-doped graphene oxide: A first-principles study. *J. Phys. Chem. C* 2013, **117**, 4337–4344.

[138] Ataca, C.; Aktürk, E.; Ciraci, S. Hydrogen storage of calcium atoms adsorbed on graphene: First-principles plane wave calculations. *Phys. Rev. B* 2009, **79**, 041406–041409.

[139] Reunchan, P.; Jhi, S. H. Metal-dispersed porous graphene for hydrogen storage. *Appl. Phys. Lett.* 2011, **98**, 093103–093105.

[140] Fair, K. M.; Cui, X. Y.; Li, L.; Shieh, C. C.; Zheng, R. K.; Liu, Z. W.; Delley, B.; Ford, M. J.; Ringer, S. P.; Stampfl, C. Hydrogen adsorption capacity of adatoms on double carbon vacancies of graphene: A trend study from first principles. *Phys. Rev. B* 2013, **87**, 014102–014108.

[141] Ao, Z. M.; Jiang, Q.; Zhang, R. Q.; Tan, T. T.; Li, S. Al doped graphene: A promising material for hydrogen storage at room temperature. *J. Appl. Phys.* 2009, **105**, 074307–074312.

[142] Ao, Z. M.; Peeters, F. M. High-capacity hydrogen storage in Al-adsorbed graphene. *Phys. Rev. B* 2010, **81**, 205406–205412.

[143] Ao, Z. M.; Tan, T. T.; Li, S.; Jiang, Q. Molecular hydrogen storage in Al-doped bulk graphite with wider layer distances. *Solid State Commun.* 2009, **149**, 1363–1367.

[144] Carrete, J.; Longo, R. C.; Gallego, L. J.; Vega, A.; Balbas, L. C. Al enhances the H₂ storage capacity of graphene at nanoribbon borders but not at central sites: A study using nonlocal van derWaals density functionals. *Phys. Rev. B* 2012, **85**, 125435–125441.

[145] Park, H. L.; Chung, Y. C. Hydrogen storage in Al and Ti dispersed on graphene with boron substitution: First-principles calculations. *Comput. Mater. Sci.* 2010, **49**, S297–S301.

[146] Sigal, A.; Rojas, M. I.; Leiva, E. P. M. Is hydrogen storage possible in metal-doped graphite 2D systems in conditions found on earth? *Phys. Rev. Lett.* 2011, **107**, 158701–158704.

[147] Dimitrakakis, G. K.; Tylianakis, E.; Froudakis, G. E. Pillared graphene: A new 3-D network nanostructure for enhanced hydrogen storage. *Nano Lett.* 2008, **8**, 3166–3170.

[148] Matsuo, Y.; Ueda, S.; Konishi, K.; Marco-Lozar, J. P.; Lozano-Castelló, D.; Cazorla-Amorós, D. Pillared carbons consisting of silsesquioxane bridged graphene layers for hydrogen storage materials. *Int. J. Hydrogen Energy* 2012, **37**, 10702–10708.

[149] Tylianakis, E.; Psofogiannakis, G. M.; Froudakis, G. E. Li-doped pillared graphene oxide: A graphene-based nanostructured material for hydrogen storage. *J. Phys. Chem. Lett.* 2010, **1**, 2459–2464.

[150] Kim, B. H.; Hong, W. G.; Moon, H. R.; Lee, S. M.; Kim, J. M.; Kang, S.; Jun, Y.; Kim, H. J. Investigation on the existence of optimum interlayer distance for H₂ uptake using pillared-graphene oxide. *Int. J. Hydrogen Energy* 2012, **37**, 14217–14222.

[151] Wu, C. D.; Fang, T. H.; Lo, J. Y. Effects of pressure, temperature, and geometric structure of pillared graphene on hydrogen storage capacity. *Int. J. Hydrogen Energy* 2012, **37**, 14211–14216.

[152] Aboutalebi, S. H.; Aminorroaya-Yamini, S.; Nevirkovets, I.; Konstantinov, K.; Liu, H. K. Enhanced hydrogen storage in graphene oxide-MWCNTs composite at room temperature. *Adv. Energy Mater.* 2012, **12**, 1439–1446.

[153]   Burress, J. W.; Gadipelli, S.; Ford, J.; Simmons, J. M.; Zhou, W.; Yildirim, T. Graphene oxide framework materials: Theoretical predictions and experimental results. *Angew. Chem. Int. Ed.* 2010, **49**, 8902–8904.

[154]   Srinivas, G.; Burress, J. W.; Fordab, J.; Yildirim, T. Porous graphene oxide frameworks: Synthesis and gas sorption properties. *J. Mater. Chem.* 2011, **21**, 11323–11329.

[155]   Chan, Y.; Hill, J. M. Hydrogen storage inside graphene-oxide frameworks. *Nanotechnology* 2011, **22**, 305403–305410.

[156]   Liu, S.; Sun, L. X.; Xu, F.; Zhang, J.; Jiao, C. L.; Li, F.; Li, Z. B.; Wang, S.; Wang, Z. Q.; Jiang, X.; Zhou, H. Y.; Yang, L. N.; Schick, C. Nanosized Cu-MOFs induced by graphene oxide and enhanced gas storage capacity. *Energy Environ. Sci.* 2013, **6**, 818–823.

INDEX

Carbon Nanomaterials for Advanced Energy Systems: Advances in Materials Synthesis and Device Applications, First Edition. Edited by Wen Lu, Jong-Beom Baek and Liming Dai.
© 2015 John Wiley & Sons, Inc. Published 2015 by John Wiley & Sons, Inc.